Atmospheric Science, Climatology and Biogeochemistry

Atmospheric Science, Climatology and Biogeochemistry

Edited by **Mary D'souza**

SYRAWOOD
PUBLISHING HOUSE
New York

Published by Syrawood Publishing House,
750 Third Avenue, 9th Floor,
New York, NY 10017, USA
www.syrawoodpublishinghouse.com

Atmospheric Science, Climatology and Biogeochemistry
Edited by Mary D'souza

International Standard Book Number: 978-1-68286-281-0 (Hardback)

Printed in China.

Contents

Preface

Over the years, the climate patterns across earth have witnessed significant changes. A major cause of this has been attributed to the greenhouse effect. Thus, these chemical, biological and atmospheric changes have been studied increasingly worldwide. This book consists of material provided by top researchers from the field of geophysics. It covers in detail various topics pertaining to atmosphere, weather and climate such as scaling, predictability, pattern formation, turbulence, time series, face transition, etc. This book will prove to be immensely beneficial to students and researchers in this field.

The information contained in this book is the result of intensive hard work done by researchers in this field. All due efforts have been made to make this book serve as a complete guiding source for students and researchers. The topics in this book have been comprehensively explained to help readers understand the growing trends in the field.

I would like to thank the entire group of writers who made sincere efforts in this book and my family who supported me in my efforts of working on this book. I take this opportunity to thank all those who have been a guiding force throughout my life.

Editor

1

Representation of model error in a convective-scale ensemble prediction system

L. H. Baker, A. C. Rudd, S. Migliorini, and R. N. Bannister

Department of Meteorology, University of Reading, Earley Gate, P.O. Box 243, Reading, RG6 6BB, UK

Correspondence to: L. H. Baker (l.h.baker@reading.ac.uk)

Abstract. In this paper ensembles of forecasts (of up to six hours) are studied from a convection-permitting model with a representation of model error due to unresolved processes. The ensemble prediction system (EPS) used is an experimental convection-permitting version of the UK Met Office's 24-member Global and Regional Ensemble Prediction System (MOGREPS). The method of representing model error variability, which perturbs parameters within the model's parameterisation schemes, has been modified and we investigate the impact of applying this scheme in different ways. These are: a control ensemble where all ensemble members have the same parameter values; an ensemble where the parameters are different between members, but fixed in time; and ensembles where the parameters are updated randomly every 30 or 60 min. The choice of parameters and their ranges of variability have been determined from expert opinion and parameter sensitivity tests. A case of frontal rain over the southern UK has been chosen, which has a multi-banded rainfall structure.

The consequences of including model error variability in the case studied are mixed and are summarised as follows. The multiple banding, evident in the radar, is not captured for any single member. However, the single band is positioned in some members where a secondary band is present in the radar. This is found for all ensembles studied. Adding model error variability with fixed parameters in time does increase the ensemble spread for near-surface variables like wind and temperature, but can actually decrease the spread of the rainfall. Perturbing the parameters periodically throughout the forecast does not further increase the spread and exhibits "jumpiness" in the spread at times when the parameters are perturbed. Adding model error variability gives an improvement in forecast skill after the first 2–3 h of the forecast for near-surface temperature and relative humidity. For precipitation skill scores, adding model error variability has the effect of improving the skill in the first 1–2 h of the forecast, but then of reducing the skill after that. Complementary experiments were performed where the only difference between members was the set of parameter values (i.e. no initial condition variability). The resulting spread was found to be significantly less than the spread from initial condition variability alone.

1 Introduction

Errors in forecasts originate from a number of sources, namely the initial conditions, the boundary conditions and the model formulation. In synoptic scale forecasts of lead times up to a day, it is thought that the first two sources dominate. However, at convective scale model errors are thought to become more important, especially for relatively short range forecasts. Here we investigate a proposed representation of model error that can influence the forecast skill at convective scale. We use an experimental convection-permitting version of the UK Met Office's 24-member Global and Regional Ensemble Prediction System (EPS) (MOGREPS) (Bowler et al., 2008): the southern UK 1.5 km EPS (Migliorini et al., 2011; Caron, 2013). We focus on the effect of model error resulting from the parameterisation of unresolved processes; specifically microphysics and turbulent boundary layer processes. In this study we modify the so-called Random Parameters (RP) scheme, used in MOGREPS (Bowler et al., 2008), by applying changes designed to make it appropriate for use in a convective-scale ensemble.

The aim of this work is to investigate how this modified version of the RP scheme affects the characteristics of the ensemble, with particular focus on the ensemble spread and the forecast skill. By evaluating these effects, we determine how useful this scheme could be as a method of representing model error in a convective-scale EPS. Although it is difficult to draw general conclusions from one case study, this work presents new results of the impact of a model error scheme on a convection-permitting model.

There are several different methods commonly used to represent model error. In the multiphysics method (also sometimes called the multimodel method), a set of different model physics parameterisation schemes is used, and generally an ensemble is constructed of members which use different combinations of schemes (Stensrud et al., 2000; Berner et al., 2011; Clark Jr. et al., 2010; Clark et al., 2011). The stochastic kinetic energy backscatter (SKEB) scheme of Shutts (2005) (developed further by Berner et al., 2009 and used by Berner et al., 2011) aims to replace upscale kinetic energy lost in the numerical integration of the model and in the parameterisation of unresolved processes. Another method is to use a stochastically perturbed physical tendencies (SPT or SPPT) scheme (Buizza et al., 1999; Bouttier et al., 2012; Fresnay et al., 2012), in which the total tendencies from parameterisation schemes are perturbed. A further method, and the one on which we focus here, is to perturb a set of individual parameters within the parameterisation schemes themselves. This method can be applied in two different ways: the first is to perturb individual or sets of parameters about their default values, and keep these perturbed values fixed throughout each forecast; the second is to use a stochastic technique to vary the parameter values periodically throughout the forecast. The fixed parameter perturbation method has been widely used for global ensembles (notably the climateprediction.net multi-thousand member ensemble project (Murphy et al., 2004; Stainforth et al., 2005) and the Met Office's Quantifying Uncertainty in Model Predictions (QUMP) ensemble (Murphy et al., 2004)), mesoscale ensemble systems (Hacker et al., 2011) and convection-permitting models (Gebhardt et al., 2011; Vié et al., 2012). In this study we use both methods of setting parameter values (see Sect. 2.2 for details of the scheme used when updating parameters periodically). A similar method was also used by Lin and Neelin (2000) and Bright and Mullen (2002), but for only one and two parameters, respectively. The advantage of updating the parameters throughout the forecast is that it increases the amount of random variability in the parameters, and increases the volume of parameter values explored during the period of the forecast. However, a disadvantage of this method is that perturbing the parameters during the forecast may introduce jumps, loss of conservation of quantites that should be conserved, and changes of state, which may lead to undesirable effects on the forecast. For the purposes of this study, the RP scheme is a logical choice since the basic RP framework already exists in the Met Office Unified Model (MetUM), and some parameters for which there is genuine uncertainty in their values had already been identified and their uncertainty quantified. We note that this method of representing model error is not guaranteed to increase the ensemble spread, due to the nonlinear dependence of the model on the parameters.

In Sect. 2 we describe the configuration of the MetUM, the formulation of the RP scheme and the evaluation tools used, and give an outline of the experiments performed in Sects. 4 and 5. In Sect. 3 we describe the meteorology of the case used in this study, and show results of parameter sensitivity tests performed for this case. In Sect. 4 we evaluate the performance of the control (no model error) ensemble, and in Sect. 5 we evaluate the effects of model error variability on the forecast skill and on the spread of the ensemble. Finally, Sect. 6 provides a summary and discussion of our results.

2 Methodology

2.1 Description of the 1.5 km EPS

This work was performed using the MetUM version 7.8. The MetUM is a finite-difference model that solves the non-hydrostatic, fully compressible, deep-atmosphere dynamical equations with a semi-Lagrangian, semi-implicit integration scheme. The equations are solved on a horizontally staggered Arakawa C grid and a terrain-following hybrid-height vertical coordinate with Charney–Phillips grid staggering (Davies et al., 2005). In all the MetUM ensemble systems discussed here, boundary-layer mixing is parameterised using the scheme of Lock et al. (2000) and large-scale precipitation is parameterised using the scheme of Wilson and Ballard (1999). In the global and regional formulations of the ensemble system (MOGREPS-G and MOGREPS-R, respectively, described in more detail below), convection is parameterised using the scheme of Gregory and Rowntree (1990); no convection scheme is used in the 1.5 km EPS. To limit the occurrence of grid-point convection in the 1.5 km EPS, a prognostic rain variable is advected with the wind field, following Sect. 2d of Lean et al. (2008). In MOGREPS-G and MOGREPS-R, gravity-wave drag is parameterised using the scheme of Webster et al. (2003); this scheme is not active in the 1.5 km EPS.

The 1.5 km EPS has a horizontal grid spacing of 1.5 km, with 360 × 288 grid points covering a domain over the southern UK (Fig. 1). The model has 70 vertical levels with variable spacing, with the highest vertical resolution in the boundary-layer, and a model lid at around 38 km. This model is run with a time step of 50 s. The 1.5 km EPS is nested within MOGREPS-R, which has a domain covering the North Atlantic and Europe and a horizontal resolution equivalent to 18 km in the mid-latitudes and 70 vertical levels, with a model lid at around 80 km. MOGREPS-R is nested within MOGREPS-G, which covers the full globe and runs with a resolution equivalent to 60 km in the mid-latitudes and the same vertical levels as MOGREPS-R[1].

A 24-member ensemble is generated in MOGREPS-G using an ensemble transform Kalman filter (ETKF) (Bishop et al., 2001). The ETKF is used to produce a set of perturbations which can be added to the Met Office 4-D-Var analysis to give the initial conditions (ICs) for the ensemble members. A detailed description of the way that the ETKF is used to generate these IC perturbations is given in Bowler et al. (2008). MOGREPS-G is run with 12 h cycling[2], starting at 00:00 UTC and 12:00 UTC. MOGREPS-R also runs with 12 h cycling starting at 06:00 UTC and 18:00 UTC, but takes its initial conditions and hourly lateral boundary conditions (LBCs) from MOGREPS-G. In this study we run a MOGREPS-G forecast starting at 00:00 UTC 20 September 2011 and extend the forecast to 19:00 UTC. We use this to provide ICs and LBCs for a MOGREPS-R forecast starting at 06:00 UTC, which outputs hourly ICs and LBCs for the 1.5 km EPS from 07:00 UTC to 18:00 UTC.

The 1.5 km EPS is described in detail by Migliorini et al. (2011). Subsequent improvements to the 1.5 km EPS setup using the so-called scale-selective ETKF to avoid ensemble perturbation mismatches between the ICs and LBCs are described by Caron (2013). In this study we run a 13 h forecast with hourly ETKF cycling for the first 6 h, followed by a 7 h forecast, forced by hourly LBCs from MOGREPS-R. The system is described schematically in Fig. 2. In each of the ETKF cycles, the initial conditions for the 1.5 km EPS control member are produced from a 3-D-Var analysis, with LBCs provided by a 4 km grid-spacing model with a domain covering the whole UK. For each 1.5 km EPS ensemble member, a perturbation is applied to the control member ICs. These IC perturbations are derived from MOGREPS-R as follows: in the first cycle, the IC perturbations for the 1.5 km EPS ensemble members are downscaled from the MOGREPS-R IC perturbations; in subsequent cycles, the 1.5 km EPS IC perturbations are partitioned into small-scale and large-scale components. The small-scale component of the IC perturbations is derived by applying the scale-selective ETKF to the 1.5 km EPS forecast perturbations from the previous cycle. The large-scale component of the 1.5 km EPS ICs is derived by using the downscaled MOGREPS-R forecast perturbations for each ensemble member to perturb the 1.5 km analysis. An inflation factor is applied to scale the perturbations. This ensures that the innovation variance is consistent with observations at $T+1$ of the forecast (Migliorini et al., 2011; Caron, 2013). A spatially and temporally fixed inflation factor value is used throughout the forecast. A suitable value for the inflation factor in the particular case studied here was determined by examining power spectra of potential temperature and horizontal wind using data from four different model levels, as was done by Caron (2013). The IC perturbations and analysis increments are added to the background forecast gradually over a one hour period ($T-30$ min to $T+30$ min) using the Incremental Analysis Update (IAU) scheme. Hourly LBCs for the control member are derived from the MOGREPS-R control member. Perturbed LBCs are provided for each ensemble member, again by MOGREPS-R. During the ETKF cycles, the perturbed LBCs are introduced using the incremental LBC update (ILBCU) method (Caron, 2013) to gradually apply the corresponding LBC perturbations over the same one-hour time period as the IAU scheme acts.

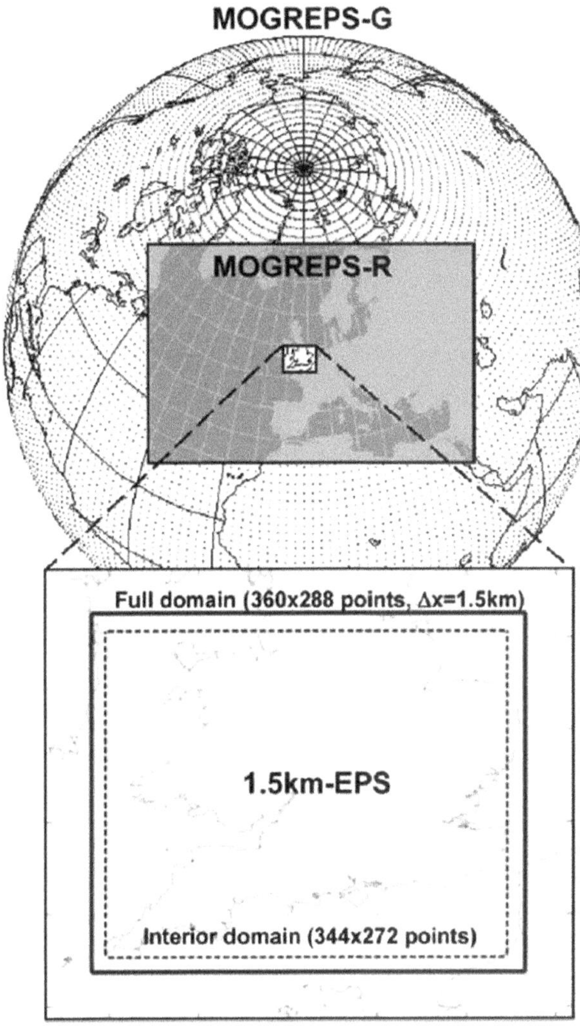

Fig. 1. Schematic diagram showing the model domains of MOGREPS-G, MOGREPS-R and the 1.5 km EPS used here. Taken from Fig. 1 of Caron (2013). ©Crown Copyright 2013, Met Office.

[1]The current operational version of MOGREPS-G runs at a higher resolution than this, with a grid length of 33 km in the mid-latitudes.

[2]The current operational version has 6 h cycling.

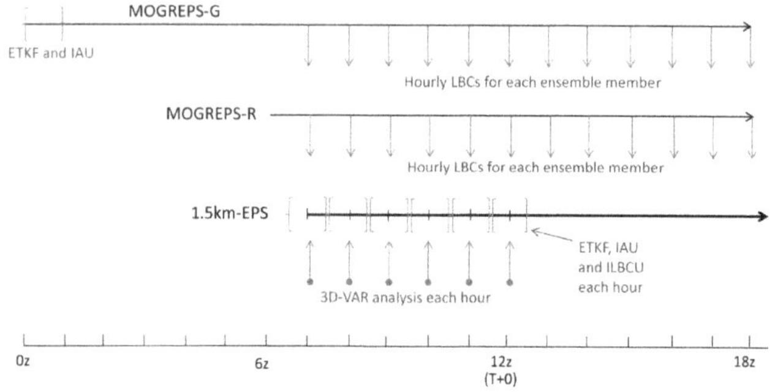

Fig. 2. Schematic diagram showing the model setup of the 1.5 km EPS used in this study. Times shown are for 20 September 2011.

2.2 The random parameters scheme as a method of representing model error

The RP scheme is used operationally in MOGREPS (Bowler et al., 2008). The scheme is designed to simulate model error due to uncertainty in the parameterisation of sub-grid scale processes. The RP scheme takes a set of parameters from various parameterisation schemes and treats them as stochastic variables. For each parameter, a physically sensible range of values has been defined as advised by experts. The parameter is allowed to vary within this range.

In the original RP scheme, two parameters from each of the boundary layer, large-scale precipitation, gravity-wave drag and convection schemes were selected (see Table 1 of Bowler et al. (2008) for the full list of parameters and values). Five additional boundary layer parameters were later added to the RP scheme following the work of Zadra and Lock (2010). The full set of boundary layer and large-scale precipitation scheme parameters used in this updated scheme is shown in the top part of Table 1.

The parameter values are calculated first using the first-order autoregression model

$$P_t^0 = \mu + r(P_{t-1} - \mu) + \epsilon, \qquad (1)$$

where P_t^0 and P_{t-1} are the intermediate and previous parameter values, μ is the mean value [3] of the parameter distribution (taken to be the default value of that parameter) and r is the auto-correlation constant (set as $r = 0.95$). ϵ is the stochastic shock term, sampled from a uniform distribution in the range $\pm(P_{\max} - P_{\min})/3$, where $P_{\max}$ and $P_{\min}$ are the maximum and minimum values for the parameter. If P_t^0 is outside of the prescribed range, it is rounded down or up to give the new parameter value, P_t:

$$P_t = \begin{cases} P_{\min} & \text{if } P_t^0 < P_{\min}, \\ P_{\max} & \text{if } P_t^0 > P_{\max}, \\ P_t^0 & \text{otherwise.} \end{cases} \qquad (2)$$

[3] μ is not necessarily the true mean of the resulting distribution of parameter values. The true mean is influenced by the rounding process described in Eq. (2).

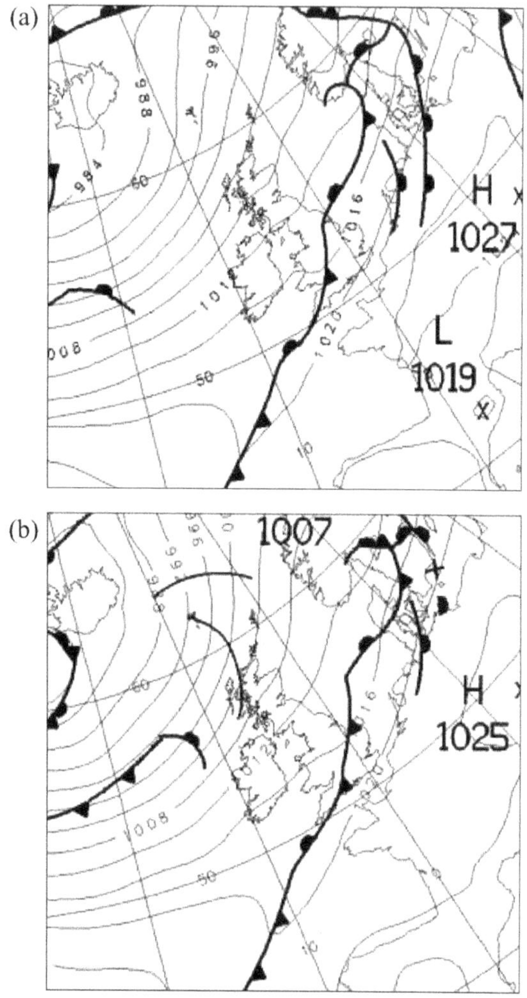

Fig. 3. Met Office mean sea level pressure analysis charts valid at **(a)** 12:00 UTC and **(b)** 18:00 UTC 20 September 2011. Grey pressure contours every 4 hPa. ©Crown Copyright 2011, Met Office.

The range of ϵ and the value of r were chosen following a series of 72 h forecasts and selecting the combination of values that gave the maximum spread and best deterministic scores for individual ensemble members (Bowler et al., 2008). These values are therefore tuned for the resolution of MOGREPS-G, which at the time had 90 km grid spacing.

In the standard formulation of the RP scheme, the parameters are updated using Eqs. (1) and (2) once every 3 h. The parameter values are then fixed for a period of 3 h until the next update time. The same parameter values are used for all grid points. Each time the scheme is applied, a new random number $k \in [0, 1]$ is generated to give a different value of ϵ defined as $\epsilon = (2k - 1)(P_{\max} - P_{\min})/3$. A different random number is used for each parameter so that they vary independently from one another. On the first application of the RP scheme, the parameter values are set to random values within the parameter range. Each ensemble member, including the

Table 1. Parameters used in the modified RP scheme. The four pairs of parameters that vary together (using the same random number to update ϵ in Eq. 1) are shown in their respective pairs in italics and bold. Parameter Ri_c varies as $Ri_c = 10/g_0$.

Scheme	Parameter	Name	Description	Min	Default	Max
BL	g_0	*flux profile parameter*	Controls stability dependence of turbulent mixing coefficients in the stable boundary-layer scheme.	5	10	20
BL	Ri_c	*critical Richardson number*	Determines the boundary-layer top (defined as the lowest level for which $Ri > Ri_c$).	0.5	1.0	2.0
BL	g_{mezcla}	**neutral mixing length**	Sets the magnitude of the turbulent mixing lengths.	0.03	0.15	0.45
BL	λ_{min}	**minimum mixing length**	Sets the magnitude of the turbulent mixing lengths.	8	40	120
BL	Charnock	Charnock parameter	Determines the magnitude of the wind-speed dependent roughness length over sea.	0.010	0.011	0.026
BL	A_1	entrainment parameter	Multiplication factor for the entrainment rate at the boundary-layer top.	0.1	0.23	0.4
BL	g_1	cloud-top diffusion parameter	Multiplication factor for the eddy diffusivity at the cloud top in the boundary layer.	0.5	0.85	1.5
LSP	RH_{crit}	critical relative humidity	Relative humidity at which cloud begins to form.	0.875	0.9	0.910
LSP	m_{ci}	ice-fall speed	Multiplication factor for the fall speed of ice crystals.	0.3	1.0	3.0
LSP	$x1r$	particle size distribution (PSD) for rain	Multiplication factor for the PSD function for rain.	2×10^6	8×10^6	2×10^9
LSP	$x1i$	*PSD for ice aggregates*	Multiplication factor for the PSD function for ice aggregates.	1×10^6	2×10^6	1×10^7
LSP	$x1ic$	*PSD for ice crystals*	Multiplication factor for the PSD function for ice crystals.	1×10^7	4×10^7	1×10^8
LSP	**ai**	**ice aggregate mass diameter**	Multiplication factor for the mass–diameter relationship for ice aggregates.	0.0222	0.0444	0.0888
LSP	**aic**	**ice crystal mass diameter**	Multiplication factor for the mass–diameter relationship for ice crystals.	0.2935	0.587	1.174
LSP	t_{nuc}	maximum ice nucleation temperature	Maximum temperature at which ice crystals can form.	−25	−10	−1
LSP	ec_{auto}	autoconversion efficiency	Collection/collision coefficient for the conversion of cloud liquid water to rain.	0.01	0.55	0.6

control member, is given a different set of random parameter perturbations by the RP scheme.

2.3 Modifications to the scheme for the convective-scale EPS

To apply the RP scheme to a forecast ensemble run in the 1.5 km EPS convective-scale setup, it was necessary to make some modifications to the existing scheme. Importantly, at this resolution, the convection scheme and gravity-wave drag schemes are not used. Therefore, only parameters in the boundary layer and large-scale precipitation schemes were perturbed. Sensitivity tests were used to determine appropriate parameters in the large-scale precipitation scheme to perturb. These tests are described in Sect. 3.3. The full set of parameters used here is given in Table 1.

Given the shorter time step of the high-resolution forecast compared with the global and NAE forecasts (50 s compared with 5–10 min) and the short forecast lead times we are interested in, it was appropriate to revise the time interval between calls to the RP scheme (3 h in the original scheme). We test update times of 30 and 60 min, which were chosen to allow the parameters to vary smoothly over time, without giving continual shocks to the model. Consideration was also given to the timescales of the processes that the parameterisation schemes represent. A further version of the RP scheme, holding parameters fixed throughout the forecasts, was also used. In all cases the first application of the RP scheme is at $T+30$ min, after the IAU and ILBCU schemes have finished. This is to allow the IC and LBC perturbations to be added completely before the RP scheme is applied.

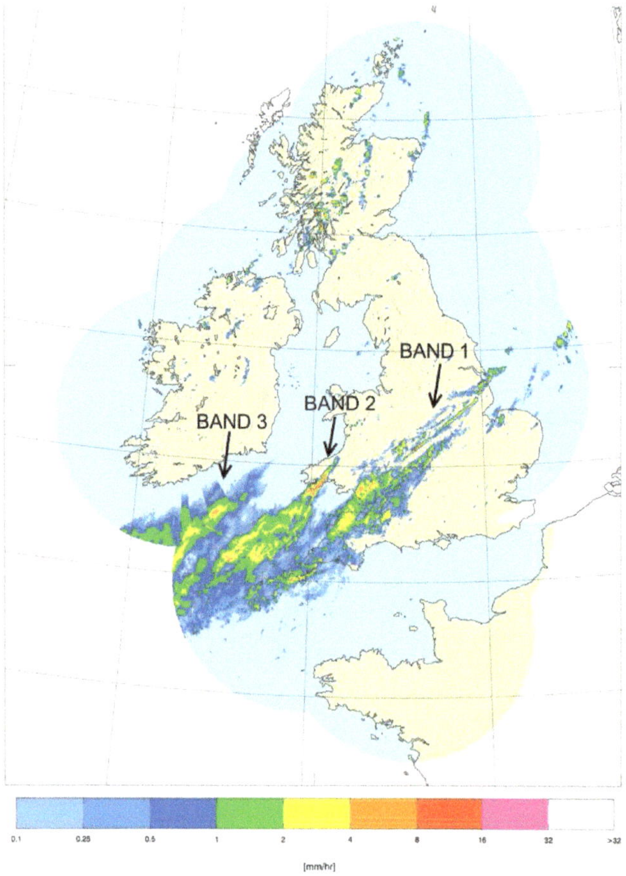

Fig. 4. Radar-derived surface precipitation rate (1 km composite) at 15:00 UTC 20 September 2011. ©Crown Copyright 2011, Met Office.

2.4 Evaluation and verification methods

We evaluate the spread and skill of the ensemble by considering near-surface fields (1.5 m temperature, 1.5 m relative humidity (RH) and 10 m wind speed), rainfall rate and rainfall accumulation. These quantities were chosen because they could be evaluated against surface and radar observations that were available to us.

To evaluate the spread of the ensemble we use two different diagnostics: the first is a domain-averaged measure of the ensemble spread at each time; the second, for the hourly rain accumulation only, is the correspondence ratio (CR), which gives a measure of spread for a whole field rather than individual grid points (as used by Gebhardt et al., 2011). We define the spread for an n-member ensemble as

$$\text{spread} = \frac{1}{nx \times ny} \sum_{i,j=1}^{nx,ny} \sigma_{i,j},$$

where

$$\sigma_{i,j} = \sqrt{\frac{1}{n-1} \sum_{k=1}^{n} (p_{i,j,k} - \overline{p_{i,j}})^2},$$

nx and ny are the numbers of grid points in the x and y directions, respectively (in our case $nx = 360$ and $ny = 288$), $p_{i,j,k}$ denotes the value of field p at point (i, j) for ensemble member k and $\overline{p_{i,j}}$ is the ensemble mean value of field p at point (i, j). CR is calculated following Gebhardt et al. (2011) as $\text{CR} = N(P_{\text{all}})/N(P_{\geq 1})$, where $N(P_{\text{all}})$ is the number of grid points where all members forecast an event, and $N(P_{\geq 1})$ is the number of grid points where at least one member forecasts the event. A low CR indicates large ensemble spread, while a high CR value indicates that the ensemble members are generally in agreement.

To evaluate the forecast skill of the ensemble by comparing with observations, we use three methods: the first, for rain and near-surface fields, is the continuous ranked probability score (CRPS); the second, for rainfall accumulation only, is the precipitation skill score (PSS); the third, for rainfall rate only, is the innovation magnitude.

The CRPS is a way of measuring the closeness between two cumulative distribution functions (CDFs). In our case the two CDFs are those of a forecast, $F_{\text{f}}(x)$, and of an observation, $F_{\text{o}}(x)$. The definition of the CRPS at point (i, j) and for quantity $x_{i,j}$ is the total squared difference between these CDFs:

$$\text{CRPS}_{i,j} = \int_{x_{i,j}=-\infty}^{\infty} \mathrm{d}x_{i,j} (F_{\text{f}}(x_{i,j}) - F_{\text{o}}(x_{i,j}))^2,$$

(Wilks, 2006), where the CDFs are $F_{\text{f/o}}(x_{i,j}) = \int_{x'_{i,j}=-\infty}^{x_{i,j}} \mathrm{d}x'_{i,j} P_{\text{f/o}}(x'_{i,j})$ and where $P_{\text{f/o}}(x_{i,j})$ is the probability density function (PDF) for the forecast/observation at the point (i, j). The forecast's PDF is described by the ensemble members, $x_{i,j}^{(k)}$, by assuming that each is equally likely: $P_{\text{f}}(x_{i,j}) = \sum_{i=1}^{N} \delta(x_{i,j} - x_{i,j}^{(k)})$, where δ is the Dirac delta-function. In previous works using the CRPS, an observation's PDF assumes that the observation is perfect, i.e. that $P_{\text{o}}(x_{i,j}) = \delta(x_{i,j} - y_{i,j}^{\text{o}})$, where $y_{i,j}^{\text{o}}$ is the observed value, which leads to the observation's CDF being a Heaviside step function. This simplifies the calculation of the CRPS as it avoids the need to do explicit quadrature (see Sect. 4 of Hersbach, 2000). In this work though we account for imperfect observations by assuming that each observation's PDFs is a normal distribution with finite width, σ_{o}. In this case the CRPS reduces to the following calculation:

$$\text{CRPS}_{i,j} = \int_{x_{i,j}=-\infty}^{\infty} \mathrm{d}x_{i,j} \left\{ \frac{1}{N} n_{\text{f}}(x_{i,j}) - \frac{1}{2}\left[1 + \text{erf}\left(\frac{x_{i,j} - y_{i,j}^{\text{o}}}{\sqrt{2}\sigma_{\text{o}}}\right)\right]\right\}^2,$$

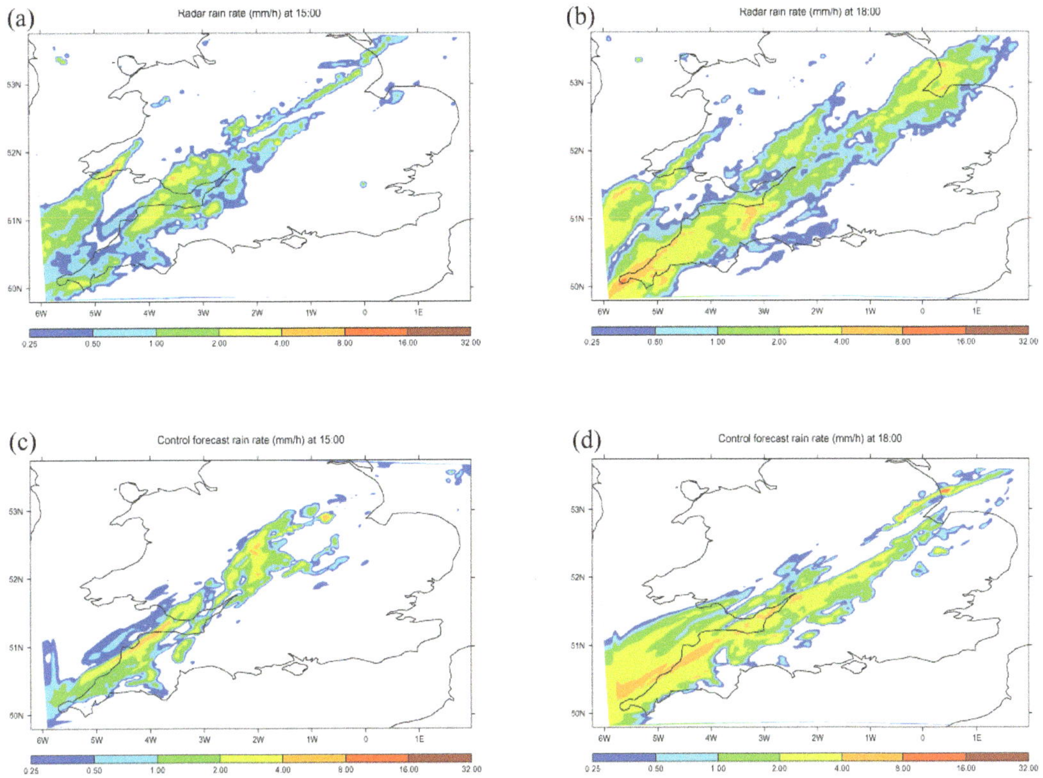

Fig. 5. Observed and model rain rate (mm h^{-1}). Left column: 15:00 UTC ($T+3$), right column: 18:00 UTC ($T+6$) 20 September 2011. Top row: radar-derived surface precipitation rates (1 km composite), bottom row: control forecast rainfall rate at the surface.

where $n_f(x_{i,j})$ is the number of ensemble members whose forecast values are less than or equal to $x_{i,j}$ and $\mathrm{erf}(x)$ is the standard error function $\mathrm{erf}(x) = \frac{2}{\sqrt{\pi}} \int_{x'=0}^{x} \exp -x^2 \mathrm{d}x$. Numerical quadrature is necessary to evaluate this integral. Note that a lower CRPS indicates a more skillful forecast.

To calculate the PSS, the hourly rainfall accumulation data for the model forecast and radar observations are first resampled to 13.5 km. This reduces the chance of penalising any small displacement errors twice, which can happen in high resolution forecasts if using grid point comparison methods. A similar resampling was done by Caron (2013) (to 15 km) when using the PSS. Next, the Brier score (BS, Wilks (2006)) is calculated, and is defined as

$$\mathrm{BS} = \frac{1}{nx \times ny} \sum_{i,j=1}^{nx,ny} (f_{i,j} - o_{i,j})^2,$$

where $f_{i,j}$ is the probability of a given event occurring in the model ensemble data, and $o_{i,j}$ is a binary indicator of the occurrence of the event in the observations. In this case the event is the exceedance of a threshold rain rate at a given grid point. The PSS for an ensemble ENS is given by

$$\mathrm{PSS} = 1 - \frac{\mathrm{BS_{ENS}}}{\mathrm{BS_{CTL}}}.$$

The PSS is therefore given with respect to a control ensemble, which here is the ensemble with IC and BC perturbations but no RP scheme applied (CTL in Table 2).

To compare the ensemble mean with the radar observations we calculate the innovation magnitude, d, defined at each grid point as the absolute difference between the mean value of the ensemble at that point and the observation value at that point:

$$d_{i,j} = \| \overline{p_{i,j}} - y^{\mathrm{o}}_{i,j} \|.$$

2.5 Description of the experiments

The various ensemble experiments discussed in Sects. 4 and 5 are listed in Table 2. The control ensemble (CTL) is composed of a set of 24 six-hour forecasts with ICs at 12:00 UTC derived from MOGREPS-R IC perturbations and previous cycles of the 1.5 km EPS, as described in Sect. 2.1. The IC and LBC perturbations are introduced gradually between $T-30$ and $T+30$ using the IAU and ILBCU schemes, respectively. This ensemble has no representation of model error variability. Experiments labelled IC + BC + RP* (where here and throughout this paper, * is shorthand for either "fix", "30" or "60") use the same IC and LBC perturbations as the control ensemble but also use the RP scheme with different lengths of time between parameter updates (once, every 30 min and every 60 min, respectively), starting from

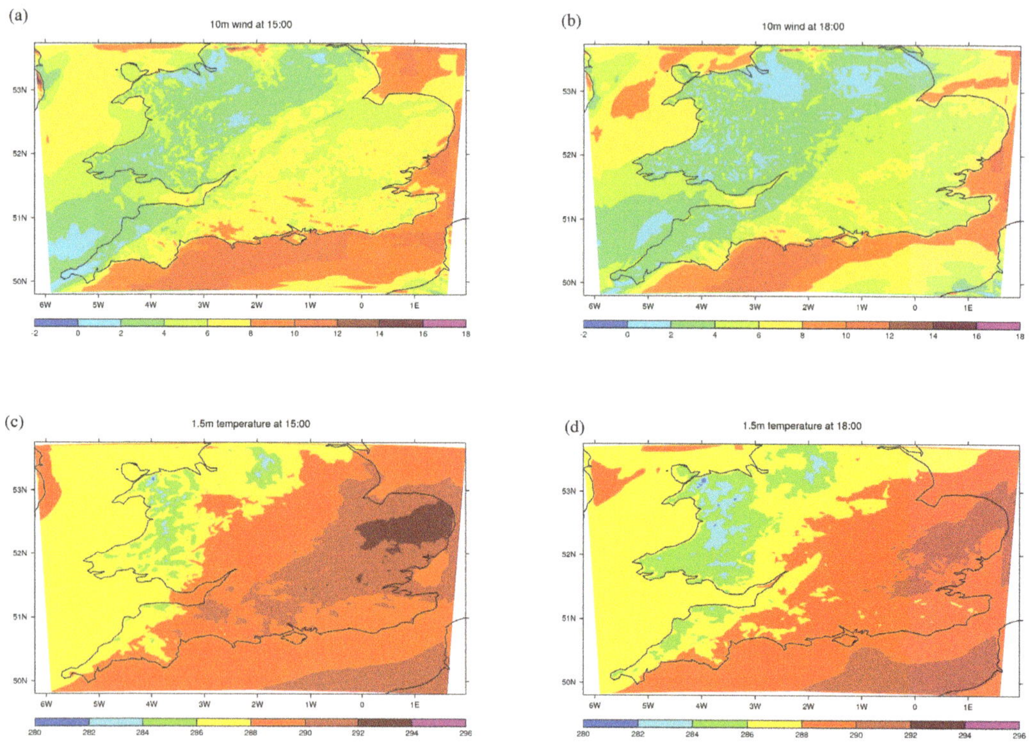

Fig. 6. Control forecast. Left column: 15:00 UTC ($T+3$), right column: 18:00 UTC ($T+6$) 20 September 2011. Top row: 10 m wind speed ($m\,s^{-1}$), bottom row: 1.5 m temperature (K).

Table 2. Summary of model configurations for the ensemble experiments.

Ensemble name	Description	IC and LBC perts	RP scheme
CTL	Control (perturbed ICs and LBCs, no RP scheme)	Yes	No
IC + BC + RPfix	perturbed ICs and LBCs, RP scheme applied once	Yes	Yes
IC + BC + RP30	perturbed ICs and LBCs, RP scheme applied every 30 min	Yes	Yes
IC + BC + RP60	perturbed ICs and LBCs, RP scheme applied every 60 min	Yes	Yes
RPfix	RP scheme applied once	No	Yes
RP30	RP scheme applied every 30 min	No	Yes
RP60	RP scheme applied every 60 min	No	Yes

$T+30$ min (after the IAU and ILBCU have finished). Experiments labelled RP* have no IC and LBC perturbations: their members all use the unperturbed ICs and LBCs of the control member, with model error represented by applying the RP scheme. These ensembles therefore show the effects of using the RP scheme to perturb parameters, without the effects of perturbing the ICs and LBCs.

3 Description of the case and sensitivity tests to inform the modifications to the RP scheme

3.1 Overview

For this study we focus on a case on 20 September 2011 characterised by the passage of a cold front over the southern UK (Fig. 3). It can be seen from the analysis charts that this is part of a long, trailing front with several weak frontal waves. This frontal structure passed across the southern UK in a north-eastward direction. There are two main reasons for selecting this particular case. First, it occurred during a three-week field campaign period of the DIAMET (DIAbatic influences on Mesoscale structures in ExTratropical storms) project (funded by the Natural Environment Research Council (NERC)), and was thus selected as a DIAMET intensive observation period (IOP-2). We therefore have access to high-resolution observations for this case, in addition to those made operationally, including in situ aircraft measurements, dropsonde profiles and extra radiosonde ascents from some UK stations. These non-operational observations are not used in the study reported in this paper, but are being

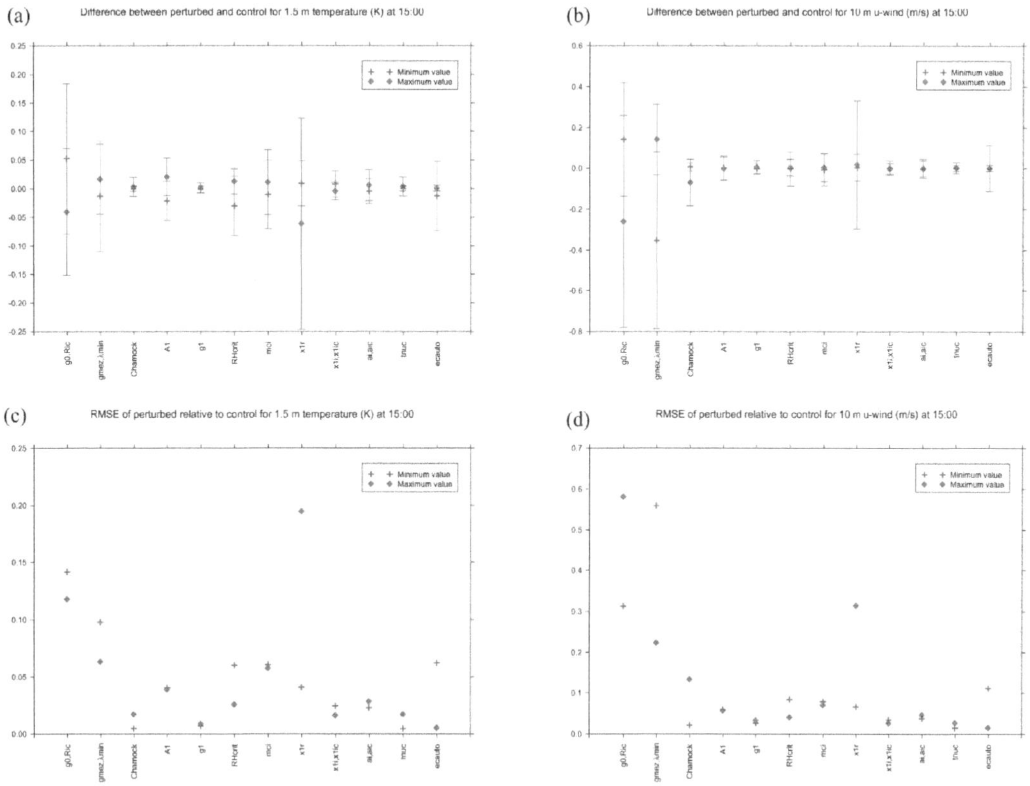

Fig. 7. Sensitivities of the forecasts at $T+3$ (15:00 UTC) to perturbing parameters to their maximum and minimum values. Top row: difference between perturbed and control forecast at each point, averaged over the domain. Error bars show the standard deviation of the differences. Bottom row: RMS difference between the perturbed and control forecast at each point, averaged over the domain. Left column: 1.5 m temperature. Right column: u component of 10 m wind.

used in ongoing related work (Migliorini, Bannister, Rudd and Baker, in preparation). The second reason for choosing this case is that it has an unusual triple banded structure in the rain band passing over the UK, which was present in the radar rain rate (Fig. 4) but was not captured by the Met Office operational high-resolution deterministic forecast (the UKV forecast; 1.5 km grid length, covering the whole UK; not shown). This triple rain band feature, and the fact that it was not captured by the operational deterministic forecast, makes it an interesting case to study using an ensemble of forecasts. It is not clear from either the model forecasts or the radar observations what the cause of this banded structure is; this is the subject of ongoing work and is beyond the scope of this paper.

We focus on the period 12:00 UTC to 18:00 UTC, and are particularly interested in evaluating the performance and spread of the ensemble at 15:00 UTC ($T+3$) as this is typically a useful timescale for convective scale forecasts. This should also allow time for model error introduced by the RP scheme to have effect before information coming through the boundaries begins to dominate (Gebhardt et al., 2011).

3.2 Control forecast from the 1.5 km EPS

The control forecast (using the standard default parameter values in each of the parameterisation schemes) captures the main rain band (Fig. 5c and d) but does not capture the second band over South Wales seen in the radar (Fig. 5a and b). Note that during the period studied here the third rain band in Fig. 4 does not reach the 1.5 km EPS domain. There is some smaller-scale banding along the rain band in the control forecast at both times shown, indicating that the model is representing the correct type of smaller-scale structure within the rain band. Figure 6a and b show stronger winds ahead of the front than behind the front, with a sharp drop in 10 m wind speeds over land compared with the wind speeds over the sea, which is likely to be caused by the different surface roughness and orographic effects. The front is relatively weak, with a temperature gradient of around 6 K across the front (Fig. 6c and d). The location of the strongest temperature gradient in Fig. 6d is consistent with the position of the analysed surface front in Fig. 3b.

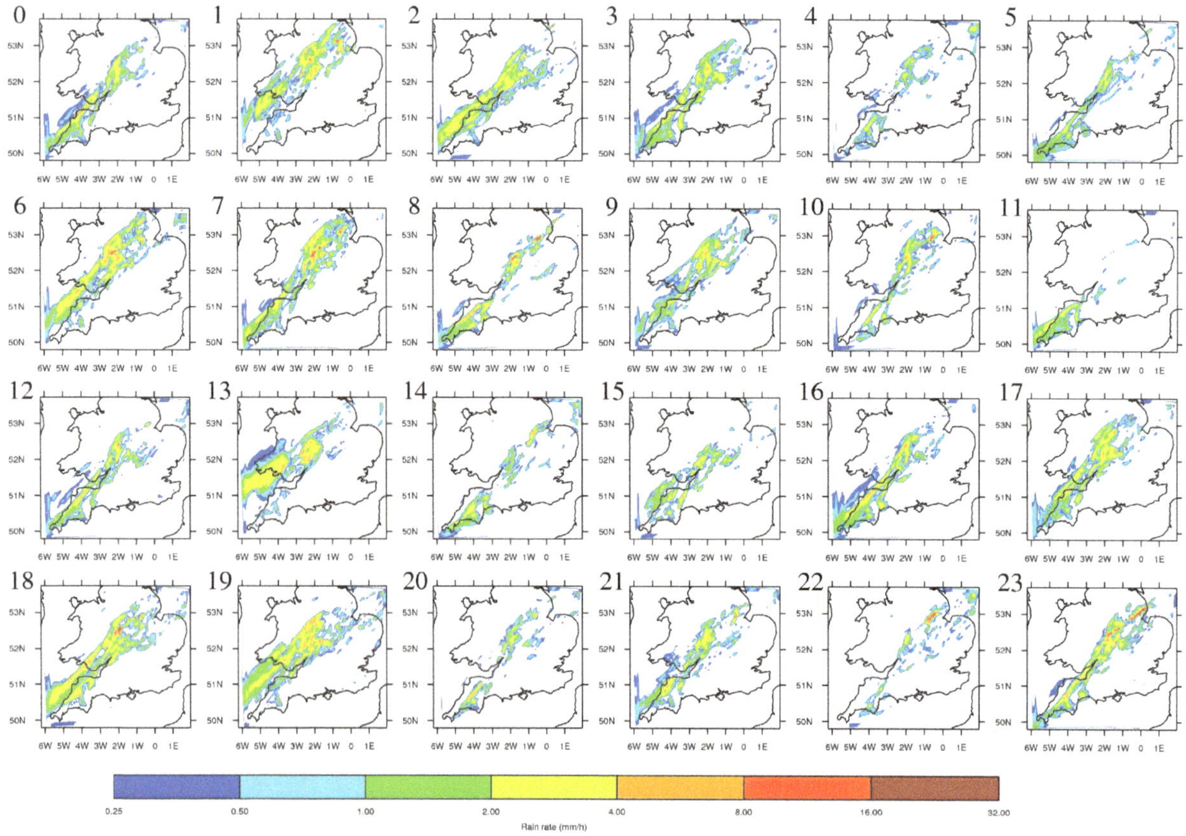

Fig. 8. Control ensemble rainfall rate ($\mathrm{mm\,h^{-1}}$) at 15:00 UTC. The control member (member 0) is in the top left-hand corner.

3.3 Parameter sensitivity tests

In order to make the RP scheme suitable to be used in the 1.5 km EPS, we first identified suitable parameters to be perturbed within the scheme, in addition to those used in the operational scheme. Since work had already been done by Zadra and Lock (2010) to identify extra parameters to be perturbed in the boundary layer scheme after the RP scheme was originally developed, we focus here on parameters in the large-scale precipitation scheme. An extensive set of around 50 parameters known to have some uncertainty was determined by consulting with experts at the Met Office. For each parameter a suitable range of values was chosen such that it was physically sensible while also reflecting the uncertainty in the value of the parameter.

From this extensive list of parameters, a smaller subset was chosen (Table 1). The first nine parameters listed in Table 1 are those used in the operational RP scheme, although some of the ranges have been changed, based on advice by experts. The last seven parameters in Table 1 are those that we have added to the RP scheme. The selection of these parameters was motivated by choosing from the 50 those with the highest sensitivity, and choosing one parameter for each physical process, in order to minimise the possibility of parameters counteracting each other. Choosing a set of parameters that collectively control a variety of different processes also has the advantage that the overall effects of the scheme may be less case-dependent. In addition, some parameters were chosen to vary together in pairs. These pairs of related parameters are indicated in Table 1. Parameters Ri_c and g_0 are related by $Ri_c = 10/g_0$; for the other parameter pairs the same random number is used to update their associated ϵ in Eq. (1), which ensures that these pairs of parameters remain correlated. In this chosen set of parameters, there are seven parameters from the boundary layer scheme and nine from the large-scale precipitation scheme, but due to the pairings of some parameters there are five independent parameters from the boundary layer scheme and seven independent parameters from the large-scale precipitation scheme.

To test the sensitivity of the forecast to each of these parameters, a series of forecasts was run in which each parameter (or pair of parameters) was perturbed to its maximum or minimum value. The perturbed forecasts were compared with the control forecast to determine the sensitivity of the forecast to each parameter, and also to test in what way the forecast was affected. In particular, it is desirable that perturbing a parameter to its maximum and minimum values gives opposite effects (e.g. one increases the temperature while the other decreases it), than for both cases to have the same result (e.g. a reduction in temperature) which would

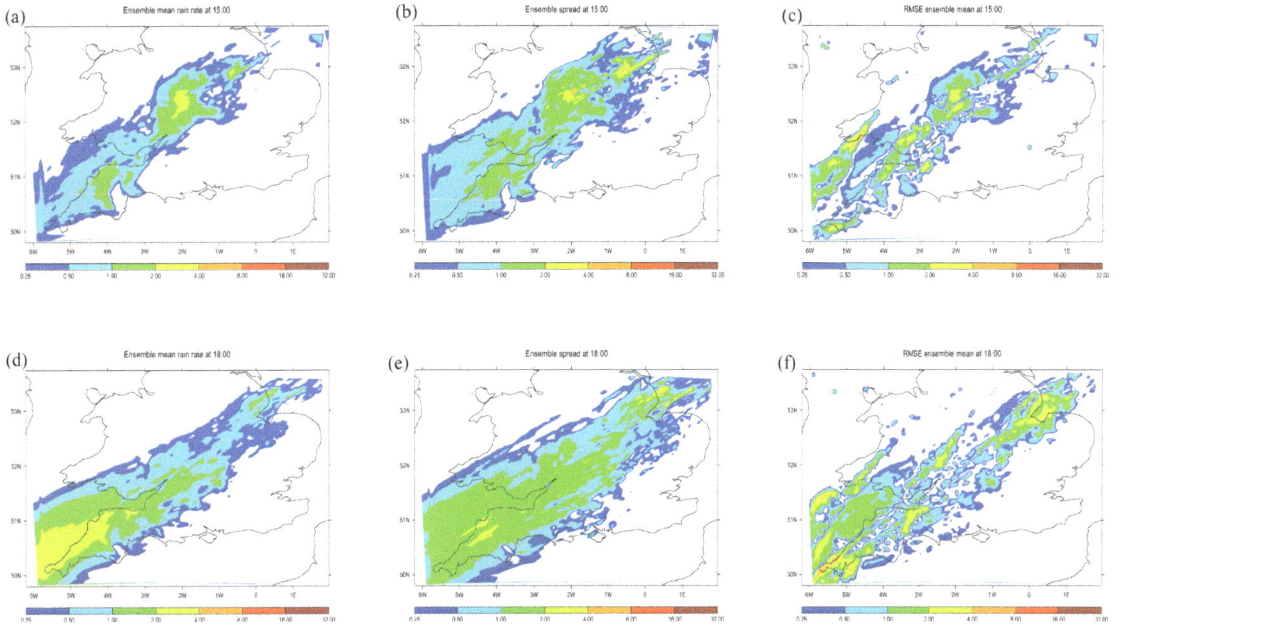

Fig. 9. Spatial distribution of rain rate diagnostics (mm h^{-1}) for the control ensemble. Left column: ensemble mean, middle column: ensemble spread and right column: innovation magnitude of the ensemble mean, at 15:00 UTC (top row) and 18:00 UTC (bottom row) 20 September 2011.

imply that perturbing this parameter randomly may introduce a bias.

The sensitivity of the forecast to perturbing each of the parameters to its maximum and minimum value was examined both by computing domain-averaged differences and root-mean-squared (RMS) differences. Difference maps of surface variables at different forecast lead times were also examined (not shown). These show that different parameters affect the forecast in different parts of the domain. Specifically, the large-scale precipitation parameters have the strongest effects in the region of the rain band, while the effects of boundary layer parameters are generally more widespread, and most strongly affect the north-east part of the domain.

Figure 7 shows the domain-averaged 3 h sensitivities with respect to each of the 12 independent parameters (or parameter pairs) for 1.5 m temperature and for the u component of the 10 m wind. Equivalent plots for 1.5 m RH and v component of 10 m wind (not shown) show qualitatively similar results to those for 1.5 m temperature and for the u component of 10 m wind (respectively), in terms of their relative sensitivities to each parameter (although note that the response for 1.5 m RH has the opposite sign to the response for 1.5 m temperature, as might be expected). Figure 7a and b show that for all parameters, the domain average effects on the temperature and u-component of wind of perturbing a parameter to its maximum value has an opposite sign to perturbing it to its minimum value. However, the bars showing the standard deviation indicate that there is a large variability in the effect of these parameters over the domain. The RMS difference plots (Fig. 7c and d) give a measure of the magnitude of the sensitivity to each of the parameters. It can be seen that the boundary layer parameter pairs (g_0, Ri_c) and (g_{mezcla}, λ_{min}), and the large-scale precipitation parameter $x1r$ (perturbed only to its maximum value) have the largest effects on both the temperature and u component of wind. The parameter pairs (g_0, Ri_c) and (g_{mezcla}, λ_{min}) affect the wind more, while $x1r$ has a large impact on the temperature. Although the effects of these parameters dominate in this case, the effects of perturbing the other parameters are not orders of magnitude smaller. The sensitivity to most of the parameters will be dependent on the atmospheric state (in particular the boundary layer stability for the boundary layer parameters, and the cloud amount and type for the large-scale precipitation parameters), and these results confirm that this is a reasonable set of parameters and ranges to choose in order to simulate model error variability.

4 Evaluation of the control ensemble

In this section we describe the properties of the control ensemble, CTL, which has IC and LBC perturbations but no representation of model error variability.

All the ensemble members capture the main rain band at 15:00 UTC (Fig. 8), although there is considerable variability in the amount of rainfall associated with this band (compare, for example, members 7 and 11). Inspection of the rain rate in the ensemble members from the driving model, MOGREPS-R (not shown), also shows the main rain band, but no member has the second band, suggesting that the mechanism causing

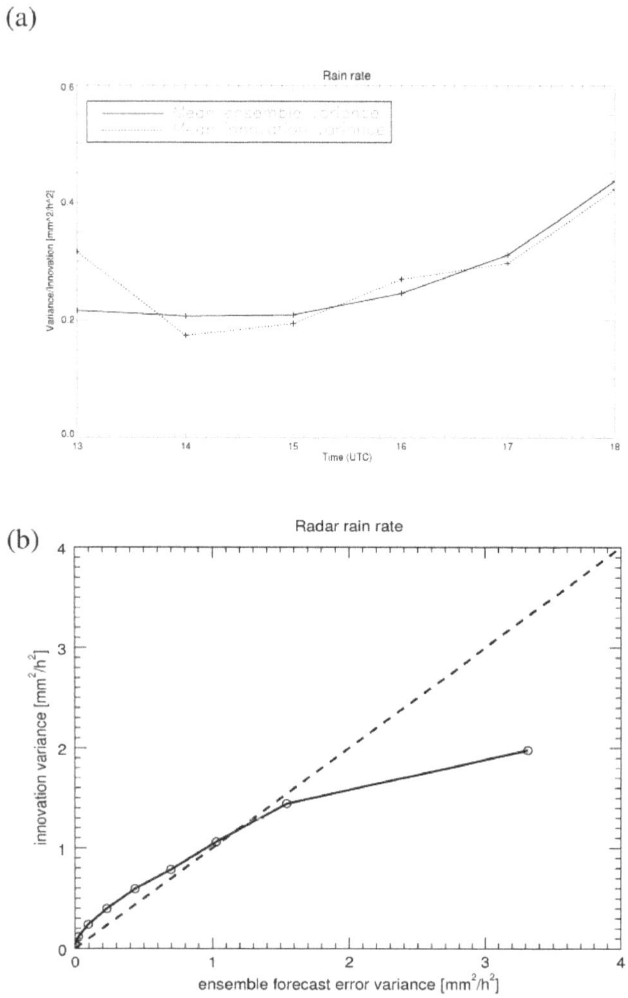

Fig. 10. **(a)** Domain averaged ensemble variance and innovation variance for rain rate every hour. **(b)** Spread-skill plot for the control ensemble. In **(b)** the data for each grid point for each hour of the forecast are binned into 10 bins by ensemble variance.

this banded structure to form in reality was not well represented in MOGREPS-R. Our hope was that the increased resolution might lead to some 1.5 km EPS members capturing the second band, but Fig. 8 shows that this ensemble also fails to capture the two bands together in any individual member. Whether this is an inevitable consequence of the lack of multiple bands in the driving model is an open question.

There is generally good agreement in the positioning of the main rain band, although three members (members 1, 13 and 19) have the rain band displaced further to the north. This means that, although the two distinct bands seen in the radar are not captured, these three members do have some rain in roughly the location of the second radar rain band. Several of the ensemble members have more intense rain in the north-east part of the rain band, and in particular some of them have one or two localised regions of heavy rain (e.g. members 7, 8 and 13) exceeding 16 mm h^{-1}, which is not seen in

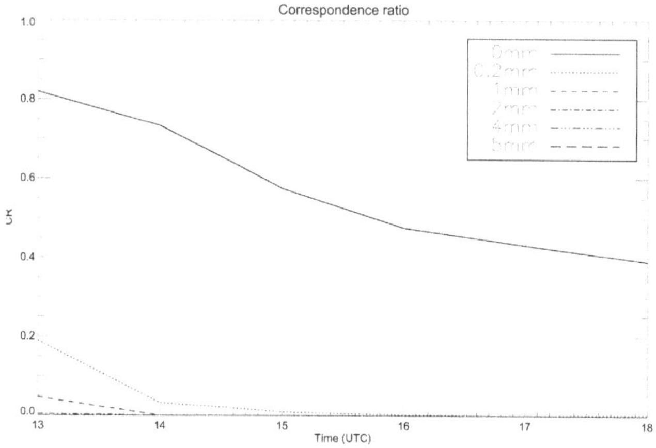

Fig. 11. Correspondence ratio for the control ensemble for hourly rain accumulation.

the radar rain rate. The locations of these localised regions of heavy rain correspond to the two regions of heaviest rain in the ensemble mean at this time (Fig. 9a). By 18:00 UTC the positioning of the rain band in the ensemble mean (Fig. 9d) compares well with the position of the main rain band in the radar (Fig. 5b), although the region of moderately heavy rain off the east coast in the radar is relatively weak in the ensemble mean.

Comparison of the ensemble spread with the innovation magnitude (middle and right columns of Fig. 9) shows that these quantities have quite different spatial distributions at both times shown. In the ensemble spread (Fig. 9b and e) there are more widespread regions of low values of spread, while in the innovation magnitude (Fig. 9c and f) there are more localised regions of higher values. Thus although the domain-averaged quantities (Fig. 10a) appear to be very similar, this is not due to these quantities having the same spatial distribution. The spread-skill plot produced by binning the data by ensemble variance (Fig. 10b) shows generally good agreement between the ensemble forecast error variance and the innovation variance. For larger ensemble forecast error variances the innovation variance is not as large, which could suggest that the ensemble is overspread in areas with higher rain rates. However, this may be at least partly due to the limited ensemble size (Bowler et al., 2008). These results are qualitatively similar to Fig. 16 in Bowler et al. (2008) and Fig. 15 in Migliorini et al. (2011).

The correspondence ratio (CR) was calculated for hourly rainfall accumulation for different accumulation thresholds (Fig. 11). At 13:00 UTC, the CR for the 0 mm threshold is initially high (0.8), but decreases throughout the forecast, reaching a minimum of around 0.4. In contrast, for the higher thresholds the CR is initially low (less than 0.2) and quickly decreases to zero or near-zero values. This indicates that after the first 1–2 h of the forecast there are no, or very few, grid points where all ensemble members agree for these thresholds. These results suggest that, while the ensemble members

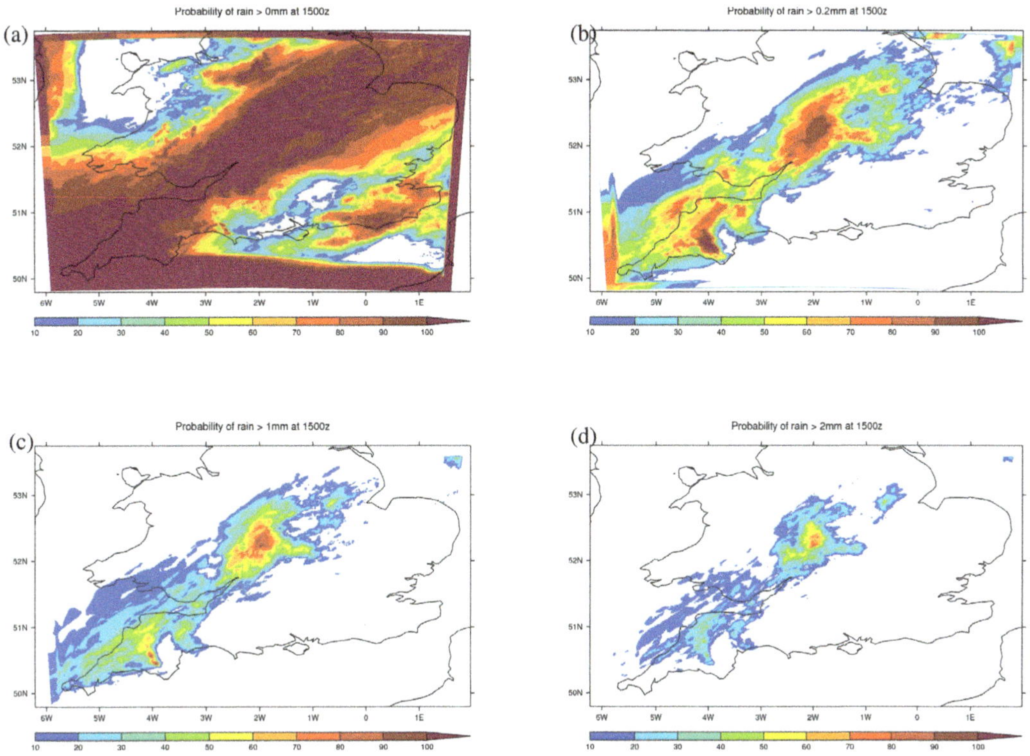

Fig. 12. Probability of rain rate exceeding given thresholds (%) for the control ensemble at 15:00 UTC. Thresholds are **(a)** 0 mm, **(b)** 0.2 mm, **(c)** 1 mm and **(d)** 2 mm.

are in generally good agreement about the location of the rain band, they disagree in the details of the location of heavier rain. This is supported by Fig. 12 which shows the probability of rain exceeding given thresholds. For a 0 mm threshold (Fig. 12a), there is a broad region across the domain with 100% probability of rain. In contrast, for higher thresholds (Fig. 12b–d) there are more areas with lower probabilities. For the three thresholds shown in Fig. 12b–d, there is a small region in the centre of the domain with high probabilities. This corresponds to some much smaller regions of rain exceeding 2 mm in the radar (Fig. 5a). In Fig. 12b and c there is some hint of a slightly separated band extending further north than the main rain band, which may be related to the second band seen in the radar (Fig. 5a), although in Fig. 12b and c it does not extend as far north.

The Brier score (BS) for hourly rain accumulation (Fig. 13a) shows an improvement in the skill over time for the 0 and 0.2 mm accumulation thresholds. This improvement over time, particularly after 15:00 UTC, suggests that the information coming through the boundaries at later times improves the forecast. There is also a significant improvement with increasing accumulation threshold, which is largely due to the decrease in the number of points exceeding the larger thresholds (Fig. 13b). This increases the number of points that are correctly forecast as not having rain exceeding the threshold, thus effectively improving the BS.

5 Evaluation of the effects of applying the RP scheme

The CTL ensemble represents just one realisation of forecast error. Here we examine the ensembles that have model error variability represented using the RP scheme. The features of each ensemble and the way that each is labelled are described in Sect. 2.5 and in Table 2. Here we show results for one run of each type of ensemble. To confirm that the results presented here are robust and not simply a result of specific combinations of random numbers used in the ϵ term in Eq. (1), these ensemble runs were repeated using different sets of random numbers. The results from these repeated ensembles were found to be qualitatively similar to those presented here.

One possibility considered is that the different parameters across the ensemble members in all IC + BC + RP* experiments may lead to some members forecasting the multiple rain bands. Once again though, all members in all of these ensembles fail to show this feature. This suggests that it may be the absence of the seed of this feature in the ICs and BCs provided by the driving model – rather than problems associated with model parameter values – that results in this forecast failure, since collectively our experiments have used so many different combinations of model parameters.

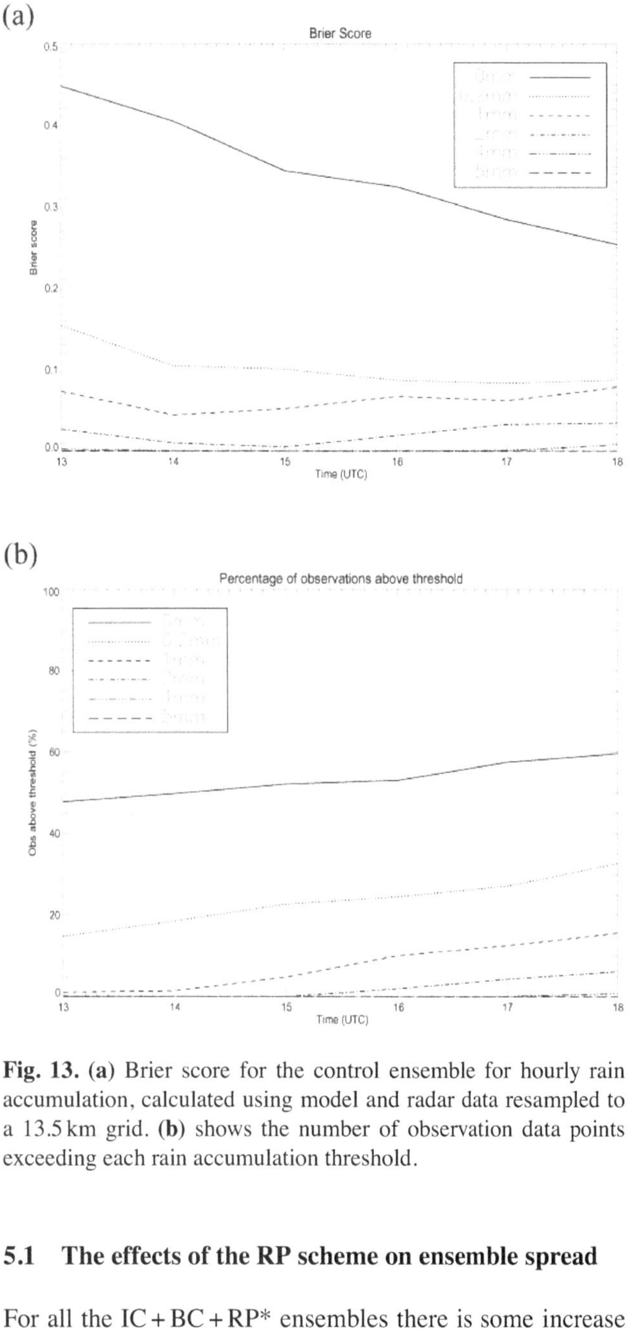

Fig. 13. **(a)** Brier score for the control ensemble for hourly rain accumulation, calculated using model and radar data resampled to a 13.5 km grid. **(b)** shows the number of observation data points exceeding each rain accumulation threshold.

5.1 The effects of the RP scheme on ensemble spread

For all the IC + BC + RP* ensembles there is some increase in the ensemble spread compared with the CTL ensemble, for the near-surface fields shown in Fig. 14b–d. For hourly rain accumulation (Fig. 14a) there is a slight decrease in spread in the IC + BC + RP* ensembles compared with the CTL ensemble. Note that the rapid increase in ensemble spread over the first hour of the simulation (until 12:30 UTC) is due to the gradual introduction of the IC perturbations by the IAU scheme over this period, which means that the ensemble members diverge throughout this time. The RP scheme is first applied at the end of the IAU period (at 12:30 UTC), causing a divergence in the spread of the four ensembles at this time. For temperature and RH (Fig. 14c and d), the three IC + BC + RP* ensembles have consistently larger spread than the CTL ensemble. For these variables there is little difference between the three model error ensembles, although IC + BC + RPfix has a slightly larger spread towards the end of the forecast. For 10 m wind speed (Fig. 14b) there is a slight increase in the spread in the IC + BC + RP* ensembles compared with the CTL ensemble. The spread for the two ensembles with periodic calls to the RP scheme (IC + BC + RP30 and IC + BC + RP60) has some small jumps at times corresponding to the times that the RP scheme updates the parameters. For hourly rain accumulation (Fig. 14a) the evolution of spread for ensembles IC + BC + RP30 and IC + BC + RP60 is less smooth than for ensemble IC + BC + RPfix, and while generally lower than the spread for the CTL ensemble, both ensembles have larger spread between around 16:30 and 17:30 UTC than the IC + BC + RPfix.

The spread gained by applying the RP scheme compared with the CTL ensemble can be compared with the results of Hacker et al. (2011). Their perturbed parameter ensemble gave an increase in spread for 2 m temperature of up to 30 % (their Fig. 6a), compared with their control ensemble. This compares well with Fig. 14c, which shows an increase in spread of 30 % in the last hour or so of the forecast for 1.5 m temperature. Their results also show an increase in spread of around 20 % for 2 m specific humidity (their Fig. 6c), compared with a 15–20 % increase in the spread for RH (Fig. 14d). In contrast to our results, Hacker et al. (2011) show a negligible change in 10 m wind speed, and an increase in spread for precipitation accumulation.

Figure 14a–d show that model error variability alone introduces much smaller spread than the IC perturbations, but still a much larger amount than the difference between the spread in the CTL and IC + BC + RP* ensembles. This is consistent with the results of Gebhardt et al. (2011) who found that the effects of their combined physics and LBC perturbations was not the sum of these two perturbations individually. For all four variables shown in Fig. 14a–d the change in spread of the RP* ensembles mirrors the shape of the spread of the IC + BC + RP* ensembles, increasing slightly throughout the forecast. It is interesting to note that the RPfix ensemble still shows an increase in spread throughout the forecast similar to the RP30 and RP60 ensembles, despite the fact that the RP scheme is only applied once. This indicates that the effect of the RP scheme on the spread is due to the members following different model manifolds from the first application, and subsequent changes to the manifolds due to repeated applications of the RP scheme in RP30 and RP60 do not increase the spread further in a substantial way. The similarity of the spreads between all RP* experiments may also be related to the high autocorrelation coefficient, r, in Eq. (1) and the nonlinear relationships between the parameter values and the forecasts.

For the 0 mm threshold (Fig. 15a) the CR for the IC + BC + RP* ensembles is higher than for the CTL ensemble, indicating a lower spread and a stronger agreement

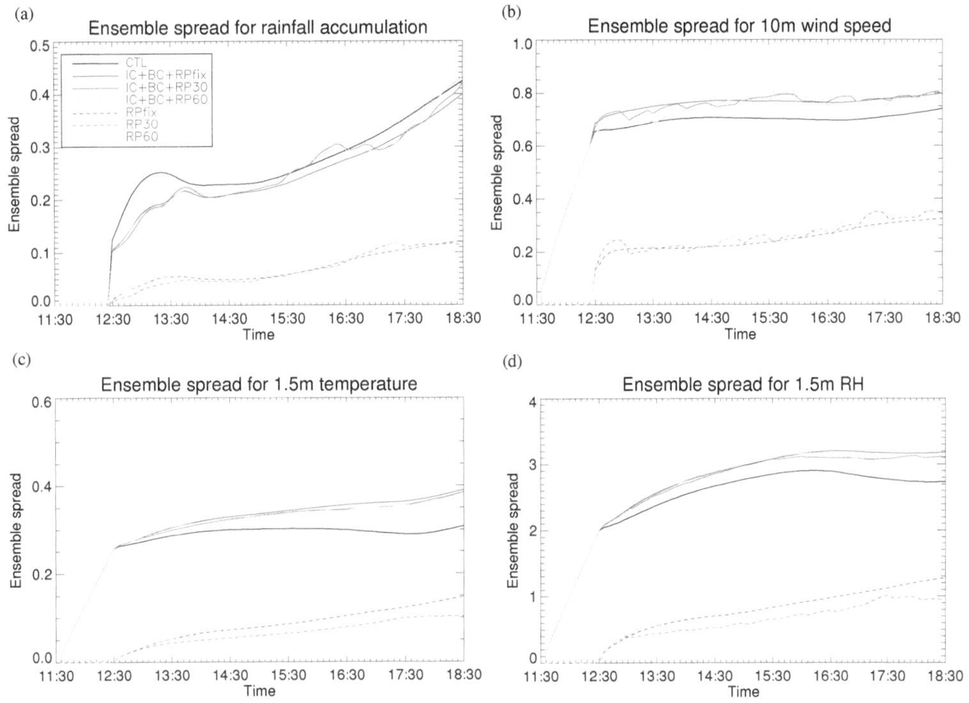

Fig. 14. Evolution of ensemble spread with forecast time for the different ensembles, for **(a)** hourly rainfall accumulation (mm), **(b)** 10 m wind speed (m s^{-1}), **(c)** 1.5 m temperature (K) and **(d)** 1.5 m RH (%).

between members in the location of the rain band. This is in agreement with plots of the probability of rain exceeding 0 mm for the IC + BC + RP* ensembles (equivalent to Fig. 12a) (not shown) which show broader regions with 100 % probability of rain than for the CTL ensemble. For higher thresholds (Fig. 15b–d) there is no significant change in the CR for the IC + BC + RP* ensembles compared with the CTL ensemble. The RP* ensembles have higher CR values than the CTL and IC + BC + RP* ensembles throughout the forecast for all thresholds (Fig. 15a–d). This is consistent with the much lower spread seen in rain accumulation for the RP* ensembles (Fig. 14a). The RP60 ensemble has consistently lower CR values than the other RP* ensembles for all thresholds, indicating more variability between ensemble members in this case. This is in agreement with Fig. 14a which shows a slightly higher spread for the RP60 ensemble than for RP30 and RPfix.

Figure 16a shows that the ensemble mean for the IC + BC + RPfix ensemble has a larger peak and a slightly narrower rain band than the ensemble mean for the CTL ensemble (Fig. 9a). The ensemble spread for the IC + BC + RPfix ensemble extends over a narrower band than the CTL ensemble (Fig. 16b compared with Fig. 9b), which is consistent with the reduction in domain-averaged spread seen in Fig. 14a. The innovation magnitude for the IC + BC + RPfix ensemble (Fig. 16c) is very similar to the innovation magnitude for the CTL ensemble (Fig. 9c), and similarly for the IC + BC + RP30 and IC + BC + RP60 ensembles (not shown). Therefore the addition of model error has not improved the relationship between the spatial distributions of ensemble spread and innovation magnitude.

5.2 The effects of the RP scheme on the forecast skill

The skill of each ensemble forecast was evaluated using the CRPS, as shown in Fig. 17 (accounting for observation error as described in Sect. 2.4). The observation error variance for each quantity was estimated from spread-skill plots for this case as the value of the skill for zero spread. The values used are specified in the figure caption. Repeating the calculations, but with zero observation error gives qualitatively similar results, which suggests that the results are not sensitive to the exact observation error variance values used.

The CRPS for 1.5 m temperature and 10 m winds (Fig. 17c, e and f) shows a general improvement in skill (reduction of CRPS) with forecast lead time, particularly in the IC + BC + RP* ensembles. This is surprising, as skill is normally expected to degrade with lead time. One hypothesis is that this improvement is due to large-scale information (and hence potentially more reliable information) entering the domain via the LBCs. If this were the case, we might expect see reductions in maps of CRPS propagating inwards, especially from the prevailing wind direction (from the south and west boundaries). Also, temporal evolutions of spread-skill relationships at different lead times were calculated to search for evidence of better spread-skill agreement at later lead times.

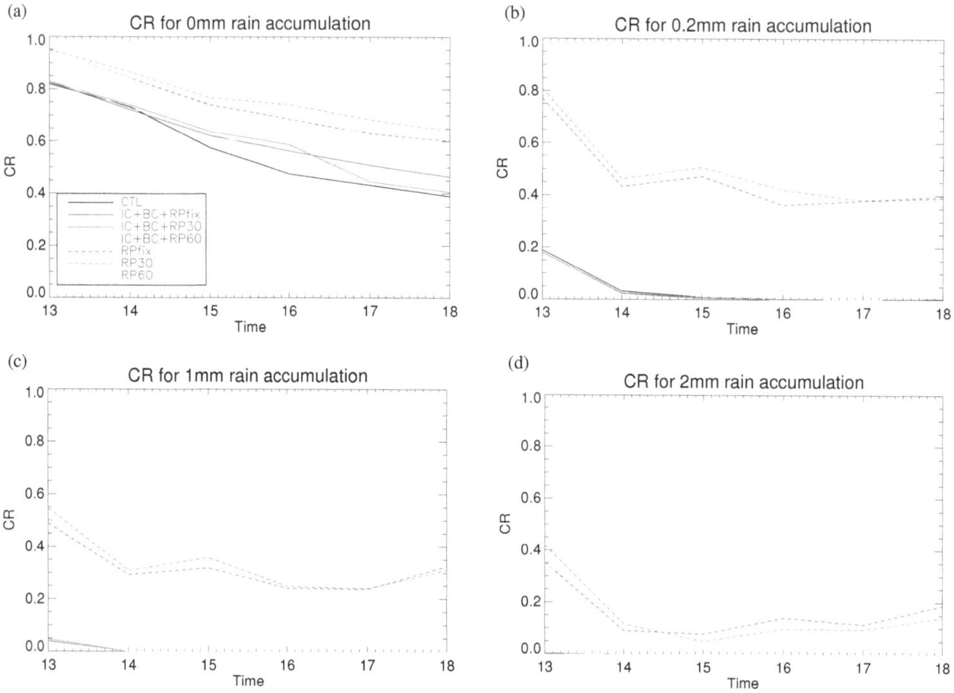

Fig. 15. Correspondence ratio for hourly rain accumulation for all the ensembles. Thresholds are **(a)** 0 mm, **(b)** 0.2 mm, **(c)** 1 mm and **(d)** 2 mm.

Both investigations, however (results not shown), showed no systematic signals to suggest that the LBC information is responsible for the decrease of CRPS (although this may be due to an insufficient observational sample at any given lead time and so this remains an open question). The improvement in skill with lead time for temperature and winds may be due to the effect of the readjustment of the forecast ensemble to a more balanced set of states after the effect of the inflation on the initial ensemble spread. As discussed in Sect. 2.1, the inflation is a spatially constant factor – determined from temperature and winds power spectra at selected levels – imposed on the ICs as determined by the ETKF. Arguably, a sub-optimal inflation on the initial ensemble spread could cause spurious gravity waves leading to loss of skill in the first hour into the forecast and subsequent improvements in skill would be achieved when the effects of the initial inflation are "forgotten" by the forecast ensemble.

The IC + BC + RP* ensembles have consistently better skill than the CTL ensemble, with a significant improvement in skill at and after the 4 h lead time. For 1.5 m RH (Fig. 17d) the skill for all ensembles gets worse over the first two hours, followed by an improvement from the 3 h lead time. As with 1.5 m temperature, the IC + BC + RP* ensembles show consistently better skill than the CTL ensemble, with the differences becoming more significant after the 2 h lead time. For surface u and v wind (Fig. 17e and f) there is large uncertainty in these results, shown by the standard error bars. There is no significant difference in the CRPS between ensembles, indicating that the RP scheme has little impact on the skill of the wind forecast. For rain rate and hourly rain accumulation (Fig. 17a and b) all ensembles show an improvement in skill between the 1 and 2 h lead times, followed by a degradation in skill over the remainder of the forecast. For the first 2 h, the IC + BC + RP* ensembles show an improvement in the skill compared with the CTL ensemble, which is particularly significant in the rain accumulation at the 1 h lead time. After this the IC + BC + RP* ensembles have a significantly worse skill than the CTL ensemble. These results are consistent with results using the PSS (Fig. 18). In the first 2 h of the forecast the PSS is positive for all thresholds, showing that the IC + BC + RP* ensembles perform slightly better than the CTL ensemble; after this time the PSS is negative, showing that the IC + BC + RP* ensembles are less skillful than the CTL ensemble according to this diagnostic. For the higher thresholds (Fig. 18c and d) the PSS is approximately zero at 15:00 UTC, indicating no change in the skill at this time for these thresholds. All three IC + BC + RP* ensembles show similar PSS values throughout much of the forecast, although for the 0 mm threshold (Fig. 18a) IC + BC + RPfix performs worse than the other ensembles towards the end of the forecast.

The CRPS for the RP* ensembles was also calculated (not shown). These ensembles have consistently worse skill than either the CTL or the IC + BC + RP* ensembles. This shows that using the RP scheme to add model error variability alone produces a less skillful ensemble than an ensemble with

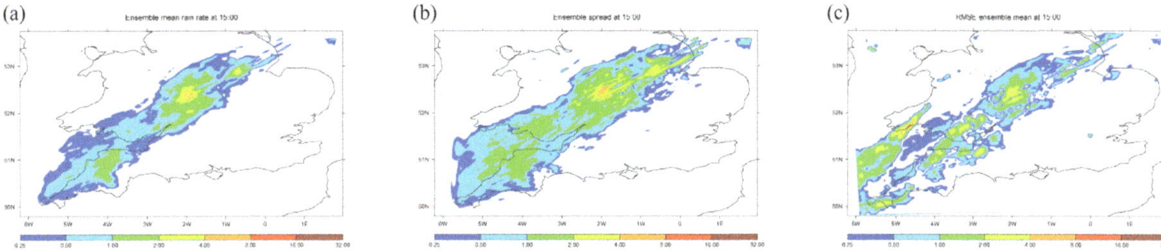

Fig. 16. Rain rate diagnostics for the IC + BC + RPfix ensemble (mm h^{-1}). **(a)** ensemble mean, **(b)** ensemble spread and **(c)** innovation magnitude of the ensemble mean, all at 15:00 UTC 20 September 2011.

perturbed ICs and LBCs. This is likely due to the smaller spread in the RP* ensembles compared with the others.

For all ensembles shown in Fig. 17 (i.e. with and without model error variability) and for the near-surface quantities (1.5 m temperature and RH and 10 m wind) the forecasts show a tendency to improve throughout the forecast. This is indicated by a decrease of the CRPS with time. This may be due to the spin-up of the model after the IAU has finished and to large-scale information coming in from the boundaries. All ensembles show an improvement in rainfall skill for the first 2 h of the forecast, with the IC + BC + RP* ensembles performing better in this period. In contrast, all ensembles show a worsening of skill in rainfall towards the end of the forecast, and the IC + BC + RP* ensembles show worse skill than the CTL ensemble for rain rate and accumulation. This may be due to inconsistencies between fields in the interior of the domain (for which some parameters have been perturbed) and the LBCs (which have not been perturbed). However, the validity of this idea is difficult to demonstrate with just one case study.

6 Conclusions

In this paper we have evaluated the effects of using the random parameters (RP) scheme as a method of representing model error in the 1.5 km EPS for a case on 20 September 2011 in which a frontal rain band crossed through the domain. Although it is difficult to draw conclusions from one case study only, we have shown that the RP scheme in this convective-scale EPS can actually reduce the ensemble variability for some variables.

6.1 Summary of results

The standard RP scheme (used operationally in MOGREPS) was used as a basis for the representation of model error uncertainty, and modifications were made to make it suitable for use in the convective-scale 1.5 km EPS. A set of extra parameters to be perturbed were identified in the large-scale precipitation scheme, in addition to the existing boundary layer and large-scale precipitation parameters (Table 1). Some pairs of related parameters were varied together (using the same random number), with the intention of controlling each physical process by only one parameter or one pair of parameters. Parameter sensitivity testing on forecasts showed that there was some sensitivity to all the chosen parameters when perturbed individually, and none of the perturbed forecasts gave physically unrealistic results.

Seven ensembles were run with different configurations: the control ensemble had IC and BC perturbations only; three ensembles had IC and BC perturbations and model error represented by the RP scheme, with parameter values updated with different periods; and three ensembles had no IC or BC perturbations but did have model error represented by the RP scheme, with the same three update periods.

The control ensemble was found to represent the main rain band reasonably well, but all members failed to capture the two separate rain bands seen in the radar in the model's domain. The position of the main rain band, however, in this ensemble was positioned in some members at locations occupied by the second rain band in the radar. This was also the case for the ensembles with model error.

The ensembles with IC and BC perturbations and the RP scheme applied were found to have higher domain-averaged ensemble spread for the near-surface temperature, RH and wind speed than the control ensemble, but a reduced spread for hourly rain accumulation. This decrease in rainfall spread may be due to the fact that the ensembles with the RP scheme applied have a narrower rain band in most members, which accounts for the lower domain averaged spread. The correspondence ratio showed a fast convergence to zero for rain accumulation thresholds greater than zero, indicating that after the first hour or so of the forecast the members disagree on the positions of regions of heavier rain. This is not a surprising result as the regions of heavier rain are found to be more localised than those of lighter rain. The ensembles with the RP scheme applied but no IC and LBC perturbations showed that model error alone introduces less spread than the IC and LBC perturbations (at least in the way that we have configured the RP system), but considerably more spread than the spread gained in the IC + BC + RP* ensembles compared with the control. According to the CRPS, applying the RP scheme gave a significant improvement in skill after the first 2–3 h of the forecast for 1.5 m temperature and

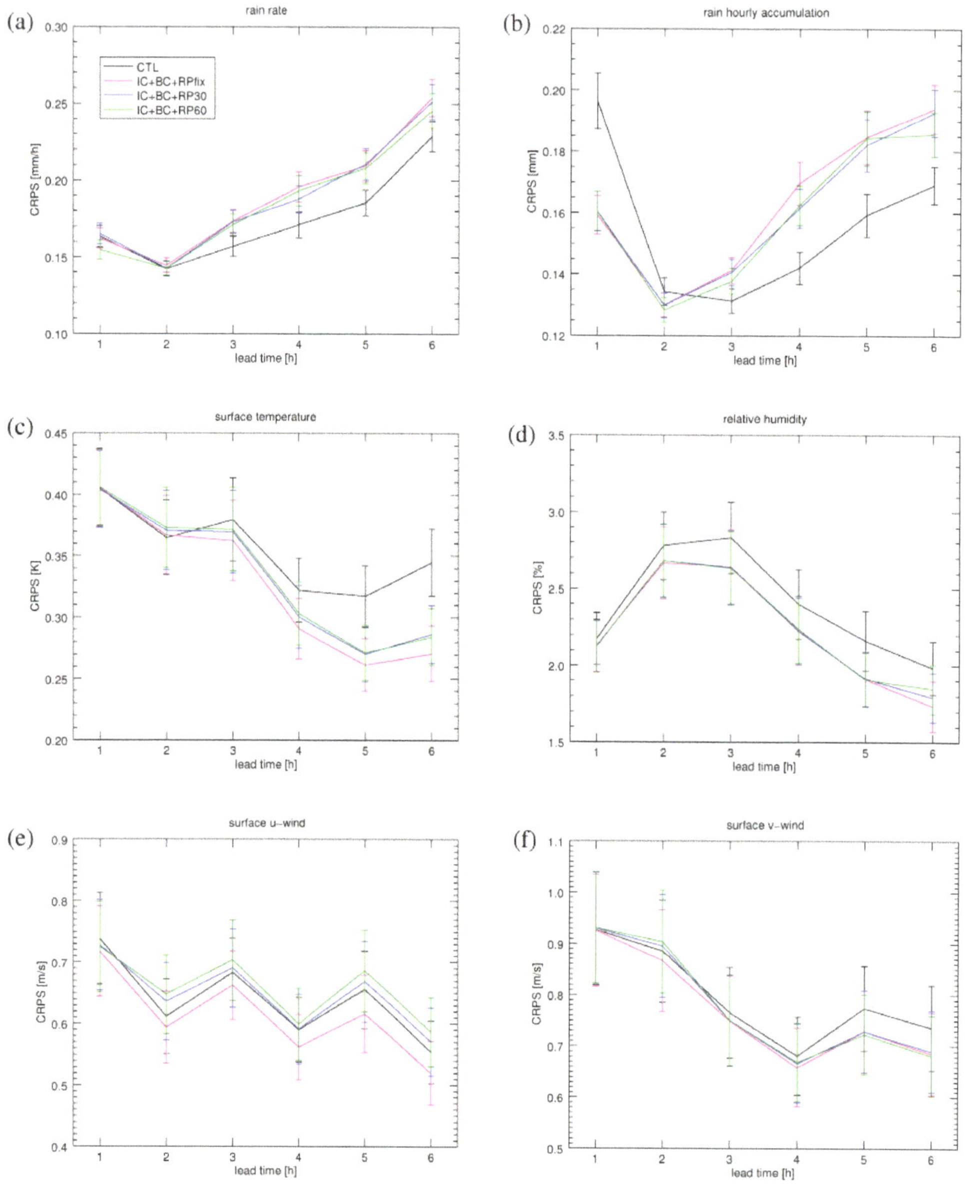

Fig. 17. CRPS for **(a)** rain rate, **(b)** hourly rain accumulation, **(c)** 1.5 m temperature, **(d)** 1.5 m RH, **(e)** u component of 10 m wind and **(f)** v component of 10 m wind. The CRPS shown is the mean CRPS over the domain for each ensemble; error bars show the standard error in the mean. The observation error standard deviations used in the calculation are **(a)** $\sigma_o = 0.47\,\mathrm{mm\,h^{-1}}$, **(b)** $\sigma_o = 0.47\,\mathrm{mm}$, **(c)** $\sigma_o = 0.69\,\mathrm{K}$, **(d)** $\sigma_o = 4.72\,\%$, **(e)** $\sigma_o = 1.54\,\mathrm{m\,s^{-1}}$ and **(f)** $\sigma_o = 1.54\,\mathrm{m\,s^{-1}}$. These values were estimated from spread-skill plots for this case.

1.5 m relative humidity, as compared with the control ensemble. For rainfall, there was an improvement in skill at the start of the forecast for the IC + BC + RP* ensembles compared with the control, but a reduced skill in the last few hours. This may be due to inconsistencies between the interior and the boundaries of the domain, which may lead to incorrect rainfall amounts later in the forecast.

The results for both the spread and the skill indicate that perturbing the parameters periodically throughout the forecast has little impact on the forecast spread or skill compared with perturbing the parameters once and holding them fixed for the rest of the forecast. However, perturbing the parameters only once means that less of the parameter space will be spanned.

Overall, for the case discussed here, the effects of applying the RP scheme to an ensemble with IC and LBC perturbations has a positive effect on the spread and skill of near-surface temperature and relative humidity, but a negative effect on the spread and skill of precipitation.

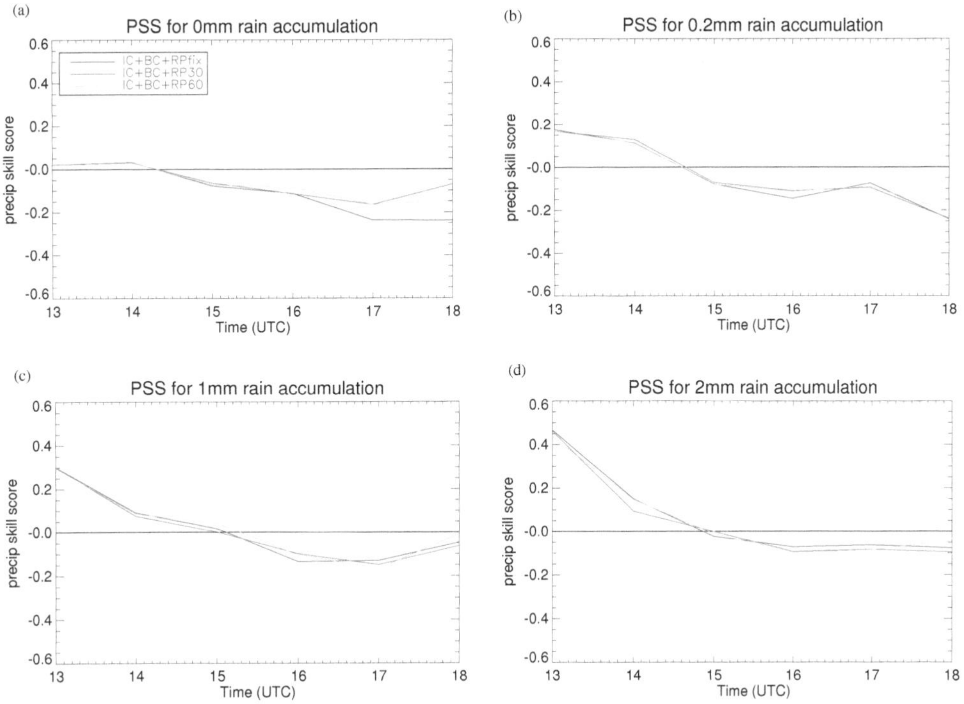

Fig. 18. PSS for hourly rain accumulation for the IC + BC + RP* ensembles with reference to the CTL ensemble for **(a)** 0 mm, **(b)** 0.2 mm, **(c)** 1 mm and **(d)** 2 mm thresholds.

6.2 Further work

An obvious limitation of this study is that we have considered only one particular case. This was due to time constraints and upgrades to the modelling system which meant that it was not possible to run further cases with the same model setup. Repeating these experiments with different cases, in particular for different synoptic situations such as convective showers rather than a synoptically forced frontal system would give a broader picture of the usefulness of the RP scheme. We expect that for different synoptic situations there would be differences in the relative sensitivity to each of the parameters. For example, we expect that in a convective case the perturbations made to the large-scale precipitation parameters may have a larger effect than in the case discussed here.

A further limitation is our constraint to using the existing formulation of the RP scheme as a basis for our study. One potential issue with this setup is that each time the parameters are perturbed the model is "kicked" into a different state corresponding to a particular set of parameter values. While the formulation of the autoregression Eq. (1) used in the RP scheme ensures that the parameters remain within a certain range of their previous values, this did still introduce some jumpiness in the wind and rain fields, which is not desirable. An alternative approach would be to update the parameter values more frequently (e.g. every time step, which we might call RPts), but change the autoregression constant r in Eq. (1) to be closer to unity, and/or reduce the range of the shock ϵ. This would allow a smooth evolution of parameters. By extrapolating the results in this study, we hypothesize that this smoother approach would give a more skillful forecast than the method used here, but may not give any additional increase in ensemble spread. It would also require tuning of r and ϵ, which would be a time-consuming process.

The possible ways in which the RP scheme may be applied raises some interesting issues. At one end of the scale RPfix allows n sets of parameters (where n is the number of ensemble members) to be applied in forecasts that each follow precisely the model equations. At the other end of the scale RPts allows more of parameter space to be explored, but no member will follow the same model equations. Strictly speaking this will lead to loss of continuity (of e.g. mass, water, energy, etc.), but this may actually be a desirable feature given that the model equations are imperfect and the conserved quantities are imperfectly known anyway. Each strategy represents a different philosophical approach to representing model error variability with variable parameters and a comparison would make an interesting piece of future work.

The issues with the degradation of forecast skill for rain rate towards the end of the forecast, which we hypothesize is due to conflicting information in the perturbed domain interior and the information coming in through the domain boundaries from the non-perturbed driving model, suggests that this method may not be ideal for forecasts of more than around a 3 h lead time. However, the improvement in skill of rainfall forecast in the first 1–2 h of the forecast show that it could be useful for nowcasting and very short forecasts. A potential application of the RP scheme could be to use it within an hourly cycling system as an alternative to the inflation factor as a method of increasing the ensemble spread. It would also be of use on a larger domain, such as that of the new MOGREPS-UK ensemble, which covers the whole of the UK, and therefore the interior of the domain would be less strongly influenced by the LBCs.

Ongoing work aims to investigate the effects of the RP scheme on forecast error covariance statistics and covariance length scales (Migliorini, Bannister, Rudd and Baker, in preparation) and on the inherent balances that are obeyed by the ensemble (Bannister, Migliorini, Baker and Rudd, in preparation). These have applications to informing the appropriate formulation of the background error covariance matrix used in convective-scale data assimilation systems.

Acknowledgements. This work was funded by the Natural Environment Research Council (NERC) under the DIAMET project (Grant ref. NE/I005234/1).The authors wish to thank Ian Boutle and Jonathan Wilkinson for their advice on parameterisation schemes; Emilie Carter, Jean-Francois Caron, Giovanni Leoncini, David Simonin, Simon Wilson and Roger Brugge for technical help; and Sue Ballard, Neill Bowler and Jonathan Flowerdew for helpful discussions on this work. We are also grateful for constructive comments from two anonymous referees and editors of NPG (Roberto Buizza and Olivier Talagrand), which helped to improve the paper. We thank NERC and the Met Office for use of the MetUM and the MONSooN supercomputing facility.

Edited by: R. Buizza

References

Berner, J., Shutts, G. J., Leutbecher, M., and Palmer, T. N.: A spectral stochastic kinetic energy backscatter scheme and its impact on flow-dependent predictability in the ECMWF Ensemble Prediction System, J. Atmos. Sci., 66, 603–629, 2009.

Berner, J., Ha, S.-Y., Hacker, J. P., Fournier, A., and Snyder, C.: Model uncertainty in a mesoscale ensemble prediction system: Stochastic versus multiphysics representations, Mon. Weather Rev., 139, 1972–1995, 2011.

Bishop, C. H., Etherton, B. J., and Majumdar, S. J.: Adaptive sampling with the ensemble tranform Kalman filter. Part I: Theoretical aspects., Mon. Weather Rev., 129, 420–436, 2001.

Bouttier, F., Vié, B., Nuissier, O., and Raynaud, L.: Impact of Stochastic Physics in a convection-permitting ensemble, Mon. Weather Rev., 140, 3706–3721, 2012.

Bowler, N. E., Arribas, A., Mylne, K. R., Robertson, K. B., and Beare, S. E.: The MOGREPS short-range ensemble prediction system, Q. J. Roy. Meteorol. Soc., 134, 703–722, 2008.

Bright, D. R. and Mullen, S. L.: Short-range ensemble forecasts of precipitation during the Southwest Monsoon, Weather Forecast., 17, 1080–1100, 2002.

Buizza, R., Miller, M., and Palmer, T. N.: Stochastic representation of model uncertainties in the ECMWF Ensemble Prediction System, Q. J. Roy. Meteorol. Soc., 125, 2887–2908, 1999.

Caron, J.-F.: Mismatching perturbations at the lateral boundaries in limited-area ensemble forecasting: A case study, Mon. Weather Rev., 141, 356–374, 2013.

Clark, A. J., Gallus Jr., W. A., Xue, M., and Kong, F.: Growth of spread in convection-allowing and convection-parameterizing ensembles, Weather Forecast., 25, 594–612, 2010.

Clark, A. J., Kain, J. S., Stensrud, D. J., Xue, M., Kong, F., Coniglio, M. C., Thomas, K. W., Wang, Y., Brewster, K., Gao, J., Wang, X., Weiss, S. J., and Du, J.: Probabilistic precipitation forecast skill as a function of ensemble size and spatial scale in a convection-allowing ensemble, Mon. Weather Rev., 139, 1410–1418, 2011.

Davies, T., Cullen, M. J. P., Malcolm, A. J., Mawson, M. H., Staniforth, A., White, A. A., and Wood, N.: A new dynamical core for the Met Office's global and regional modelling of the atmosphere, Q. J. Roy. Meteorol. Soc., 131, 1759–1782, 2005.

Fresnay, S., Hally, A., Garnaud, C., Richard, E., and Lambert, D.: Heavy precipitation events in the Mediterranean: sensitivity to cloud physics parameterisation uncertainties, Nat. Hazards Earth Syst. Sci., 12, 2671–2688, doi:10.5194/nhess-12-2671-2012, 2012.

Gebhardt, C., Theis, S. E., Paulat, M., and Bouallègue, Z. B.: Uncertainties in COSMO-DE precipitation forecasts introduced by model perturbations and variation of lateral boundaries, Atmos. Res., 100, 168–177, 2011.

Gregory, D. and Rowntree, P. R.: A mass flux convection scheme with representation of cloud ensemble characteristics and stability-dependent closure, Mon. Weather Rev., 118, 1483–1506, 1990.

Hacker, J. P., Snyder, C., Ha, S.-Y., and Pocernich, M.: Linear and non-linear response to parameter variations in a mesoscale model, Tellus, 63A, 429–444, 2011.

Hersbach, H.: Decomposition of the Continuous ranked probability score for ensemble prediction systems, Weather Forecast., 15, 559–570, 2000.

Lean, H. W., Clark, P. A., Dixon, M., Roberts, N. M., Fitch, A., Forbes, R., and Halliwell, C.: The Met Office global three-dimensional variational data assimilation scheme, Q. J. R. Meteorol. Soc., 131, 1759–1782, 2008.

Lin, J. W.-B. and Neelin, J. D.: Influence of a stochastic moist convective parameterization on tropical climate variability, Geophys. Res. Lett., 27, 3691–3694, 2000.

Lock, A. P., Brown, A. R., Bush, M. R., Martin, G. M., and Smith, R. N. B.: A new boundary layer mixing scheme. Part I: scheme description and single-column model tests, Mon. Weather Rev., 128, 3187–3199, 2000.

Migliorini, S., Dixon, M., Bannister, R., and Ballard, S.: Ensemble prediction for nowcasting with a convection-permitting model – I: description of the system and the impact of radar-derived surface precipitation rates, Tellus, 63, 468–496, doi:10.1111/j.1600-0870.2010.00503.x, 2011.

Murphy, J. M., Sexton, D. M. H., Barnett, D. N., Jones, G. S., Webb, M. J., Collins, M., and Stainforth, D. A.: Quantification of modelling uncertainties in a large ensemble of climate change simulations, Nature, 430, 768–772, 2004.

Shutts, G.: A kinetic energy backscatter algorithm for use in ensemble prediction systems, Q. J. Roy. Meteorol. Soc., 131, 3079–3102, 2005.

Stainforth, D. A., Aina, T., Christensen, C., Collins, M., Faull, N., Frame, D. J., Kettleborough, J. A., Knight, S., Martin, A., Murphy, J. M., Pianl, C., Sexton, D., Smith, L. A., Spicer, R. A., Thorpe, A. J., and Allen, M. R.: Uncertainty in predictions of the climate response to rising levels of greenhouse gases, Nature, 433, 403–406, 2005.

Stensrud, D. J., Bao, J.-W., and Warner, T. T.: Using initial condition and model physics perturbations in short-range ensemble simulations of mesoscale convective systems, Mon. Weather Rev., 128, 2077–2107, 2000.

Vié, B., Molinié, G., Nuissier, O., Vincendon, B., Ducrocq, V., Bouttier, F., and Richard, E.: Hydro-meteorological evaluation of a convection-permitting ensemble prediction system for Mediterranean heavy precipitating events, Nat. Hazards Earth Syst. Sci., 12, 2631–2645, doi:10.5194/nhess-12-2631-2012, 2012.

Webster, S., Brown, A. R., Cameron, D. R., and Jones, C. P.: Improvements to the representation of orography in the Met Office Unified Model, Q. J. Roy. Meteorol. Soc., 129, 1989–2010, 2003.

Wilks, D. S.: Statistical methods in the atmospheric sciences, Elsevier Academic Press, 2nd Edn., 2006.

Wilson, D. R. and Ballard, S. P.: A microphysically based precipitation scheme for the Meteorological Office Unified Model, Q. J. Roy. Meteorol. Soc., 125, 1607–1636, 1999.

Zadra, A. and Lock, A.: Sensitivity tests with respect to random parameters in the boundary layer scheme, Tech. rep., Met office, 2010.

2

Nonlinear dynamics approach to the predictability of the Cane–Zebiak coupled ocean–atmosphere model

L. Siqueira and B. Kirtman

Rosenstiel School of Marine and Atmospheric Science, University of Miami, Miami, USA

Correspondence to: L. Siqueira (lsiqueira@rsmas.miami.edu)

Abstract. The predictability of the Cane–Zebiak coupled ocean–atmosphere model is investigated using nonlinear dynamics analysis. Newer theoretical concepts are applied to the coupled model in order to help quantify maximal prediction horizons for finite amplitude perturbations on different scales. Predictability analysis based on the maximum Lyapunov exponent considers infinitesimal perturbations, which are associated with errors in the smallest fastest-evolving scales of motion. However, these errors become irrelevant for the predictability of larger scale motions. In this study we employed finite-size Lyapunov exponent analysis to assess the predictability of the Cane–Zebiak coupled ocean–atmosphere model as a function of scale. We demonstrate the existence of fast and slow timescales, as noted in earlier studies, and the expected enhanced predictability of the anomalies on large scales. The final results and conclusions clarify the applicability of these new methods to seasonal forecasting problems.

1 Introduction

The fundamental question regarding the predictability limits of the coupled ocean–atmosphere system has not been well established yet, although recent coupled models are able to forecast phenomena like El Niño–Southern Oscillation (ENSO) up to about one year in advance (Kirtman and Min, 2009). The atmospheric predictability has been extensively investigated in the past, and classical predictability studies (Lorenz, 1965, 1969; Charney et al., 1966; Smagorinsky, 1969; Shukla, 1985) have established that the instantaneous state of the atmosphere cannot be predicted beyond a few weeks. However, only a few studies have examined the predictability of the coupled ocean–atmosphere system. Uncertainties in the atmospheric state are passed to the ocean through coupling processes. The predictability of the coupled system depends critically on how uncertainties in the atmospheric state reach nonlinear equilibrium, and how this relatively high-frequency noise interacts with the low-frequency evolution of the ocean model (Goswami et al., 1997). Within the atmosphere, a similar reasoning for the predictability dependence on 2-D turbulence interacting with 3-D turbulence also applies. The tropical Pacific coupled ocean–atmosphere system with its primary oscillatory instability, i.e., ENSO, is an interesting example of a coupled fast–slow system of crucial importance for seasonal forecasting, and is the focus of this study.

In the early 90s, Goswami and Shukla (1991) and Blumenthal (1991), following different procedures but using the same model, showed that the growth of errors in the Cane–Zebiak coupled model (Zebiak and Cane, 1987) is governed by two different timescales. However, none of these studies involving the predictability of the coupled ocean–atmosphere system made use of methods from nonlinear dynamics analysis available today. Newer theoretical concepts are available to reveal the detailed structure of the observed or modeled variability in a condensed way, providing additional conceptual insight as well as quantifying maximal prediction horizons that have useful applicability to seasonal forecasting problems like ENSO.

Classical predictability analysis based on the maximum Lyapunov exponent (MLE) considers infinitesimal perturbations (Rosenstein et al., 1993; Kantz, 1994), which are associated with errors in the smallest fastest-evolving scales of motion as in Eq. (1). However, the Lyapunov exponent and its association with prediction horizons (Boffetta et al., 1998a) are of very limited use in multiple timescale systems like geophysical flows, and predictions can actually be made far beyond the characteristic Lyapunov timescale. The MLE can be defined as

$$\lambda = \lim_{t\to\infty} \lim_{\delta_x^0\to 0} \frac{1}{t} \ln \frac{|\boldsymbol{\delta x}(t)|}{|\boldsymbol{\delta}_x^0|}, \tag{1}$$

where $\boldsymbol{\delta}_x^0 = \boldsymbol{\delta x}(0)$ represents an infinitesimal initial uncertainty in the model state. Because the initial conditions can be measured only with a finite uncertainty $\boldsymbol{\delta x}(0)$, we can forecast the future state of the system with a given error tolerance level Δ only to a maximum time T_p (prediction horizon)

$$T_\mathrm{p} = \frac{1}{\lambda} \ln \frac{\Delta}{\boldsymbol{\delta x}(0)}. \tag{2}$$

Equation (2) gives the shortest prediction horizon, which is associated with the smallest, low energy-containing scales that, as soon as their error reaches nonlinear saturation, do not play a role in the error growth. Larger errors, which typically occur at larger scales, will grow at a different rate. This is usually due to the small-scale instabilities growing faster but becoming nonlinearly saturated at a much smaller level than large-scale instabilities. In general, large errors will grow with the characteristic timescale of the largest slow-evolving high-energy containing scales. It is worth noting at this point that the largest energy-containing scales are associated with instabilities that obtain energy from the mean flow and also from each other. Thus these are the energy production instabilities that contain most of the energy, but are not the same as the largest scales in space and time. Therefore, the effect of the smallest fast-evolving scales (considered by the MLE analysis) becomes irrelevant for the predictability of larger scale motions.

In order to overcome this problem associated with the infinitesimally small amplitude perturbations, Aurell et al. (1997) have generalized the Lyapunov exponent concept to non-infinitesimal finite amplitude perturbations. The finite-size Lyapunov exponent (FSLE) is particularly useful when there exist characteristic scales; this allows the computation of the average exponential error growth at a given scale, say, δ. This generalization has the advantage that both the nonlinear dynamical evolution of these perturbations as well as the predictability of multiple timescale systems can be treated appropriately (Boffetta et al., 1998a). The finite-size Lyapunov exponent $\lambda(\delta)$ is based on the concept of error growing time $\tau_q(\delta)$ (error q-pling time), which is the smallest time it takes for a perturbation of initial size δ to grow by a factor q.

To practically compute the FSLE, we assume that a system has been evolved for long enough that the transient dynamics has lapsed, e.g., for dissipative systems the motion has settled onto the attractor. Then we consider at $t = 0$ a "reference" trajectory $\boldsymbol{x}(0)$ supposed to be on the attractor, and generate a "perturbed" trajectory $\boldsymbol{x}'(0) = \boldsymbol{x}(0) + \delta \boldsymbol{x}(0)$. We need the perturbation to be initially very small (essentially infinitesimal) in some chosen norm $\delta(t=0) = \| \delta \boldsymbol{x}(t=0) \| = \delta_{\min} \ll 1$. Typically, in numerical experiments with "toy models" $\delta_{\min} = \mathcal{O}(10^{-6} – 10^{-8})$, but for intermediate complexity models like the Cane–Zebiak coupled model and in a state-of-the-art coupled general circulation model (CGCM), $\delta_{\min}$ is limited by the precision of the finite data set that leads to a minimum length scale below which the attractor structure cannot be resolved (later shown in Sect. 4). Then, in order to study the perturbation growth through different scales, we can define a series of thresholds, $\delta_n = q^n \delta_0$, and to measure the time $\tau_q(\delta)$ that a perturbation with size δ_n takes to grow up to δ_{n+1}, where $n = 0, \ldots, N_\mathrm{s}$ represents an increasing scale and N_s is the scale at which the error saturates. The ratio q should not be taken too large in order to avoid interference due to the growth through different scales before reaching the next threshold. On the other hand, the rate q cannot be too close to one; typically a sensible threshold rate is $q = 2$ (which we take throughout this work) or $q = \sqrt{2}$. The choice of $q = 2$ leads to the "doubling time" τ_2, while $q = e$ yields the so-called "e-folding time". Both doubling time and e-folding time are frequently used in the literature to characterize error growth; thus, we pick the doubling time as a convenient number for comparison with previous work on CZ predictability (Goswami and Shukla, 1991; Blumenthal, 1991; Goswami et al., 1997). The time $\tau_q(\delta)$ is obtained by integrating two trajectories of the system that start at an initial distance $\delta_{\min}$ and this process is illustrated in Fig. 1.

The algorithm schematized in Fig. 1 proceeds as follows. After time t_0, the perturbation has grown from $\delta_{\min}$ up to δ_n, ensuring that the perturbed trajectory relaxes on the attractor and aligns along the maximally expanding direction. Then, we measure the time $\tau_1(\delta_n)$, the error needed to grow up to δ_{n+1}, i.e., the first time such that $\delta(t_0) = \| \delta \mathbf{x}(t_0) \| = \delta_n$ and $\delta(t_0 + \tau_1(\delta_n)) = \delta_{n+1}$. The perturbation is thus rescaled to δ_n, along the direction $\mathbf{x}' - \mathbf{x}$ (see Fig. 1). The procedure is repeated $\mathcal{N}_\mathrm{d}$ times for each threshold, obtaining the set of the "doubling" times (strictly speaking the name applies for $q = 2$ only) $\{\tau_i(\delta_n)\}$ for $i = 1, \ldots, \mathcal{N}_\mathrm{d}$ error-doubling experiments. Note that $\tau(\delta_n)$ also depends on q. The doubling rate is defined as

$$\gamma_i(\delta_n) = \frac{1}{\tau_i(\delta_n)} \ln q. \tag{3}$$

The error growing time is a fluctuating quantity and one has to take the average along the trajectory as in Eq. (1), therefore the FSLE $\lambda(\delta)$ is computed by taking the average over the natural measure along the trajectory (invariant measure of the dynamics), which is approximated by the ensemble average $\langle \cdot \rangle_\mathrm{ens}$ over many realizations.

$$\lambda(\delta_n) = \langle \gamma(\delta_n) \rangle_t = \frac{1}{T} \int_0^T \mathrm{d}t\, \gamma = \frac{\sum_i \gamma_i \tau_i}{\sum_i \tau_i} \tag{4}$$

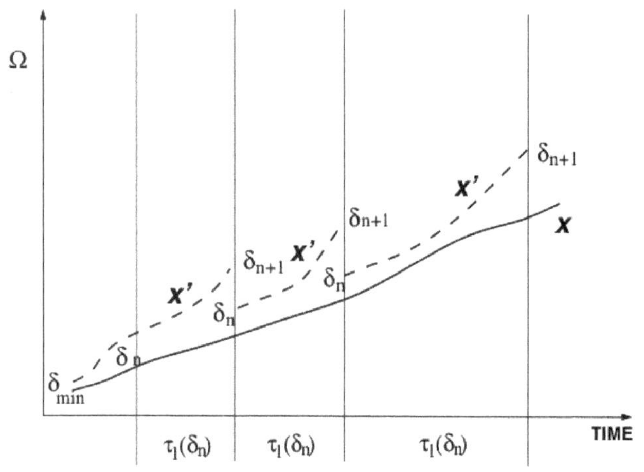

Fig. 1. Finite-size Lyapunov exponent algorithm schematized after Boffetta et al. (1998a).

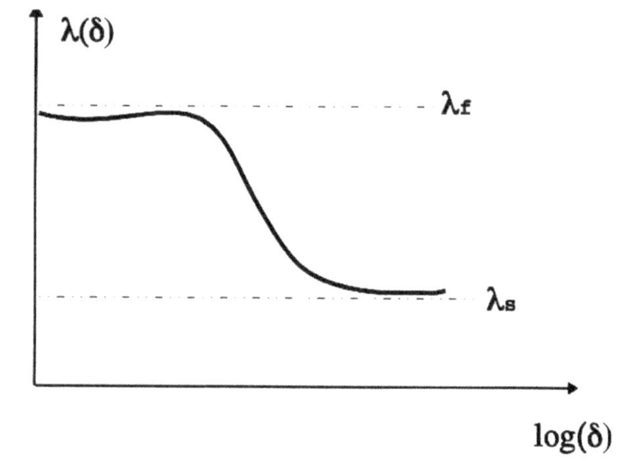

Fig. 2. Finite-size Lyapunov exponent of two coupled (slow–fast) Lorenz models computed from the slow variables. The two horizontal lines represent the uncoupled Lyapunov exponents. Figure schematized after Boffetta et al. (1998a).

using Eq. (3) yields

$$\lambda(\delta_n) = \frac{1}{\langle \tau(\delta_n)\rangle_{\mathrm{d}}} \ln q, \tag{5}$$

where $\langle \tau(\delta)\rangle_{\mathrm{d}} = \sum_i \tau_i / \mathcal{N}_{\mathrm{d}}$ is the average over the doubling experiments and the total duration of the trajectory is $T = \sum_i \tau_i$. When δ_n is infinitesimal, $\lambda(\delta_n)$ recovers the MLE, in Eq. (1), and we have $\lim_{\delta\to 0} \lambda(\delta_n) = \lambda_1$. We must underline two things: (i) unlike the standard LE, when δ is finite $\lambda(\delta)$ depends on the chosen norm and this dependence is a manifestation of the fact that in the nonlinear regime the predictability time depends on the chosen observable; (ii) since the FSLE is based on "q-pling" times, it cannot detect negative LEs.

Boffetta et al. (1998a) illustrated this new concept by studying two coupled sets of nonlinear Lorenz (1963) aligns, one characterized by a slow and the other one characterized by a fast timescale. They showed that the FSLE is indeed an adequate quantity to describe the predictability of systems with multiple timescales. The FSLE is expected to converge to the leading Lyapunov exponent λ_{f} of the fast decoupled system for small scales, whereas for large scales it asymptotically approaches the Lyapunov exponent λ_{s} of the slow decoupled system (Fig. 2).

It is worth noting that Fig. 2, schematized after Boffetta et al. (1998a), does not illustrate the fact that if the phase space is bounded, then $\lambda(\delta)$ goes to zero for sufficiently large δ, simply because the distance between two trajectories cannot grow without bound on a bounded phase space. This is later shown for the CZ model in Figs. 7 and 8 in Sect. 4 (Results).

One interesting fact is that the MLE defined in Eq. (1) does not depend on the initial state, but for the FSLE one has to take an average over initial states in Eq. (5). Since the FSLE depends on the initial state, one can consider the application of the FSLE for a particular season (ENSO spring predictability barrier) or regime of the system as in Mitchell and Gottwald (2012). In order to consider the FSLE seasonal or regime dependence, the algorithm can be modified to avoid the rescaling at finite δ_n as follows. The thresholds $\{\delta_n\}$ and the initial perturbation ($\delta_{\min} \ll \delta_0$) are chosen as before, but now the perturbation growth is followed from δ_0 to $\delta_{N_{\mathrm{s}}}$ without rescaling back the perturbation once the threshold is reached. In practice, after the system reaches the first threshold δ_0, we measure the time $\tau_1(\delta_0)$ to reach δ_1, then following the same perturbed trajectory we measure the time $\tau_1(\delta_1)$ to reach δ_2, and so forth up to $\delta_{N_{\mathrm{s}}}$: we thus record the time $\tau(\delta_n)$ for going from δ_n to δ_{n+1} for each value of n. One important difference in the modified algorithm is that the evolution of the error from the initial value $\delta_{\min}$ to the largest threshold $\delta_{N_{\mathrm{s}}}$ carries out a single error-doubling experiment. Therefore, it relies on the realization of an ensemble of simulations in order to take the average over many realizations with different initial conditions for a particular season or regime of the system.

In order to quantify the typical predictability time for a trajectory with initial uncertainty δ, we introduce the average error growing time for a trajectory with initial uncertainty δ and a given tolerance Δ as

$$T_{\mathrm{p}}(\Delta, \delta) = \int_{\delta}^{\Delta} \frac{\mathrm{d}\ln\delta'}{\lambda(\delta')}, \tag{6}$$

which reduce to Eq. (2) in the case of constant λ. Furthermore, from the considerations mentioned above, one expects that $\lambda(\delta)$ is a decreasing function of δ and thus Eq. (6) gives longer predictability time than Eq. (2).

In the case studied by Boffetta et al. (1998a) the prediction horizon T_{p} of the slow component of the coupled fast-slow Lorenz (1963) system becomes orders of magnitude larger

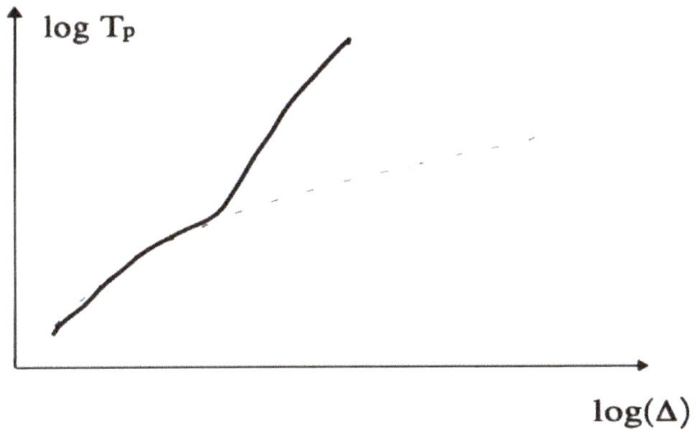

Fig. 3. Average error growing time T_p for the slow component of the two coupled Lorenz models as a function of error tolerance Δ. The dashed line represents the classical Lyapunov estimate. Figure schematized after Boffetta et al. (1998a).

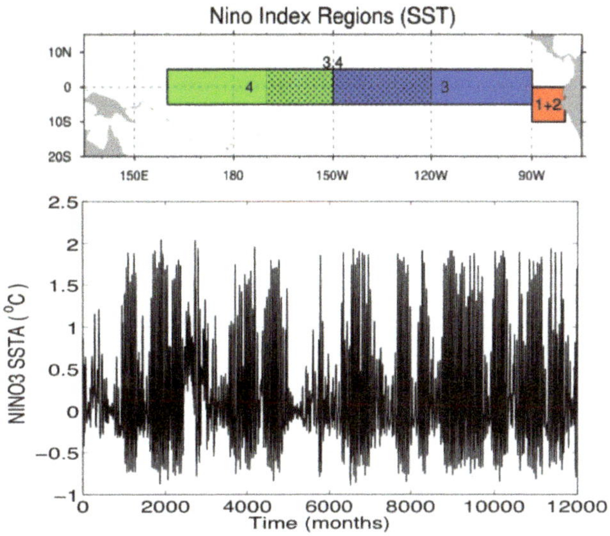

Fig. 4. Sea surface temperature anomalies in the NINO3 (5° S–5° N; 150–90° W) region from a 50 000 yr run of the CZ model (monthly means from the first thousand years).

than the one based on the classical Lyapunov estimate, when large error tolerances are allowed, which is shown in Fig. 3.

The interest in error growth rates and their influence on the predictability has also promoted the development of breeding vectors, which are optimal finite-size perturbations that represent the directions of growing analysis errors to produce optimal perturbations for ensemble forecasting (Toth and Kalnay, 1993, 1997; Vikhliaev et al., 2007). The average growth rate of the breeding vectors, which are also scale dependent, is closely related to the FSLE (Boffetta et al., 2002; Peña and Kalnay, 2004). An interesting property of the FSLE methodology as presented here is that it also applies to systems that are forced or partly forced by noise.

2 The model

In this study we used the standard version of the Cane–Zebiak (CZ) model that has been used in a large number of studies related to ENSO (Zebiak and Cane, 1987; Goswami and Shukla, 1991, 1993; Blumenthal, 1991). This is an anomaly coupled model, i.e., the governing aligns represent oceanic and atmospheric perturbations about the mean climatological state, with monthly climatology prescribed from observation. The nonlinearity enters through the thermodynamic energy align for the ocean and the model does not contain the high-frequency internal variability in the atmosphere since it considers a steady-state atmosphere. It is worth noting that this is in contrast with the real coupled ocean–atmosphere system discussed in the Introduction, since in the CZ model the nonlinearities and high-frequency variability do not enter through the atmosphere.

In this paper, we performed a 50 000 yr-long run of the CZ model to quantify its predictability. The coupled model run is able to simulate important aspects of interannual variations in the tropical Pacific reasonably realistically, contain nonlinearity, and shows a timescale similar to the observed system. The model is able to simulate the characteristic large-scale spatial (not shown) and temporal structure of ENSO (see Zebiak and Cane, 1987). The evolution of mean monthly sea surface temperature anomalies in the NINO3 region of the eastern tropical Pacific (5° S–5° N; 150–90° W) is shown in Fig. 4. The choice of this region to represent the system's evolution is based on the fact that this is the region of the tropical Pacific that has the largest variability in sea surface temperature on interannual timescales and is highly correlated with the dominant mode of variability, i.e., the first EOF of the system (not shown). The evolution is clearly irregular, with decades of little activity and decades of irregular cycles of warm and cold phases with an average period of 3–4 yr. The chaotic dynamics of the model is due to the fact that the SST align and the atmospheric heating are nonlinear.

3 Methodology

Previous predictability experiments involving the CZ model were conducted introducing small perturbations distributed randomly in space in Goswami and Shukla (1991), which actually introduces perturbations in all scales of motion, or by representing the fields in terms of its EOFs, and then studying the evolution of uncertainties that have the spatial structure of one of them (Blumenthal, 1991).

The methodology employed here differs from the previous studies, since it is based on newer theoretical concepts from nonlinear dynamics analysis and follows the following steps. First, we reconstruct the system dynamics in phase space using the Takens embedding theorem (Takens, 1981; Sauer et al., 1991) and then apply the finite-size Lyapunov exponent algorithm in order to infer the CZ model predictability.

A similar procedure could be done using an EOF reduction to reconstruct the system's phase space, since the NINO3 index is highly correlated with the first EOF.

4 Results

The choice of optimal embedding is crucial for reconstructing the system dynamics in phase space as well as for the application of the FSLE algorithm in order to infer the CZ model predictability. Given the NINO3 index, which is a scalar series, a suitable delay time and an embedding dimension were chosen following the methods of mutual information and correlation dimension, respectively (Kantz and Schreiber, 2004).

We determined a suitable delay time based on the autocorrelation function and mutual information, since one criterion may be superior for one dynamical system and poor for another. The optimal delay is estimated by the first minimum of mutual information (when it has a minimum) or when the autocorrelation drops below $1/e$ or has a point of inflection (Albano et al., 1988; Fraser and Swinney, 1986). Figure 5 shows the autocorrelation and mutual information as a function of the time lags for the CZ model, suggesting a delay time of 9–10 months. Note that the autocorrelation for $\tau = 50$ months is well above the $1/e$ threshold. This indicates that the Cane–Zebiak model is predictable at least 50 months ahead, much longer than the predictability of real ENSO by current systems.

We also need to estimate the dimensionality of the system to perform the phase space reconstruction. The curves for the correlation dimension (Grassberger and Procaccia, 1983) of the CZ model NINO3 index in Fig. 6 show a scaling range where all curves saturate (plateau) slightly below 4, giving the estimated dimensionality of the signal. The low value of saturation of the slopes reveals that the underlying dynamics of the CZ model correspond to a low-dimensional deterministic process. According to the embedding theory, the dynamical degrees of freedom cannot be higher than $2D+1 = 9$ and lower than $D_{\text{int}} \approx 4$ where D_{int} is the first integer value after the saturation value. The size of the 50 000 yr simulation, with monthly outputs of the CZ model, implies that the highest dimension attractor that can be discerned using this time series is between 9 and 10 dimensions according to Tsonis (1992).

Our result is in agreement with the dimensionality estimate of 3.5 for the CZ model obtained by Tziperman et al. (1994). It is worth noting that the ocean component of this model is essentially made up of two prognostic aligns, one for the mixed layer thermodynamics align for SST, a thermocline depth align for upper ocean dynamics, along with SST-gradient-wind relation, and the wind-equatorial upwelling relation. This implies roughly 4 degrees of freedom and therefore this data set is expected to be low dimensional.

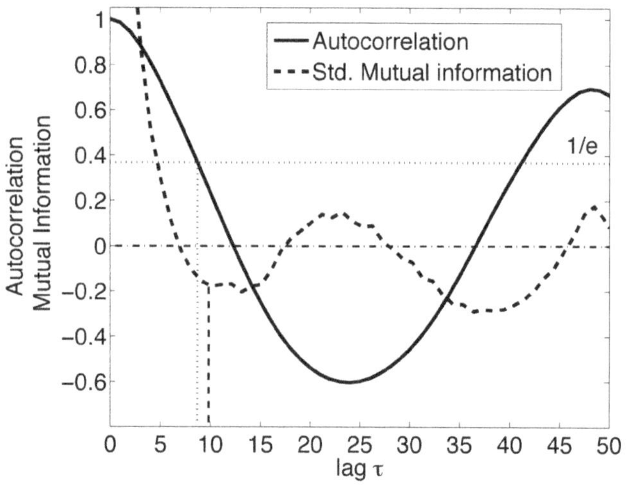

Fig. 5. Autocorrelation function of the NINO3 index (solid line) and mutual information (dashed line) at different time lags for the CZ model.

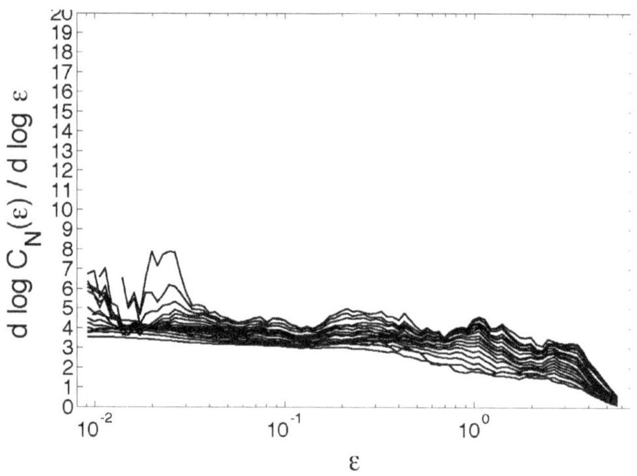

Fig. 6. The correlation dimension (local slopes of the correlation integral) of the NINO3 index of the CZ model. Embedding is done with a delay of 10 months. The different curves represent $m = 4, \ldots, 20$ dimensions (from bottom to top).

We reconstruct the CZ model phase space using the Takens embedding theorem (Takens, 1981; Sauer et al., 1991) applied to the NINO3 index time series. The dynamics in phase space appears as a chaotic attractor (fractal) where the interannual oscillations clearly display low-order deterministic chaos (Fig. 7). The attractor also shows the ENSO dynamics, characterized by El Niño positive loops and normal La Niña conditions (neutral and negative values).

In the final step, the NINO3 time series of the CZ model were analyzed by the FSLE algorithm to infer its predictability. For the sake of brevity a detailed description of the computation of FSLE is not covered here. Interested readers are referred to the paper by Boffetta et al. (1998b).

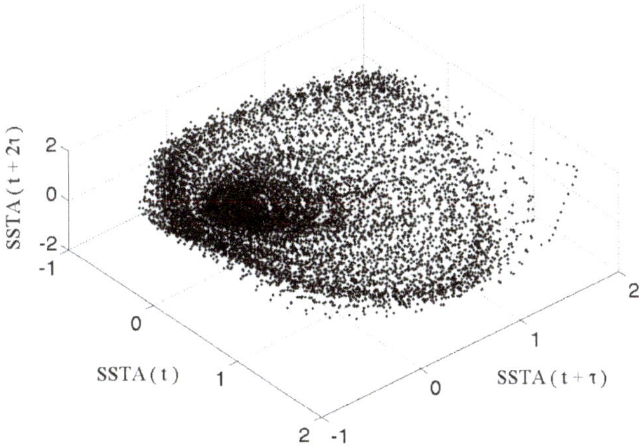

Fig. 7. Reconstructed phase space (three-dimensional projection) of SSTA of the NINO3 region from the CZ model.

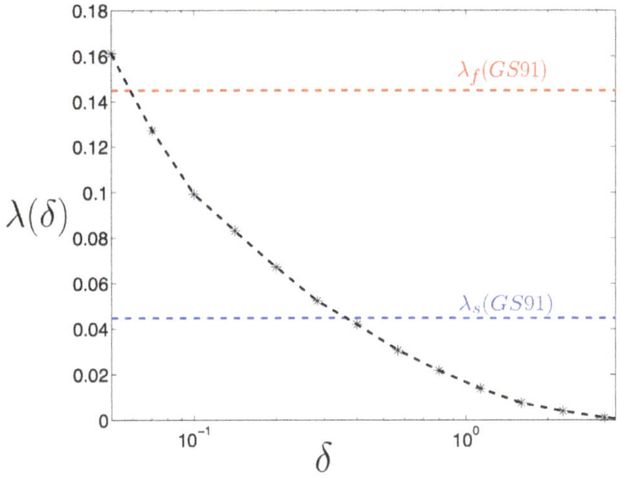

Fig. 8. Scale-dependent Lyapunov exponents $\lambda(\delta)$ of the CZ model as a function of scale δ in the range [0.05, 1.5]. The dashed red line shows the Goswami and Shukla (1991) estimation for the fast mode growth rate λ_f and the dashed blue line the slow mode growth rate λ_s.

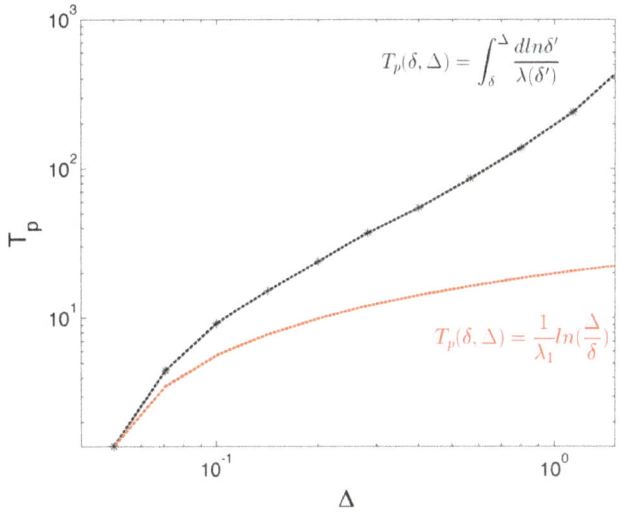

Fig. 9. Average error growing time T_p (months) of the CZ model as a function of the error tolerance level Δ based on: FSLE (black line) and the maximum Lyapunov exponent λ_{fast} (red line). Note that predictability inferred by assuming "infinitesimal" perturbations (dashed red line) is much smaller than that inferred by finite-size perturbations (black line).

Figure 8 shows the plot of the FSLE $\lambda(\delta)$ as a function of the scale δ. The scales δ are defined on an embedded phase space with dimension $m=9$, and a delay time $\tau_{corr}=10$ months as suggested by the autocorrelation and mutual information. The metric used for delta is $\delta = \| \mathbf{x}-\mathbf{x}' \|$, see Fig. 1, and the choice for the initial size of δ_{min} is limited by the precision of a finite data set of N points that leads to a minimum length scale ϵ_0 below which the attractor structure cannot be resolved. The scales at which the correlation dimension curves, shown in Fig. 6, start to fluctuate tremendously due to the lack of neighbors indicate the minimum length scale. These scales are smaller than the average interpoint distance in the attractor in embedding space and δ_{min} can be approximated by $\epsilon_0 \approx L/N^{(1/d)}$, where L is the extent of the attractor and d is its correlation dimension (Farmer et al., 1983). In order to compute the FSLE, we performed a 50 000 yr-long run, which roughly gives a minimum length scale $\epsilon_0 \approx 0.05$ if we consider the attractor dimension of $d=4$, which is in agreement with the behavior of the correlation dimension curves for scales $\epsilon < 0.04$ in Fig. 6. For values $\delta > 3$ the perturbation size is comparable to the variance of the NINO3 SSTA, rendering the FSLE meaningless (saturation value δ_{N_s}).

One noticeable feature in Figs. 8 and 9 is that they show very gradually changing curves for $\lambda(\delta)$ and T_p, which is rather different from the abrupt changes in the simple systems of Boffetta et al. (1998a) that have only two different timescales. As pointed out by Lorenz (1969), the errors in different scales in systems possessing many scales of motion tend to grow at different rates and saturate at different levels. First, let us consider very small amplitude perturbations $\delta = 0.05$ associated with errors in the smallest fastest-evolving scales of motion we can discern in this work. It is seen in Fig. 8 that for very small δ the magnitude of $\lambda(\delta=0.05) \approx 0.16\,\text{month}^{-1}$ (doubling time of 4.3 months).

Goswami and Shukla (1991), in a predictability study of the CZ model, found the fast growth rate to be $\lambda_f \approx 0.145\,\text{month}^{-1}$, corresponding to a doubling time of 4.8 months. This matches the growth rate found here and we must underline that, like us, they do not consider the dependence of predictability on season or regime. Blumenthal (1991), following a predictability study involving principal oscillation patterns (POPs) but using the same model, found that there is a seasonality to error growth in the

CZ model, with spring starts showing the largest error growth. Blumenthal (1991) found that the doubling time for February and May starts is 4.2 months, while integrations starting in August and November show much less error growth with doubling times of 7 months and 9.5 months, respectively. Later on, Goswami et al. (1997) used a decomposition of model fields in terms of the model's EOFs, allowing them to introduce errors with a given spatial structure and to study growth of errors in all scales. Using a similar approach to Goswami and Shukla (1991) to compute error growth, but examining the seasonal and scale dependence, they found that growth of small or moderate error in the initial condition is always slow in winter forecasts and fast in forecasts started from spring of certain years. Furthermore, they concluded that the slow growth rate of errors (doubling time of about 15 months) is intrinsic to the model, whereas the fast growth rate (doubling time of about 5 months) comes from the coupling with the atmospheric winds and its interaction with the tropical wave dynamics.

The fast growth rate seems therefore to be closely related to the growth-phase predictability barrier also known as the "spring predictability barrier" (Moore and Kleeman, 1996; Chen et al., 2004; Yu et al., 2009). The spring predictability barrier has been explained in terms of the seasonal dependence of the air-sea coupling process, due to the seasonal movement of the Intertropical Convergence Zone, which is closest to the Equator, and SST in the east is warmest. The spring is also, of course, the period when the El Niño growth phase tends to occur, since the coupling between model atmosphere and ocean is such that small perturbations in the initial state are easily amplified. Samelson and Tziperman (2001) suggest that there is a predictability barrier associated with the growth phase of El Niño conditions instead of a spring barrier. That is, ENSO's predictability is smallest during the growth phase of El Niño conditions in the equatorial eastern Pacific, indicating that the growth mechanism of the disturbance is likely to be similar to that of the mechanism due to which the mature El Niño develops.

The mechanism that leads to a mature El Niño lies in the coupling with the atmospheric winds and its interaction with the tropical wave dynamics leading to the growth of coupled instabilities. In the CZ model, the model run is initiated with an imposed westerly wind anomaly that is held fixed for a given period and then removed. The westerly wind anomaly produces an initial thermocline depression that leads to an ocean heating anomaly, atmospheric convergence, latent heat release, then further downwelling and deepening of the thermocline (positive coupled feedback). The disequilibrium is carried by the ocean wave dynamics as eastward propagating equatorial Kelvin waves and westward propagating off-equatorial Rossby waves. In the central and eastern Pacific the Kelvin waves grow unstable if they have large wavelengths and are slow enough to stay within the coupling region so that the coupling process feeds back on the perturbation. The CZ model then shows the recurrence of warm events, deriving solely from self-interactions of the coupled system since there is no external forcing.

The CZ model was designed specifically to capture interannual variability of the large spatial scales of motion in the ocean. Therefore, the slow growth rate of errors is intrinsic to the ocean dynamics (slow model). On larger scales the growth rate is much slower and is associated with equatorial trapped Rossby wave timescales (ENSO mode). The perturbations of the thermocline also produce negative off-equatorial depth anomalies (that is, a shallowing signal of the thermocline) in the central Pacific that excite westward propagating Rossby waves. These Rossby waves are reflected off the western boundary as large-scale equatorial Kelvin waves (long wave approximation) and eventually arrive at the eastern Pacific months later, shallow the thermocline there and cause cooling of the SST. It is worth noting that by making a long wave approximation, one is assuming that all the energy associated with reflected Rossby waves goes into the largest Kelvin modes. Therefore, reflection becomes too efficient as it happens in the CZ model and is also enhanced by the use of unrealistic solid boundaries in this model. That means that the detection of the slow growth rate may be harder when considering more realistic models like CGCMs or in the observations.

Although a lot of attention has been given to the fast growth rate, the existence of slow growth in the coupled system as shown by Goswami and Shukla (1991) and Goswami et al. (1997) has been the basis for optimism for ENSO prediction at longer lead times. The slow growth rate found by Goswami and Shukla (1991) is $\lambda_s = 0.045\,\text{month}^{-1}$, corresponding to a doubling time of 15.3 months. The slow timescale proposed by Goswami and Shukla (1991) actually comes from fitting their error curves, based on 151 initial conditions and 180 months long runs, with a linear combination of two processes having two timescales of error growth. Here we cannot point out a specific value for the slow growth rate, but a spectra of slow growth rates for the large scales. An equivalent slow growth rate to the one found by Goswami and Shukla (1991) corresponds to perturbation scales of $\delta \approx 0.4\,°\text{C}$ as shown in Fig. 8. To put these growth rates into perspective, an initial error of 0.4 °C in SST would take about 7 months to grow to 1.5 °C (approximately the standard deviation of NINO3 SST anomalies) at the fast rate, and about 2 yr at the slow rate.

Another fact that can affect the estimation of growth rates is that, in the presence of regimes like in the CZ model (Fig. 4), the error growth rate given by the FSLE can be a highly fluctuating function of initial state. In particular, troughs in the large scales of FSLE spectra are shown to be a signature of slow regimes (in the slow system – ocean), whereas regimes occurring in the fast system (atmosphere) are shown to cause large peaks in the FSLE spectra where error growth rates far exceed those estimated from the maximal Lyapunov exponent (Mitchell and Gottwald, 2012).

Figure 9 shows the average error growing time T_p (in months) as a function of the error tolerance level Δ for a small uncertainty of fixed size $\delta = 0.05\,°C$ to grow to a value $\Delta = 1.5\,°C$, and is obtained by Eq. (6). If one considers a constant $\lambda(\delta)$, the integral reduces to Eq. (2). The black line in Fig. 9 corresponding to T_p is based on the FSLE results shown in Fig. 8. The red line in Fig. 9 represents the estimation of T_p given by Eq. (2) for a hypothetical system with a single (fast mode) Lyapunov exponent $\lambda_{fast} = 0.16\,\text{month}^{-1}$ in red. Observe that the predictability inferred by assuming infinitesimal perturbations (red line) and fast growth rate is much smaller than that inferred by finite-size perturbations (black line), illustrating the fact that large errors evolve in general with large-scale characteristic time, which thus rules large-scale predictability.

In the past, Goswami and Shukla (1991) and Blumenthal (1991) performed predictability studies of the CZ model using different methods from the FSLE. Goswami and Shukla (1991) found that the error growth has two timescales, the shorter timescale having a doubling time of approximately 4.8 months, while Blumenthal (1991) found a doubling time of 4.2 months for the spring season. The slow mode doubling time found by Goswami and Shukla (1991) is 15 months and is somewhat closer to the doubling time of 10 months found by Blumenthal (1991) for the winter season. Here we found, using the FSLE, a similar doubling time for the fast mode (~4.3 months). These earlier estimates of the growth rates are not very different from the fast growth rates found here. Although we cannot establish a specific doubling time for the slow mode (we have a spectra of slow growth rates for large scales), we believe that the ultimate limit on predictability in the CZ model is governed by the slow growth rates of errors intrinsic to the ENSO mode.

Although the FSLE is a relatively simple and very useful method for estimating the model predictability, it does not offer ways to infer the source of the fast and slow timescales in terms of spatial patterns of these finite-size perturbations that represent the directions of growing errors. However, it may provide tools to study the predictability of the real coupled ocean–atmosphere in the tropical Pacific, since it does not rely on the realization of twin experiments in order to compute the error growth like in the previous studies.

5 Conclusions

In this study we could identify different intrinsic timescales in the Cane–Zebiak intermediate model of the coupled ocean–atmosphere system of the tropical Pacific, with the shorter timescale showing a doubling time of approximately 5 months. The results also confirm that for systems with different intrinsic timescales, the slow varying quantities (i.e., large-scale features) have a longer predictability than the fast-evolving ones. Present-day CGCMs have useful ENSO prediction skill at lead times of ~6 months, but the skill also depends strongly on season, ENSO phase, and ENSO intensity as shown in Jin et al. (2008). It seems that the skill of current models is limited by rapid growth of error, at a rate comparable to the fast rate identified in here and in the previous work on the predictability of the tropical Pacific coupled system.

Futhermore, assuming that the ENSO in a state-of-the-art CGCM is governed by chaotic nonlinear dynamics, we expect that the high-frequency noise or regime changes from the atmospheric internal dynamics can strongly influence the estimation of the λ_{fast} using the FSLE approach. However, we do not expect the predictability of the slow ENSO mode (λ_{slow}) to be significantly affected by the relatively high-frequency noise from the atmosphere, since the slow growth rate of errors is intrinsic to the ocean dynamics. Of course, we recognize that this assumption is the subject of considerable debate (e.g., Kirtman et al., 2005 for a discussion) and needs further investigation. The literature offers external stochastic forcing and chaos as two distinct possibilities for the irregularity of ENSO. This debate of stochastic forcing versus chaotic dynamics has important implications for understanding predictability, since the theoretical upper limit of predictability depends heavily on the mechanism for irregularity. Nevertheless, the approach may help diagnose the role of chaos in various CGCMs, and the applicability of this technique in higher complexity models like state-of-the-art CGCMs deserves further studies. We also expect that ENSO in most state-of-the-art CGCMs may also exhibit low-order dimensionality and that the essence of the method used here can be applied. The approach presented here in conjunction with the atmospheric noise reduction technique introduced by Kirtman and Shukla (2002) can be used to isolate the role of high-frequency atmospheric noise in the predictability of ENSO.

The FSLE is a relatively simple concept and very useful method for estimating the model predictability. However, it does not allow for the inference of the source of different timescales in terms of spatial patterns, nor does it provide answers to questions regarding the cascade of errors from one scale of motion to another. The main advantages of this technique relative to the others is that it is relatively simple to perform and allows the study of the predictability of the real coupled ocean–atmosphere in the tropical Pacific, since it does not rely on the realization of twin experiments in order to compute error growth as in the previous studies, only possible with coupled models.

Acknowledgements. The authors acknowledge support from NSF grants OCI0749165, ATM0754341, NOAA grant NA08OAR432088 and CAPES/FULBRIGHT (BEX2837/06-4). We are also grateful to G. Burgers for many stimulating suggestions and constructive comments.

Edited by: O. Talagrand

References

Albano, A. M., Muench, J., Schwartz, C., Mees, A. I., and Rapp, P. E.: Singular value decomposition and the Grassberger-Procaccia algorithm, Phys. Rev. A, 38, 3017–3026, 1988.

Aurell, E., Boffetta, G., Crisanti, A., Paladin, G., and Vulpiani, A.: Predictability in the large: An extension of the concept of Lyapunov exponent, J. Phys. A-Math. Gen., 30, 1–26, 1997.

Blumenthal, M. B.: Predictability of a coupled ocean–atmosphere model, J. Climate, 4, 766–784, 1991.

Boffetta, G., Crisanti, A., Paparella, F., Provenzale, A., and Vulpiani, A.: Slow and fast dynamics in coupled systems: A time series analysis view, Physica D, 116, 301–312, 1998a.

Boffetta, G., Giuliani, P., Paladin, G., and Vulpiani, A.: An extension of the Lyapunov analysis for the predictability problem, J. Atmos. Sci., 55, 3409–3412, 1998b.

Boffetta, G., Cencini, M., Falcioni, M., and Vulpiani, A.: Predictability: a way to characterize complexity, Phys. Rep., 356, 367–474, 2002.

Charney, J. G., Fleagle, R. G., Riehl, H., Lally, V. E., and Wark, D.: The feasibility of a global observation and analysis experiment, B. Am. Meteorol. Soc., 47, 200–220, 1966.

Chen, D., Cane, M. A., Kaplan, A., Zebiak, S. E., and Huang, D. J.: Predictability of El Niño over the past 148 years, Nature, 428, 733–736, 2004.

Farmer, J. D., Ott, E., and Yorke, J. A.: The dimension of chaotic attractors, Physica D, 7, 153–180, 1983.

Fraser, A. M. and Swinney, H. L.: Independent coordinates for strange attractors from mutual information, Phys. Rev. A, 33, 1134–1140, doi:10.1103/PhysRevA.33.1134, 1986.

Goswami, B. N. and Shukla, J.: Predictability of a coupled ocean atmosphere model, J. Climate, 3, 2–22, 1991.

Goswami, B. N. and Shukla, J.: Aperiodic variability in the Cane–Zebiak model: A diagnostic study, J. Climate, 5, 628–638, 1993.

Goswami, B. N., Rajendran, K., and Sengupta, D.: Source of Seasonality and Scale Dependence of Predictability in a Coupled Ocean-Atmosphere Model, Mon. Weather Rev., 125, 846–858, 1997.

Grassberger, P. and Procaccia, I.: Characterization of strange attractors, Phys. Rev. Lett., 50, 346–349, 1983.

Jin, E. K., Kinter, J. L., Wang, B., Park, C.-K., Kang, I.-S., Kirtman, B. P., Kug, J.-S., Kumar, A., Luo, J.-J., Schemm, J., Shukla, J., and Yamagata, T.: Current status of ENSO prediction skill in coupled ocean atmosphere models, Clim. Dynam., 31, 647–664, doi:10.1007/s00382-008-0397-3, 2008.

Kantz, H.: A robust method to estimate the maximal Lyapunov exponent of a time series, Phys. Lett. A, 185, 77–87, 1994.

Kantz, H. and Schreiber, T.: Nonlinear Time Series Analysis, 2nd edn., Cambridge University Press, Cambridge, UK, 2004.

Kirtman, B. P. and Min, D.: Multimodel Ensemble ENSO Prediction with CCSM and CFS, Mon. Weather Rev., 1372, 2908–2930, 2009.

Kirtman, B. P. and Shukla, J.: Interactive coupled ensemble: A new coupling strategy for GCMs, Geophys. Res. Lett., 29, 1029–1032, 2002.

Kirtman, B. P., Pegion, K., and Kinter, S. M.: Internal atmospheric dynamics and tropical Indo-Pacific climate variability, J. Atmos. Sci., 62, 2220–2233, 2005.

Lorenz, E. N.: Deterministic Nonperiodic Flow, J. Atmos. Sci., 20, 130–141, 1963.

Lorenz, E. N.: A study of the predictability of a 28-variable atmospheric model, Tellus, 17, 321–33, 1965.

Lorenz, E. N.: The predictability of a flow which possesses many scales of motion, Tellus, 21, 289–307, 1969.

Mitchell, L. and Gottwald, G. A.: On finite-size Lyapunov exponents in multiscale systems, Chaos, 22, 1–10, 2012.

Moore, A. M. and Kleeman, R.: The dynamics of error growth and predictability in a coupled model of ENSO, Q. J. Roy. Meteor. Soc., 122, 1405–1446, 1996.

Peña, M. and Kalnay, E.: Separating fast and slow modes in coupled chaotic systems, Nonlin. Processes Geophys., 11, 319–327, doi:10.5194/npg-11-319-2004, 2004.

Rosenstein, M. T., Collins, J. J., and De Luca, C. J.: A practical method for calculating largest Lyapunov exponents from small data sets, Physica D, 65, 117–134, 1993.

Samelson, R. M. and Tziperman, E.: Instability of the Chaotic ENSO: The Growth-Phase Predictability Barrier, J. Atmos. Sci., 58, 3613–3625, 2001.

Sauer, T., Yorke, J. A., and Casdagli, M.: Embedeology, J. Stat. Phys., 65, 579–616, 1991.

Shukla, J.: Predictability, in: Issues in atmospheric and oceanic modeling, Part II. Weather Dynamics, Academic Press, Inc., 28B 87–122, 1985.

Smagorinsky, J.: Problems and promises of deterministic extended range forecasting, B. Am. Meteorol. Soc., 50, 286–311, 1969.

Takens, F.: Detecting strange attractors in turbulences, in: Dynamical Systems and Turbulence, Springer-Verlag, Berlin, 898, 366–381, 1981.

Toth, Z. and Kalnay, E.: Ensemble forecasting at NMC: The generation of perturbations, B. Am. Meteorol. Soc., 74, 2317–233, 1993.

Toth, Z. and Kalnay, E.: Ensemble forecasting at NCEP and the breeding method, Mon. Weather Rev., 125(12), 3297–3319, 1997.

Tsonis, A. A.: Chaos: from theory to applications, Plenum, NY, 1992.

Tziperman, E., Stone, L., Cane, M. A., and Jarosh, H.: El-Niño chaos: overlapping of resonances between the seasonal cycle and the Pacific ocean–atmosphere oscillator, Science, 264, 72–74, 1994.

Vikhliaev, Y., Kirtman, B., and Schopf, P.: Decadal North Pacific Bred Vectors in a Coupled GCM, J. Climate, 20, 5744–5764, 2007.

Yu, Y. S., Duan, W. S., and Xu, H.: Dynamics of nonlinear error growth and season-dependent predictability of El Niño events in the Zebiak-Cane model, Q. J. Roy. Meteor. Soc., 135, 2146–2160, 2009.

Zebiak, S. E. and Cane, M. A.: A model El Niño/Southern Oscillation, Mon. Weather Rev., 115, 2262–2278, 1987.

3

Sensibility to noise of new multifractal fusion methods for ocean variables

A. Turiel[1], J. Isern-Fontanet[2], and M. Umbert[1]

[1]Institut de Ciencies del Mar, CSIC, Barcelona, Spain
[2]Institut Catala de Ciencies del Clima (IC3), Barcelona, Spain

Correspondence to: A. Turiel (turiel@icm.csic.es)

Abstract. The repeated observation of the same signatures of mesoscale and submesoscale features in different ocean variables indicates that some common, non-linear processes affect them to a significant extent. A new method to exploit these common signatures to improve the quality of a noisy variable (i.e. increasing the signal-to-noise ratio) using another variable as template has recently been introduced. The method is based on superimposing the multifractal structure of singularity exponents from the template variable to the variable to be enhanced. In this paper, we will discuss the sensitivity of this method to the presence of noise of different types and amplitude. Our results indicate that multifractal methods can be a key to enhancing the existing databases of remote sensing images and give hints about non-linear dynamics of the ocean.

1 Introduction

The introduction of new remote sensing platforms for the observation of oceans has posed more questions than the answers they have provided. Previously unobserved processes can now be seen by repeated observations (as for instance with the SMOS – Soil Moisture and Ocean Salinity – mission (Font et al., 2012) and the large-scale deployment of the Argo floater system) and improved numerical models. Making sense of the large amount of new data and at the same time understanding the associated processes is a pressing challenge.

On the other hand, ocean structures associated with many different processes (frontogenesis and eddy generation Hallberg and Gnanadesikan, 2006, eddy propagation Isern-Fontanet et al., 2003; Chelton et al., 2007, variations in sea roughness, upwelling Mason et al., 2011, etc.) have been shown to leave a recognisable footprint in many different variables. This common footprint is the result of similar non-linear terms within the equations defining the evolution of ocean scalars (such as sea surface temperature – SST, sea surface salinity – SSS, chlorophyll concentration – CC, etc.), even if those non-linear terms are not the dominant contribution to the dynamics of the particular scalars. This common signature in ocean scalars indicates that a part of the information conveyed by the different ocean variables is in fact redundant. Understanding this redundancy is useful in two different ways. On one hand it can serve to assess and even to measure the magnitude of the associated terms in the evolution equations. On the other hand, that redundancy can be exploited to code information in a more compact way and even to enhance data quality when one of the data sources is corrupted or otherwise contaminated.

The introduction of the microcanonical multifractal formalism (Isern-Fontanet et al., 2007; Turiel et al., 2008b) provides a unified formalism to model and quantify this redundancy in a consistent mathematical model. The common signature in all ocean variables is the multifractal structure underlying all of them, which is dimensionless and independent of the signal amplitude of any particular scalar. Evidence is now extensive about the validity of this approach to deal with ocean variables (Turiel et al., 2005, 2008a, 2009). Recently, a non-parametric, non-linear method to fuse the information coming from different scalars using their common multifractal structure has been proposed (Umbert et al., 2014) for dealing with a specific application (improving the quality of SMOS maps).

The goal of this paper is to analyse the performance of this novel approach to fuse the information from different scalars as a data quality enhancer. To that goal, we have started from controlled situations, in which the amplitude and characteristics of noise are known and a ground truth is available. The consequences of this study go beyond a simple denoising algorithm, as this method settles the basis for the implementation of new dynamic equations based on a new dynamic, more robust entities coming from multifractal analysis.

2 Data

For the analysis developed in this paper the output of a global circulation model has been used. Our data set consists of output from the Ocean general circulation model For the Earth Simulator (OFES) (Masumoto et al., 2004; Masumoto, 2010). The simulation has been carried out with a model developed starting from Princeton GFDL MOM-3 (a z coordinate, eddy resolving, explicit free-surface model) by adapting it to the parallel architecture of the Earth Simulator. The simulation has been run on the near-global ocean, spun up for 50 yr under climatological forcing taken from monthly mean NCEP (United States National Centers for Environmental Prediction) atmospheric data. After that period the OFES is forced by the daily mean NCEP reanalysis for 48 yr from 1950 to 1998. See Masumoto et al. (2004) for additional details on the forcings.

The output of the model corresponds to daily data for the last 8 yr of simulation. Horizontal angular resolution is the same in both the zonal, ϕ, and meridional, θ, directions, with values of $\Delta\theta = \Delta\phi = 1/10°$. The calculations are restricted to variables in surface layer, namely temperature and salinity, to emulate remote sensing data. The results shown in this paper refer to 1 January 1990 (see Fig. 1), although similar ones are obtained for any other day.

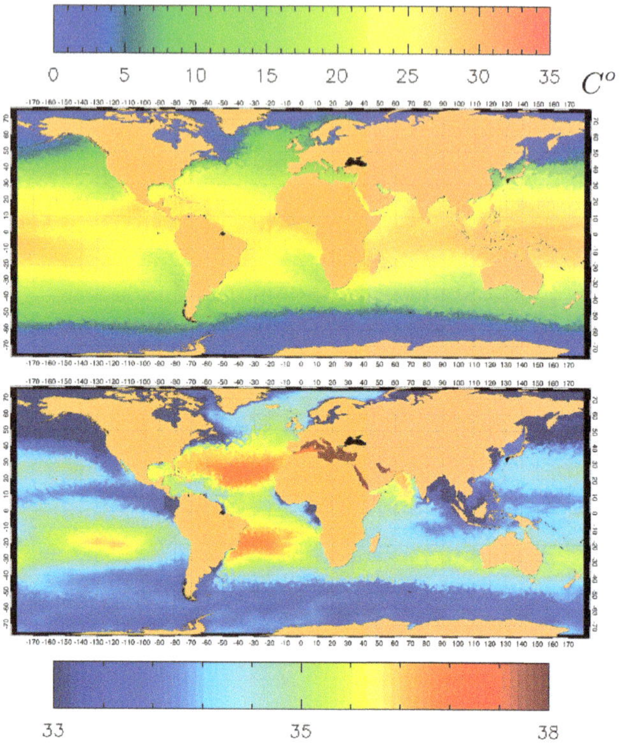

Fig. 1. OFES output for 1 January 1990. Top: SST map. Bottom: SSS.

2.1 Noise sources

All noise sources $n_\beta(\boldsymbol{x})$ in this paper are Gaussianly distributed with zero mean and marginal standard deviation σ_n which is the same at all the points. The spatial correlation matrix of the noise is given by the exponent β of its power spectrum, namely:

$$\left|\hat{n}_\beta\right|^2(\boldsymbol{k}) \propto \| \boldsymbol{k} \|^{-\beta} \qquad (1)$$

The value of the exponent can be taken between $\beta = 0$ (purely decorrelated noise) and $\beta = 2$ (strongly correlated noise). In this study we take β as $\beta = 0$, 1 and 2.

3 Non-linear methods used in this study: theoretical review

3.1 Singularity analysis

The existence of a multifractal structure in synoptic maps of ocean scalars has been studied for several years now and can be given a geometrical meaning by means of the appropriate analysis. A number of studies have shown that singularity fronts forming a multifractal hierarchy can be extracted from SST maps of different types and resolutions (Turiel et al., 2005, 2008a). It has also been observed that other scalars, such as chlorophyll concentration maps (Nieves et al., 2007) and even brightness temperature maps (Isern-Fontanet et al., 2007), have exactly the same singularity fronts that can be put into correspondence among the different scalars. The emergence of such a singular structure is the result of the advective forces acting on a quasi-2-D turbulence regime, and can be thus observed for scales ranging from kilometres to the planetary scale. Notice that 2-D turbulence is not a requirement for having singularity fronts: 3-D turbulence also generates singularity fronts, although they form compiles 2-D manifolds in general (Kestener and Arnéodo, 2003). Quasi-2-D turbulence is required to be able to identify singularity fronts derived from maps of surface variables with dynamic structures.

Singularity analysis is the keystone of the so-called microcanonical multifractal formalism (MMF); in essence, the fluid can be understood as a hierarchical arrangement of different fractal components, each one with its own fractal dimensional and characterized by a particular value of the singularity exponent. The existence of scaling properties of scalars submitted to specific types of turbulent flow can be tracked back to Kolmogorov's seminal works (Kolmogorov, 1941a, b). Later works showed that structure functions in fully developed turbulence (2- and 3-D) present multiscale dependence on the scale parameter (Dubrulle, 1994; She and Leveque, 1994; She and Waymire, 1995). The intuition that such multiscale behaviour is the result of the hierarchy of multiple fractal components is due to Parisi and Frisch (Parisi and Frisch, 1985), although the explicit extraction of the fractal components of an ocean variable (just for the surface layer) was not done until Turiel et al. (2005). With the explicit calculation of the singularity exponents of an ocean scalar MMF can be applied to provide a dynamic interpretation of the resulting singular fronts. For a comprehensive presentation of MMF the reader is referred to Turiel et al. (2008b).

Obtaining the singularity exponents associated with a given point $\boldsymbol{x}$ of a scalar map s should be easy by just studying the neighbourhood of the point, as for simple scalars:

$$|s(\boldsymbol{x}+\boldsymbol{r}) - s(\boldsymbol{x})| = \alpha_s(\boldsymbol{x})\, r^{d+h(\boldsymbol{x})} + o\left(r^{d+h(\boldsymbol{x})}\right) \tag{2}$$

where d is the dimension of the domain of s ($d = 2$ in our case) and $\boldsymbol{r}$ is any displacement vector, with r denoting its modulus. In practical terms, real signals s are affected by many artifacts, as noise, discretization errors and spurious long-range correlations. Hence, the calculation of singularity exponents requires the use of an appropriate interpolation scheme, the most usual one being wavelet projections. Let s be an arbitrary 2-D scalar signal. It will be said that s has a singularity exponent $h(\boldsymbol{x})$ at the point $\boldsymbol{x}$ if, for any wavelet function Ψ (Mallat, 1999), the following relation holds:

$$\begin{aligned} T_\Psi |\nabla s|(\boldsymbol{x}, r) &\equiv \int d\boldsymbol{x}' \, |\nabla s|\left(\boldsymbol{x}'\right) \frac{1}{r^2} \Psi\left(\frac{\boldsymbol{x}-\boldsymbol{x}'}{r}\right) \\ &= \alpha(\boldsymbol{x})\, r^{h(\boldsymbol{x})} + o\left(r^{h(\boldsymbol{x})}\right) \end{aligned} \tag{3}$$

where r stands for an arbitrary scale parameter (normalized by the integral scale so it is dimensionless and smaller than 1) and $o\left(r^{h(\boldsymbol{x})}\right)$ is a term that becomes negligible compared to $r^{h(\boldsymbol{x})}$ when r goes to zero. The amplitude function $\alpha(\boldsymbol{x})$ does not depend on the particular scale at which the wavelet projection is calculated and has the same units as the gradient of the scalar.

The left-hand side, i.e. $T_\Psi|\nabla s|(\boldsymbol{x}, r)$, is called the wavelet projection of the gradient modulus of s over the wavelet Ψ, and represents a local zoom of variable size around the point $\boldsymbol{x}$. What is important in Eq. (3) is the function $h(\boldsymbol{x})$, which is called the singularity exponent of the function s at the point $\boldsymbol{x}$. The singularity exponent is, by construction, a dimensionless measure of the regularity or irregularity of the function s around the point $\boldsymbol{x}$, extending the usual concepts of differentiability (integer values of $h(\boldsymbol{x})$ have the regular meaning of the amount of continuous derivatives accepted by the function) and of Hölder exponents (Rudin, 1987; Mallat and Huang, 1992).

If a singularity exponent can be assigned at each point of a signal s, the signal is said to be multifractal in the microcanonical sense (although strictly speaking some additional conditions should hold, see Turiel et al., 2008b). All the scalar variables studied in this paper are multifractal in the microcanonical sense, which can be verified in a standard way (Turiel et al., 2006).

Singularity exponents are related to ocean circulation and thus they are not specific to any particular scalar under study (as has been shown in the case of remote sensed chlorophyll concentration and SST in Nieves et al., 2007). In fact, it has been verified that to a good approximation singularity lines coincide with the streamlines of the flow (Turiel et al., 2009), confirming that singularity exponents are characterized by the flow and are not scalar-specific, as shown in Figs. 2 and 3. Singularity analysis is useful to track fronts or streamlines similarly to the methods described in Cayula and Cornillon (1995); Ullman and Cornillon (2000).

However, the coincidence of the isolines of singularity exponents associated with two different scalars does not imply that the values of singularity exponents of those two scalars coincide. It turns out that, at least with our output of OFES, the singularities from SST and from SSS coincide within the numerical accuracy of the singularity analysis methods and the uncertainties associated with the model dynamics. In Fig. 4 we show the conditioned histogram of the singularity exponents of SSS conditioned by the value of the singularity exponents of SST. The maximum probability line is given by $h_{\mathrm{SSS}} = h_{\mathrm{SST}} - 0.08$, while more than 95 % of the probability is concentrated within ± 0.15 of this line, so empirically $h_{\mathrm{SSS}} = h_{\mathrm{SST}} - 0.08 \pm 0.15$ with a good approximation.

3.2 Fusion method

The complete theoretical discussion of the fusion method can be found in Umbert et al. (2014); we summarize the main results here.

Let us suppose that we have two scalar variables, s and θ, that are multifractal according to MMF. It is additionally assumed that both scalars have the same singularity exponents at all points. We have hence:

$$\begin{aligned} T_\Psi |\nabla s|(\boldsymbol{x}, r) &= \alpha_s(\boldsymbol{x})\, r^{h(\boldsymbol{x})} + o\left(r^{h(\boldsymbol{x})}\right) \\ T_\Psi |\nabla \theta|(\boldsymbol{x}, r) &= \alpha_\theta(\boldsymbol{x})\, r^{h(\boldsymbol{x})} + o\left(r^{h(\boldsymbol{x})}\right) \end{aligned} \tag{4}$$

By using some theorems from functional analysis (extensions of Riesz's representation theorem) we conclude from Eq. (4) that the gradients of θ and s must be related by a 2×2 smooth matrix $\boldsymbol{\rho}$, namely:

$$\nabla s(\boldsymbol{x}) = \boldsymbol{\rho}(\boldsymbol{x})\, \nabla\theta(\boldsymbol{x}) . \tag{5}$$

The matrix $\boldsymbol{\rho}(\boldsymbol{x})$ must be slowly varying, meaning that it has a small gradient; otherwise it will introduce new singularity exponents in s not present in θ, which contradicts our first hypothesis, Eq. (4).

Equation (5), although appealing, does not provide a constructive rule for calculating the matrix $\boldsymbol{\rho}$ from the data. In

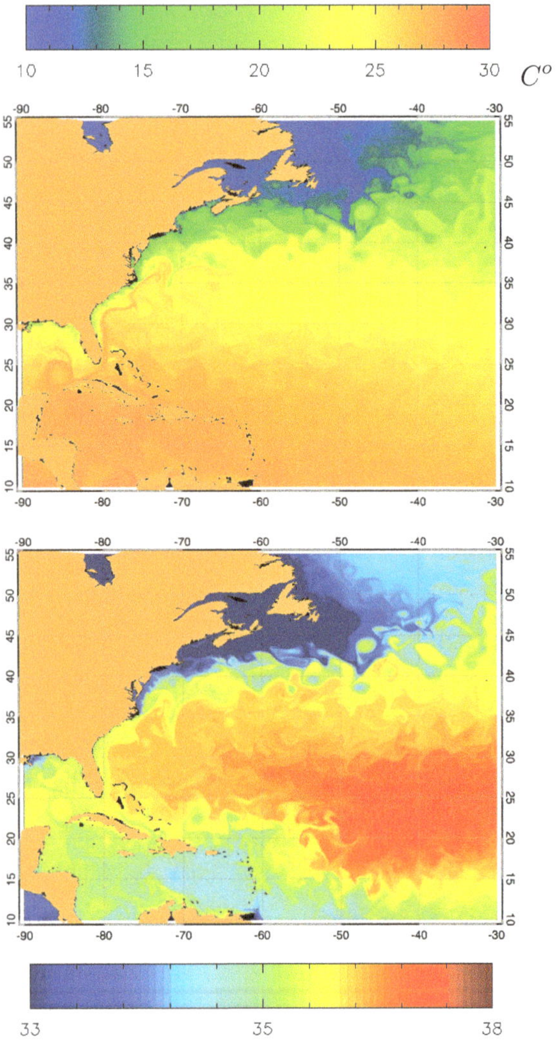

Fig. 2. OFES outputs for 1 January 1990 in the Gulf Stream area. Top: SST, bottom: SSS.

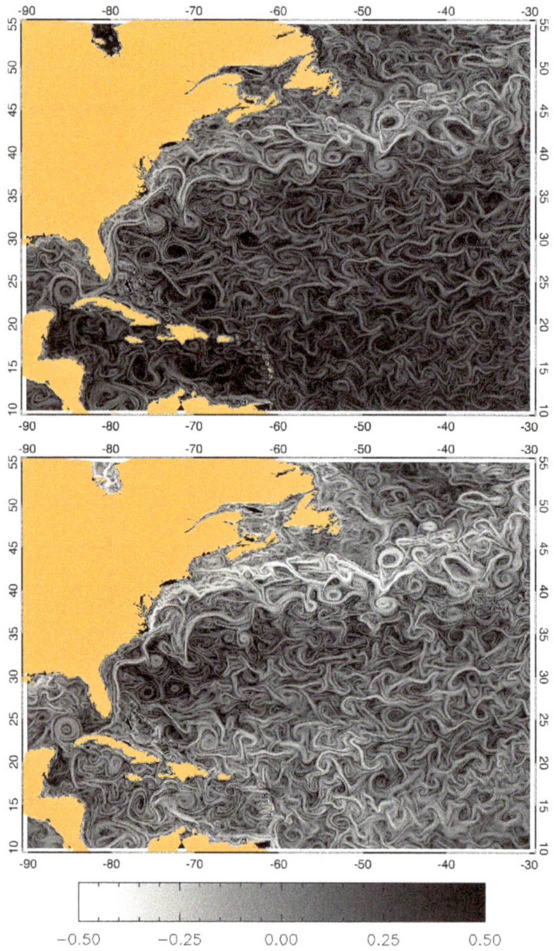

Fig. 3. Singularity exponents associated with the maps shown in Fig. 2. Top: SST, bottom: SSS.

Umbert et al. (2014) a simplifying hypothesis is discussed: if this matrix is almost everywhere proportional to the identity matrix we can conclude that:

$$s(\boldsymbol{x}) \approx a(\boldsymbol{x})\,\theta(\boldsymbol{x}) + b(\boldsymbol{x})\,, \tag{6}$$

where, as for $\boldsymbol{\rho}$, the functions a and b must have small gradients.

Let us now suppose that the function s is contaminated by some source of noise n, so in fact our data are $\theta(\boldsymbol{x})$ (assumed to be noiseless) and $s'(\boldsymbol{x}) = s(\boldsymbol{x}) + n(\boldsymbol{x})$. We can construct a filtered version of s, denoted by s_{f}, by applying the following relation:

$$s_{\mathrm{f}}(\boldsymbol{x}) = \hat{a}(\boldsymbol{x})\theta(\boldsymbol{x}) + \hat{b}(\boldsymbol{x})\,, \tag{7}$$

where $\hat{a}(\boldsymbol{x})$ and $\hat{b}(\boldsymbol{x})$ are estimates of the actual parameters $a(\boldsymbol{x})$ and $b(\boldsymbol{x})$ in Eq. (6). These estimates must be obtained from the values of s' and θ by an appropriate method which takes into account the value of both variables and which enforces that both $\hat{a}$ and $\hat{b}$ must have small gradients almost everywhere. In Umbert et al. (2014) one such method was proposed, in which $\hat{a}$ and $\hat{b}$ are calculated by linear regressions weighted around each point (Nieves et al., 2007).

The total weight of a point $\boldsymbol{x}$, $N(\boldsymbol{x})$, is defined as follows:

$$N(\boldsymbol{x}) \equiv \sum_{\boldsymbol{x} \neq \boldsymbol{x}' \in \mathrm{sea}} \frac{1}{\left|\boldsymbol{x}' - \boldsymbol{x}\right|^2}. \tag{8}$$

For any function f we define its local average around the point $\boldsymbol{x}$, $\langle f \rangle_x$, as:

$$\langle f \rangle_x \equiv \frac{1}{N(\boldsymbol{x})} \sum_{\boldsymbol{x} \neq \boldsymbol{x}' \in \mathrm{sea}} \frac{f\left(\boldsymbol{x}'\right)}{\left|\boldsymbol{x}' - \boldsymbol{x}\right|^2}. \tag{9}$$

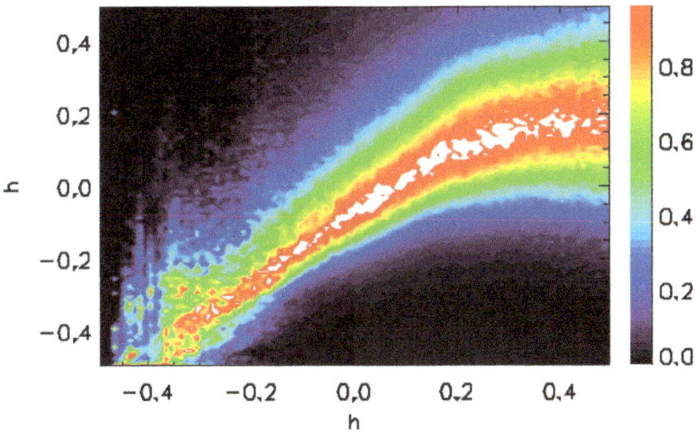

Fig. 4. Histograms of h_{SSS} conditioned by h_{SST} as derived from OFES simulation for the whole globe on 1 January 1990. For each column of the histogram the value of the SST-related variable is fixed, so the values in that column represent the distribution of probability of values of the SSS-related variable for that fixed value of the SST-related variable. To enhance visualization, in this figure the histogram is normalized by its maximum probability in each row (so the brightest colour in each row represents the mode or maximum of probability along that row).

We have thus:

$$\sigma_s'^2(\boldsymbol{x}) = \langle s'^2 \rangle_x - \langle s' \rangle_x^2 , \tag{10}$$

$$\sigma_\theta^2(\boldsymbol{x}) = \langle \theta^2 \rangle_x - \langle \theta \rangle_x^2 , \tag{11}$$

$$\sigma_{s'\theta}(\boldsymbol{x}) = \langle s'\theta \rangle_x - \langle s' \rangle_x \langle \theta \rangle_x , \tag{12}$$

and

$$\hat{a}(\boldsymbol{x}) \equiv \frac{\sigma_{s'\theta}(\boldsymbol{x})}{\sigma_s'^2(\boldsymbol{x})} , \tag{13}$$

$$\hat{b}(\boldsymbol{x}) \equiv \langle s' \rangle_x - \hat{a}(\boldsymbol{x}) \langle \theta \rangle_x , \tag{14}$$

then Eq. (7) is applied to obtain the filtered signal s_{f}. This algorithm was shown in Umbert et al. (2014) to lead to a large increase in the quality of SMOS SSS maps using OSTIA SST maps as templates, and a partial restoration of the structure of singularity exponents. The method, however, is far from perfect (the singularity exponents of the filtered signal s_{f} and those of the template θ differ significantly), so further research must be conducted to improve the method.

The method was introduced directly to process remote sensing data, without any prior study on its sensitivity to noise and on the expected quality of the results. This paper intends to fill this gap by analysing in detail the performance of the method in front of several sources of noise of controlled spatial correlation and amplitude.

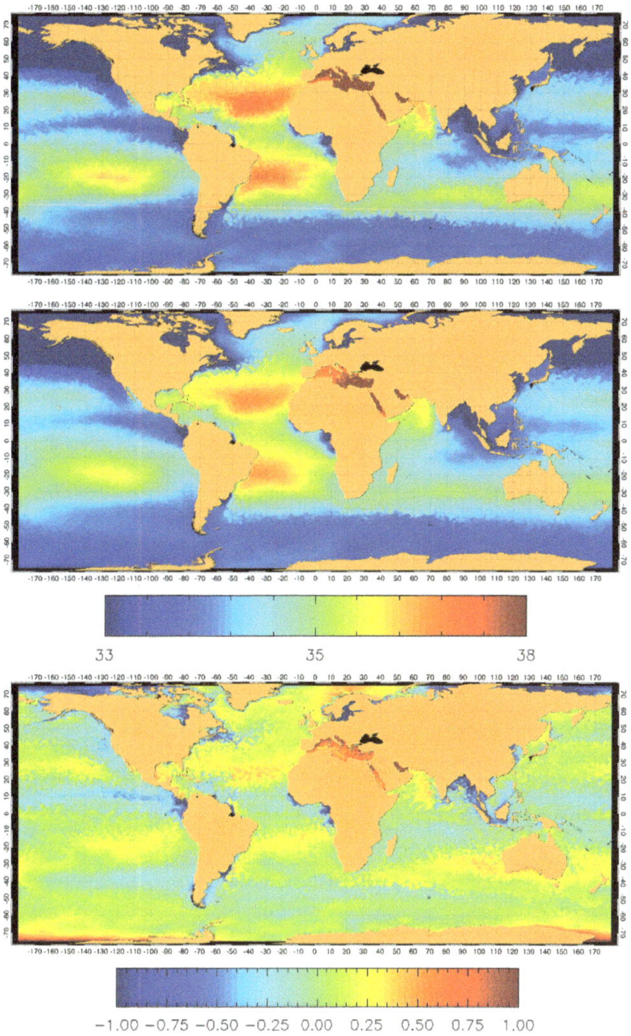

Fig. 5. Top: Original SSS. Middle: SSS map after filtering using the SST map as a template. Bottom: Map of differences.

4 Sensibility of the fusion algorithm to different types of noise

4.1 No noise

In the absence of noise, the fusion algorithm creates a map which slightly deviates from the original, as can be observed in Fig. 5.

The map of singularity exponents, Fig. 6 shows that some structures have become a bit blurred. This is probably a consequence of the approximation implied by Eq. (7). The error map at the bottom of Fig. 5 has a standard deviation of 0.3. This is an indication that the error correction by our multifractal fused method cannot go below this threshold within the present formulation.

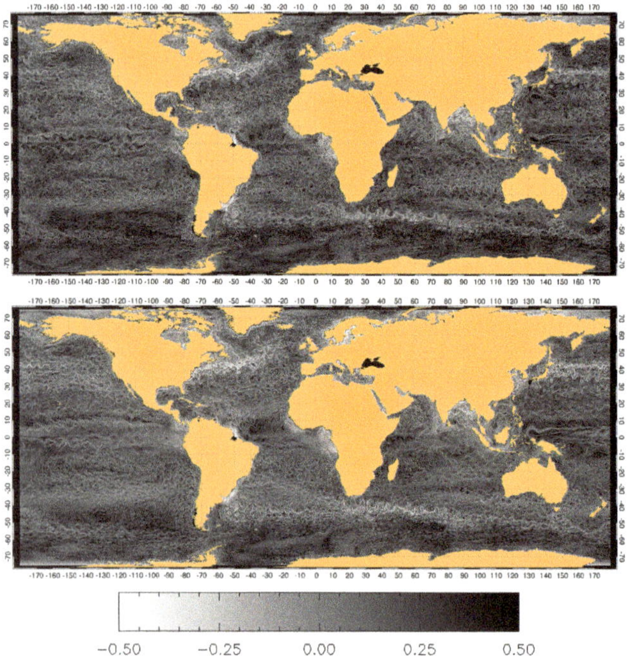

Fig. 6. Top: Singularity exponents associated with SSS. Bottom: Singularity exponents derived from the fused SSS map.

4.2 Decorrelated noise ($\beta = 0$)

Decorrelated noise is the easiest to filter away by our fusion algorithm. In Fig. 7 we show an example of white noise added to the original SSS map and of the result of applying the fusion algorithm to it, as well as the difference map.

Significantly, the average error is negligible (0.02) and the standard deviation of the error map (obtained by averaging over all spatial locations) is 0.33 for input noise of unity standard deviation, a considerable reduction in the error, to very nearly the same level occurring when no noise is introduced. Nevertheless, the restoration of singularity exponents, Fig. 8, is very incomplete: just the singularity exponents associated with the main currents can be recognised. Notice however that the level of input noise is rather high.

4.3 Correlated noise ($\beta = 1$)

Correlated noise is still very well filtered away by our fusion algorithm. In Fig. 9 we show an example of noise added to the original SSS map with the spatial correlation of the noise proportional to k^{-1} and of the result of applying the fusion algorithm to it, as well as the difference map.

The error map has a spatial average of 0.02 and a standard deviation of 0.36 for input noise of unity standard deviation, a considerable error reduction slightly worse than in the previous case. The error map is still associated with significant geophysical structures. The restoration of singularity exponents, Fig. 10, is worse than in the previous case and centered on the main currents.

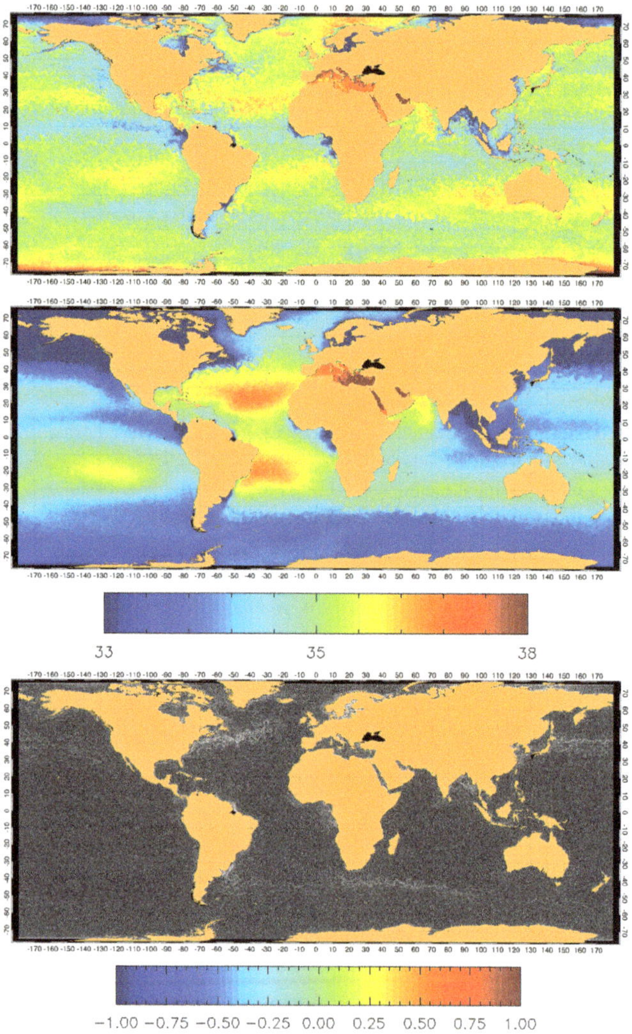

Fig. 7. Top: SSS after the introduction of decorrelated noise with input noise of standard deviation $\sigma_i = 1$ psu. Middle: SSS map after filtering the map on top using the SST map as a template. Bottom: Differences between the original (no noise) SSS map and the fused SSS map.

4.4 Strongly correlated noise ($\beta = 2$)

The correction on strongly correlated noise is very challenging for our fusion algorithm, with an important deformation of large-scale oceanic features. In Fig. 11 we show an example of noise added to the original SSS map with the spatial correlation of the noise proportional to k^{-2} and of the result of applying the fusion algorithm to it, as well as the difference map.

Strongly correlated noise introduces large-scale structures, which are not of geophysical origin, but even so the error map has a negligible spatial average of 0.014. Submitted to such a strong perturbation, the standard deviation of the error is larger in this case, arriving at a value 0.66 for input noise of unity standard deviation; significant but not quite

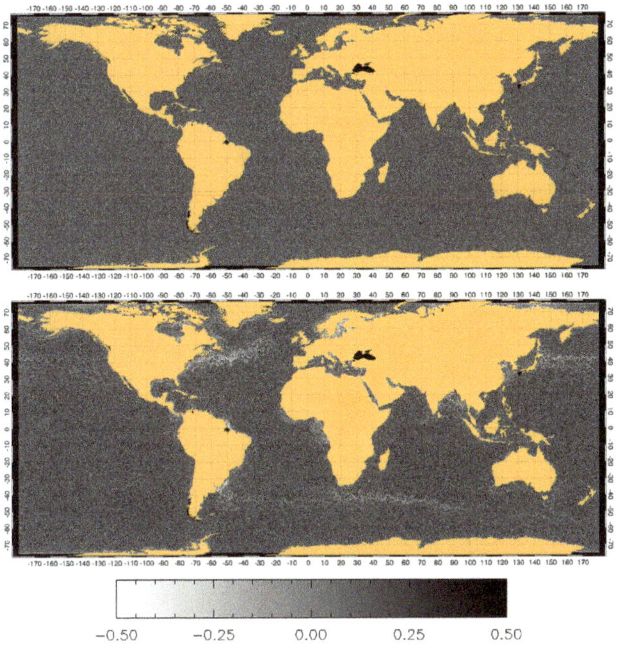

Fig. 8. Top: Singularity exponents associated with SSS after noise of unity amplitude and $\beta = 0$. Bottom: Singularity exponents derived from the fused SSS map.

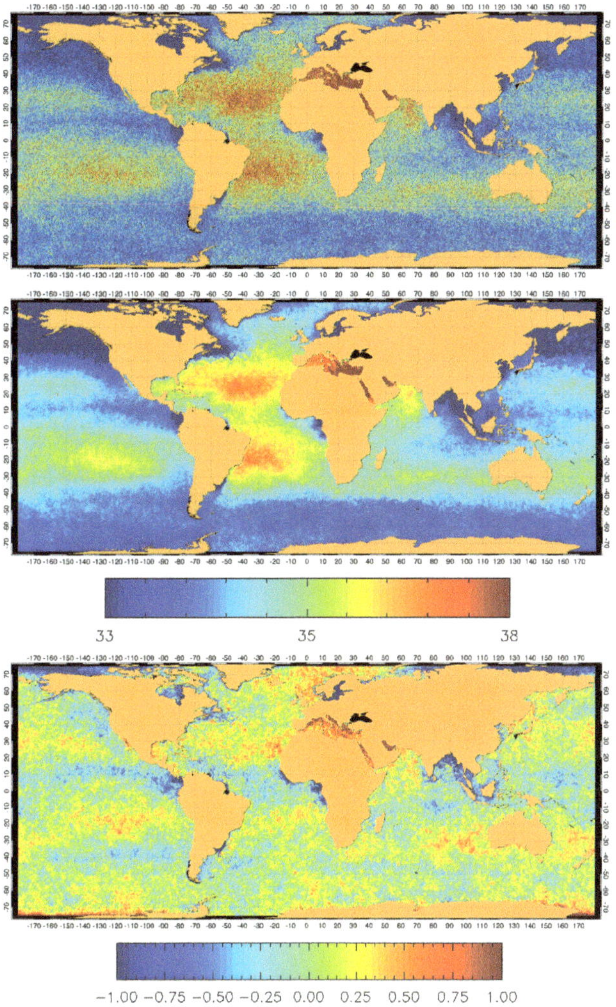

Fig. 9. Top: SSS after the introduction of correlated noise with input noise of standard deviation $\sigma_i = 1$ psu. Middle: SSS map after filtering the map on top using the SST map as a template. Bottom: Differences between the original (no noise) SSS map and the fused SSS map.

satisfactory. The error map has a fuzzy structure, and the main resulting structures are void of geophysical meaning. The restoration of singularity exponents, Fig. 10, is even worse than in the previous case and even the main currents are hard to recognise.

4.5 Dependence of the reconstruction quality on noise intensity

To quantify the quality of our fusion algorithm, we define the output noise level, σ_o, as the standard deviation of the error image, calculated over the whole image (in fact the average error is always lower than 0.02 and negligible by comparison, so the standard deviation and the root squared error are almost identical). We define the input noise, σ_i, as the standard deviation of the injected noise.

In Fig. 13 we show the evolution of the ratios of output noise to input noise as the size of input noise increases for the three types of noise studied. We observe that there is a crossover value σ_c below which fusion is not useful: for any input noise $\sigma_i < \sigma_c$ the output noise σ_o is equal to σ_c (and then the ratio σ_o/σ_i diverges). On the other hand, when the level of input noise is large enough the capability of our fusion algorithm to remove noise stagnates at a fixed percentage of input noise (the asymptotic maximum percentages of noise removal are 91 % ($\beta = 0$), 83 % ($\beta = 1$), and 43 % ($\beta = 2$). Those ratios are really very good, implying that fusion is always a good strategy even for very noisy images.

5 Comparison with median filters

Fusion methodologies have the appeal of invoking the use of synergistic approaches to combine the information coming from different variables; however, it is a relatively sophisticated methodology to filter noise which does not depend on signal amplitude, as the one used in this paper. Many conventional filters, as for instance median filters or average filters, could lead to better or comparable results and are easier to implement. We have thus compared the results of fusion filters with median filters (average filters always have worse performance than median filters).

To define our median filters, we will consider square neighbourhoods of $(2N+1) \times (2N+1)$ points (at the given resolution) centered at each point of the map. For each square neighbourhood we compute the median of the points in the

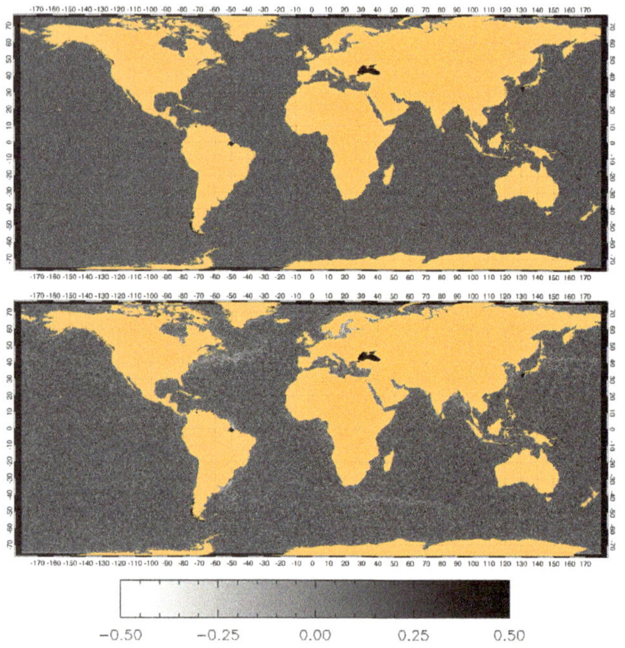

Fig. 10. Top: Singularity exponents associated with SSS after noise of unity amplitude and $\beta = 1$. Bottom: Singularity exponents derived from the fused SSS map.

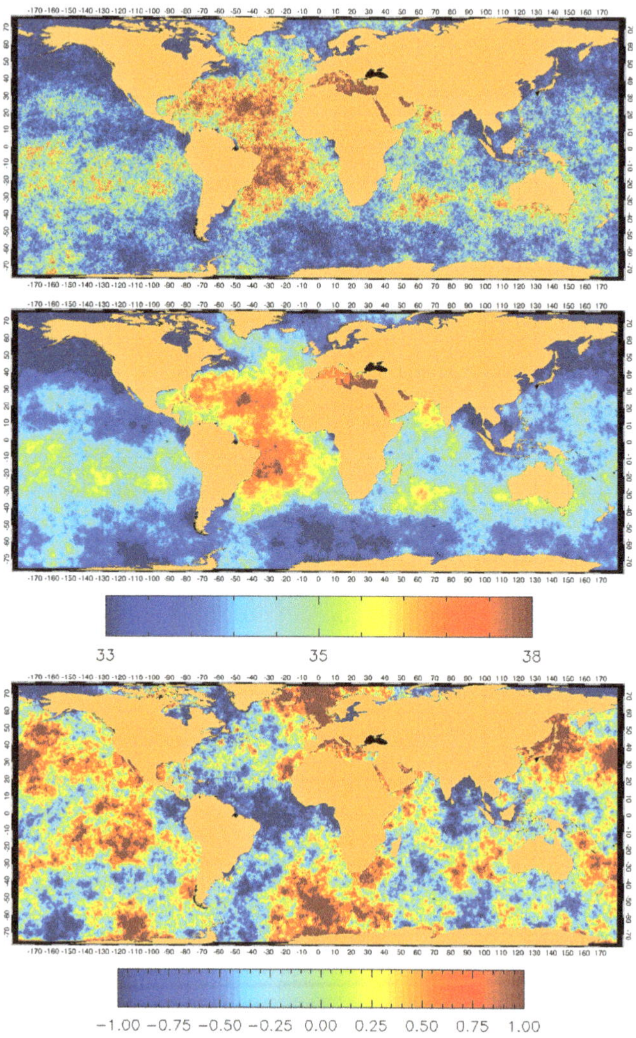

Fig. 11. Top: SSS after the introduction of strongly correlated noise with input noise of standard deviation $\sigma_i = 1$ psu. Middle: SSS map after filtering the map on top using the SST map as a template. Bottom: Differences between the original (no noise) SSS map and the fused SSS map.

neighbourhood (the value of the signal which has exactly 50 % probability of occurrence, that is, the one exactly in the middle position if the list of $(2N+1) \times (2N+1)$ values were ordered), and we assign this local median value to that point to obtain the median filtered map. Obviously the quality of this map depends on the size of the neighbourhood, which is parametrized by N. Larger Ns tend to reduce the impact of the overimposed Gaussian noise, but at the cost of increasing the error associated with smoothening of the original signal. In Fig. 14 we show an example of the application of median filters of different sizes to a map of SSS corrupted with spatially decorrelated Gaussian noise.

As in Sect. 4.5 we have calculated the evolution of the ratio of output to input noise as the amplitude of input noise increases for median filters of different sizes, compared to the performance of fusion. As shown in Figs. 15, 16, and 17, median filters achieve good reconstruction only when the amplitude of noise is small enough. In general, median filters have better performance when noise is less correlated and when the filter size increases, at the cost of sacrificing resolution. At large sizes (5 × 5 or greater) median and average filters become very similar for input noises greater than 0.5 (figures not shown) because the median and the average of the Gaussian noise coincide. Taking this into account we can predict that the output noise for a median filter applied to a map contaminated with decorrelated noise with $N \geq 2$ and $\sigma_i > 0.5$ behaves as:

$$\sigma_o^{\text{median}} = \sigma_g(N) + \sigma_i/(2N+1), \tag{15}$$

where $\sigma_g(N)$ is the inner geophysical variability of SSS in the $(2N+1) \times (2N+1)$ neighbourhoods. SSS is a variable with relatively low variability to scales of less than 1 degree (Jorda et al., 2011), so up to $N = 5$ the first term in Eq. (15) can be neglected as compared to σ_i for $\sigma_i > 0.5$. We obtain thus

$$\frac{\sigma_o^{\text{median}}}{\sigma_i} \approx \frac{1}{2N+1}. \tag{16}$$

This explains why the curves in Fig. 15 asymptotically tend to straight lines, and why the asymptotical value is proportional to $\frac{1}{2N+1}$. When noise is correlated, it is possible to infer that the formula above can be generalized to:

$$\frac{\sigma_o^{\text{median}}}{\sigma_i} \approx \frac{1}{(2N+1)^\gamma}, \tag{17}$$

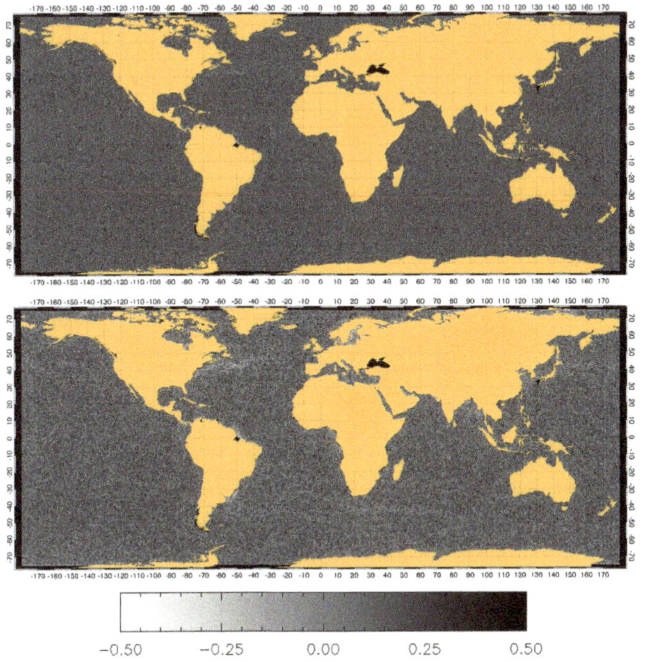

Fig. 12. Top: Singularity exponents associated with SSS after noise of unity amplitude and $\beta = 2$. Bottom: Singularity exponents derived from the fused SSS map.

where $\gamma \to 0$ as $\beta \to \infty$. This explains why median filters have lower performance with correlated noise. In contrast, fusion offers a much greater reduction of noise, probably because it is a non-linear method.

6 Summary and future work

In this paper we have shown that ocean scalars of different types possess a common multifractal structure. This correspondence has been highlighted in previous works (Nieves et al., 2007; Isern-Fontanet et al., 2007), but it has only been used recently (Umbert et al., 2014) to derive a new algorithm for data fusion.

Our multifractal fusion method can be used to enhance the signal quality of a variable contaminated with noise by using another variable as a template. The method is constructed to reintroduce the multifractal structure of the noised variable based on that of the template. This guarantees the reestablishment of the structure of short-range spatial correlations in the signal, but it seems to impact also long-range correlations.

We have shown the effect of the fusion algorithm when a source of uncorrelated noise of given amplitude is added. Our results indicate that until a threshold in noise amplitude of about 0.3 is reached the application of our denoising algorithm does not improve the quality of the signal; even a noiseless image degrades in quality to 0.3 after the application of the fusion algorithm. The results of the application of

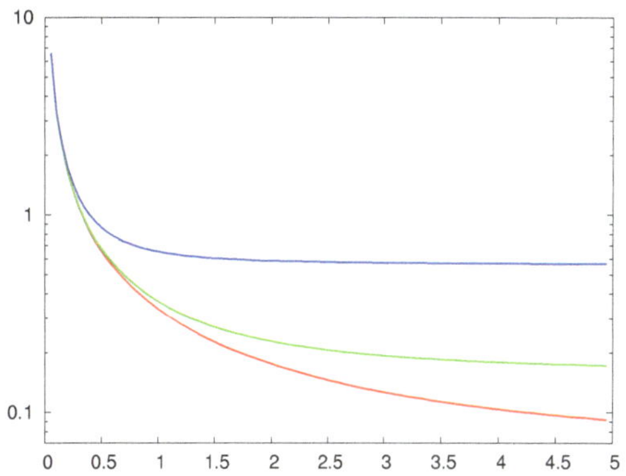

Fig. 13. Evolution of the ratio output noise/input noise depending on the value of input noise. Red curve: decorrelated noise ($\beta = 0$); green curve: correlated noise ($\beta = 1$); blue curve: strongly correlated noise ($\beta = 2$); a constant level of 1 is also drawn (in black) for reference. The estimated asymptotic values for the input/output ratios are 0.09, 0.17, and 0.57, respectively. The crossovers (passings by 1) occur for input noise levels of 0.33, 0.33, and 0.40, respectively.

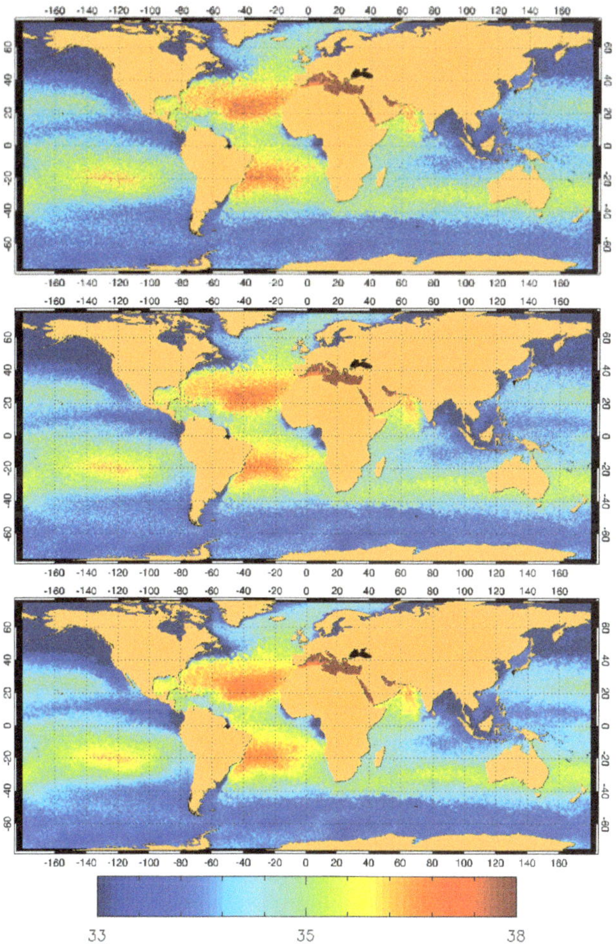

Fig. 14. Results of applying median filters to a noisy salinity with no spatial correlation and amplitude 1 (same as in Fig. 7, top). The filters are calculated over boxes of variable sizes centered around each point. Results are for 3×3 boxes (top), 5×5 boxes (middle), and 11×11 (bottom).

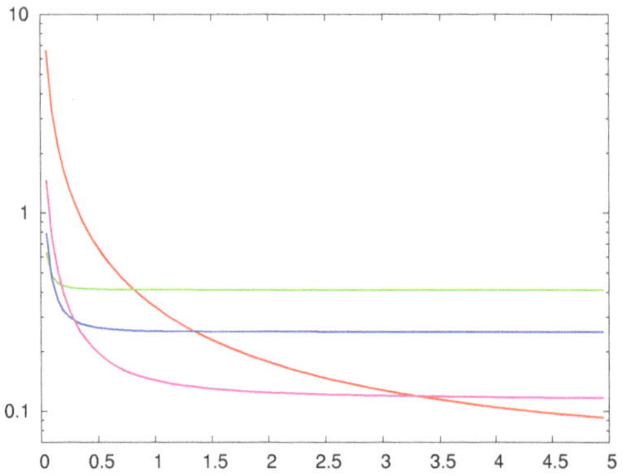

Fig. 15. Evolution of the ratio output noise/input noise depending on the value of input noise for decorrelated noise ($\beta = 0$). Red curve: fusion; green curve: median filter with 3×3 neighbourhoods; blue curve: median filter with 5×5 neighbourhoods; purple curve: median filter with 11×11 neighbourhoods. For large input noises fusion performs better than the others; the crossover points occur for input noise levels of about 0.75, 1.25, and 3.25, respectively.

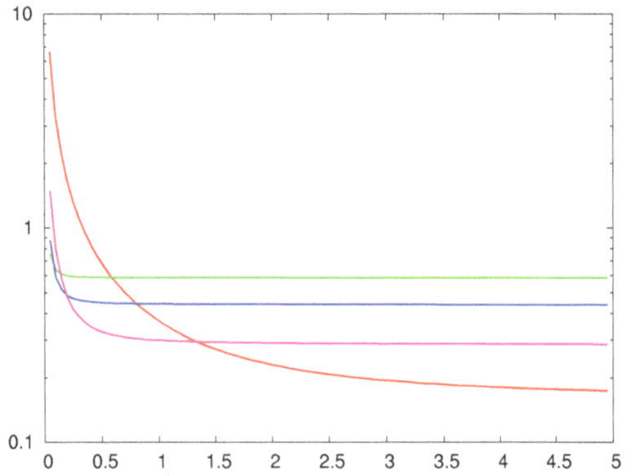

Fig. 16. Evolution of the ratio output noise/input noise depending on the value of input noise for slightly correlated noise ($\beta = 1$). Red curve: fusion; green curve: median filter with 3×3 neighbourhoods; blue curve: median filter with 5×5 neighbourhoods; purple curve: median filter with 11×11 neighbourhoods. For large input noises fusion performs better than the others; the crossover points occur for input noise levels of about 0.6, 0.8, and 1.3, respectively.

the fusion algorithm to remote sensing data (SMOS SSS maps using OSTIA SST templates) discussed in Umbert et al. (2014) show that in fact the quality could be even better, with post-fusion errors of around 0.20 and even smaller. This difference may indicate that the correspondence in the multifractal structure of the maps of SSS and SST resulting from OFES is not as good as in real data from the oceans. The result indicates that reinforcing the correspondence between ocean scalars in numerical models of the ocean could potentially improve their assessment quality about the actual ocean state, which is important both for operational applications and for the study of ocean processes. The introduction of an assimilation scheme accounting for the singularity structure of the scalars should be studied.

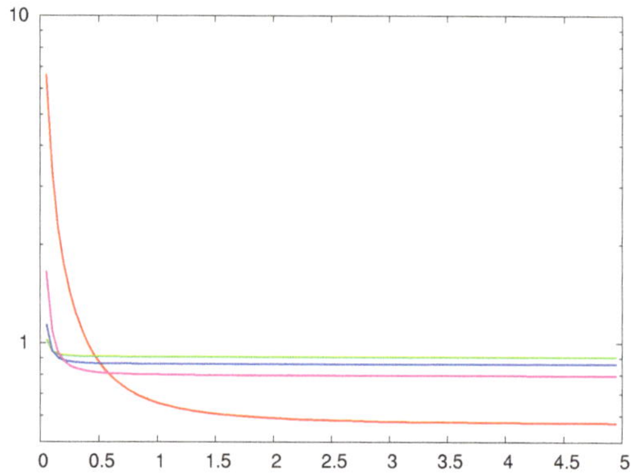

Fig. 17. Evolution of the ratio output noise/input noise depending on the value of input noise for strongly correlated noise ($\beta = 2$). Red curve: fusion; green curve: median filter with 3×3 neighbourhoods; blue curve: median filter with 5×5 neighbourhoods; purple curve: median filter with 11×11 neighbourhoods. For large input noises fusion performs better than the others; the crossover points occur for input noise levels of about 0.5, 0.55, and 0.65, respectively.

The existence of a common singularity structure is a central hypothesis of our algorithm, which seems to be confirmed for more and more examples from remote sensing and from numerical models. The development of an appropriate theoretical framework to explain the creation of singularities is one of the major challenges in this field of research, but will help to unify the evolution equations for different scalars and most probably control their numerical behaviour and improve their quality.

The method presented in Umbert et al. (2014) and analysed here is just a first step in the theory of multifractal methods for data fusion. From the experiences shown in this paper it is clear that the method is not fully able to restore the multifractal structure using that of the template. Most likely this limitation comes from the approximation used (the matrix issued from Riesz's representation theorem considered to be proportional to the identity matrix). The extension for a representation with a non-trivial matrix and a better inference strategy for the smooth matrix elements should be actively sought. This matrix will not only serve to improve the quality of existing remote sensing maps, but also will serve to extrapolate maps in missing areas and will provide geophysical insight into the dynamic relations governing the relation between the different variables (in particular, providing canonical classification of regions according to the physical and chemical properties of their water masses – mostly reminiscent of the biogeochemical provinces).

Acknowledgements. We acknowledge ICMAT Severo Ochoa project SEV-2011-0087 for its support in the publication of this work. M. Umbert is funded by an FPI grant from the Spanish Ministry of Economy. This work has been funded by the Spanish Ministry of Economy through the National R+D Plan by means of MIDAS-7 project AYA2012-39356-C05-03, MIDAS-6 project AYA2010-22062-C05 and previous grants. The OFES simulation was conducted on the Earth Simulator with the support of JAMSTEC. We thank the JAMSTEC Earth Simulator Center team for providing these data.

Edited by: E. Hernández-García

References

Cayula, J.-F. and Cornillon, P.: Multi-Image Edge Detecion for SST Images, J. Atmos. Ocean. Tech., 12, 821–829, 1995.

Chelton, D., Schlax, M., Samelson, R., and de Szoeke, R.: Global observations of large oceanic eddies, Geophys. Res. Lett., 34, L15606, doi:10.1029/2007GL030812, 2007.

Dubrulle, B.: Intermittency in fully developed turbulence: Log-Poisson statistics and generalized scale covariance, Phys. Rev. Lett., 73, 959–962, 1994.

Font, J., Boutin, J., Reul, N., Spurgeon, P., Ballabrera-Poy, J., Chuprin, A., Gabarro, C., Gourrion, J., Guimbard, S., Henocq, C., Lavender, S., Martin, N., Martinez, J., McCulloch, M., Meirold-Mautner, I., Mugerin, C., Petitcolin, F., Portabella, M., Sabia, R., Talone, M., Tenerelli, J., Turiel, A., Vergely, J.-L., Waldteufel, P., Yin, X., Zine, S., and Delwart, S.: SMOS first data analysis for sea surface salinity determination, Int. J. Remote Sens., 34, 3654–3670, 2012.

Hallberg, R. and Gnanadesikan, A.: The Role of Eddies in Determining the Structure and Response of the Wind-Driven Southern Hemisphere Overturning: Results from the Modeling Eddies in the Southern Ocean (MESO) Project, J. Phys. Oceanogr., 36, 2232–2252, 2006.

Isern-Fontanet, J., García-Ladona, E., and Font, J.: Identification of Marine eddies from Altimetry, J. Atmos. Ocean. Tech., 20, 772–778, 2003.

Isern-Fontanet, J., Turiel, A., Garcia-Ladona, E., and Font, J.: Microcanonical Multifractal Formalism: application to the estimation of ocean surface velocities, J. Geophys. Res., 112, C05024, doi:10.1029/2006JC003878, 2007.

Jorda, G., Gomis, D., and Talone, M.: The SMOS L3 Mapping Algorithm for Sea Surface Salinity, IEEE T. Geosci. Remote, 49, 1032–1051, doi:10.1109/TGRS.2010.2068551, 2011.

Kestener, P. and Arnéodo, A.: Three-Dimensional Wavelet-Based Multifractal Method: The Need for Revisiting the Multifractal Description of Turbulence Dissipation Data, Phys. Rev. Lett, 91, 194501, doi:10.1103/PhysRevLett.91.194501, 2003.

Kolmogorov, A. N.: The local structure of turbulence in an incompressible fluid with very large Reynolds number, Dokl. Akad. Nauk. SSSR, 309, 301–305, 1941a.

Kolmogorov, A. N.: Dissipation of energy in a locally isotropic turbulence, Dokl. Akad. Nauk. SSSR, 32, 16–18, 1941b.

Mallat, S.: A Wavelet Tour of Signal Processing, 2nd Edn., Academic Press, 1999.

Mallat, S. and Huang, W. L.: Singularity detection and processing with wavelets, IEEE T. Inform, Theory, 38, 617–643, 1992.

Mason, E., Colas, F., Molemaker, J., Shchepetkin, A. F., Troupin, C., McWilliams, J. C., and Sangra, P.: Seasonal variability of the Canary Current: A numerical study, J. Geophys. Res., 116, C06001, doi:10.1029/2010JC006665, 2011.

Masumoto, Y.: Sharing the results of a high-resolution ocean general circulation model under a multi-discipline framework – a review of OFES activities, Ocean Dynam., 60, 633–652, 2010.

Masumoto, Y., Sasaki, H., Kagimoto, T., Komori, N., Ishida, A., Sasai, Y., Miyama, T., Motoi, T., Mitsudera, H., Takahashi, K., Sakuma, H., and Yamagata, T.: A fifty-year eddy-resolving simulation of the world ocean – Preliminary outcomes of OFES (OGCM for the Earth Simulator), Journal of the Earth Simulator, 1, 35–56, 2004.

Nieves, V., Llebot, C., Turiel, A., Solé, J., García-Ladona, E., Estrada, M., and Blasco, D.: Common turbulent signature in sea surface temperature and chlorophyll maps, Geophys. Res. Lett., 34, L23602, doi:10.1029/2007GL030823, 2007.

Parisi, G. and Frisch, U.: On the singularity structure of fully developed turbulence, in: Turbulence and Predictability in Geophysical Fluid Dynamics, edited by: Ghil, M., Benzi, R., and Parisi, G., Proceedings of the International School of Physics "Enrico Fermi", North Holland, Amsterdam, 84–87, 1985.

Rudin, W.: Real and Complex Analysis, Mc Graw Hill, New York, USA, 1987.

She, Z. S. and Leveque, E.: Universal scaling laws in fully developed turbulence, Phys. Rev. Lett., 72, 336–339, 1994.

She, Z. S. and Waymire, E. C.: Quantized energy cascade and log-poisson statistics in fully developed turbulence, Phys. Rev. Lett., 74, 262–265, 1995.

Turiel, A., Isern-Fontanet, J., García-Ladona, E., and Font, J.: Multifractal method for the instantaneous evaluation of the stream function in geophysical flows, Phys. Rev. Lett., 95, 104502, doi:10.1103/PhysRevLett.95.104502, 2005.

Turiel, A., Pérez-Vicente, C., and Grazzini, J.: Numerical methods for the estimation of multifractal singularity spectra on sampled data: a comparative study, J. Comput. Phys., 216, 362–390, 2006.

Turiel, A., Solé, J., Nieves, V., Ballabrera-Poy, J., and García-Ladona, E.: Tracking oceanic currents by singularity analysis of Micro-Wave Sea Surface Temperature images, Remote Sens. Environ., 112, 2246–2260, 2008a.

Turiel, A., Yahia, H., and Pérez-Vicente, C.: Microcanonical Multifractal Formalism: a geometrical approach to multifractal systems. Part I: Singularity analysis, J. Phys. A, 41, 015501, doi:10.1088/1751-8113/41/1/015501, 2008b.

Turiel, A., Nieves, V., Garcia-Ladona, E., Font, J., Rio, M.-H., and Larnicol, G.: The multifractal structure of satellite sea surface temperature maps can be used to obtain global maps of streamlines, Ocean Sci., 5, 447–460, doi:10.5194/os-5-447-2009, 2009.

Ullman, D. and Cornillon, P.: Evaluation of Front Detection Methods for Satellite-Derived SST Data Using In Situ Observations, J. Atmos. Ocean. Tech., 17, 1667–1675, 2000.

Umbert, M., Hoareau, N., Turiel, A., and Ballabrera-Poy, J.: Multifractal synergy among ocean scalars: applications to the blending of remote sensing data, Remote Sens. Environ., accepted, doi:10.1016/j.rse.2013.09.018, 2014.

4

Tidally induced internal motion in an Arctic fjord

E. Støylen[1,*] and I. Fer[2]

[1]Department of Geosciences, University of Oslo, Oslo, Norway
[2]Geophysical Institute, University of Bergen, Bergen, Norway
[*]now at: Norwegian Meteorological Institute, Oslo, Norway

Correspondence to: I. Fer (ilker.fer@gfi.uib.no)

Abstract. The internal response in a stratified, partially enclosed basin subject to semi-diurnal tidal forcing through a narrow entrance is investigated. The site is located above the critical latitude where linear internal waves of lunar semi-diurnal frequency are not permitted to propagate freely. Generation and propagation of tidally induced internal Kelvin waves are studied, for baroclinically sub- and supercritical conditions at the mouth of the fjord, using a non-linear 3-D numerical model in an idealized basin and in Van Mijenfjorden, Svalbard, using a realistic topography. The model results are compared to observations of hydrography and currents made in August 2010. Results from both the model and measurements indicate the presence of internal Kelvin waves, even when conditions at the fjord entrance are supercritical. The entrance of Van Mijenfjorden is split into two sounds. Sensitivity experiments by closing each sound separately reveal that internal Kelvin waves are generated at both sounds. When the conditions are near supercritical, a wave pulse propagates inward from the fjord entrance at the beginning of each inflow phase of the tidal cycle. The leading crest is followed by a series of smaller amplitude waves characterized as non-linear internal solitons. However, higher model resolution is needed to accurately describe the influence of small-scale mixing and processes near the sill crest in establishing the evolution of the flow and internal response in the fjord.

1 Introduction

Internal waves in the Arctic regions have received increasing scientific interest recently because of their role in vertical mixing and influence on the regional and large-scale heat budget and ice cover. Wind-induced internal waves are important during ice-free conditions in seasonally ice-covered regions, as demonstrated in the northern Chukchi Sea by Rainville and Woodgate (2009). The positive effect of reduced ice cover on internal wave forcing, however, may be offset by increased stratification by meltwater which amplifies the negative effect of boundary layer dissipation on internal wave energy (Guthrie et al., 2013). Another source of internal waves is the action of tidal flow over topography; the Yermak Plateau is noted as an important region for enhanced internal wave activity (e.g. Padman and Dillon, 1991; Fer et al., 2010). In Arctic fjords short period internal waves are observed under ice (Marchenko et al., 2010; Morozov and Marchenko, 2012), and longer period internal Kelvin waves are documented in the Kongsfjorden–Krossfjorden system (Svendsen et al., 2002).

Internal waves in fjords are typically forced by changing winds and the barotropic tide. For fjords that are wide with respect to the internal (baroclinic) Rossby radius, internal Kelvin waves may arise, propagating cyclonically around the fjord. Such waves induce a mean current in the wave propagation direction (Støylen and Weber, 2010), which may lead to an accumulation or deposition of pollutants and biological material in certain areas along the coastline. As discussed in Cottier et al. (2010), fjords in the Arctic are typically wide with respect to the baroclinic Rossby radius. Thus, given sufficient forcing, internal Kelvin waves are to be expected in many stratified Arctic fjords.

In the present work we consider a particular wide, tidally forced Arctic fjord, namely Van Mijenfjorden in Svalbard (Fig. 1). The entrance of this fjord is partly covered by an island, which restricts water inflow to two narrow sounds, and thus makes the fjord a good "laboratory" for process studies (e.g. Widell, 2006; Fer and Widell, 2007). The energy extracted from the barotropic tide, partitioning to tidal jet flux and baroclinic jet flux, as well as the modal contributions to kinetic energy and horizontal shear are discussed

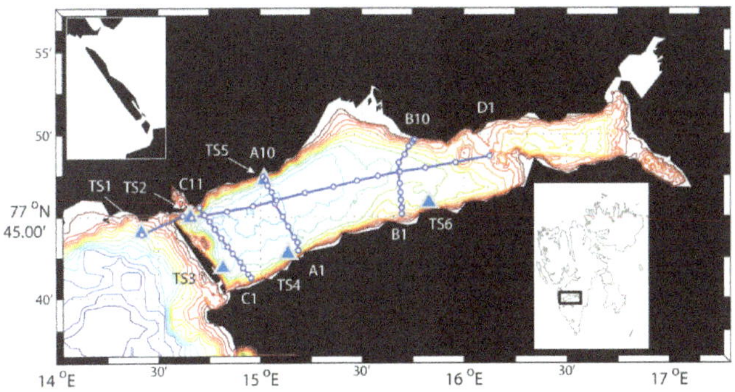

Fig. 1. Map of Van Mijenfjorden in Svalbard, situated at 77.8° N, 15.5° E. The insets show (top left) a blow-up of the fjord mouth with the island of Akseløya, and (bottom right) the location of Van Mijenfjord in Svalbard. Lines and dots denote CTD measurement sections (A, B, C, and D) with start and end stations indicated. Triangles marked TS1–TS6 are the time series stations (Table 1).

in Fjellsbø (2013). Our aim is to describe the generation and propagation of internal waves induced by tides in this fjord, by use of recent observations and a 3-D numerical model. In particular, we investigate the possibility of internal Kelvin waves, as suggested in Støylen and Weber (2010) and Skarðhamar and Svendsen (2010). The sensitivity of internal wave generation to hydraulic conditions through the fjord entrance is investigated numerically. Finally, as the fjord entrance consists of two sounds we demonstrate numerically the effect of closing each sound on the respective wave field.

This study is organized as follows: in Sect. 2 we present a theoretical background for internal Kelvin waves in Arctic fjords. Observations from Van Mijenfjorden are given in Sect. 3. In Sect. 4 we treat the numerical problem in an idealized geometry, before applying realistic bottom topography of Van Mijenfjorden. Finally, we discuss our findings and provide concluding remarks in Sect. 5.

2 Theory

In this section we briefly discuss what types of internal motion we may expect in an Arctic fjord, and describe a basic theory regarding the internal Kelvin wave. Where appropriate, we will assume a two-layer system with constant densities ρ_1 and ρ_2 for the upper and lower layer, respectively. Other variables have similar subscripts. The Cartesian coordinate system (x, y, z) with z as upward vertical direction has corresponding current components (u, v, w).

Internal waves span a broad range of spatial and temporal scales. In a fjord basin there may be propagating Poincaré waves (Brown, 1973; Farmer and Freeland, 1983) and soliton trains (Helfrich and Melville, 2006). Along the coast there may possibly be propagating edge waves (Llewellyn Smith, 2004; Weber and Støylen, 2011) trapped by the sloping bottom, or the internal Kelvin wave trapped by rotation.

The theory behind internal Kelvin waves is well established in the literature (e.g. Gill, 1982). Nonlinearity modifies this solution, as demonstrated in laboratory experiments by Maxworthy (1983), noting a curvature in the cross-wall wave front as well as a modification of the transverse scale c_i/f, where f is the Coriolis parameter and c_i is the baroclinic phase velocity. Later, others confirmed these findings (e.g. Renouard et al., 1987); see Helfrich and Melville (2006) for a comprehensive review. Full 3-D numerical efforts on fjord scale internal Kelvin waves arose during the late 1990s, with a primary focus on closed inland lakes. Beletsky et al. (1997) were among the first to conduct 3-D simulations on this scale, concerning the response in a closed stratified basin (the Great Lakes) to changes in the large-scale wind field. Later numerical studies with increased spatial resolution include Hodges et al. (2000) and Gómez-Giraldo et al. (2006) for Lake Kinneret, Israel.

There are several mechanisms for internal wave generation. In an Arctic fjord, the dominant sources of forcing are the changing wind fields and the barotropic tide interacting with topographic features at the fjord entrance. We will restrict our attention to the tidal case which is particularly relevant for an Arctic fjord that is ice covered. When the cross-sectional area of the fjord entrance is narrow, as is the case in Van Mijenfjorden, the tidal current is intensified here. If the water is stratified, one can determine whether conditions are sub- or supercritical with respect to the first baroclinic mode. For a two-layer system, a densimetric Froude number is defined as $F_D = |u_S|/c_i$, where u_S is the upper layer current. $F_D < 1$ indicates subcritical conditions, and favours generation of long internal waves. When $F_D > 1$, the current enters the fjord as a jet, and there will be a hydraulic jump just inside the entrance. In this case the generation of long waves may be prohibited; we will discuss this further in Sect. 5. For completeness, a derivation of the linear internal Kelvin wave in a two-layer reduced gravity system is given in the following. If the fjord width is large compared to the baroclinic Rossby radius, $a = c_i/f$, wave solutions of Kelvin type may exist. In a two-layer reduced gravity system with a deep lower layer, the velocities in the lower layer are neglected. The balance of forces in the lower layer may thus be written as

$$g\eta_x + P_{Sx}/\rho_1 = -g'\xi_x$$
$$g\eta_y + P_{Sy}/\rho_1 = -g'\xi_y. \quad (1)$$

Here $g' = g(\rho_2 - \rho_1)/\rho_2$ is the reduced gravity, P_S is surface pressure, η and ξ are surface and interface displacements, respectively, and subscripts x and y denote partial derivation. The baroclinic phase velocity is (Gill, 1982)

$$c_i = \sqrt{\frac{\rho_2 - \rho_1}{\rho_2} g \frac{H_1 H_2}{H_1 + H_2}}, \quad (2)$$

where H_1 and H_2 are the upper and lower layer thicknesses, respectively. If we take a straight coast at $y = 0$, the

Table 1. Overview of observation sections and stations where CTD and LADCP profiles were taken. Names A–D denote horizontal sections, TS1–TS6 time series. A–1 to A–3, B–1 and B–2, and C–1 and C–2 are the re-occupations of the corresponding sections. Time is date of August, and hour [DD/HH]. Depth [m] is min–max for sections, average for time series.

Name	B-1	A-1	C-1	TS1	TS2	C-2	TS3
Time	09/20–10/00	10/02–10/04	10/06–10/09	10/10–10/23	11/00–11/13	11/13–11/16	11/17–12/06
No. of samples	10	10	11	18	27	11	27
Depth	42–87	61–117	41–110	60	60	42–114	84
Name	TS4	A-2	TS5	A-3	B-2	TS6	D
Time	12/07–12/20	12/20–12/23	12/23–13/12	13/12–13/15	13/17–13/19	13/20–14/09	14/09–14/15
No. of samples	27	10	27	10	11	27	15
Depth	79	31–112	79	32–113	34–88	66	33–113

first-order current component v_1 is zero everywhere. Inserting from Eq. (1), the linear first order momentum balance in the upper layer becomes

$$u_{1t} = g'\xi_x + \nu\nabla_h^2 u_1 + \nu u_{1zz}$$
$$f u_1 = g'\xi_y, \tag{3}$$

where ν is the kinematic viscosity coefficient, and ∇_h^2 is the horizontal Laplacian operator. The corresponding linearized continuity equation is obtained by assuming $|\eta| \ll |\xi|$:

$$\xi_t = \frac{\partial}{\partial x}\int_0^{H_1} u_1 \mathrm{d}z. \tag{4}$$

For the sake of simplicity we neglect the effect of friction here. We consider waves of constant frequency ω in accordance with steady tidal forcing. A solution to Eqs. (3)–(4) is obtained by assuming internal motion of the form $\xi = \xi_0 e^{-y/a} e^{i(kx-\omega t)}$ with wave number k, and near-coast wave amplitude ξ_0. Letting the real part denote the physical solution, we obtain

$$\xi = \xi_0 e^{-y/a} \cos(kx - \omega t) \tag{5}$$

with the corresponding dispersion relation $\omega^2 = g' H_1 k^2$. Equation (5) describes the linear internal Kelvin wave, in accordance with Støylen and Weber (2010). The wave is travelling along the positive x axis on the Northern Hemisphere, the interface displacement is largest near the coast, and is exponentially damped seaward. The frequency of the wave is the forcing frequency; in our case the lunar semi-diurnal M_2 tidal component. Retaining the friction term leads to the resulting wave being further damped along the coast (Støylen and Weber, 2010).

Linear wave solutions for free internal Poincaré waves are restricted at the critical latitude where $\omega = f$ for the respective forcing frequency (Vlasenko et al., 2003). For the M_2 tidal component the critical latitude is $\phi = 74.5°$. This restriction does not apply to internal Kelvin waves, however, and they may exist in Arctic regions above the critical latitude (Farmer and Freeland, 1983).

3 Observations

Our study site is Van Mijenfjorden in Svalbard; see Fig. 1. This fjord is an interesting "laboratory" for the study of propagating baroclinic waves of tidal periodicity. At the outer region of the fjord the near-coast bathymetry is quite steep. The entrance of the fjord is partly covered by an island, Akseløya. This restricts the water exchange into two narrow sounds, Akselsundet in the north, and Mariasundet in the south. The typical tidal amplitude outside Akseløya is in the range from 0.3 to 0.8 m (see Fig. 2), and currents measured in Akselsundet may exceed $2\,\mathrm{m\,s^{-1}}$ (Bergh, 2004). During summer and autumn the water in the fjord is stratified as a result of glacial melting. It is during this period that we expect baroclinic tidal activity to be most pronounced. Indeed, using a 22 h CTD time series near Blixodden in July 1996, Skarðhamar and Svendsen (2010) observed a vertical displacement of the pycnocline of 20 m, which they argued was caused by a passing internal Kelvin wave.

Our measurements were performed in the period between 9 and 14 August 2010 during a cruise of the Research Vessel (R/V) *Håkon Mosby*. Profiles of hydrography and horizontal currents were collected using a rosette equipped with a Sea-Bird Electronics 911+ CTD (conductivity, temperature, depth) system and a set of up- and down-looking 300 kHz, RD Instruments LADCP (lowered acoustic Doppler current profiler). Data were processed using well-established routines and averaged vertically in 1 and 4 m-thick bins for CTD and LADCP, respectively. Weather data were collected from the ship's meteorological mast. Figure 1 shows a map of the measurement locations; see also Table 1. Labels A–D denote horizontal sections, and TS1–TS6 are time series. For each time series station, a profile was obtained approximately every 30 min for a station occupation period of 13 h, thus encompassing the M_2 tidal period of 12.4 h. The location of the stations and the sampling frequency in time were chosen so as to best capture a potentially propagating internal Kelvin wave, which is expected to have its largest amplitude near the coast (see Sect. 2), and propagating cyclonically around the basin. Due to the steep topographic slope near the fjord

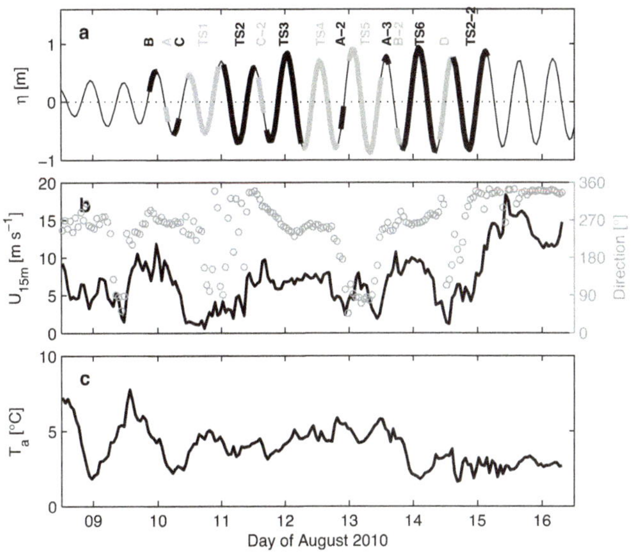

Fig. 2. **(a)** Tidal surface amplitude outside Van Mijenfjorden inferred at the location of TS1 from the AOTIM-5 model (Padman and Erofeeva, 2004). Alternating thick grey and black portions represent the time of occupation of the sections and the time series stations indicated by letters. Several sections were repeated, e.g. A-2 corresponds to the second occupation of section A. Time series of hourly averaged **(b)** wind direction and speed, and **(c)** air temperature, measured from the meteorological mast onboard R/V *Håkon Mosby* at 15 m height.

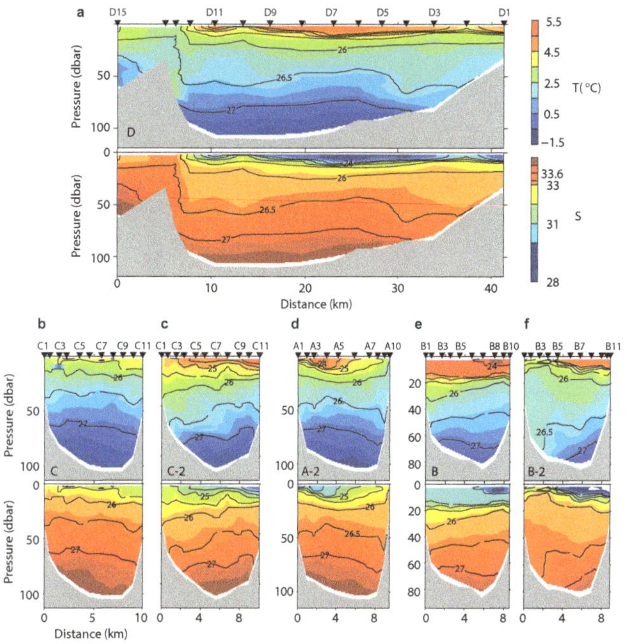

Fig. 3. Temperature and salinity (colour) and density (black contours) for CTD sections **(a)** D, **(b)** C-1, **(c)** C-2, **(d)** A-2, **(e)** B-1, and **(f)** B-2. Orientation of the figures is west–east in **(a)**, and south–north in **(b)–(f)**. See Fig. 1 and Table 1 for reference.

entrance and strong tidal forcing, in addition to the long internal Kelvin waves, short non-linear solitary waves are expected (see e.g. Farmer and Armi, 1999; Grue et al., 1999; Cummins et al., 2003). These short wavelength and high-frequency phenomena are, however, not resolved with our sampling scheme, and will be treated further in the numerical analysis (Sect. 4). The narrow sound combined with swift currents posed severe navigational difficulties and hindered a detailed, high-time and spatial resolution sampling near the sill region.

A general view of the hydrography in the basin is presented in Fig. 3. The stratification is typical for this season. Surface water is relatively fresh as a result of summer melt and runoff from glaciers, and warmer than the water below due to surface heating; see Fig. 2 for air temperature measurements. The colder and saltier deep water seen in Fig. 3a) originates from outside the fjord. During high tide this water is lifted over the sill as can be inferred from the D-section; the outmost D-profile was the last sample in the section, and was thus captured during high tide; see Fig. 2. The repeated sections B and C indicate a strong variability on relatively short time scales throughout the entire fjord basin (Fig. 3). This is in accordance with Skarðhamar and Svendsen (2010), where a more detailed description of the short-term hydrographic variability in Van Mijenfjorden can be found. In the present work we restrict our attention to the internal waves.

Time–depth contours of velocity and density for each time series station are presented in Fig. 4. Again we refer to Fig. 2 for comparison with the tide. Outside Akseløya (TS1) the dominant features of the tide entering and leaving Akselsundet are distinguishable in the upper 40 m. Just inside Akselsundet (TS2) we observe intense mixing associated with the peak tidal inflow. The small-scale overturning and mixing associated with this hydraulic jump will influence the generation of long waves in the lee of the sill; see discussion in Sect. 5. An intensified inflow may also contribute to generation of short non-linear solitons, as we observe in the numerical experiments reported in Sect. 4. Further south, along Akseløya north of Mariasundet (TS3), the flow field is dominated by a southward current that is related to the time of maximum tidal inflow through Akselsundet. Assuming a typical current velocity of 35 cm s^{-1} as measured at TS3 and a distance between Askelsundet and the TS3 station of 7.5 km, we obtain a travel time of about 6 h in agreement with the time of tidal inflow inside Akselsundet. We note the displacements of the isopycnals which weakly indicate oscillations of tidal periodicity. Along the southern coast (TS4) the currents are in-fjord during the entire tidal period. This station is potentially influenced by inflow from both sounds, thus a more complex current structure is expected. Interestingly, the isopycnals are displaced vertically by more than 30 m during the time series, accompanied by a clear vertical shift in the horizontal current. A similar displacement is not observed along the northern coast (TS5). The observations from the

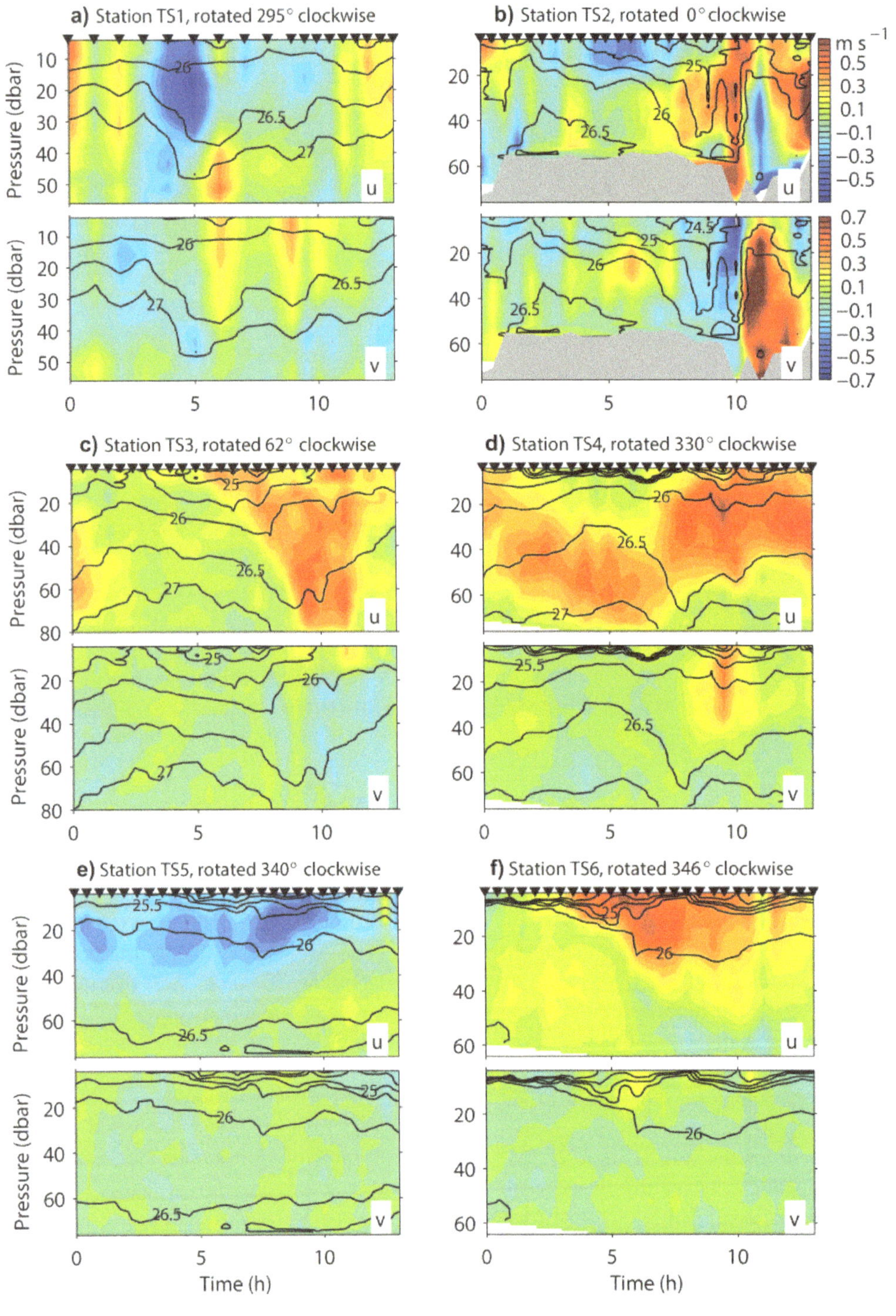

Fig. 4. u and v velocities (colour) and density (σ_{theta}, black contours) for each time series station. **(a)–(f)** correspond to TS1–TS6 respectively. Axes are rotated so that u is aligned eastward along the coast where applicable (TS3–TS6), or along a direct line inwards through Akselsundet (TS1–TS2).

time series stations suggest the presence of an internal Kelvin wave propagating in-fjord from Akseløya along the southern coast, and dissipating before returning outward along the northern side. The role of Mariasundet in this process is difficult to assess (Sect. 4). The main outflow along TS5, and the dominant eastward flow along the southern slope at TS4 and TS6 are consistent with Kelvin waves and the mean circulation in Van Mijenfjorden induced by the tidal forcing (Bergh, 2004).

Time series of horizontal velocity and density profiles collected at stations TS1 to TS6 are used to calculate the baroclinic energy flux for the semi-diurnal signal

$$\boldsymbol{F}_E(z) = \langle \boldsymbol{u}' p' \rangle_\varphi \quad (6)$$

where $\boldsymbol{u}'$ is the baroclinic perturbation velocity vector, p' is the pressure perturbation, and averaging is over the M_2 phase, φ. The calculations are made following the methods detailed in Nash et al. (2005). The pressure anomaly is inferred from the density profiles assuming hydrostatic balance, after removing the full depth average to satisfy baroclinicity. The perturbation velocity is calculated from the LADCP profiles after removing the depth and time average. The semi-diurnal components for the pressure and velocity perturbations are then isolated by harmonic analysis of time series at each depth level. Using the amplitude and phase obtained from the harmonic analysis, one cycle of a sinusoidal semi-diurnal wave is constructed, and time averaging is done over one complete phase. The depth integrated fluxes are shown in Fig. 5. The semi-diurnal baroclinic energy is generated at the fjord sill, in both in-fjord and out-fjord directions. The in-fjord energy propagates cyclonically along the slope, decaying in magnitude, presumably due to friction. The observations of the baroclinic semi-diurnal energy flux are consistent with Kelvin waves.

The depth of the isopycnal $\sigma_\theta = 26$ at TS2 gradually decreases from about 11 m to approximately 59 m, close to the seabed, which then abruptly rises to 19 m in 2 h. The details of this transition are not resolved by our observations. The evolution of the stratification at TS2 is complex, and a two-layer approximation is not possible. Nevertheless, between 2 and 7 h into the record, when the isopyncals are relatively smooth, the depth of the pycnocline, inferred from the depth of the maximum in vertical density gradient, is 15 ± 4 m ($\pm$one standard deviation); the depth of the $\sigma_\theta = 26$ surface during this period is approximately 22 m. Assuming that the pynocline depth is representative of the upper layer thickness in a two-layer flow, this vertical displacement corresponds to a normalized excursion (vertical displacement divided by the upper layer thickness) of 2.7, or 1.8 when normalized by the mean depth of the isopycnal.

At TS4, large vertical isopycnal displacements occur in the relatively weakly stratified portion of the water column. This is expected through the WKB scaling, since less energy is required to vertically displace the isopycnals in weaker stratification. The vertical excursion of $\sigma_\theta = 26.5$ surface is 37 m in approximately 2 h. This isopycnal is located at 40 m depth on average, corresponding to a normalized excursion of 0.9. The stratification is strongest in the upper layers, and the pycnocline depth is estimated at 6 ± 2 m using all the profiles. The typical vertical displacement in the pycnocline, using the 25 isopycnal, is 4–6 m, yielding a normalized excursion close to 1. The exact period of the wave at TS4 (or at other stations) cannot be inferred from the time series with confidence because of the short duration of the record. While the

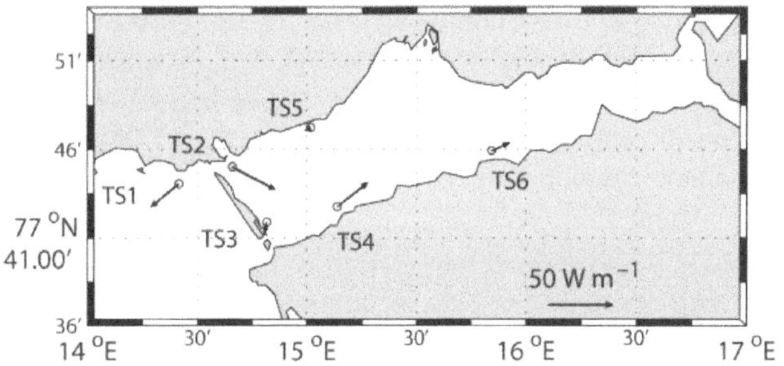

Fig. 5. Depth-integrated semi-diurnal baroclinic energy flux vectors inferred from the time series stations. The scale is shown on the bottom right.

semi-diurnal period fits explain up to 50 % of the total variance, the wave period inferred from the first zero crossing of the autocovariance function is 8 h for the $\sigma_\theta = 26$ and 26.5 isopycnals, and increases to 5 h for the isopycnals in the pycnocline ($\sigma_\theta = 25$ and 25.5). A similar analysis for the alongshore component of the baroclinic velocity suggests consistent periods (5.2 h at 8 m depth, and 9 ± 0.6h between 12 and 68 m). The pattern is similar at TS3; the $\sigma_\theta = 26.5$ isopycnal oscillates at 8 h, and the oscillation of the rotated baroclinic velocity component is between 7 and 10 h throughout the water column.

4 Numerical simulations

4.1 Model and the setup

The model utilized is the MITgcm model (Marshall et al., 1997; Adcroft et al., 2004). This is a finite volume, nonlinear z coordinate model with non-hydrostatic capabilities. The model has been widely used for study of internal waves (Legg and Adcroft, 2003; Vlasenko and Stashchuk, 2007; Xing and Davies, 2007; Boegman and Dorostkar, 2011). Bottom topography in MITgcm is represented by the use of shaved cells (Adcroft et al., 1997). For our model setup we disregard the effect of diffusion by setting small constant values for the horizontal and vertical diffusivities of temperature and salt[1]. We use a vertical turbulent viscosity of $A_z = 0.001\,\mathrm{m^2\,s^{-1}}$. The horizontal viscosity is of Smagorinsky type with value 2.2 along with a small biharmonic viscosity factor (viscC4smag = 1) as suggested by Griffies and Hallberg (2000). No-slip conditions are employed at side walls, along with a quadratic bottom drag coefficient of 0.0025. The time step used is 5 s, the horizontal grid size is 100 m, and we employ 32 non-uniformly spaced vertical levels with the lowest spacing of 0.75 m where the density gradient is largest, and up to 14 m spacing below 100 m depth. We let the model run hydrostatically, as the grid should be too coarse for nonhydrostatic effects to be observable (Berntsen et al., 2009).

[1] $K_{hT} = 1\times10^{-6}$, $K_{zT} = 1\times10^{-7}$, $K_{hS} = 1\times10^{-8}$, $K_{zS} = 1\times10^{-9}\,\mathrm{m^2\,s^{-1}}$

In our simulations we ignore the effect of wind and freshwater runoff. Initially, the system is at rest, with horizontally constant hydrography. The hydrography of Van Mijenfjorden varies considerably on short time scales (Sect. 3). Performing a realistic simulation of the entire fjord system requires detailed information on local winds and freshwater runoff, and detailed initial and boundary conditions on hydrography and currents. In this study, we focus on the generation and propagation of internal waves. Thus, it is instructive to idealize the problem. In the following we describe the idealized box runs, and subsequently, the relatively realistic Van Mijenfjord runs.

4.2 Idealized box runs

We shall consider the Van Mijenfjord topography in Sect. 4.3, but in this section we use a simplified semi-enclosed box with the topography shown in Fig. 6. The the long side, x axis, is aligned in the west–east direction, and the short side, y axis, along south–north. The western boundary is open, whereas all other boundaries are closed. At 20 km in-fjord, we place a constriction resembling Akseløya. In this simplified setup we only consider one sound (recall that Van Mijenfjorden has two sounds), and place it on the southern side in order to avoid the complicating factor of the wave making a 90°turn in the southwestern corner. The sound is 20 m deep and 1300 m wide, giving a cross-sectional area similar to Akselsundet. The sloping bottom toward the eastern coast is introduced as a crude wave damper to minimize reflections. The open boundary condition is of the form $u = u_0 \sin(\omega t)$, where u_0 is constant across the open boundary and $\omega = 1.4 \times 10^{-4}\,\mathrm{s}^{-1}$ corresponds to the M_2 semi-diurnal tide. The interior solution is relaxed toward the boundary over a 64-grid point wide sponge layer. The relaxation time scale increases linearly with distance from the open boundary, up to 16 h at the inner border of the sponge layer. The model is initiated with a forcing amplitude of $u_0/2$ in the first 6 h to ensure a smooth spin-up, and run for 48 h.

The solution is dependent on the choice of the forcing amplitude u_0 and the vertical hydrography profile. In the following, we use our idealized setup to test the internal response of the system by varying these parameters.

4.2.1 Box run 1: Van Mijenfjord hydrography

In the first test, we set forcing and hydrography to resemble the conditions in Van Mijenfjorden. From Fig. 2, we want surface tidal amplitudes close to 0.5 m. If we integrate over the domain this leads approximately to a boundary forcing of $u_0 = 3\,\mathrm{cm\,s^{-1}}$ (verified a posteriori). Regarding the hydrography, we neglect the effect of horizontal variation, and consider only the outer part of the fjord. Representative profiles of temperature, salinity, and density are obtained by averaging the CTD measurements from the time series stations TS2–TS5, and the sections A and C (Fig. 7).

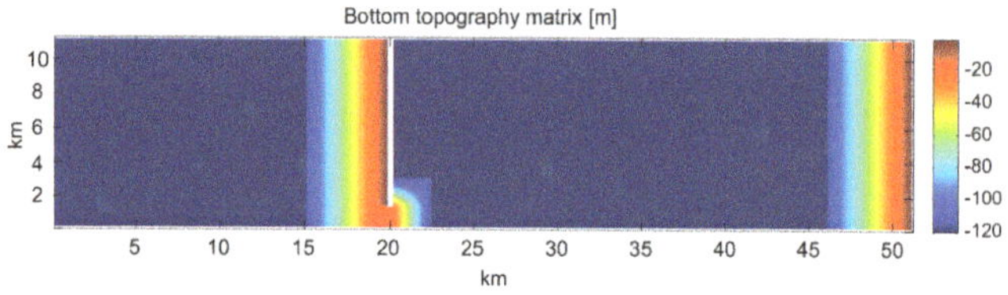

Fig. 6. Bottom topography test box run. The western boundary is the only open boundary.

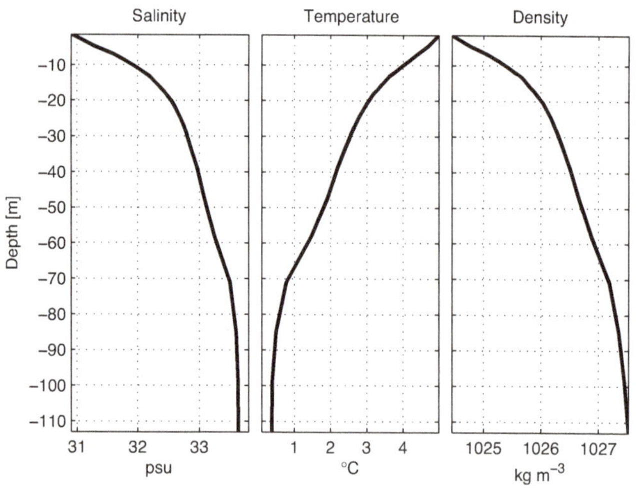

Fig. 7. Initial hydrography for box run 1.

In order to visualize the horizontal distribution of the vertical isopycnal displacements, we introduce a perturbation potential energy per unit area, defined as

$$\Delta\mathrm{PE}(t) = \mathrm{PE}(t) - \mathrm{PE}_0 = \int_{-H}^{\eta(t)} \rho(t) g z \mathrm{d}z - \int_{-H}^{0} \rho(t=0)\, g z \mathrm{d}z. \quad (7)$$

From this definition PE is negative, so a positive ΔPE indicates a mean depression in the water column. Plot of the normalized ΔPE 41 h into the simulation is shown in Fig. 8. The region shown is from 20 to 45 km in-fjord, i.e. the sound is immediately in the southwestern corner, and the eastern-most region with the sloping bottom is omitted. We clearly see that the most energetic displacements occur near the boundaries. These displacements propagate cyclonically around the basin, similar to internal Kelvin waves. The radius of the displacement signal is 2–3 km, and by comparing similar plots at different times we obtain a propagation velocity of 40–45 $\mathrm{cm\,s^{-1}}$ (not shown). A similar velocity can be inferred from $\lambda/T_{M2} = 42\,\mathrm{cm\,s^{-1}}$, using a wavelength of $\lambda = 19\,\mathrm{km}$ obtained from Fig. 8, and the wave period $T_{M2} = 12.42\,\mathrm{h}$. Here, $c_i = 42\,\mathrm{cm\,s^{-1}}$ is taken as the typical internal long wave speed for this model configuration.

In addition to the internal Kelvin wave signal, a second feature is evident in Fig. 8. In front of the coastal wave, there is a narrow pulse, a non-linear soliton that propagates radially

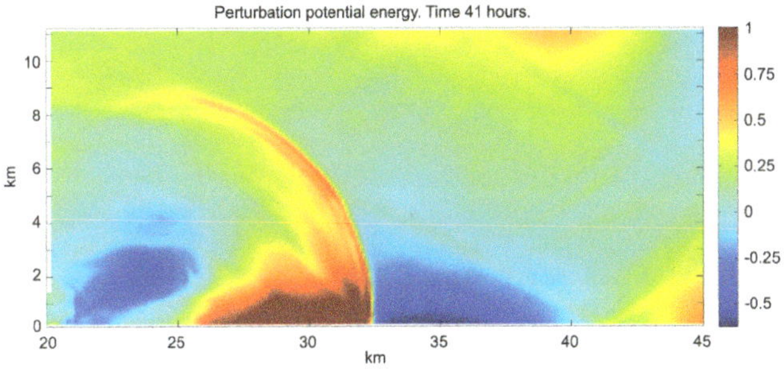

Fig. 8. Horizontal plot of normalized perturbation potential energy ΔPE after 41 h, box run 1. The plotted region corresponds to the first 25 km of the inner basin; see Fig. 6. Red and blue indicate depression and elevation respectively; see Fig. 9.

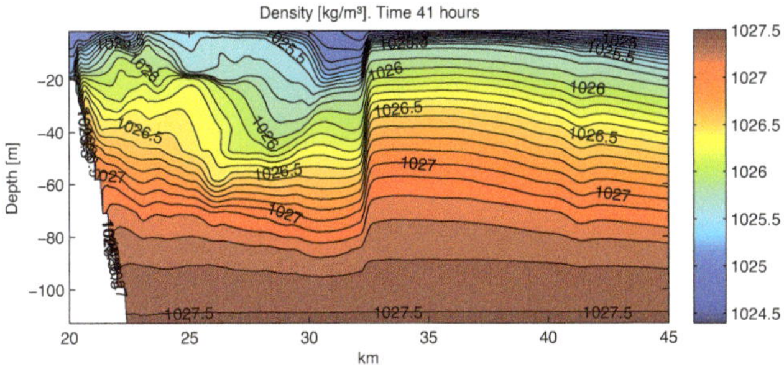

Fig. 9. Vertical section of density near the southern coast between 20 and 45 km east after 41 h simulation, box run 1. Contour interval is $0.1\,\mathrm{kg\,m^{-3}}$.

into the basin in all directions from the sound. By inspection of the circular shape of the leading wave crest, we infer that the propagation velocity is approximately constant in all directions. Model output from successive time steps indicates that the velocity of the pulse is similar to the internal long wave speed c_i (not shown). From a vertical section of density near the southern wall, at $y = 100\,\mathrm{m}$ (Fig. 9), we see more clearly the steep front of the depression, coinciding with the pulse. The vertical displacement exceeds 20 m over a 600 m horizontal distance. We discuss this pulse more thoroughly in Sect. 5.

We now turn our attention to the velocities through the sound. A vertical cross section of normalized maximum velocity u_s/c_i across the sound is presented in Fig. 10. During inflow the current peaks at 3.69 ($1.55\,\mathrm{m\,s^{-1}}$) in a jet at approximately 16 m depth. The outflow is relatively homogenous across the sound, peaking at −3.07 ($-1.29\,\mathrm{m\,s^{-1}}$). This asymmetry is expected in accordance with tidal choking theory (Stigebrandt, 1980). Further, these values relate directly to the definition of the Froude number (Sect. 2). $F_D \approx 3$ is a clear indication of supercritical conditions, reinforcing the jet-type behaviour shown in Fig. 10.

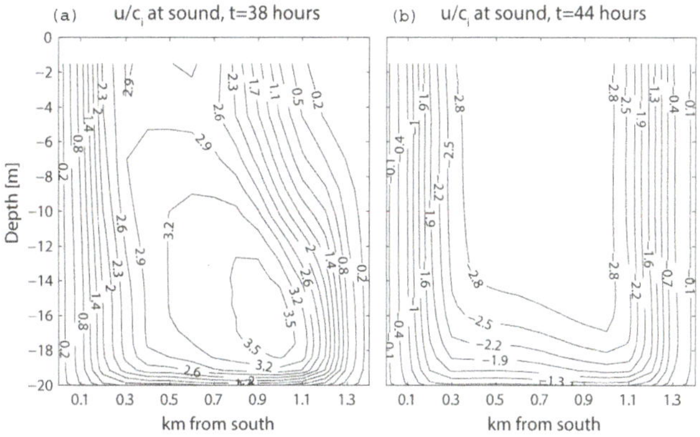

Fig. 10. Cross-sectional velocity $[\mathrm{m\,s^{-1}}]$ at time of maximum inflow **(a)** and outflow **(b)** through the sound at 20 km east for box run 1. Left–right in the figures correspond to south–north.

4.2.2 Box run 2: Two-layer stratification

Having obtained the expected baroclinic phase velocities (c_i) from a realistic stratification, we set up a case with two-layer stratification. We preserve the approximate location of the pycnocline while keeping the typical value of c_i. For simplicity, we take temperature constant at 2° C, and a two-layer salinity profile as shown in Fig. 11. The resulting density is also shown, in which a small correction for pressure can be seen. The phase velocity c_i is calculated according to Eq. (2), which yields $c_i = 0.42\,\mathrm{m\,s^{-1}}$ using the appropriate values from Fig. 11.

Plot of ΔPE for the two-layer stratification after 41 h simulation is shown in Fig. 12. The structure is comparable to the previous case, with most of the energy near the south coast, and a similar pulse propagating into the basin. A vertical section near the south coast (Fig. 13) reveals that most of the displacement occurs near the pycnocline, and a steepened structure similar to Fig. 9 is seen. The normalized velocities through the sound (not shown) are very similar to the previous case in structure, with the maximum values of 3.61 ($1.52\,\mathrm{m\,s^{-1}}$) and −3.07 ($-1.29\,\mathrm{m\,s^{-1}}$), during the inflows and outflows, respectively.

4.2.3 Box run 3: Test forcing

In the next set of tests we consider the sub-critical scenario by reducing the forcing amplitude. We perform two runs using the two-layer hydrography from box run 2, with $u_0 = 0.45\,\mathrm{cm\,s^{-1}}$ and $0.9\,\mathrm{cm\,s^{-1}}$, or 15 % (a) and 30 % (b) of the original forcing, respectively. ΔPE for the two runs are presented in Fig. 14. We clearly see the difference between the two cases; at case a, the pulse in front of the coastal wave is almost absent and the shape is quite similar to what we expect from a Kelvin wave (i.e. Eq. 5). At case b, however, the pulse is more visible, and the shape is reminiscent of the result for the box run 2.

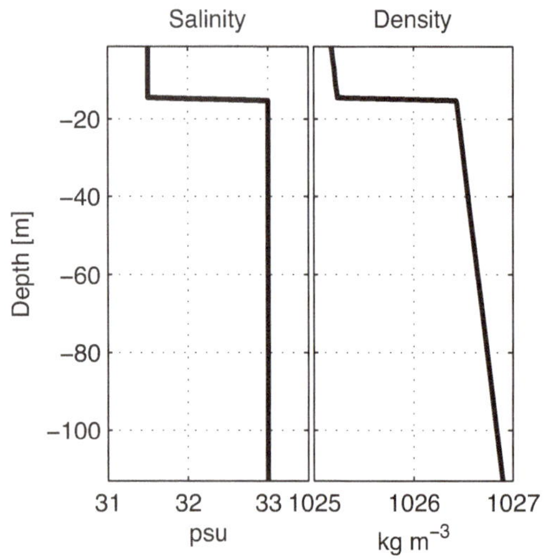

Fig. 11. Initial hydrography for box run 2. Temperature is constant at 2 °C and not shown.

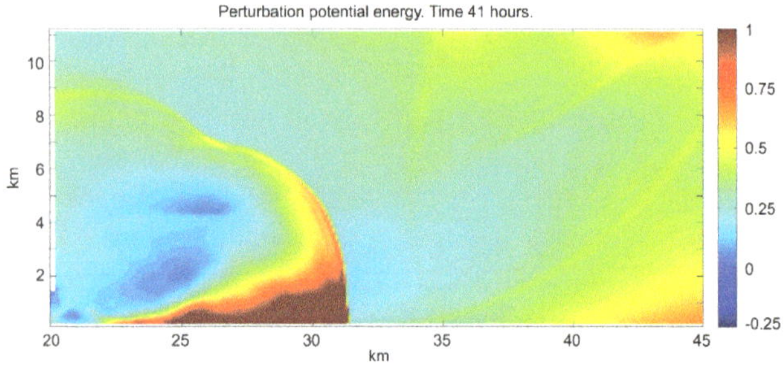

Fig. 12. Normalized ΔPE (similar to Fig. 8) after 41 h simulation, box run 2.

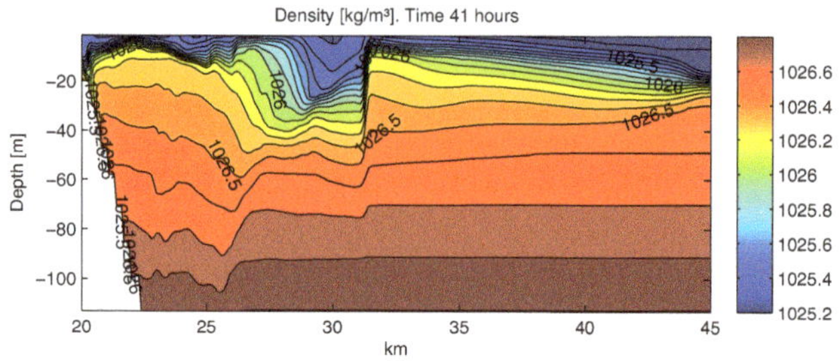

Fig. 13. Vertical section of density near southern coast after 41 h simulation, box run 2. Contour interval is 0.1 kg m^{-3}.

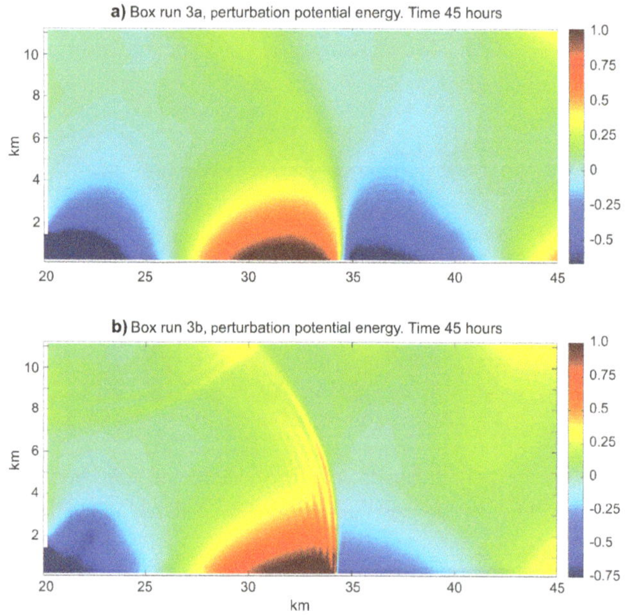

Fig. 14. Normalized ΔPE (similar to Fig. 8) after 45 h simulation, box run 3. Forcing at 15 % **(a)** and 30 % **(b)** of original forcing.

The normalized velocities through the sound have maximum values of -0.57 ($-0.24\,\mathrm{m\,s^{-1}}$) and 0.67 ($0.28\,\mathrm{m\,s^{-1}}$) for case a, and of -1.07 ($-0.45\,\mathrm{m\,s^{-1}}$) and 1.33 ($0.56\,\mathrm{m\,s^{-1}}$) for case *b*. The velocity structure across the sound in the latter case is similar to box run 2 with a jet during inflows and a more barotropic distribution during outflows. The jet is not as visible in case *a*, however; here the structure is rather similar for in- and outflow, in accordance with the subcritical conditions reflected by the normalized velocities.

4.3 Van Mijenfjord runs and comparison with observations

Having obtained some experience with the idealized model runs, we now apply a topography similar to Van Mijenfjorden. As this fjord has two sounds, it is of interest to isolate the effect of each sound. We also investigate the influence of the realistic hydrography in comparison with the two-layer structure we applied in Sect. 4.2.2.

We set up four model runs with different topography and initial hydrography; see Table 2. The bottom topography matrix is shown in Fig. 15. The leftmost part is the sponge layer, and the shallow, secondary inner basin on the east of Van Mijenfjorden is omitted from the simulation. As our point of interest is close to Akseløya, this should not influence our solution significantly. When a sound is closed, we simply set the depth to zero in a three grid point-wide band across the respective sound. Model parameters are similar to what is used in Sect. 4. As a result of varying depth, ΔPE (Eq. 7) is not easily visualized; instead, contours of constant density are shown.

Results from the first run with realistic hydrography and topography (VMrun1) are shown in Fig. 16. The left column shows the density distribution at three different depths after 45 h simulation, whereas the right column shows the structure at -16.9 m depth at different times. As before, we identify the internal Kelvin wave pattern along the southern coast. It is quite visible at all three depths, but most pronounced at -16.9 m. Snapshots at different times suggest that the primary internal wave generation occurs at Akselsundet. We also note the pulse at all depths. Current magnitude and

Table 2. Setup of the four Van Mijenfjorden model runs.

	VMrun1	VMrun2	VMrun3	VMrun4
Topography:	Realistic	Mariasundet closed	Akselsundet closed	Realistic
Hydrography:	Realistic	Realistic	Realistic	Two-layer

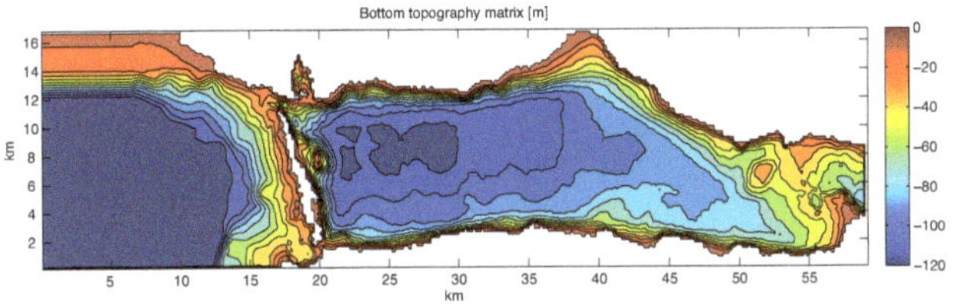

Fig. 15. Bottom topography of Van Mijenfjorden.

direction at −16.9 m depth are shown in Fig. 17 at the time of maximum inflow (a) and maximum outflow (b) through Akselsundet. We see the wave along the southern coast propagating inward at all times through the tidal cycle. Near Akselsundet during inflow, the current has a dominantly southeastern direction along Akseløya. During outflow the current is toward the sound from a much wider region, in accordance with tidal choking theory discussed earlier. Current amplitudes through Akselsundet reach 1.4–1.6 $\mathrm{m\,s^{-1}}$, consistent with observations from moored current meters (Bergh, 2004; Fer and Widell, 2007; Fjellsbø, 2013).

As a means of comparing the realistic model run with the measurements, we plot model density and along-coastal current at the time series locations TS4 and TS5 (Fig. 18). For the station near the southern boundary (TS4), the largest currents coincide with depression of the isopycnals (low-density values) in the upper layer (see also Figs. 16 and 17). The excursion of the 25.5 isopycnal, located at 14 m on average, is 13 m over the tidal period, corresponding to a normalized excursion of 0.9. We also see a local current maximum around −60 m depth along with elevation of the isopycnals. This is also observed in the measurements (Fig. 4d). The current magnitude and the maximum isopycnal displacement are lower than what was measured. At the northern station TS5 (Fig. 18b), we see only small oscillations, consistent with the observations. The strong mean westward flow in the upper layer in the measurements at TS5 (Fig. 4e) is not present in the model because of model shortcomings and simplifications; see Sect. 5.

The next two model runs are presented in Fig. 19. Here we experiment with closing Mariasundet (VMrun2, left column) and Akselsundet (VMrun3, right column) separately. The result when the Mariasundet is closed is quite similar to what was found in the realistic run (Fig. 16), suggesting that Akselsundet is the dominant generator for internal waves. However, as we see in VMrun3, closing off Akselsundet reveals a substantial contribution from Mariasundet. Comparing the location of the largest depression from the two runs, we infer that waves in VMrun2 are only about 2–3 km ahead of those in VMrun3. Thus the full picture in VMrun1 is close to a superposition of two internal Kelvin waves. Regarding the pulse we note that it is clearly visible in all the plotted depths, propagating radially from each respective sound.

As a last test, we apply the two-layer hydrography with the realistic topography (VMrun4). The result is shown in Fig. 20 for −16.9 m depth. Again, a Kelvin-type signal can be identified, along with several pulses. At −10.3 and −28 m depths, the density surfaces are approximately undisturbed (not shown).

5 Discussion and concluding remarks

Through a series of simulations and measurements, we have shown strong indications of tidally induced internal wave motion in Van Mijenfjorden, an Arctic fjord in Svalbard. As expected, a major part of the internal wave energy is in the form of an internal Kelvin wave, propagating cyclonically around the fjord. Numerical tests using a two-layer stratification show that the propagating internal wave emerges when conditions at the sound are baroclinically sub- or supercritical. This is in agreement with observations from Loch Etive, Scotland (Inall et al., 2004; Stashchuk et al., 2007), which is a typical "jet-type" fjord following the definition of Stigebrandt and Aure (1989). According to Stigebrandt (1980), generation of long waves in a jet fjord is prohibited, although he notes that for weakly supercritical conditions wave generation cannot be ruled out.

Our results indicate that both sounds in Van Mijenfjorden serve as internal wave generators; the dominant waves propagate from Akselsundet, whereas the contribution from Mariasundet can be substantial. According to the model results, a realistic summer hydrography leads to waves generated from the two sounds that are approximately in phase. As propagation velocity is directly affected by the stratification, waves from the two sounds are likely to be in phase occasionally, because of the considerable variability in hydrography. Consequently, the contribution from each respective sound may be masked in measurements. It is important to note that Mariasundet is relatively narrow (0.7 km wide) compared to the grid size (100 m), and the representation of this sound in the topography matrix is coarse. Increasing the resolution and applying a more accurate representation of the topography would thus contribute to a more realistic wave field.

Increasing the resolution and including non-hydrostatic terms in the model equations can help resolve the

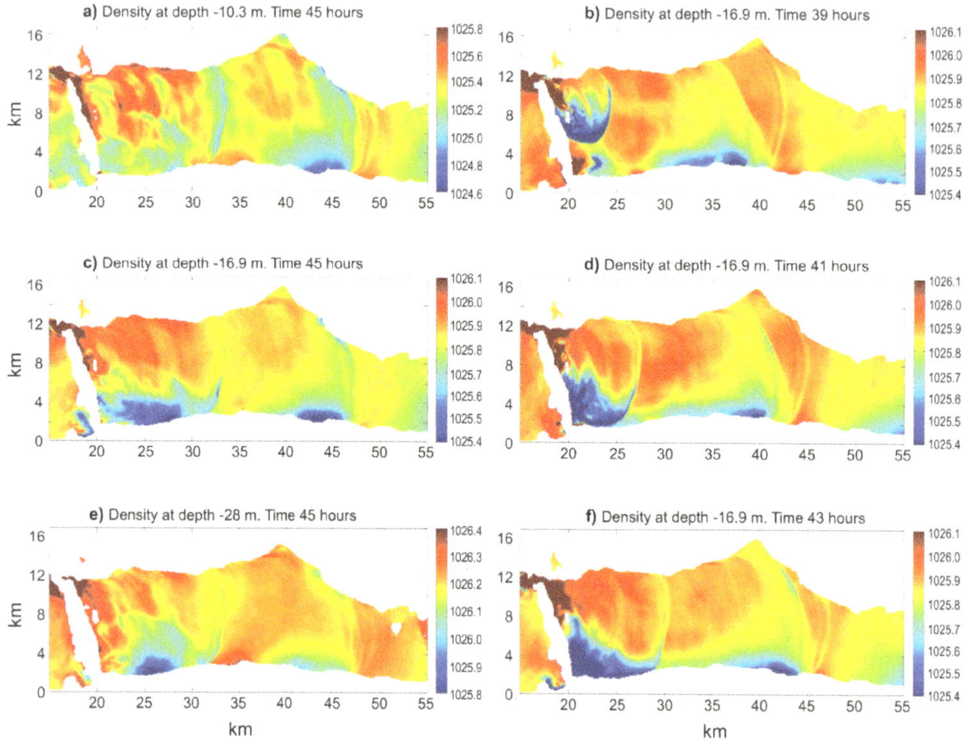

Fig. 16. Horizontal plot of density [kg m^{-3}], Van Mijenfjorden (VMrun1). Left: 45 h simulated; depth 10.3 m **(a)**, 16.9 m **(c)**, 28 m **(e)**. Right: depth 16.9 m; 39 h **(b)**, 41 h **(d)**, 43 h **(f)**.

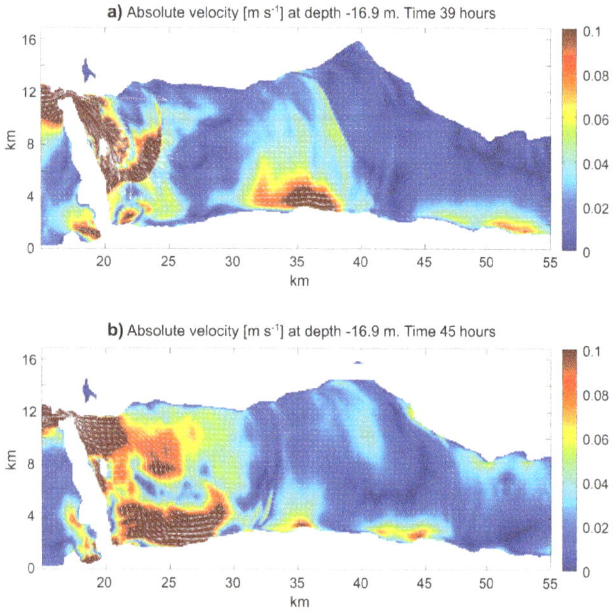

Fig. 17. Current magnitude [m s^{-1}] and direction for VMrun1 at 16.9 m depth. Simulated 39 h **(a)** and 45 h **(b)**.

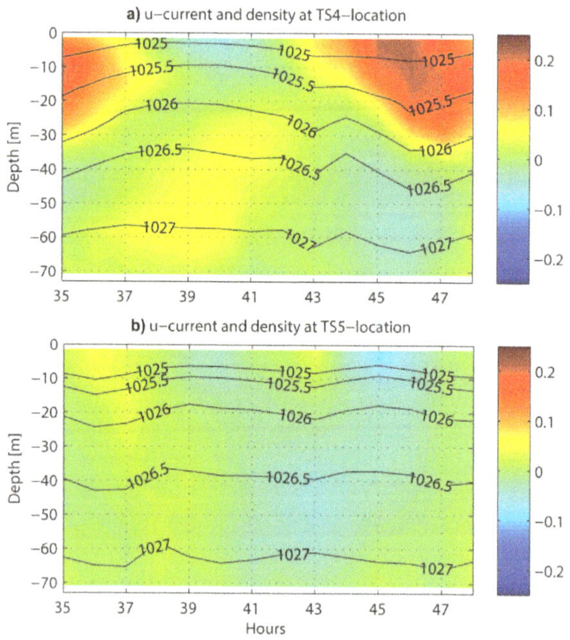

Fig. 18. Along-coastal current [m s^{-1}] (colour) and density [kg m^{-3}] (black lines) at locations TS4 **(a)** and TS5 **(b)** from VMrun1, 35–48 h simulated.

non-hydrostatic, small-scale overturning and mixing near the sills (Berntsen et al., 2009), as observed just inside Akselsundet. Our observations do not resolve the near field mixing processes in the vicinity of the sill. It is expected that the energy lost to mixing will be unavailable for generation of the long internal wave, influencing the resulting in-fjord wave field. Farmer and Armi (1999, 2001), using detailed

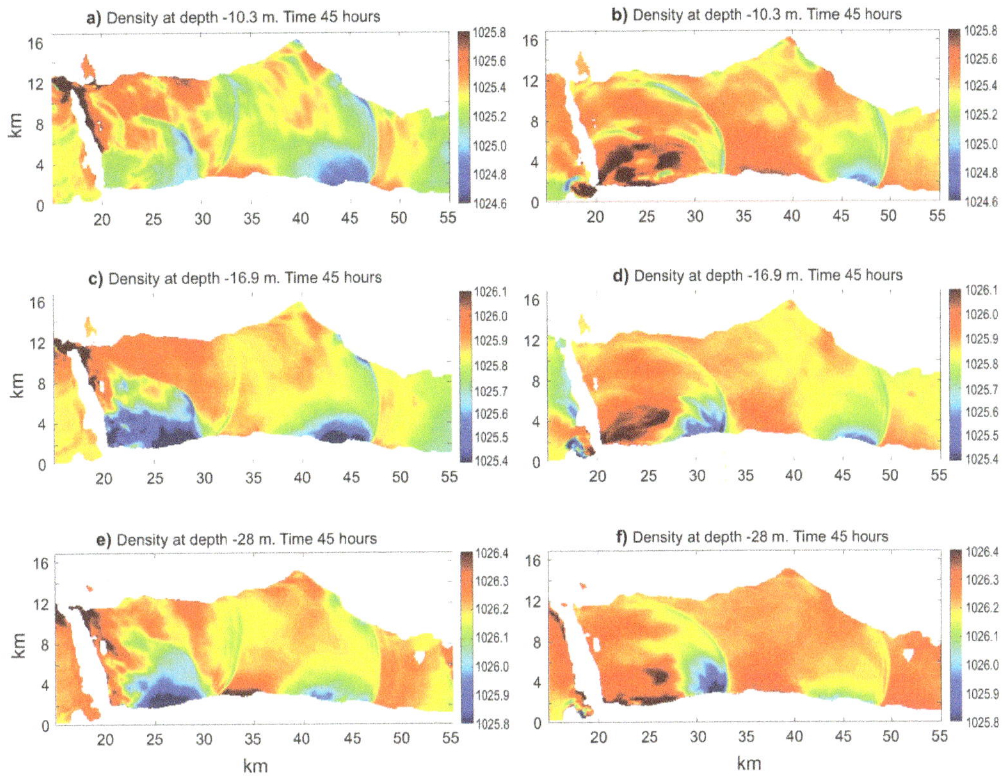

Fig. 19. Horizontal plot of density [$kg\,m^{-3}$]. Left: Mariasundet closed (VMrun2). Right: Akselsundet closed (VMrun3). **(a)** and **(b)** 10.3 m, **(c)** and **(d)** 16.9 m, **(e)** and **(f)** 28 m.

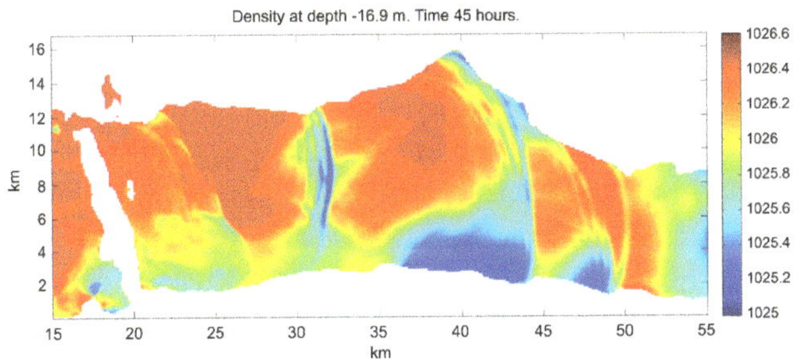

Fig. 20. Horizontal plot of density [$kg\,m^{-3}$] after 45 h simulation, 16.9 m depth. Two-layer hydrography profile (VMrun4).

observations, explain the time-dependent evolution of the stratified flow response over a sill that is typically not captured by the numerical simulations because of unresolved mixing. Rapid flow over the sill separates the streamlines into a fast deep current immediately downstream of the crest, beneath a spreading, weakly stratified, nearly stagnant intermediate wedge susceptible to overturning and further mixing. Small scale instability and boundary layer separation act in concert to determine the time evolution of the flow. Numerical studies, e.g. Berntsen et al. (2009), examine the resulting lee waves as a result of varying grid sizes, and thus the ability of the model to resolve the aforementioned mixing. They report a distinct difference in the lee wave field moving from 100 to 12.5 m resolution; as the resolution increases, waves become shorter in scale and their amplitudes increase. For the smaller scales, the vortices in the wedge zone increase in intensity. In all their experiments, Berntsen et al. (2009) reported boundary layer separation, a well-mixed wedge and internal waves in the lee of the sill; however, those in the hydrostatic and relatively coarser resolution runs were a result of artificial mixing, whereas the high-resolution non-hydrostatic models resolved the mixing processes. In our simulations, it would be of interest to apply a nested grid, where resolution is increased near the sounds, in order to resolve the mixing.

One must take care when directly comparing the observations to the model results, due to the numerous and crude assumptions involved. Firstly, we neglected wind. Thus, the modelled near-surface turbulence distribution and currents are bound to differ from the measured data. Another important aspect of the wind is that it may set up internal waves upon changing direction. Wind measurements during the cruise (Fig. 2) show that the wind direction in Van Mijenfjorden is unidirectional for long periods of time, which sets up a tilting interface that when released (as a result of wind changes), will propagate along the coast as an internal Kelvin wave. A second aspect is that our model is started from rest and is run for 48 h in an attempt to isolate the effect of internal wave propagation on an otherwise undisturbed system. In order for a mean circulation pattern to be set up, one would need a spin-up time of several days, possibly months. In addition, the model does not resolve the estuarine circulation properly, and large-scale pressure gradients from outside the fjord region, bar the tidal contribution, are also absent in the simulation. These shortcomings are all likely candidates for explaining why the observed mean currents in Fig. 4e) are not captured in the model. However, this might just as well be another example of insufficient resolution not resolving small-scale overturning and mixing from e.g. wind and background shear.

The wave pulse observed in the model results demands some discussion. When the conditions are near super-critical, a wave pulse propagates inward from the fjord entrance at the beginning of each inflow phase of the tidal cycle. This pulse of steep front is not dominantly affected by rotation, propagates with the long internal wave speed, and the leading crest is followed by a series of smaller amplitude waves. The train of waves develops as a result of dispersion, since the group velocity of the shorter waves is smaller than the long wave speed. The waves are characterized as non-linear internal solitons. The process is similar to that for free surface waves which are comparatively less non-linear. The latter has been illustrated comparing fully nonlinear computations and KdV-formulation (Grue et al., 2008), and similar generation process for internal waves was computed using fully nonlinear-dispersive calculations in Grue (2005). The vertical velocity profiles of the two-layer solitary wave are described in detail in Grue et al. (1999). The upper and lower layer velocities are in opposite directions, with an abrupt change across the interface. This may be compared to our observations, i.e Fig. 4d), where a similar separation of velocities is seen. It should be kept in mind, however, that the KdV-theory is applicable to two-layer solitons with non-dimensional excursions of up to 0.4 (Grue et al., 1999; Grue, 2005). In our study, results from both observations and the numerical simulations indicate normalized excursions up to 1, exceeding the validity of the KdV-formulation. It must be emphasized that the hydrography in Van Mijenfjorden varies considerably, and the profile at TS4 cannot be characterized by a two-layer system. Thus, caution must be exercised in the idealized theory-observation comparison.

The generation process for internal solitons as explained in Cummins et al. (2003) and Grue (2005) is as follows: during intense outflow, an internal depression forms over the sill crest, accompanied by an elevation more upstream. When the outflow relaxes, the depression propagates upstream into the fjord. Interestingly, as flow is supercritical during both in- and outflow through Akselsundet, we might expect solitons emerging on both sides of the sound at opposite phases of the tidal cycle. The dispersive degeneration of the solitary front into smaller waves is not properly resolved. An additional test with non-hydrostatic dynamics turned on yielded little difference in the results, which is expected due to insufficient horizontal resolution.

Acknowledgements. The authors are grateful to Jarle Berntsen and Jiuxing Xing for valuable input on the model configuration, and the captain and crew of RV Håkon Mosby during the field work. This study is partly supported by the Research Council of Norway, through the "Internal hydraulic processes in an Arctic fjord" project. The comments from two reviewers helped to improve an earlier version of the manuscript.

Edited by: Y. Troitskaya

References

Adcroft, A., Hill, C., and Marshall, J.: Representation of Topography by Shaved Cells in a Height Coordinate Ocean Model, Mon. Weather Rev., 125, 2293–2315, doi:10.1175/1520-0493(1997)125<2293:rotbsc>2.0.co;2, 1997.

Adcroft, A., Hill, C., Campin, J.-M., Marshall, J., and Heimbach, P.: Overview of the formulation and numerics of the MITgcm, in: Proceedings of the ECMWF seminar series on Numerical Methods, Recent developments in numerical methods for atmosphere and ocean modelling, ECMWF, 139–149, 2004.

Beletsky, D., O'Connor, W. P., Schwab, D. J., and Dietrich, D. E.: Numerical simulation of internal Kelvin waves and coastal upwelling fronts, J. Phys. Oceanogr., 27, 1197–1215, 1997.

Bergh, J.: Measured and modelled tidally driven mean circulation under ice cover in Van Mijenfjorden, M.Sc. thesis, Department of Oceanography, Göteburg University, Göteburg, Sweden, 65 pp., 2004.

Berntsen, J., Xing, J., and Davies, A.: Numerical studies of flow over a sill: sensitivity of the non-hydrostatic effects to the grid size, Ocean Dynam., 59, 1043–1059, doi:10.1007/s10236-009-0227-0, 2009.

Boegman, L. and Dorostkar, A.: Three-dimensional simulation of NLIW generation, propagation and breaking in Cayuga Lake, in: 7th Int. Symp. on Stratified Flows, Rome, Italy, 22–26 August 2011.

Brown, P. J.: Kelvin-wave reflection in a semi-infinite canal, J. Mar. Res., 31, 1–10, 1973.

Cottier, F. R., Nilsen, F., Skogseth, R., Tverberg, V., Skarðhamar, J., and Svendsen, H.: Arctic fjords: a review of the oceanographic environment and dominant physical processes, Geological Society, London, Special Publications, 344, 35–50, doi:10.1144/sp344.4, 2010.

Cummins, P. F., Vagle, S., Armi, L., and Farmer, D. M.: Stratified flow over topography: upstream influence and generation of nonlinear internal waves, P. Roy. Soc. Lond. A Mat., 459, 1467-1487, 10.1098/rspa.2002.1077, 2003.

Farmer, D. and Armi, L.: Stratified flow over topography: the role of small-scale entrainment and mixing in flow establishment, P. Roy. Soc. Lond. A Mat., 455, 3221–3258, 1999.

Farmer, D. M. and Armi, L.: Stratified flow over topography: models versus observations, P. Roy. Soc. Lond. A Mat., 457, 2827–2830, doi:10.1098/rspa.2001.0802, 2001.

Farmer, D. M. and Freeland, H. J.: The physical oceanography of fjords, Prog. Oceanogr., 12, 147–220, 1983.

Fer, I. and Widell, K.: Early spring turbulent mixing in an ice-covered Arctic fjord during transition to melting, Cont. Shelf Res., 27, 1980–1999, 2007.

Fer, I., Skogseth, R., and Geyer, F.: Internal waves and mixing in the Marginal Ice Zone near the Yermak Plateau, J. Phys. Oceanogr., 40, 1613–1630, 2010.

Fjellsbø, T. K. B.: Circulation, mixing and tidally-induced energy fluxes in Van Mijenfjorden, M.Sc. thesis, Geophysical Institute, University of Bergen, Bergen, Norway, 98 pp., 2013.

Gill, A. E.: Atmosphere-Ocean Dynamics, Academic Press, London, 662 pp., 1982.

Gómez-Giraldo, A., Imberger, J., and Antenucci, J. P.: Spatial structure of the dominant basin-scale internal waves in Lake Kinneret, Limnol. Oceanogr., 51, 229–246, 2006.

Griffies, S. M. and Hallberg, R. W.: Biharmonic Friction with a Smagorinsky-Like Viscosity for Use in Large-Scale Eddy-Permitting Ocean Models, Mon. Weather Rev., 128, 2935–2946, doi:10.1175/1520-0493(2000)128<2935:BFWASL>2.0.CO;2, 2000.

Grue, J.: Generation, propagation, and breaking of internal solitary waves, Chaos, 15, 037110, doi:10.1063/1.2047907, 2005.

Grue, J., Jensen, A., Riså s, P.-O., and Sveen, J. K.: Properties of large-amplitude internal waves, J. Fluid Mech., 380, 257–278, doi:10.1017/S0022112098003528, 1999.

Grue, J., Pelinovsky, E. N., Fructus, D., Talipova, T., and Kharif, C.: Formation of undular bores and solitary waves in the Strait of Malacca caused by the 26 December 2004 Indian Ocean tsunami, J. Geophys. Res., 113, C05008, doi:10.1029/2007jc004343, 2008.

Guthrie, J. D., Morison, J. H., and Fer, I.: Revisiting internal waves and mixing in the Arctic Ocean, J. Geophys. Res., 118, 3966–3977, doi:10.1002/jgrc.20294, 2013.

Helfrich, K. R. and Melville, W. K.: Long nonlinear internal waves, Annu. Rev. Fluid. Mech., 38, 395–425, doi:10.1146/annurev.fluid.38.050304.092129, 2006.

Hodges, B. R., Imberger, J., Saggio, A., and Winters, K. B.: Modeling basin-scale internal waves in a stratified lake, Limnol. Oceanogr., 45, 1603–1620, 2000.

Inall, M., Cottier, F., Griffiths, C., and Rippeth, T.: Sill dynamics and energy transformation in a jet fjord, Ocean Dynam., 54, 307–314, doi:10.1007/s10236-003-0059-2, 2004.

Legg, S. and Adcroft, A.: Internal Wave Breaking at Concave and Convex Continental Slopes*, J. Phys. Oceanogr., 33, 2224–2246, doi:10.1175/1520-0485(2003)033<2224:IWBACA>2.0.CO;2, 2003.

Llewellyn Smith, S. G.: Stratified rotating edge waves, J. Fluid Mech., 498, 161–170, doi:10.1017/S002211200300702X, 2004.

Marchenko, A. V., Morozov, E. G., Muzylev, S. V., and Shestov, A. S.: Interaction of Short Internal Waves with the Ice Cover in an Arctic Fjord, Oceanology, 50, 18–27, doi:10.1134/S0001437010010029, 2010.

Marshall, J., Adcroft, A., Hill, C., Perelman, L., and Heisey, C.: A finite-volume, incompressible Navier Stokes model for studies of the ocean on parallel computers, J. Geophys. Res., 102, 5753–5766, doi:10.1029/96jc02775, 1997.

Maxworthy, T.: Experiments on solitary internal Kelvin waves, J. Fluid Mech., 129, 365–383, 1983.

Morozov, E. G. and Marchenko, A. V.: Short-period internal waves in an arctic Fjord (Spitsbergen), Izv. Atmos. Ocean. Phys., 48, 401–408, doi:10.1134/s0001433812040123, 2012.

Nash, J. D., Alford, M. H., and Kunze, E.: Estimating internal wave energy fluxes in the ocean, J. Atmos. Ocean. Tech., 22, 1551–1570, 2005.

Padman, L. and Dillon, T. M.: Turbulent mixing near the Yermak Plateau during the coordinated Eastern Arctic Experiment, J. Geophys. Res., 96, 4769–4782, 1991.

Padman, L. and Erofeeva, S.: A barotropic inverse tidal model for the Arctic Ocean, Geophys. Res. Lett., 31, L02303, doi:10.1029/2003gl019003, 2004.

Rainville, L. and Woodgate, R. A.: Observations of internal wave generation in the seasonally ice-free Arctic, Geophys. Res. Lett., 36, L23604, doi:10.1029/2009GL041291, 2009.

Renouard, D. P., D'Hières, G. C., and Zhang, X.: An experimental study of strongly nonlinear waves in a rotating system, J. Fluid Mech., 177, 381–394, 1987.

Skarðhamar, J. and Svendsen, H.: Short-term hydrographic variability in a stratified Arctic fjord, Geological Society, London, Special Publications, 344, 51–60, doi:10.1144/sp344.5, 2010.

Stashchuk, N., Inall, M., and Vlasenko, V.: Analysis of supercritical stratified tidal flow in a Scottish Fjord, J. Phys. Oceanogr., 37, 1793–1810, doi:10.1175/jpo3087.1, 2007.

Stigebrandt, A.: Some aspects of tidal interaction with fjord constrictions, Estuar. Coast. Mar. Sci., 11, 151–166, doi:10.1016/S0302-3524(80)80038-7, 1980.

Stigebrandt, A. and Aure, J.: Vertical mixing in basin waters of fjords, J. Phys. Oceanogr., 19, 917–926, 1989.

Støylen, E. and Weber, J. E. H.: Mass transport induced by internal Kelvin waves beneath shore-fast ice, J. Geophys. Res., 115, C03022, doi:10.1029/2009jc005298, 2010.

Svendsen, H., Beszczynska-Møller, A., Hagen, J. O., Lefauconnier, B., Tverberg, V., Gerland, S., Ørbøk, J. B., Bischof, K., Papucci, C., Zajaczkowski, M., Azzolini, R., Bruland, O., Wiencke, C., Winther, J.-G., and Dallmann, W.: The physical environment of Kongsfjorden–Krossfjorden, an Arctic fjord system in Svalbard, Polar Res., 21, 133–166, doi:10.1111/j.1751-8369.2002.tb00072.x, 2002.

Vlasenko, V. and Stashchuk, N.: Three-dimensional shoaling of large-amplitude internal waves, J. Geophys. Res., 112, C11018, doi:10.1029/2007jc004107, 2007.

Vlasenko, V., Stashchuk, N., Hutter, K., and Sabinin, K.: Nonlinear internal waves forced by tides near the critical latitude, Deep-Sea Res. Pt. I, 50, 317–338, doi:10.1016/s0967-0637(03)00018-9, 2003.

Weber, J. E. H. and Støylen, E.: Mean drift velocity in the Stokes interfacial edge wave, J. Geophys. Res., 116, C04002, doi:10.1029/2010jc006619, 2011.

Widell, K.: Ice-ocean interaction and the under-ice boundary layer in an Arctic fjord, Ph.D. thesis,, Geophys. Inst., University of Bergen, Bergen, Norway, 2006.

Xing, J. and Davies, A. M.: On the importance of non-hydrostatic processes in determining tidally induced mixing in sill regions, Cont. Shelf Res., 27, 2162–2185, doi:10.1016/j.csr.2007.05.012, 2007.

5

Distinguishing the effects of internal and forced atmospheric variability in climate networks

J. I. Deza[1], **C. Masoller**[1], **and M. Barreiro**[2]

[1]Departament de Física i Enginyeria Nuclear, Universitat Politècnica de Catalunya, Colom 11, 08222, Terrassa, Barcelona, Spain

[2]Instituto de Física, Facultad de Ciencias, Universidad de la República, Iguà 4225, Montevideo, Uruguay

Correspondence to: J. I. Deza (juan.ignacio.deza@upc.edu)

Abstract. The fact that the climate on the earth is a highly complex dynamical system is well-known. In the last few decades great deal of effort has been focused on understanding how climate phenomena in one geographical region affects the climate of other regions. Complex networks are a powerful framework for identifying climate interdependencies. To further exploit the knowledge of the links uncovered via the network analysis (for, e.g., improvements in prediction), a good understanding of the physical mechanisms underlying these links is required. Here we focus on understanding the role of atmospheric variability, and construct climate networks representing internal and forced variability using the output of an ensemble of AGCM runs. A main strength of our work is that we construct the networks using MIOP (mutual information computed from ordinal patterns), which allows the separation of intraseasonal, intraannual and interannual timescales. This gives further insight to the analysis of climatological data. The connectivity of these networks allows us to assess the influence of two main indices, NINO3.4 – one of the indices used to describe ENSO (El Niño–Southern oscillation) – and of the North Atlantic Oscillation (NAO), by calculating the networks from time series where these indices were linearly removed. A main result of our analysis is that the connectivity of the forced variability network is heavily affected by "El Niño": removing the NINO3.4 index yields a general loss of connectivity; even teleconnections between regions far away from the equatorial Pacific Ocean are lost, suggesting that these regions are not directly linked, but rather, are indirectly interconnected via El Niño, particularly at interannual timescales. On the contrary, on the internal variability network – independent of sea surface temperature (SST) forcing – the links connecting the Labrador Sea with the rest of the world are found to be significantly affected by NAO, with a maximum at intraannual timescales. While the strongest non-local links found are those forced by the ocean, the presence of teleconnections due to internal atmospheric variability is also shown.

1 Introduction

The existence of long-range teleconnections in climate is an established fact, as the atmosphere connects far-away regions through waves and advection of heat and momentum. This long-range coupling makes the complex network approach (Albert and Barabási, 2002) of the earth's climate very attractive (Tsonis et al., 2006, 2008; Donges et al., 2009b). Climate networks are constructed by considering the earth as a regular grid of nodes and assigning links connecting two different nodes via an analysis of their similarity over a particular field. This approach has been used in the literature both on local and on global scales and for analyzing climate phenomena considering both linear and nonlinear interdependencies.

For example, the network approach has been recently used to analyze patterns of extreme monsoonal rainfall over South Asia (Malik et al., 2012), to infer early warning indicators for the Atlantic Meridional Overturning Circulation collapse (van der Mheen et al., 2013), to gain insight into the origin of decadal climate variability (Tsonis and Swanson, 2012) and to study the El Niño phenomenon as an autonomous component of the climate network (Gozolchiani et al., 2011).

Various methods for constructing climate networks have been proposed (computing information measures from temperature or geopotential fields, from daily or monthly data, etc.) and the reliability and robustness of the networks uncovered have also been analyzed in terms of a critical comparison of the networks found with the various methods used (Paluš et al., 2011; Hlinka et al., 2013; Martin et al., 2013; Tirabassi and Masoller, 2013). A main conclusion of these studies is that it is crucial to analyze the robustness of the method used to quantify climate similarities because trends and serial correlations in the time series, as well as time lags, can significantly affect the topology of the network obtained.

This paper is focused on understanding the atmospheric variability by means of networks constructed from monthly averaged surface air temperature (SAT) anomalies.

Atmospheric variability can be considered, to first order, as a superposition of an internal part due to intrinsic dynamics, and an external part due to the variations of the boundary conditions, primarily given by the sea surface temperature (SST) forcing. These two components can be distinguished by using atmospheric general circulation models (AGCMs) forced with prescribed historical SSTs (Straus and Shukla, 2000; Barreiro et al., 2002; Molteni, 2003; Bracco et al., 2004) – see also the accompanying paper (Arizmendi et al., 2014) in this Special Issue.

The separation between internal and forced atmospheric variability is a standard procedure to study the impact of the oceans on the atmosphere and has led to important advances regarding our understanding of the dynamics involved. James (1995) and Trenberth (1997) provide two excellent summaries of the processes involved, mainly based on the propagation of Rossby waves and the generation of teleconnection patterns. Although there are some nonlinear secondary effects, the theory asserts that, to first order, the observed propagation and establishment of teleconnection patterns are linear.

Separating forced from internal atmospheric variability is also important because it can allow for improvements in climate prediction. In many geographical regions, the atmosphere is strongly influenced by SST variations that force persistent anomalies (Shukla, 1998). Because the evolution of the tropical oceans presents some predictability at timescales longer than the atmosphere, prediction of atmospheric variables beyond the chaotic timescale of 7–10 days is possible, provided that the atmospheric dynamics have been forced by the ocean (Shukla, 1998).

The usual modeling strategy to study predictability consists in forcing AGCMs with idealized or observed SST anomalies. This allows for investigating the response of the atmosphere to different boundary conditions and different initial conditions. If the time series of anomalies of a climatic field (e.g., SAT anomalies) is considered as a combination of internal and forced variability (e.g., $x = x_{\mathrm{for}} + x_{\mathrm{int}}$) the output of several numerical experiments initialized differently but forced with the *same* boundary conditions (i.e., same SST) can be used to separate the internal and forced variability. For each run i it results in

$$x^i = x^i_{\mathrm{for}} + x^i_{\mathrm{int}} = x_{\mathrm{for}} + x^i_{\mathrm{int}}$$

(as x_{for} does not depend on the initial conditions).

Averaging over N runs yields

$$\bar{x} = x_{\mathrm{for}} + (1/N)\sum_i x^i_{\mathrm{int}}.$$

If N is large enough, the second term will be small, as each model run will have a different value. Thus, to the first order $\bar{x} \approx x_{\mathrm{for}}$.

In other words, each time series x^i can be separated into a part that changes from run to run, x^i_{int}, and a part that does not depend on the initial conditions (is forced by the boundary conditions only and is the same for all runs), $x_{\mathrm{for}} \approx \bar{x}$.

This method allows for the construction of two types of networks, those in which the links represent similarities in internal atmospheric variability (referred to as *internal variability* network), and those in which the links represent similarities in forced atmospheric variability (the *forced variability* network).

The connectivity of these networks allows us to assess the influence of two main phenomena: El Niño (characterized by the NINO3.4 index) and the North Atlantic Oscillation (characterized by the NAO index). This was done by calculating the networks from time series where either the NINO3.4 index or the NAO index was linearly removed.

The forced variability networks is found to be intimately related to El Niño phenomenon and that linearly removing its evolution yields a breakdown of the long-range teleconnections of the climate network, particularly at interannual timescales. A similar result is observed for the internal variability network in the Northern Hemisphere when NAO is removed, with the maximum effect at intra-annual timescales.

The paper is organized as follows. In Sect. 2 the method used for constructing climate networks is shown. The data and the model, employed as well as the NINO3.4 and NAO indices, are discussed in Sect. 3. Section 4 presents the results. Here the internal and forced variability networks, and the effects of NAO and El Niño are analyzed. Section 5 presents a summary and the conclusions.

2 Methods for network construction

2.1 Measure of statistical interdependence

The mutual information (MI) is computed from the probability density functions (PDFs) that characterize two time series in two nodes, p_i and p_j, as well as their joint probability function, p_{ij} (Cover and Thomas, 2006; Amigó, 2010; Paluš, 2007):

$$M_{ij} = \sum_{m,n} p_{ij}(m,n) \log \frac{p_{ij}(m,n)}{p_i(m)p_j(n)}.$$

M_{ij} is a symmetric measure ($M_{ij} = M_{ji}$) of the degree of statistical interdependence of the time series in nodes i and j; if they are independent $p_{ij}(m,n) = p_i(m)p_j(n)$, and thus $M_{ij} = 0$.

In this paper the PDFs p_i, p_j, and p_{ij} are computed in two ways: by histograms of values (this case will be referred to as MIH) and by using a symbolic transformation, in terms of probabilities of *ordinal patterns* (this case will be referred to as MIOP) (Amigó, 2010; Pompe and Runge, 2011; Bandt and Pompe, 2002; Barreiro et al., 2011).

The ordinal patterns are calculated from time series by comparing the value of a given data point relative to its neighbors (Fig. 1). When a value (v_2) is higher than the previous one (v_1) and lower than the next one (v_3) ($v_1 < v_2 < v_3$), the ordinal transformation gives pattern "a" (see inset in Fig. 1); when $v_1 > v_2 > v_3$, it gives pattern "f", and so forth. Considering patterns of length D, then there are $D!$ possible patterns.

A significant advantage of MIOP for climate data analysis over other methods is that the ordinal transformation allows us to select the timescale of the analysis, not only by considering shorter or longer patterns, but also, by comparing data points in the time series which are not consecutive, but separated by a time interval.

This symbolic transformation keeps the information about correlations present in a time series at the selected timescale, but does not keep information about the absolute values of the data points. Therefore, the mutual information computed from ordinal patterns (MIOP) can be expected to provide complementary information with respect to the standard method of computing the mutual information (MIH) with monthly averaged data and zero-lag regressions, adequate for many geographical regions of earth. However, there are some exceptions, and the analysis of lagged responses could be an interesting extension of the present study that is left for future work.

Monthly data in the period January 1948–December 2006 is analyzed. Due to the short length of the time series (708 data points), in order to compute the probabilities of the patterns with good statistics we have considered ordinal patterns of length three. Since for $D = 3$ there are six possible patterns, for the sake of consistency, the MIH is computed using histograms with 6 equi-sized bins.

We varied the timescale of the MIOP analysis by constructing the patterns in three ways (see Fig. 1): (1) by comparing temperature anomalies in 3 consecutive months (constructing patterns with three consecutive data points), (2) by comparing anomalies in 3 equally spaced months that cover a 1-year period (by taking one data point every four points), and (3) by comparing anomalies in the same month of 3 consecutive years (by taking one data point every twelve points). The MIOP computed in these ways is referred to as *intraseasonal*, *intra-annual*, and *interannual*, respectively. While constructing the patterns with one data point per season (i.e., every three points) could seem more useful – as seasons cover 3 months– we decided to use 3 data points per year because seasons are not well-defined worldwide (for example, in the tropics), but the seasonal cycle is.

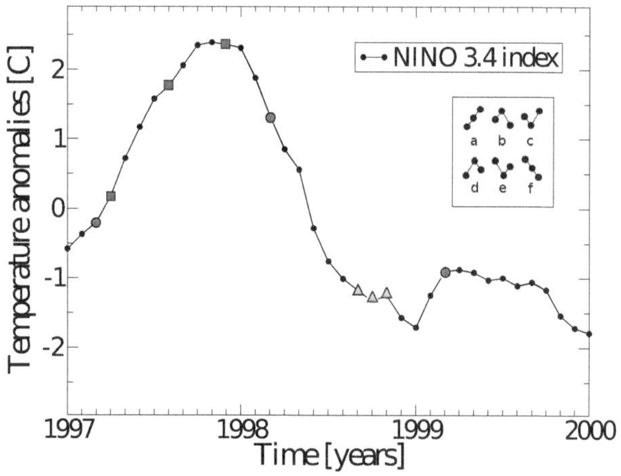

Figure 1. An example of three ordinal patterns in the time series of the NINO3.4 index (monthly averaged). Green triangles: intraseasonal pattern, blue squares: intra-annual pattern and red circles: interannual pattern. The possible patterns for $D = 3$ are shown in the inset. In this example, the intraseasonal pattern corresponds to an "e", the intra-annual, to an "a" and the interannual, to a "b".

2.2 Definition of links

To construct the network, a link between nodes i and j is defined if M_{ij} is above an appropriate threshold, which is calculated in terms of *surrogate*-shuffled data as in Deza et al. (2013), where the data are shuffled before we calculate the histograms or the ordinal patterns. When defining the significance criterion for the links in climate networks there is always a degree of arbitrariness. As we have shown in our previous work, the distribution of MI values computed from surrogate data is approximately Gaussian (see Fig. 1 of Deza et al., 2013). Therefore, here we use a significance criterion computed in terms of the mean, μ, and the standard deviation, σ, of the distribution of MI values computed from surrogate data, and accept links whose MI value, computed from the original data, is above $\mu + 3\sigma$.

While this is the simplest approach, it has well-known drawbacks (see, e.g., Schreiber and Schmitz, 1996; Paluš, 2007), and the use of block surrogates and/or quantile thresholds, as significance criterion will provide a valuable and computationally not too demanding improvement to the method used here.

2.3 Network representation

To represent the network we plot the *area-weighted connectivity* (AWC) of the nodes, which is the fraction of the total area of earth to which each node is connected, that is,

$$\mathrm{AWC}_i = \frac{\sum_j^N A_{ij}\cos(\lambda_j)}{\sum_j^N \cos(\lambda_j)},$$

where λ_i is the latitude of node i and $A_{ij} = 1$ if nodes i and j are connected, and 0 otherwise (Tsonis et al., 2008). It is of particular interest to identify significant weak links, as the strongest links are usually of shorter spatial range. In all the AWC maps presented in Sect. 4, the color scale has been set from 0 to a fixed value (0.4), and any node with stronger connectivity is shown with the color code of 0.4. This allows a clearer visualization the weakest part of the accepted significant links. It also allows for a direct comparison of all the AWC maps.

In Sect. 4 the significant connections of some selected geographical regions (represented by individual network nodes) are explored. In these connectivity maps the value of the interdependency measure (MIH or MIOP) will be displayed using a color scale which is also fixed, from 0 up to 0.3; MI values larger than this will be shown using the same color code as 0.3.

3 Data sets and model used

3.1 Climate indices

A climate index describes the state and changes of a particular region of the ocean or the atmosphere. Indices can be determined from monitoring station or reanalysis data, or identified by means of empirical orthogonal functions (EOF) analysis. In the latter case, they result in the principal component (PC) related to an EOF (generally the leading mode) over a chosen area, calculated for a predetermined variable (e.g., temperature or pressure). As explained below, the average of SST on the NINO3.4 area has been used for calculating the NINO3.4 index, and the leading PC over the North Atlantic region has been used to calculate NAO. The indices have been linearly detrended.

3.1.1 NINO 3.4

The NINO3.4 index (Trenberth, 1997) was calculated as the average of SST anomalies in the equatorial Pacific bounded by latitudes 5° S–5° N and by longitudes 120° W–170° W, using the SST data incorporated by the model as a boundary condition for all the runs. The index, obtained in this way, has been compared with the monthly index from NOAA (2013), updated monthly, obtaining an excellent agreement. As this index is based on SST – a boundary condition for the AGCM – this phenomenon can be expected to affect mainly the *forced* part of the atmospheric variability.

3.1.2 NAO

The North Atlantic Oscillation (NAO) has been shown to be mainly an *atmospheric* phenomenon only weakly forced by the ocean (Hurrel, 1995). The NAO index is calculated as the leading EOF of surface pressure over the North Atlantic region (20° –80° N and 90° W–40° E) for each model run. Comparison among indices from different model runs and between these and the observed NAO index from NCEP/NCAR reanalysis data yielded different time series modulating essentially the same spatial pattern. The NAO properties have been studied elsewhere (see Lind et al., 2005, and references therein), but for our purposes it is enough to assume that these series have the same power spectra as low frequency noise.

3.2 Model used

In this study the AGCM from the International Centre for Theoretical Physics (ICTP AGCM) has been used. It consists of a full atmospheric model with simplified physics and a horizontal resolution of T30 (3.75° × 3.75°, which gives $N = 608$ grid points or network nodes) with eight vertical levels (Molteni, 2003). The model is forced with historical global sea surface temperatures (ERSSTv.2) (Smith and Reynolds, 2004). In order to separate forced from internal atmospheric variability, nine runs using the same boundary (SSTs) conditions (but slightly different initial conditions) were performed.

In our experiment design SST is taken as a boundary condition and it is not changed by the atmospheric flow. In the real world there is a two-way interaction between the ocean and the atmosphere. This limitation is especially important in the extratropics where the SST evolution strongly depends on the atmospheric forcing (Frankignoul and Hasselmann, 1977; Barsugli and Battisti, 1998). However, current understanding indicates that the atmosphere is most sensitive to SST anomalies in the tropics and thus the forced atmospheric variability will be related to the evolution of the tropical oceans (Trenberth et al., 1998). This model setup allows, as explained in the Introduction, to separate the *forced* and *internal* components of the atmospheric variability. While an ensemble of only nine model runs might seem insufficient for a robust estimation of the forced response, as it could be contaminated by noise due to the relatively small ensemble size, it will be shown that the results found here are consistent with well-known climate phenomena, indicating that, at least at the "first order" description of the network via AWC, nine model runs are enough to separate forced and internal variability. This is consistent with previous works that show that an ensemble of about 10 runs is enough to separate internal and forced variability in most places (e.g., Barreiro, 2009; Barreiro and Díaz, 2011; Pohlmann and Latif , 2005; Seager et al , 2010). More sophisticated methods for identifying the forced variability despite the small-ensemble noise contamination are discussed in Allen and Smith (1997), Venzke et al. (1999), Barreiro et al. (2002, 2005), and Ting et al. (2009).

Monthly averaged air surface temperature (SAT) in the period January 1948–December 2006 was analyzed. This results in a total of 708 data points per node. For each node, the time series were linearly detrended and the anomalies of these series were computed by subtracting the long term average to each monthly data point.

The influence of NINO3,4 or NAO indices was assessed by computing time series where one of these indices was linearly removed from the original time series, respectively. This was done in three steps:

1. calculate the indices, as explained above;
2. perform a zero-lag regression of the time series of each node with respect to the time series of the index (see Fig. 2);
3. subtract the linear regression from the original data.

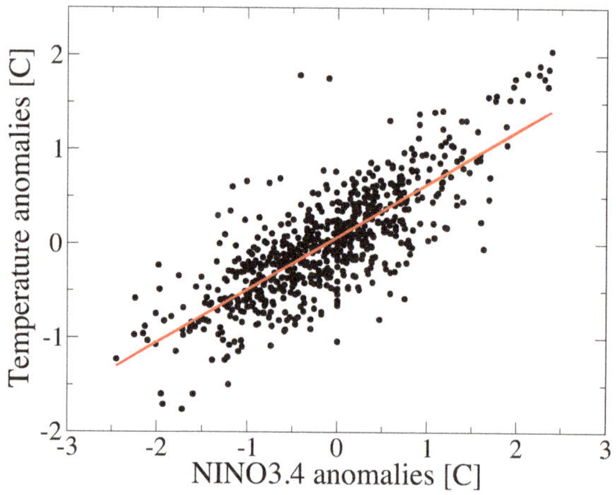

Figure 2. Graphical representation of the linear index-removal procedure. The SAT anomalies are compared at zero lag with the index (in this case NINO3.4 SST anomalies) and a linear regression is performed (in red).

In the following section it is shown that this procedure effectively removes the linear contribution of the given index in the evolution of each node (Rodwell et al., 1999; Barreiro, 2009). However, this simple approach for assessing the influence of an index could be improved in two ways: on the one hand, nonlinear methods for calculating the index could be considered (see e.g., Gámez et al., 2004), on the other hand, lagged regressions could be considered.

To validate the model (see Sect. 4.1) we considered reanalysis data from NCEP/NCAR (Kalnay, et al., 1996) in the same time period (1948–2006). Since NCEP/NCAR reanalysis data are given on a 2.5° × 2.5° grid, for easier comparison it was resampled using bilinear interpolation of the gridded data to fit the grid of the ICTP-AGCM data. The detrended and normalized anomalies were computed as stated with the model data.

4 Results

4.1 Model validation

While the ICTP-AGCM model has been used extensively in the literature (see, e.g., Bracco et al., 2004; Kucharski et al., 2005; Molteni, 2003; Barreiro, 2009, and references therein), the model has not yet been validated in the context of climate networks. Therefore, the first step of our study is to validate the model by comparing the networks obtained from one model run with the networks obtained from reanalysis data (Deza et al., 2013).

This can be done by comparing Fig. 3 with Fig. 4. Figure 3a displays the AWC map computed from reanalysis data using MIH as an interdependency measure; Fig. 4a displays the AWC map computed from one model run, also using MIH. Clearly, the model is able to capture the same overall pattern of global connectivity with a maximum in the central tropical Pacific, relative maxima in the tropical Atlantic and Indian oceans and over Alaska, the Labrador Sea and the Southern Ocean. Differences are mainly in the magnitude of the AWC, with the model underestimating the connectivity in most places. Similar observation applies to the comparison between Fig. 3b, c, and d and the corresponding panel in Fig. 4, where the network was built by using the MIOP as an interdependency measure.

Figure 4a shows the AWC using MIH and thus reveals global interdependencies of all the time series; Fig. 4b–d show the AWC using MIOP in intraseasonal, intraannual and interannual timescales, respectively. Clearly, the connectivity increases as the timescale increases, in good agreement with the results found in Deza et al. (2013) using reanalysis data. Many other features of the AWC maps are also qualitatively well reproduced by the model.

While the networks obtained from AGCM and reanalysis data, when visualized via the AWC, look qualitatively very similar, quantitative differences are seen, for example, with respect to the spatial extent of the structures. These differences might be relevant, especially if more sophisticated network measures were to be used. Nevertheless, the good qualitative agreement between networks constructed from model and reanalysis data lets us focus on using model output to distinguish the networks associated with intrinsic and forced atmospheric variability.

4.2 AWC maps

4.2.1 Forced variability

The AWC maps presented in Fig. 4 for one run of the model contain information on both forced and internal variability. To analyze *forced variability* only, we have constructed the

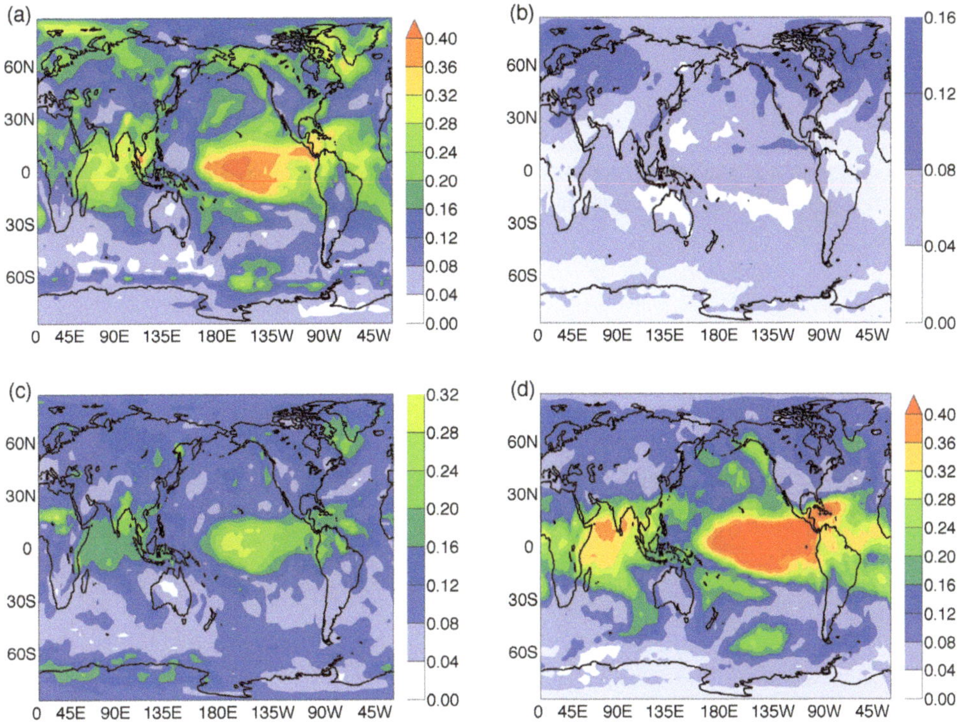

Figure 3. Maps of AWC constructed from reanalysis NCEP/NCAR data. The statistical interdependencies are quantified via **(a)** MIH, **(b)** MIOP intraseasonal, **(c)** intra-annual, and **(d)** interannual timescales (see Sect. 2.1 for details). The color scale is the same for all panels and for all the following AWC maps.

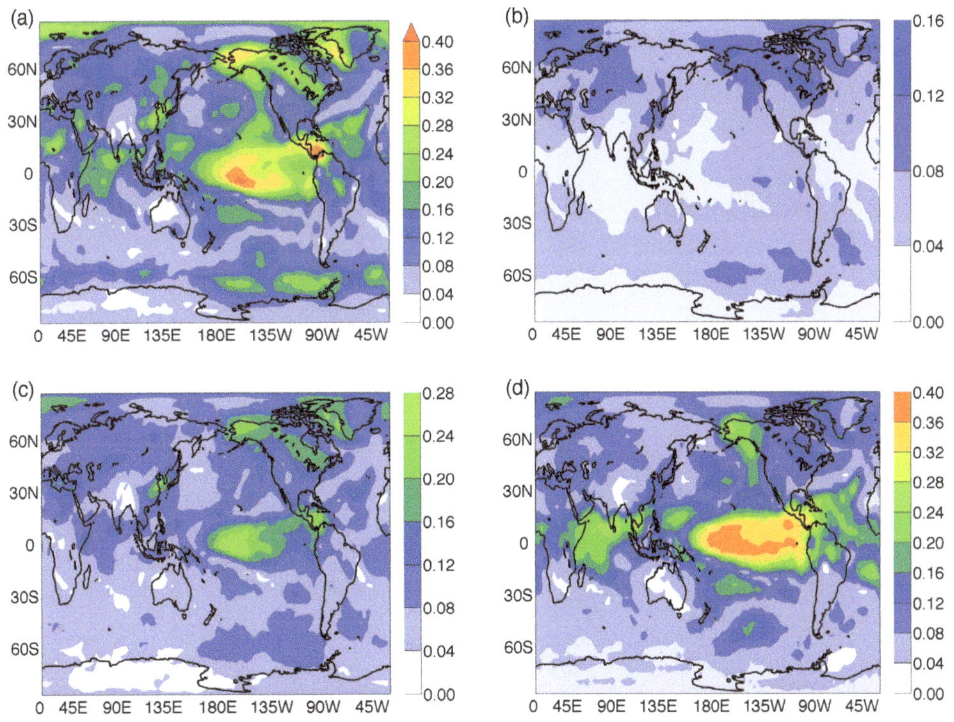

Figure 4. Maps of AWC obtained from single model run. The statistical interdependencies are quantified via **(a)** MIH, **(b)** MIOP intraseasonal, **(c)** intra-annual, and **(d)** interannual timescales (see Sect. 2.1 for details). Comparing **(a)** with Fig. 3a and **(d)** with Fig. 3b we observe that the main features of the maps are the same, providing a visual validation of the model.

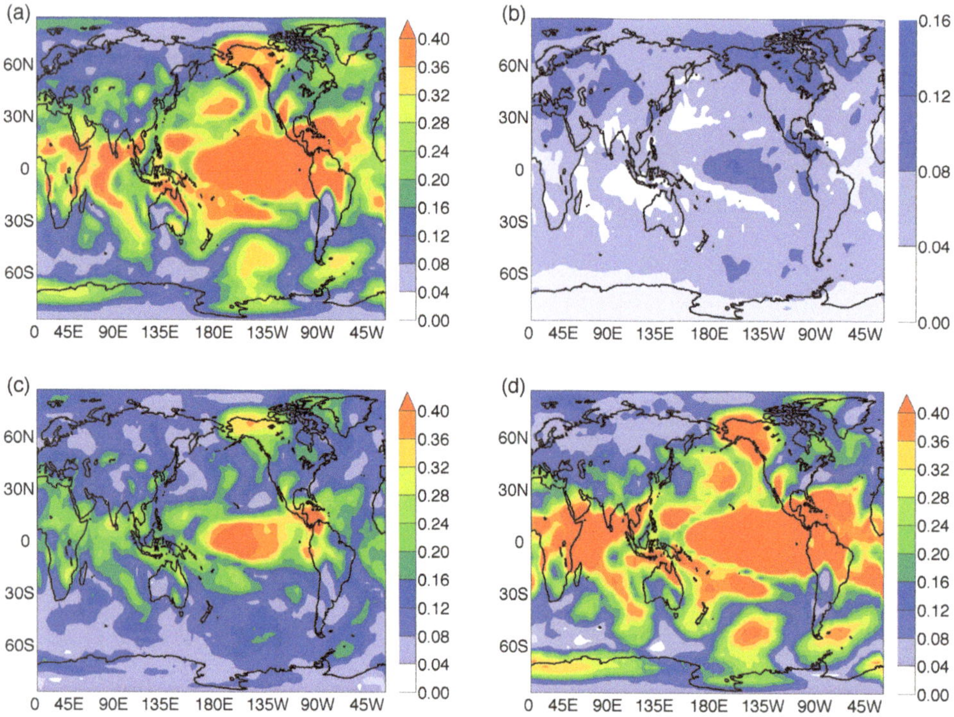

Figure 5. Maps of AWC computed from averaged time series, and thus containing information only of the forced component of atmospheric variability. The quantifiers of statistical similarity are as in Fig. 4: **(a)** MIH, **(b)** MIOP intraseasonal, **(c)** intra-annual, and **(d)** interannual. It can be noticed that in the shorter timescale the tropical area, especially the Pacific Ocean has a weak influence, and it grows stronger with increasing timescale. The fact that the maps in **(a)** and **(d)** are similar suggests that most of the links uncovered by the MIH, **(a)**, actually reflect interdependencies in the longer timescale and thus, are seen in **(d)**.

network from averaged time series (over nine model runs), as explained in Sect. 1.

The results are presented in Fig. 5. Figure 5a displays the AWC map when the MIH is used to quantify statistical interdependencies. Here, connectivity is higher in the tropics and on the Pacific, Indian and Atlantic basins than in the rest of the world. It is worth noting that while tropical connectivity is relatively symmetrical around the equator for the Pacific and Indian oceans, the North Atlantic is significantly more connected than in the south of the equator. Figure 5b–d show that the connectivity of the forced variability increases with the timescale. At intraseasonal timescales connectivity is very low compared with the connectivity from Fig. 5a. If we increase the timescale to intra-annual – as in Fig. 5c – the entire tropical area becomes more connected than the extratropics, indicating a better longitudinal energy and momentum exchange. Forced by the tropical Pacific SST anomalies, a long-range strong teleconnection is found in Alaska (Ropelewski and Halpert, 1987). For interannual timescales (3 years) within the period of the El Niño events (from 2 to 7 years) many very connected areas are found, especially in the tropics but also in the extratropics. The presence of highly connected spots is observed in the extratropics, especially in the Pacific Basin, but also in the Indian and Atlantic oceans. Comparing these three maps with that in Fig. 5a which, as explained before, was computed via MIH and thus contains information from all the time series, it can be inferred that most of the connections seen in Fig. 5a occur at long timescales, because they are clear only in Fig. 5d, and are weak or not seen in Fig. 5b, c.

Figure 6 represents the same maps as Fig. 5 but after removing the NINO3.4 index, as explained in Sect. 3.2. Figures 5a and 6a show large differences. It is clear that the signal of El Niño in the tropical Pacific was successfully removed, and moreover, the connection hotspots in the extratropics were also removed, indicating that they were mainly forced by El Niño. However, a few small well-connected areas remain over the equatorial Pacific, indicating that a linear regression is not sufficient to fully eliminate the ENSO effect on the network connectivity.

The Caribbean and North Atlantic are the largest regions that maintain a similar AWC, even after Niño has been removed. Note, however, that the instantaneous regression does not completely remove the ENSO signal if there is a lag in the response. This is so in the tropical North Atlantic (Chang et al., 2000), where El Niño affects sea surface temperature through heat flux changes that, given the ocean's heat capacity, take a few months to induce an anomaly. Thus, this

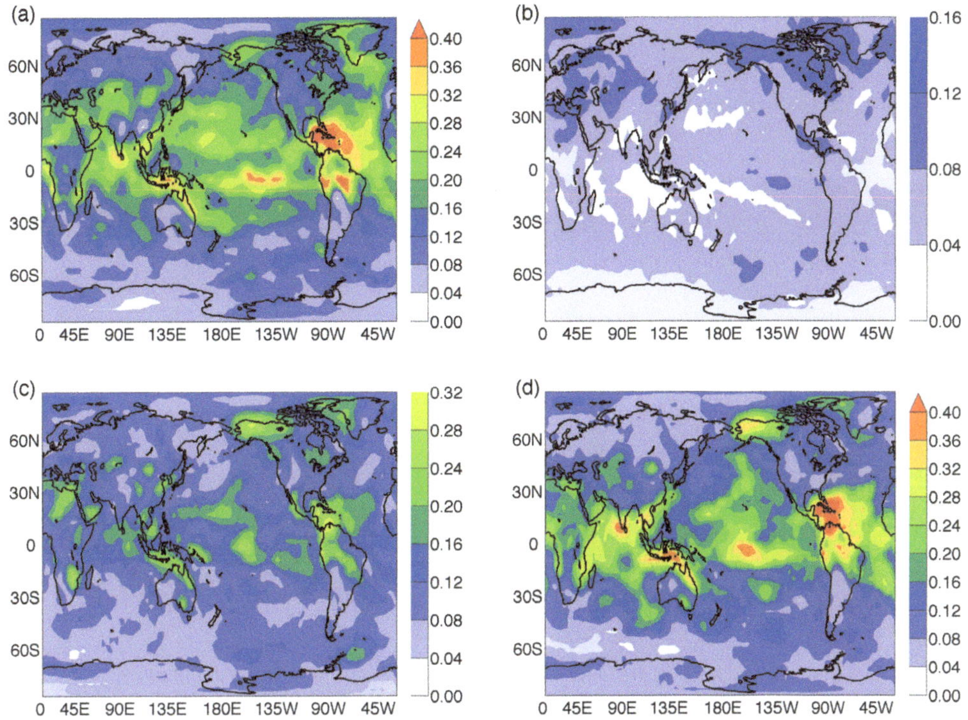

Figure 6. Maps of AWC of the forced component of the network when the index ENSO3.4 is removed from the time series (for the description of the index and for the removal procedure, see Section 3.2). The statistical interdependencies are quantified as in Fig. 4: **(a)** MIH, MIOP **(b)** intraseasonal, **(c)** intra-annual and **(d)** interannual. A comparison with Fig. 5 allows for an assessment of the influence of El Niño phenomenon over the network connectivity.

might be a reason for the still large connectivity observed in the Caribbean in Fig. 6a.

Other areas, like over China and central Asia, which are weakly connected to the El Niño phenomenon, show the same connectivity in Figs. 5 and 6. The fact that areas not related to ENSO do not change when removing the index hints that the statistical test used to fix the network density is robust and allows us to compare maps with and without the index.

Figure 6b is very similar to Fig. 5b except on the absence of a connected (dark blue) area on the Pacific Ocean, suggesting that the influence of El Niño at these timescales is very low and restricted to the tropical Pacific. At intra-annual timescales, Fig. 6c shows the disappearance of many links from the corresponding Fig. 5c. This suggests that at this timescale, even if the El Niño signal is not as strong as on interannual scales, it is already connecting far-away tropical and extratropical areas such as Alaska (Chiang and Sobel, 2002). Thus, removing the El Niño signal very heavily affects the connectivity of the network. For longer timescales – shown in Fig. 6d – the scenario is similar as for Fig. 6a with only a remnant of connectivity in the tropical region.

4.2.2 Internal variability

Figure 7 shows AWC maps of internal variability, computed by averaging the nine AWC maps obtained from the individual model runs, where in each time series, the forced signal (the average of the nine runs) was removed as explained in the Introduction. Contrary to the forced variability case presented before, in this case the most connected areas are on the extratropics. This is consistent with the results of previous figures, and indicates that in the tropics the ocean forces the largest portion of atmospheric variability. As the tropical atmosphere cannot sustain horizontal gradients generated by SST anomalies, it induces vertical movements of air, convection and release of latent heat, thus giving rise to atmospheric circulation anomalies.

In the extratropics internal atmospheric variability is larger, leading to stronger connections. The larger connectivity in the Northern Hemisphere suggests that the large landmasses affect atmospheric variability, which is consistent with our current understanding of storm track dynamics and low-frequency transients (James, 1995).

The most connected spot in Fig. 7a is over the Labrador Sea. The rest of the highly connected areas (in green) are present mostly in the Northern Hemisphere. In the Southern Hemisphere connectivity is largest over the Southern Ocean. Investigation over this well-connected area near Antarctica

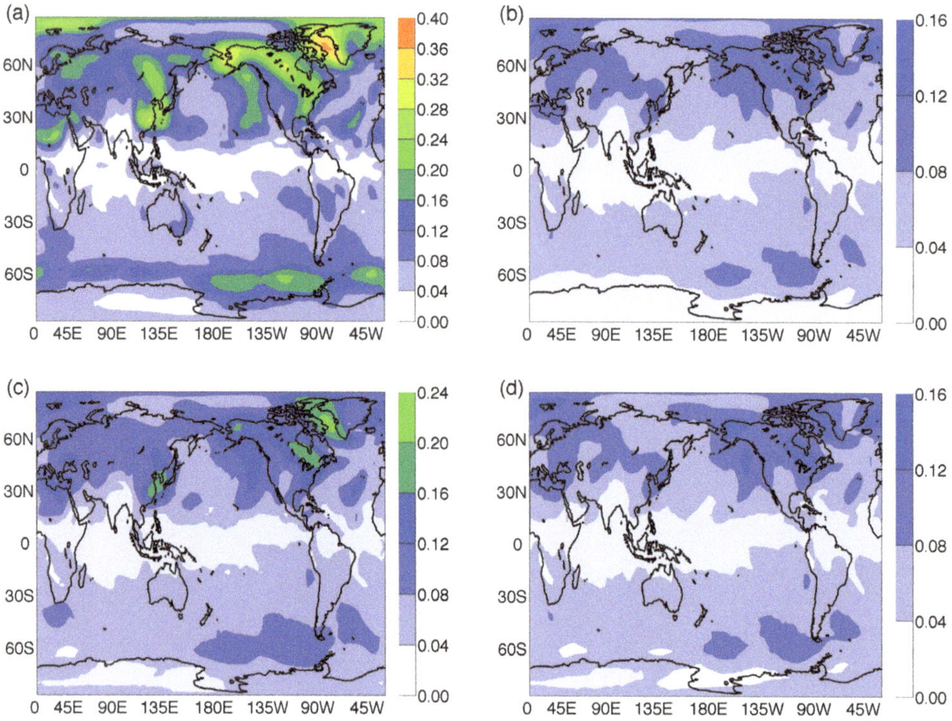

Figure 7. Maps of averaged AWC, revealing the internal variability network (see text for details). The statistical interdependencies are quantified as in Fig. 4 **(a)** MIH, MIOP **(b)** intraseasonal, **(c)** intra-annual, and **(d)** interannual. It can be noticed that in this network, the timescale showing more connectivity is the intra-annual timescale. This is consistent with the shorter memory of the atmosphere compared with the ocean.

– only found using MIH to quantify interdependencies – showed that in this area histograms have a higher skewness than in the rest of the nodes, an effect that has also been reported and discussed in Hlinka et al. (2014). This effect is found on the internal-plus-forced AWC map of Fig. 4a, and using reanalysis data as shown in Fig. 3a. When considering other measures to quantify interdependencies, such as Pearson cross correlation or MIOP, the AWC maps do not show high connectivity in this region (Deza et al., 2013).

With respect to the AWC maps computed by using MIOP, in contrast to the forced case, the intraseasonal, intra-annual and interannual maps are very similar to each other. This is a sign of "multiscale variability"; in other words, variability distributed over many timescales. Internal variability cycles are less well defined, with spectra similar to "red" noise. It can be seen that the most connected AWC map is the intra-annual one, stronger than both the intraseasonal and the interannual, consistent with the fact that atmospheric anomalies are less persistent than oceanic ones (Hasselmann, 1976; Barsugli and Battisti, 1998).

The fact that the most connected area in Fig. 7a is over the Labrador Sea suggests that it is related to NAO. In order to verify this, we have removed NAO from the time series using the same procedure as with NINO3,4, explained above. The results are shown in Fig. 8. Here, indeed the Labrador connected area disappears in all the panels, while the connectivity unrelated to NAO (i.e., over the Southern Hemisphere or China) remains almost unchanged.

4.3 Node connectivity maps

AWC maps provide information about the connectivity of the geographical regions, but no information about the nature – spatial range or distribution – of the links. It is expected that nearby points behave similarly and this leads to high values of correlation between nearby places (Radebach et al., 2013; Donges et al., 2009a). The distance over which the climate variables are well connected is related with the Rossby radius of deformation (RRD) (James, 1995), which is the distance that a particle or wave travels before being significantly affected by the earth's rotation. Also, in the tropics, this proximity effect can be greatly enhanced, as there the information is propagated very fast longitudinally. Here we are interested in unveiling the presence of teleconnections, that is, connections between regions separated by more than the RRD.

The following plots display the connections of a node, indicated with "X". Figures display MIH in the left column and MIOP in the right column. The timescale of the MIOP maps is interannual for the forced variability network and intra-annual for the internal variability network, following the above results. As explained in Sect. 2.3, since we are interested in unveiling weak but long-range significant links, we have saturated the color scale for nearby links. In this

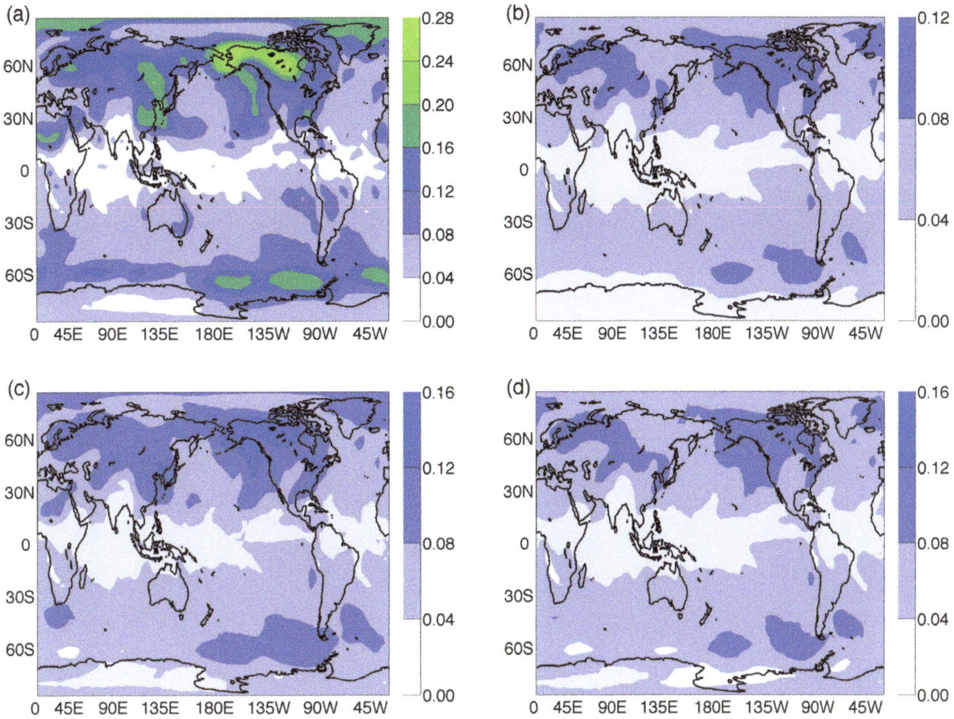

Figure 8. Maps of averaged AWC, revealing the internal variability network when the NAO index is removed from the time series (see Sect. 3.2 for details). The statistical interdependencies are quantified as in Fig. 4 **(a)** MIH, MIOP **(b)** intraseasonal, **(c)** intra-annual, and **(d)** interannual. It can be noticed that in this network the timescale showing more connectivity is the intra-annual timescale. This is consistent with the shorter memory of the atmosphere compared with the ocean.

way we are able to see the weak links with good resolution, losing information for the stronger links (stronger than 0.3 on the arbitrary scale of MI, where the highest links have values of 1 or 2 on the same scale, as shown in Deza et al., 2013) which will be all represented with the same color.

4.3.1 Forced variability

Figure 9 shows the connections of a point in the central Pacific Ocean in the forced variability network. It is clear from the comparison of the maps in the first row that most links are interannual links.

Figure 9c and d display the same node connectivity maps as in Fig. 9a and b respectively, however, in this case the NINO3.4 index has been removed from the time series and thus (to first order) they do not contain links due to El Niño phenomenon. The differences between Fig. 9a and Fig. 9c and between Fig. 9b and Fig. 9d are evident. First, after eliminating the effects of El Niño the tropical and extratropical teleconnection patterns associated to the spot in the Pacific disappear independently of the methodology used to quantify interdependencies (MIH or MIOP): the connectivity becomes restricted to the tropical Pacific Basin. Even inside this region the connectivity is greatly decreased as seen by a much smaller red spot of links over 0.3, although the remaining connections indicate that, either a linear regression is not enough to fully remove the influence of El Niño, or the ENSO dynamics is not fully represented by the NINO3.4 index.

According to Fig. 9a and b, Alaska is an area well connected to the equatorial Pacific Ocean. To further investigate, Fig. 10 shows global connections to a point near Alaska. It can be seen in Fig. 10a and b that it indeed presents connections to the equatorial Pacific Ocean with a maximum close to the dateline.

Furthermore, connections to the southern Pacific Ocean, Central Africa, Indian Ocean and even the Drake Passage are found. These connections are stronger in Fig. 10b, especially those linking Alaska with the Indian and southern Atlantic Ocean and the Drake Passage. If we remove NINO3.4 we find a dramatic change in the maps. Connections become almost local and all the north–south teleconnections are lost; only connections probably associated with an imperfect removal of the El Niño signal remain. This indicates that there are no direct teleconnections between Alaska and (for example) the Drake Passage, but both are strongly connected to El Niño. As these networks are constructed using symmetrical measures of dependency, calculated directly from the data, they are unable to distinguish between a direct connection and an indirect one.

Figure 11 is as Figs. 9 and 10, but for a node in the Southern Hemisphere extratropics. We chose southern New

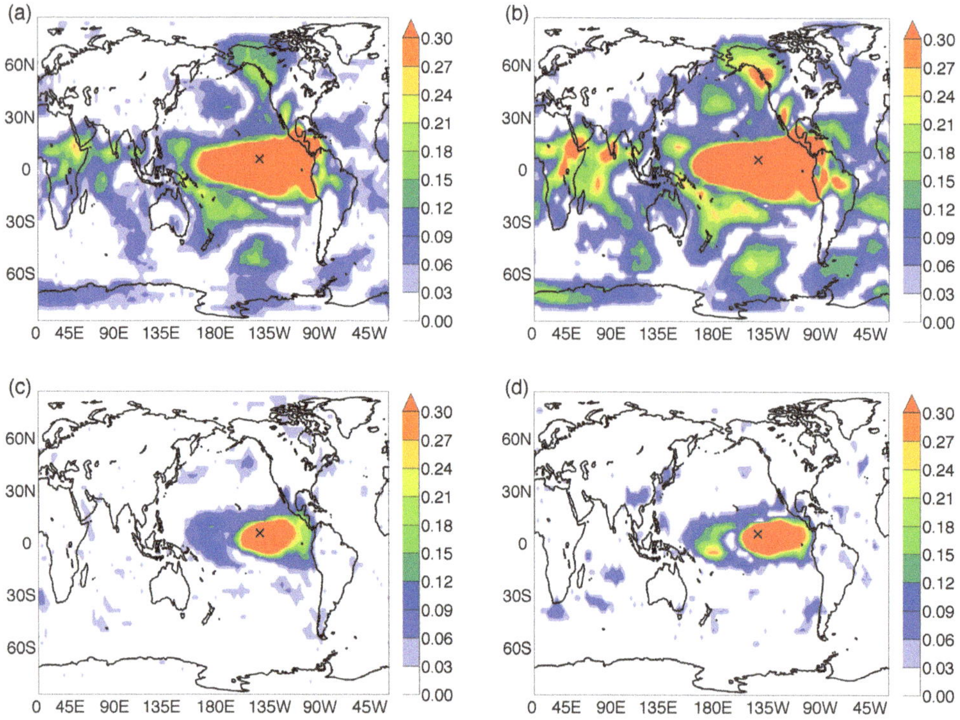

Figure 9. Connectivity map of a node in the central Pacific (indicated with X). Panels **(a)** and **(b)** are computed from forced time series (averaging over nine model realizations); panels **(c)** and **(d)** are also computed from forced time series, but with ENSO3.4 linearly removed and thus not containing – to the first order – contributions due to El Niño. In **(a)**, **(c)** interdependencies are quantified via MIH; in **(b)**, **(d)** via MIOP interannual timescale.

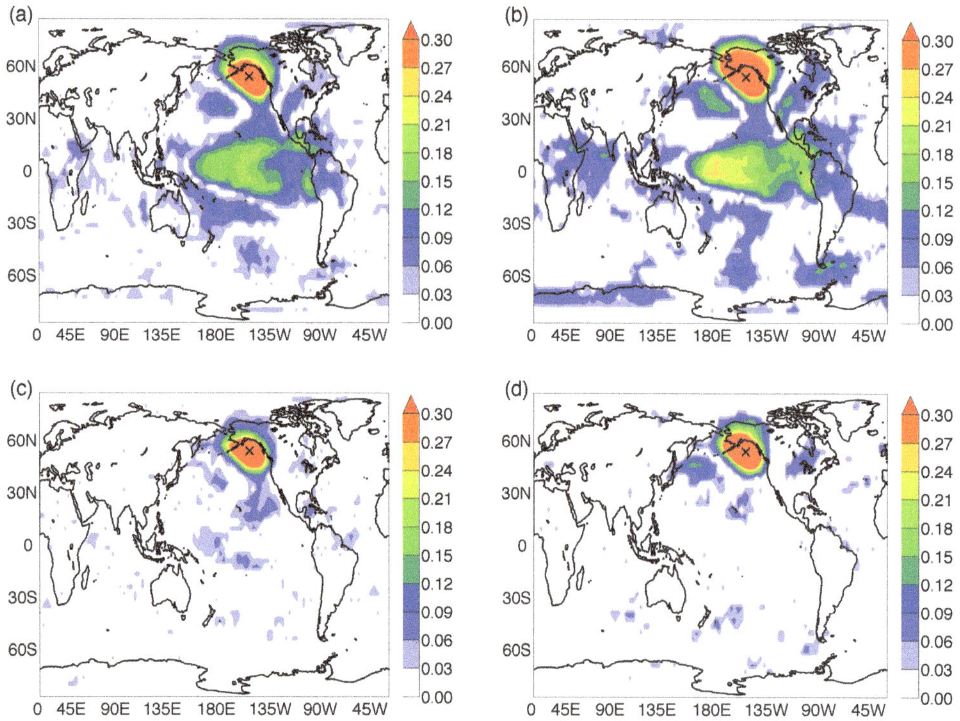

Figure 10. As Fig. 9 but regarding a node near Alaska (indicated with X). Compared to Fig. 9 one can notice that the teleconnection between this region and the Pacific is due mainly to El Niño.

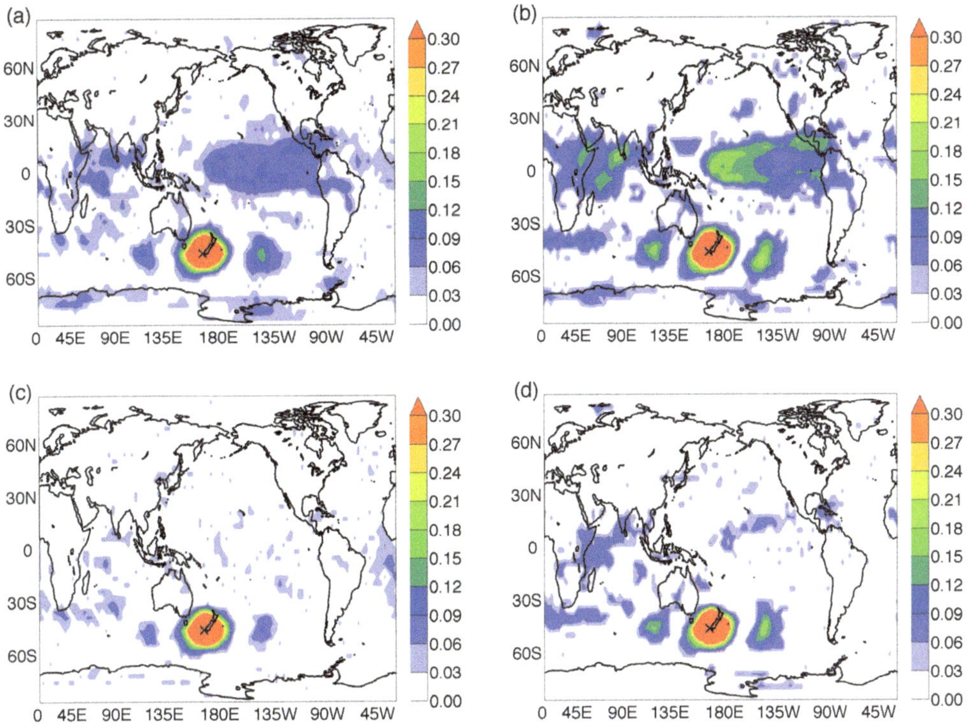

Figure 11. As Fig. 9 but regarding a node near New Zealand (indicated with X). In panels **(b)** and **(d)** the MIOP is adjusted to interannual timescale. Compare with Figs. 9 and 10.

Zealand because it shows a relatively high forced density (seen in Fig. 5a, b) and it is connected to the selected point over the tropical Pacific of Fig. 9a, b. Figure 11a shows connectivity between the chosen point and the Pacific and Indian oceans, as well as wave patterns (probably a Rossby wave train) along the extratropics. Figure 11b adds information to Fig. 11a, showing that these teleconnections are of interannual type. If we remove NINO3.4 (Fig. 11c, d) not surprisingly the links to the tropical Pacific disappear, but also some of the connectivity to the Indian Ocean, suggesting that part of the links with the Indian Ocean are indirect. Nevertheless, the extra-tropical wave train remains, and Fig. 11d suggests that the wave train may be forced by the Indian Ocean at interannual timescales. As in the previous figure, some weak north–south teleconnections are found, but they disappear if we remove the NINO3.4 index, indicating again an indirect connection between the extratropics through the Pacific Ocean.

4.3.2 Internal variability

Figure 12 displays the *internal* variability connections of a node over the most connected area of Fig. 7. The average of the resulting nine different connectivity maps is shown. In the left column the connectivity computed using MIH is displayed, while in the right column, the intra-annual scale is shown, using MIOP. This timescale shows the strongest response for internal variability. In Fig. 12a the original internal variability connections are shown, revealing teleconnections extending over the Northern Hemisphere, especially over Scandinavia, Mediterranean Europe, the east coast of North America and the tropical North Atlantic. Figure 12a also shows connections to eastern China and the Aleutian islands. The pattern shown in Fig. 12b mainly corresponds to the known influence of the North Atlantic Oscillation. This is further substantiated in Fig. 12c and d of the same figure, where the NAO influence is removed and the connections of the Labrador Sea, particularly in the northern Atlantic Basin, are strongly weakened.

5 Summary and conclusions

The monthly variability of the surface air temperature field has been decomposed into a part forced by the ocean temperature, and another due to internal atmospheric variability. This has been performed using an ensemble of nine AGCM runs forced with the same SST data, and starting from slightly different initial conditions. The model data was firstly validated by observing a qualitative agreement between the networks constructed from one model run and those constructed from reanalysis data. Afterwards, climate networks were constructed from model data, for the forced and for the internal variability components, using mutual information to assess the interdependencies between the time

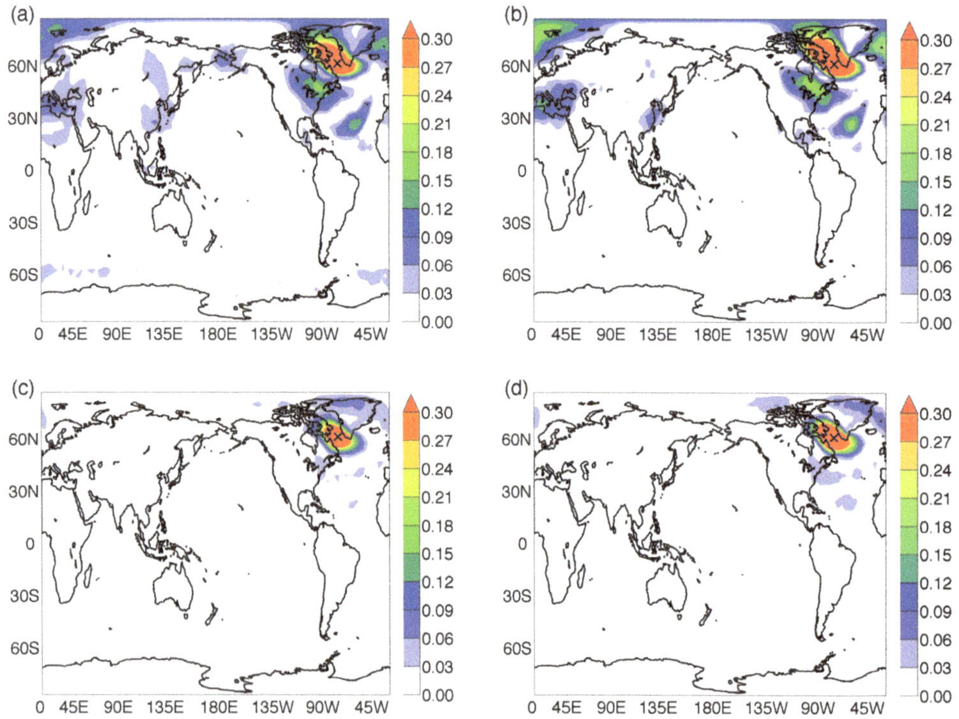

Figure 12. Maps of internal variability showing the connectivity of a node in the Labrador Sea (indicated with X). Panels **(a)** and **(b)** correspond to the original internal time series as in Fig. 7; in panels **(c)** and **(d)** the NAO was linearly removed and thus the links do not contain – to the first order – contributions due to the North Atlantic Oscillation. In **(a)** and **(c)** interdependencies are quantified via MIH; in **(b)** and **(d)** via MIOP intra-annual timescale.

series. Ordinal patterns have been used in order to separate and determine the strength of the links at different timescales.

While the main conclusions of our analysis (the connectivity of the forced variability network is heavily affected by El Niño, whereas that of the internal variability network is significantly affected by the NAO) are not new, new information has been uncovered, as ordinal analysis allows us to study these phenomena on different timescales. This has revealed that most of the links detected in the forced variability proceed from long timescales, while the contributions of intra-annual timescales to the internal variability are the most important. This work also opens the possibility of studying how various network measures, such as the average path length, assortativity, clustering coefficient, betweeness, etc. depend on the timescale considered for quantifying statistical interdependencies.

Another conclusion of this work is that forced and internal atmospheric variability are characterized by very different networks. Because the separation of internal and forced variability done here requires averaging over several model runs, the networks obtained here could not have been obtained from observational/reanalysis data only. It is shown that the forced variability is stronger in the tropics, while the internal variability peaks in the mid-latitudes. The network of forced variability has the strongest connections at interannual timescales. *Long-range teleconnections* from the tropics to the extratropics and even from different hemispheres in the forced network were observed and explained by the influence of El Niño. On the other hand, the network of internal atmospheric variability has the strongest connections in the extratropics, and it was found that connections to the Labrador Sea are heavily affected by the North Atlantic Oscillation.

This study is focused on the lowest levels of the atmosphere. A complementary analysis is performed in the companion paper by Arizmendi et al. (2014), devoted to the study of the evolution of the upper atmosphere during the 20th century and aiming at distinguishing the oceanically forced component from the atmospheric internal variability on different timescales. The methodology proposed here for distinguishing links in spatial range (short and long), timescale (intraseasonal, intra-annual and interannual) and type of variability (forced vs. internal) is a novel approach for the study of climate networks that provides new insight into the climatological meaning of the links found, and their connection to physical phenomena.

Acknowledgements. The research leading to these results has received funding from the European Community's Seventh Framework Programme FP7/2007-2011 under grant agreement no. 289447 (ITN LINC). C. Masoller also acknowledges financial support of grant FIS2012-37655-C02-01 of the Spanish Ministerio de Ciencia e Innovación, grant 2009 SGR 1168 of the Agencia de

Gestió d'Ajuts Universitaris i de Recerca (AGAUR), Generalitat de Catalunya, and the ICREA ACADEMIA award.

Edited by: R. Donner

References

Albert, R. and Barabási, A. L.: Statistical mechanics of complex networks, Rev. Mod. Phys., 74, 47–97, doi:10.1103/RevModPhys.74.47, 2002.

Allen, M. R., and Smith L. A.: Optimal filtering in singular spectrum analysis, Phys. Lett. A, 234, 419–428, doi:10.1016/S0375-9601(97)00559-8, 1997.

Amigó, J. M.: Topological permutation entropy, Springer Berlin Heidelberg, 2010.

Arizmendi, F., Martí, A., and Barreiro, M.: Evolution of atmospheric connectivity in the 20th century, accepted, Nonlin. Processes Geophys., 2014.

Bandt, C. and Pompe, B.: Permutation entropy: A natural complexity measure for time series, Phys. Rev. Lett., 88, 174102, doi:10.1103/PhysRevLett.88.174102, 2002.

Barreiro, M.: Influence of ENSO and the South Atlantic Ocean on climate predictability over Southeastern South America, Clim. Dynam., 35, 1493–1508, doi:10.1007/s00382-009-0666-9, 2009.

Barreiro, M. and Díaz, N.: Land-atmosphere coupling in El Niño influence over South America, Atmos. Sci. Lett., 12, 351–355, doi:10.1002/asl.348, 2011.

Barreiro, M., Chang, P., and Saravanan, R.: Variability of the South Atlantic Convergence Zone as simulated by an atmospheric general circulation model, J. Climate, 15, 745–763, doi:10.1175/1520-0442(2002)015<0745:VOTSAC>2.0.CO;2, 2002.

Barreiro, M., Chang, P., Ji, L., Saravanan, R., and Giannini, A.: Dynamical Elements of Predicting Boreal Spring Tropical Atlantic Sea-Surface Temperatures, Dynam. Atmos. Oceans, 39, 69–85, doi:10.1016/j.dynatmoce.2004.10.013, 2005.

Barreiro, M., Martí, A. C., and Masoller, C.: Inferring long memory processes in the climate network via ordinal pattern analysis, Chaos, 21, 013101, doi:10.1063/1.3545273, 2011.

Barsugli, J. J. and Battisti, D. S.: The Basic Effects of Atmosphere–Ocean Thermal Coupling on Midlatitude Variability, J. Atmos. Sci., 55, 477–493, doi:10.1175/1520-0469(1998)055<0477:TBEOAO>2.0.CO;2, 1998.

Bracco, A., Kucharski, F., Kallummal, R., and Molteni, F.: Internal variability, external forcing and climate trends in multi-decadal AGCM ensembles, Clim. Dynam., 23, 659–678, doi:10.1007/s00382-004-0465-2, 2004.

Chang, P., Saravanan, R., Ji, L., and Hegerl, G. C.: The effect of local sea surface temperatures on atmospheric circulation over the tropical Atlantic sector, J. Climate, 13, 2195–2216, doi:10.1175/1520-0442(2000)013<2195:TEOLSS>2.0.CO;2, 2000.

Chiang, J. C. H. and Sobel, A. H.: Tropical tropospheric temperature variations caused by ENSO and their influence on the remote tropical climate, J. Climate, 15, 2616–2631, doi:10.1175/1520-0442(2002)015<2616:TTTVCB>2.0.CO;2, 2002.

Cover, T. M. and Thomas, J. A.: Elements of information theory, John Wiley & Sons, 2006.

Deza, J. I., Barreiro, M., and Masoller, C.: Inferring interdependencies in climate networks constructed at interannual, intra–season and longer time scales, Eur. Phys. J.-Spec. Top., 222, 511–523, doi:10.1140/epjst/e2013-01856-5, 2013.

Donges, J. F., Zou, Y., Marwan, N., and Kurths, J.: Complex networks in climate dynamics, Eur. Phys. J.-Spec. Top., 174, 157–179, doi:10.1140/epjst/e2009-01098-2, 2009a.

Donges, J. F., Zou, Y., Marwan, N., and Kurths, J.: The backbone of the climate network, EPL-Europhys. Lett., 87, 48007, doi:10.1209/0295-5075/87/48007, 2009b.

Frankignoul, C. and Hasselmann, K.: Stochastic climate models, Part II Application to sea-surface temperature anomalies and thermocline variability, Tellus, 29, 289–305, doi:10.1111/j.2153-3490.1977.tb00740.x, 1977.

Gámez, A. J., Zhou, C. S., Timmermann, A., and Kurths, J.: Nonlinear dimensionality reduction in climate data, Nonlin. Processes Geophys., 11, 393–398, doi:10.5194/npg-11-393-2004, 2004.

Gozolchiani, A., Havlin, S., and Yamasaki, K.: Emergence of El Niño as an autonomous component in the climate network, Phys. Rev, Lett., 107, 148501, doi:10.1103/PhysRevLett.107.148501, 2011.

Hasselmann, K.: Stochastic climate models part I. Theory, Tellus, 28, 473–485, doi:10.1111/j.2153-3490.1976.tb00696.x, 1976.

Hlinka, J., Hartman, D., Vejmelka, M., Runge, J., Marwan, N., Kurths, J., and Paluš, M.: Reliability of inference of directed climate networks using conditional mutual information, Entropy, 15, 2023–2045, doi:10.3390/e15062023, 2013.

Hlinka, J., Hartman, D., Vejmelka, M., Novotná, D., and Paluš, M.: Non-linear dependence and teleconnections in climate data: sources, relevance, nonstationarity, Clim. Dynam., 42, 1873–1886, doi:10.1007/s00382-013-1780-2, 2014.

Hurrell, J. W.: Decadal trends in the North Atlantic oscillation: Regional temperatures and precipitation, Science, 269, 676–679, doi:10.1126/science.269.5224.676, 1995.

James, I. N.: Introduction to circulating atmospheres, Cambridge University Press, 1995.

Kalnay E., Kanamitsu, M., Kistler, R., Collins, W., Deaven, D., Gandin, L., Iredell, M., Saha, S., White, G., Woollen, J., Zhu, Y., Leetmaa, A., Reynolds, R., Chelliah, M., Ebisuzaki, W., Higgins, W., Janowiak, J., Mo, K. C., Ropelewski, C., Wang J., Jenne, R., and Joseph, D.: The NCEP/NCAR 40-Year Reanalysis Project, B. Am. Meteorol. Soc., 77, 437–471, doi:10.1175/1520-0477(1996)077<0437:TNYRP>2.0.CO;2, 1996.

Kucharski, F., Molteni, F., and Bracco, A.: Decadal interactions between the Western Tropical Pacific and the North Atlantic Oscillation, Clim. Dynam., 26, 79–91, doi:10.1007/s00382-005-0085-5, 2005.

Lind, P. G., Mora, A., Gallas, J. A., and Haase, M.: Reducing stochasticity in the North Atlantic Oscillation index with coupled Langevin equations, Phys. Rev. E, 72, 056706, doi:10.1103/PhysRevE.72.056706, 2005.

Malik, N., Bookhagen, B., Marwan, N., and Kurths, J.: Analysis of spatial and temporal extreme monsoonal rainfall over South Asia using complex networks, Clim. Dynam., 39, 971–987, doi:10.1007/s00382-011-1156-4, 2012.

Martin, E. A., Paczuski, M., and Davidsen, J.: Interpretation of link fluctuations in climate networks during El Niño periods, EPL-Europhys. Lett., 102, 48003, doi:10.1209/0295-5075/102/48003, 2013.

Molteni, F.: Atmospheric simulations using a GCM with simplified physical parametrizations. I: model climatology and variability in multi-decadal experiments, Clim. Dynam., 20, 175–191, doi:10.1007/s00382-002-0268-2, 2003.

NOAA: website available at: http://www.esrl.noaa.gov/psd/data/climateindices/list/, last access: 31 October 2013, 2013.

Paluš, M.: From Nonlinearity to Causality: Statistical testing and inference of physical mechanisms underlying complex dynamics, Contemp. Phys., 48, 307–348, doi:10.1080/00107510801959206, 2007.

Paluš, M., Hartman, D., Hlinka, J., and Vejmelka, M.: Discerning connectivity from dynamics in climate networks, Nonlin. Processes Geophys., 18, 751–763, doi:10.5194/npg-18-751-2011, 2011.

Pohlmann, H. and Latif, M.: Atlantic versus Indo-Pacific influence on Atlantic-European climate, Geophys. Res. Lett., 32, 05707, doi:10.1029/2004GL021316, 2005.

Pompe, B. and Runge, J.: Momentary information transfer as a coupling measure of time series, Phys. Rev. E, 83, 051122, doi:10.1103/PhysRevE.83.051122, 2011.

Radebach, A., Donner, R. V., Runge, J., Donges, J. F., and Kurths, J.: Disentangling different types of El Niño episodes by evolving climate network analysis, Phys. Rev. E, 88, 052807, doi:10.1103/PhysRevE.88.052807, 2013.

Rodwell, M. J., Rowell, D. P., and Folland, C. K.: Oceanic forcing of the wintertime North Atlantic Oscillation and European climate, Nature, 398, 320–323, doi:10.1038/18648, 1999.

Ropelewski, C. F. and Halpert, M. S.: North American precipitation and temperature patterns associated with the El Niño–Southern Oscillation (ENSO), Mon. Weather Rev., 114, 2352–2362, doi:10.1175/1520-0493(1986)114<2352:NAPATP>2.0.CO;2, 1986.

Schreiber, T. and Schmitz, A.: Improved Surrogate Data for Nonlinearity Tests, Phys. Rev. Lett., 77, 635–638, doi:10.1103/PhysRevLett.77.635, 1996.

Seager, R., Naik, N., Baethgen, W., Robertson, A., Kushnir, Y., Nakamura, J., and Jurburg, S.: Tropical Oceanic Causes of Interannual to Multidecadal Precipitation Variability in Southeast South America over the Past Century, J. Climate, 23, 5517–5539, doi:10.1175/2010JCLI3578.1, 2010.

Shukla, J.: Predictability in the midst of chaos: a scientific basis for climate forecasting, Science, 282, 728–731, doi:10.1126/science.282.5389.728, 1998.

Smith, T. M. and Reynolds, R. W.: Improved extended reconstruction of SST (1854–1997), J. Climate, 17, 2466–2477, doi:10.1175/1520-0442(2004)017<2466:IEROS>2.0.CO;2, 2004.

Straus, D. M. and Shukla, J.: Distinguishing between the SST-forced variability and internal variability in mid latitudes: analysis of observations and GCM simulations, Q. J. Roy. Meteor. Soc., 126, 2323–2350, doi:10.1002/qj.49712656716, 2000.

Ting, M., Kushnir, Y., Seager, R., and Li, C.: Forced and internal twentieth-century SST Trends in the North Atlantic, J. Climate, 22, 1469–1481, doi:10.1175/2008JCLI2561.1, 2009.

Tirabassi, G. and Masoller, C.: On the effects of lag-times in networks constructed from similarities of monthly fluctuations of climate fields, EPL-Europhys. Lett., 102, 59003, doi:10.1209/0295-5075/102/59003, 2013.

Trenberth, K. E.: The Definition of El Niño, B. Am. Meteorol. Soc., 78, 2771–2777, doi:10.1175/1520-0477(1997)078<2771:TDOENO>2.0.CO;2, 1997.

Trenberth, K. E., Branstator, G. W., Karoly, D., Kumar, A., Lau, N. C., and Ropelewski, C.: Progress during TOGA in understanding and modeling global teleconnections associated with tropical sea surface temperatures, J. Geophys. Res., 103, 14291–14324, doi:10.1029/97JC01444, 1998.

Tsonis, A. A. and Swanson, K. L.: Review article "On the origins of decadal climate variability: a network perspective", Nonlin. Processes Geophys., 19, 559–568, doi:10.5194/npg-19-559-2012, 2012.

Tsonis, A. A., Swanson, K. L., and Roebber, P. J.: What do networks have to do with climate?, B. Am. Meteorol. Soc., 87, 585–595, doi:10.1175/BAMS-87-5-585, 2006.

Tsonis, A. A., Swanson, K. L., and Wang, G.: On the role of atmospheric teleconnections in climate, J. Climate, 21, 2990–3001, doi:10.1175/2007JCLI1907.1, 2008.

van der Mheen, M., Dijkstra, H. A., Gozolchiani, A., den Toom, M., Feng, Q. Y., Kurths, J., and Hernandez Garcia, E.: Interaction network based early warning indicators for the Atlantic MOC collapse, Geophys. Res. Lett., 40, 2714–2719, doi:10.1002/grl.50515, 2013.

Venzke, S., Allen, M. R., Sutton, R. T., and Rowell, D. P.: The atmospheric response over the North Atlantic to decadal changes in sea surface temperature, J. Climate, 12, 2562–2584, doi:10.1175/1520-0442(1999)012<2562:TAROTN>2.0.CO;2, 1999.

Zheng, Y., Waliser, D. E., Stern, W. F., and Jones, C.: The role of coupled sea surface temperatures in the simulation of the tropical intraseasonal oscillation, J. Climate, 17, 4109–4134, doi:10.1175/JCLI3202.1, 2004.

6

Scale invariant events and dry spells for medium-resolution local rain data

A. Deluca[1,2] and Á. Corral[1,2]

[1]Centre de Recerca Matemàtica, Edifici Cc, Campus Bellaterra, 08193 Bellaterra, Barcelona, Spain
[2]Departament de Matemàtiques, Universitat Autònoma de Barcelona, 08193 Bellaterra, Barcelona, Spain

Correspondence to: A. Deluca (adeluca@crm.cat) and Á. Corral (acorral@crm.cat)

Abstract. We analyze distributions of rain-event sizes, rain-event durations, and dry-spell durations for data obtained from a network of 20 rain gauges scattered in a region of the northwestern Mediterranean coast. While power-law distributions model the dry-spell durations with a common exponent 1.50 ± 0.05, density analysis is inconclusive for event sizes and event durations, due to finite size effects. However, we present alternative evidence of the existence of scale invariance in these distributions by means of different data collapses of the distributions. These results demonstrate that scaling properties of rain events and dry spells can also be observed for medium-resolution rain data.

1 Introduction

The complex atmospheric processes related to precipitation and convection arise from the cooperation of diverse non-linear mechanisms with different temporal and spatial characteristic scales. Precipitation combines, for instance, the $\mathcal{O}(100\mu\text{m})$ microphysics effects as evaporation with $\mathcal{O}(1000\,\text{km})$ planetary circulation of masses and moisture (Bodenschatz et al., 2010). Rain fields also present high spatial and temporal intermittency as well as extreme variability, in such a way that their intensity cannot be characterized by its mean value. Despite the complexity of the processes involved, surprising statistical regularities have been found: numerous geometric and radiative properties of clouds present clear scaling or multiscaling behavior (Lovejoy, 1982; Cahalan and Joseph, 1989; Peters et al., 2009; Wood and Field, 2011); also, raindrop arrival times and raindrop sizes, are well characterized by power-law distributions over several of orders of magnitude (Olsson et al., 1993; Lavergnat and Golé, 2006).

The concept of self-organized criticality (SOC) aims for explaining the origin of the emergence of structures across many different spatial and temporal scales in a broad variety of systems (Bak, 1996; Jensen, 1998; Sornette, 2004; Christensen and Moloney, 2005). Indeed, it has been found that for diverse phenomena that take place intermittently, in terms of bursts of activity interrupting larger quiet periods, the size, s, of these bursty events or avalanches follows a power-law distribution,

$$P_S(s) \propto \frac{1}{s^{\tau_s}}, \tag{1}$$

over a certain range of s, where $P_S(s)$ is the probability density of the event size and τ_s its exponent (and the sign $\propto$ indicating proportionality). The size, s, can be understood as a measure of energy dissipation. If durations of events are measured, a power-law distribution also holds. These power-law distributions provide an unambiguous proof of the absence of characteristic scales within the avalanches, as power laws are the signature of scale invariance (Christensen and Moloney, 2005).

The main idea behind SOC is the claim that such scale invariance is achieved because of the existence of a non-equilibrium continuous phase transition whose critical point is an attractor of the dynamics when the system is slowly driven (Tang and Bak, 1988; Dickman et al., 1998, 2000). When the system settles at the critical point, scale invariance and power-law behavior are ensured, as these peculiarities are the defining characteristics of critical phenomena (Christensen and Moloney, 2005). SOC has had a big impact in the geosciences, in particular earthquakes (Bak, 1996; Sornette and Sornette, 1989), landslides and rock avalanches (Malamud, 2004), or forest fires (Malamud et al., 1998). Due to the existence of power-law distributed events in them, these systems have been proposed as realizations of SOC in the natural world; nevertheless, no evidence of an underlying critical point has been presented. On the other hand, it is also possible to understand SOC in a broader sense, as the spontaneous emergence of scale invariance; therefore, SOC is not considered a mechanism based on an attraction towards a continuous phase transition, but arising from a different phenomenology, where stochasticity and non-vanishing fluxes play a key role, as it happens in multifractal cascades (Schertzer and Lovejoy, 1994, 2011; Hooge et al., 1994; Lovejoy and Schertzer, 2013).

SOC observables have been explored in rainfall while looking at precipitation as an avalanche process, and paying attention to the properties of these avalanches, called rain events. To our knowledge, the first works following this approach are those of Andrade et al. (1998) and Peters et al. (2002) (Peters and Christensen, 2002, 2006). These authors defined, independently, a rain event as the sequence of rain occurrence with a rain rate (i.e., the activity) always greater than zero. Therefore, the focus of these approaches is not on the total amount of rain recorded in a fixed time period (for instance, one hour, one day, or one month), but on the rain event, which is what defines in each case the time period of rain-amount integration. In this way, the event size is the total amount of rain collected during the duration of the event.

Andrade et al. (1998) studied long-term daily local (i.e., zero dimensional) rain records from weather stations in Brazil, India, Europe, and Australia, with observation times ranging from a decade to a century approximately, with detection threshold 0.1 mm day^{-1}. Although the dry spells (the times between rain events) seemed to follow in some case a steep power-law distribution, the rain-event size distributions were not reported, and therefore the connection between SOC and rainfall could not really be checked. Later, Peters et al. (2002) analyzed high resolution rain data from a vertically pointing Doppler radar situated on the Baltic coast, which provided rates at an altitude between 250 and 300 m a.s.l., covering an area of 70 m^2, with detection threshold 0.0005 mm h^{-1} and temporal resolution of 1 min. Power-law distributions for event sizes and for dry-spell durations over several orders of magnitude were reported, with exponents $\tau_s \simeq \tau_q \simeq 1.4$. For the event–duration distribution the results were unclear, although a power law with an exponent $\tau_d \simeq 1.6$ was fit to the data. More recently, a study covering 10 sites across different climates has checked the universality of rain-event statistics using rain data from optical gauges (Peters et al., 2010). The data had a detection threshold of 0.2 mm h^{-1}, and were collected at intervals of 1 min. The results showed unambiguous power-law distributions of event sizes, with apparent universal exponents $\tau_s = 1.17 \pm 0.03$. Power-law distributions were also found for the dry spell durations, but for event durations the behavior was not conclusive.

Nevertheless, scale-free distributions of the avalanche-like observables are insufficient evidence for "classical" SOC dynamics, as there are many alternative mechanisms of power-law genesis (Sornette, 2004; Dickman, 2003). In other words, SOC implies power laws, but the reciprocal is not true, power laws are not a guarantee of SOC. Of particular relevance is the multifractal approach, which can also reproduce scale invariance but using different observables. When applied to rainfall, this approach focus on the rain rate field, which is hypothesized to have multifractal support as a result from a multiplicative cascade process. From this point of view, alternative statistical models and new forecasting and downscaling methods have emerged (Schertzer and Lovejoy, 1992; Lovejoy and Schertzer, 1995; Deidda, 2000; Veneziano and Lepore, 2012).

We will study the scaling of the distributions of event observables without ascribing this scaling to SOC or any other mechanism. As we have seen, the number of studies addressing this kind of scaling is rather limited, mostly due to the supposed requirement that the data has to be of very high time and rate resolution. The goal of the paper is to extend the evidence for power laws in rainfall, in particular in the distribution of the observables related to rain events, studying the applicability of this paradigm when the rain data available is not of high resolution. With this purpose, we perform an in-depth analysis of local rainfall records in a representative region of the northwestern Mediterranean. For this lower (in comparison with previous studies) resolution, the range in which the power law holds can be substantially decreased. This may require the application of more refined fitting techniques and scaling methods. Thus, as a byproduct, we explore different scaling forms and develop a collapse method based on minimizing the distance between distributions that also gives an estimation of the power-law exponent. With these tools we will be able to establish the existence of scale-invariant behavior in the medium-resolution rain data analyzed.

We proceed as follows: Sect. 2 describes the data used in the present analysis and defines the rain event, its size and duration, and the dry spell. Section 3 shows the corresponding probability densities and describes and applies an accurate fitting technique for evaluating the power-law existence. Section 4 introduces two collapse methods (parametric and non-parametric) in order to establish the fulfillment of scaling, independently of power-law fitting. Discussion and conclusions are presented in Sect. 5.

2 Data and definitions

2.1 Data

We have analyzed 20 stations in Catalonia (NE Spain) from the database maintained by the Agència Catalana de l'Aigua (ACA, 2014). These data come from a network of rain gauges, called SICAT (Sistema Integral del Cicle de l'Aigua al Territori, formerly SAIH, Sistema Automàtic d'Informació

Table 1. Characteristics of all the sites for the 9-year period 2000–2008. Every site is named by the corresponding river basin or subbasin (the municipality is included in ambiguous cases); Ll. stands for Llobregat river; f_M is the fraction (in %) of missing records (time missing divided by total time); f_D is the fraction (in %) of discarded times; f_r is the fraction (in %) of rainy time (time with $r > 0$ divided by total non-discarded time, for a time resolution $\Delta t = 5$ min); a. rate is the annual rain rate in mm year^{-1}, calculated only over non-discarded times; c. rate is the rain rate in mm h^{-1} conditioned to rain, i.e., calculated over the (non-discarded) rainy time; N_s is the number of rain events and N_q the number of dry spells (the differences between N_s and N_q are due to the missing records); the rest of symbols are explained in the text. $\langle s \rangle$ is measured in mm, and $\langle d \rangle$ and $\langle q \rangle$ in min. Sites are ordered by increasing annual rate. The table shows a positive correlation between f_r, the annual rate, N_s and N_q, and that these variables are negatively correlated with $\langle q \rangle$. In contrast, the rate conditioned to rain is roughly constant, taking values between 3.3 and 3.8 mm h^{-1}.

Site	f_M	f_D	f_r	a. rate	c. rate	N_s	N_q	$\langle s \rangle$	$\langle d \rangle$	$\langle q \rangle$
1 Gaià	0.08	3.71	1.6	470.9	3.3	5021	5014	0.81	14.9	894
2 Foix	0.07	3.38	1.6	500.6	3.6	4850	4844	0.90	15.0	929
3 Baix Ll. S.J. Despí	0.07	2.28	1.7	505.8	3.3	5374	5369	0.83	15.0	847
4 Garraf	0.09	3.30	1.6	507.8	3.7	4722	4716	0.94	15.2	956
5 Baix Ll. Castellbell	0.06	2.81	1.7	510.7	3.4	4950	4947	0.90	15.8	914
6 Francolí	0.44	13.37	1.8	528.2	3.4	4539	4540	0.91	16.1	887
7 Besòs Barcelona	0.15	4.17	1.7	531.8	3.5	4808	4803	0.95	16.2	928
8 Riera de La Bisbal	0.07	3.66	1.6	540.0	3.8	4730	4724	0.99	15.8	950
9 Besòs Castellar	4.34	13.59	2.0	633.3	3.6	4918	4970	1.00	16.9	806
10 Ll. Cardener	0.07	3.33	2.1	652.4	3.5	6204	6197	0.92	15.7	723
11 Ridaura	0.12	2.41	2.0	674.2	3.8	5780	5774	1.02	16.1	784
12 Daró	0.06	2.09	2.2	684.5	3.6	5553	5547	1.09	18.0	818
13 Tordera	0.08	2.04	2.3	688.8	3.4	7980	7977	0.76	13.6	568
14 Baix Ter	0.07	2.71	2.3	710.2	3.6	6042	6036	1.03	17.4	746
15 Cap de Creus	0.07	2.92	2.3	741.5	3.7	5962	5955	1.09	17.7	754
16 Alt Llobregat	3.12	5.82	2.6	742.8	3.3	6970	6988	0.90	16.7	621
17 Muga	0.06	2.56	2.4	749.3	3.6	6462	6457	1.02	16.9	698
18 Alt Ter Sau	0.08	2.43	2.5	772.1	3.6	6966	6961	0.97	16.3	647
19 Fluvià	3.09	4.74	2.3	772.4	3.8	6287	6319	1.05	16.7	697
20 Alt Ter S. Joan	0.07	1.98	2.8	795.1	3.3	8333	8327	0.84	15.5	452

Hidrològica), used to monitor the state of the drainage basins of the rivers that are born and die in the Catalan territory. The corresponding sites are listed in Table 1 and have longitudes and latitudes ranging from 1°10′51″ to 3°7′35″ E and from 41°12′53″ to 43°25′40″ N. All data sets cover a time period starting on 1 January 2000, at 00:00 LT, and ending either on 30 June or on 1 July 2009 (spanning roughly 9.5 years), except the Cap de Creus, which ends on 19 June 2009.

In all the stations, rain is measured by the same weighing precipitation gauge, the device called *Pluvio* from OTT Hydromet GmbH, either with a capacity of 250 or 1000 mm and working through the balance principle. It measures both liquid or/and solid precipitation. The precipitation rate is recorded in intervals of $\Delta t = 5$ min, with a resolution of 1.2 mm h^{-1} (which corresponds to 0.1 mm in 5 min). This precipitation rate can be converted into an energy flux through the latent heat of condensation of water, which yields 1 mm h$^{-1} \simeq 690$ W m^{-2}, nevertheless, we have not performed such conversion. Figure 1a shows a subset of thc time series for site 17 (Muga).

In order to make the files more manageable, the database reports zero-rain rates only every hour; then we consider time voids larger than 1 h as operational errors. The ratio of these missing times to the total time covered in the record is denoted as f_M in Table 1, where it can be seen that this is usually below 0.1 %. However, there are three cases in which its value is around 3 or 4 %. Other quantities reported in the table are the fraction of time corresponding to rain (or wet fraction), f_r, the annual mean rate, and the mean rate conditioned to rain periods. Nevertheless, note that for a fractal point process a quantity as f_r depends on the time resolution; so, f_r only makes sense for a concrete time division, in our case, $\Delta t = 5$ min.

2.2 Rain event sizes, rain event durations, and dry spell durations

Following Andrade et al. (1998) and Peters et al. (2002), we define a rain event as a sequence of consecutive rain rates bigger than zero delimited by zero rates, i.e., $\{r(t_n), r(t_{n+1}), \dots r(t_m)\}$, such that $r(t_i) > 0$ for $i = n, n+1, \dots m$ with $r(t_{n-1}) = r(t_{m+1}) = 0$. Due to the resolution of the record, this is equivalent to take a threshold with a value below $1.2\,\text{mm}\,\text{h}^{-1}$. It is worth mentioning that this simple definition of rain events may be in conflict with those used by the hydrologist community, so caution is required in order to make comparisons between the different approaches (Molini et al., 2011).

The first observable to consider is the duration, d, of the event, which is the time that the event lasts (a multiple of 5 min, in our case). The size of the event is defined as the total rain during the event (i.e., the rate integrated over the event duration),

$$s \equiv \sum_{i=n}^{m} r(t_i)\Delta t \simeq \int_{t_n}^{t_m} r(t)dt,$$

measured in mm (and multiple of 0.1 mm in our case, $1.2\,\text{mm}\,\text{h}^{-1} \times 5\,\text{min}$). Notice that this event size is not the same as the usual rain depth, due to the different definition of the rain event in each case. Figure 1b shows as an illustration the evolution of the rate for the largest event in the record, which happens at the Muga site, whereas Fig. 1c displays the sequence of all event sizes in the same site for the year 2002. It is important to realize that this quantity is different to the one at Fig. 1a. Regarding event durations, the time series have a certain resemblance to Fig. 1c, as usually they are (nonlinearly) correlated with event sizes (Telesca et al., 2007). Further, the dry spells are the periods between consecutive rain events (then, they verify $r(t) = 0$); we denote their durations by q. When a rain event, or a dry spell, is interrupted due to missing data, we discard that event or dry spell, and count the recorded duration as discarded time; the fraction of these times in the record appears in Table 1, under the symbol f_D. Although in some cases the duration of the interrupted event or dry spell can be bounded from below or from above (as in censored data), we have not attempted to use that partial information.

3 Power-law distributions

3.1 Probability densities

Due to the enormous variability of the three quantities just defined, the most informative approach is to work with their probability distributions. Taking the size as an example, its probability density, $P_S(s)$, is defined as the probability that the size is between s and $s + ds$ divided by ds, with $ds \to 0$. Subsequently, $\int_0^\infty P_S(s)ds = 1$. This implicitly assumes that s is considered as a continuous variable (but this will be corrected later, see more details on Appendix A). In general, we illustrate all quantities with the event size s, the analogous for d and q are obtained by replacing s with the symbol of each observable. The corresponding probability densities are denoted as $P_\text{D}(d)$ and $P_Q(q)$. Note that the annual number densities (Peters et al., 2002; Peters and Christensen, 2002,

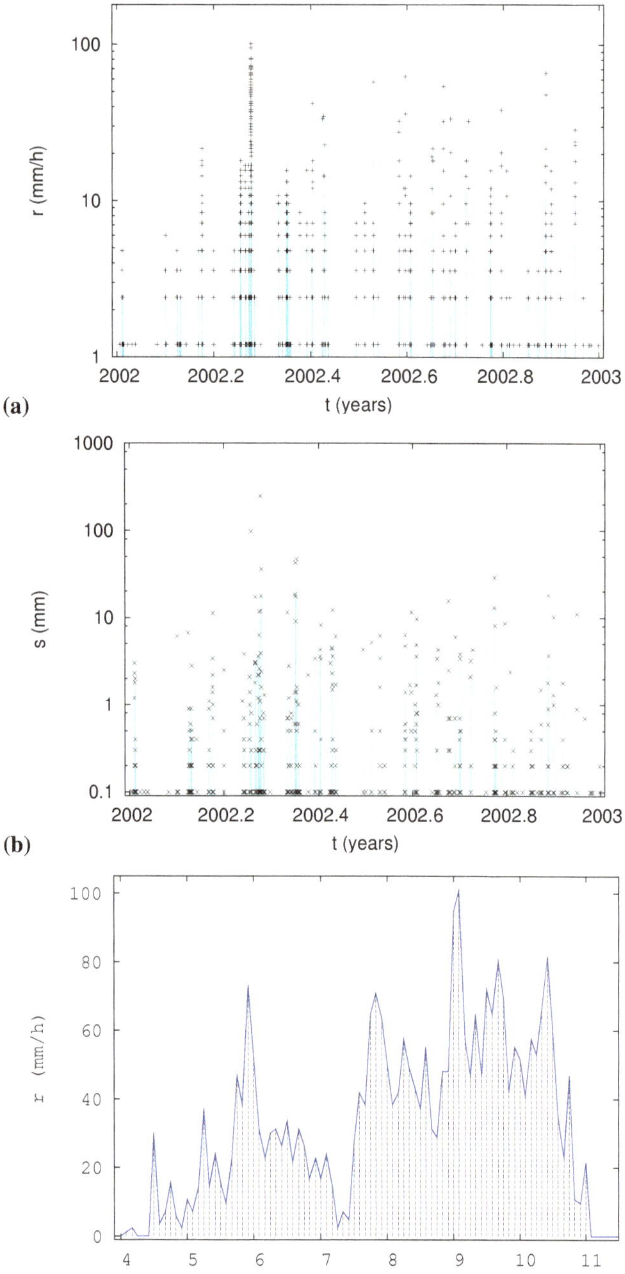

Fig. 1. **(a)** Subset of the rain rate time series for site 17 (Muga) for year 2002. **(b)** More reduced subset of the rain rate time series for the same site, corresponding to the largest rain event on the record, with $s = 248.7$ mm, on 11 April 2002. Time refers to hours since midnight. A very small rain event is also present at the beginning, with $s = 0.3$ mm and separated to the main event by a dry spell of duration $q = 15$ min. **(c)** Corresponding event-size time series for the same site, for year 2002.

2006) are trivially recovered multiplying the probability densities by the total number of events or dry spells and dividing by total time.

The results for the probability densities $P_S(s)$, $P_D(d)$, and $P_Q(q)$ of all the sites under study are shown in Fig. 2a, b, and c, respectively. In all cases the distributions show a very clear behavior, monotonically decreasing and covering a broad range of values. However, to the naked eye, a power-law range is only apparent for the distributions of dry spells, $P_Q(q)$ (remember that a power law turns into a straight line in a log–log plot). Moreover, the $P_Q(q)$ are the broadest distributions, covering a range of more than 4 orders of magnitude (from 5 min to about a couple of months), and present in some cases a modest daily peak (in comparison to Peters et al., 2002, with 1 day = 1440 min). In the opposite side, we find the distributions of durations, $P_D(d)$, whose range is the shortest, from 5 min to about 1 day (two and a half orders of magnitude), and for which no straight line is visible in the plot; rather, the distributions appear as convex. The size distributions, $P_S(s)$, defined for about 3 orders of magnitude (from 0.1 to 200 mm roughly), can be considered in between the other two cases, with a visually shorter range of power-law behavior.

3.2 Fitting and testing power laws

A quantitative method can put more rigor into these observations. The idea is based on the recipe proposed by Clauset et al. (2009) – see also Corral et al. (2011) – but improved and generalized to our problem. Essentially, an objective procedure is required in order to find the optimum range in which a power law may hold. Taking again the event size for illustration, we report the power-law exponent fit between the values of $s_{\min}$ and $s_{\max}$ which yield the maximum number of data in that range but with a p value greater than 10 %. The method is described in Peters et al. (2010) and in more detail in Deluca and Corral (2013), but we summarize it in the following paragraphs.

For a given value of the pair $s_{\min}$ and $s_{\max}$, the maximum-likelihood (ML) power-law exponent is estimated for the events whose size lies in that range. This exponent yields a fit of the distribution, and the goodness of such a fit is evaluated by means of the Kolmogorov–Smirnov (KS) test (Press et al., 1992). The purpose is to get a p value, which is the probability that the KS test gives a distance between true power-law data and its fit larger than the distance obtained between the empirical data and its fit.

For instance, $p = 20\,\%$ would mean that truly power-law distributed data were closer than the empirical data to their respective fits in 80 % of the cases, but in the rest 20 % of the cases a true power law were at a larger distance than the empirical data. Therefore, in such a case the KS distance turns out to be somewhat large, but not large enough to reject that the data follow a power law with the ML exponent.

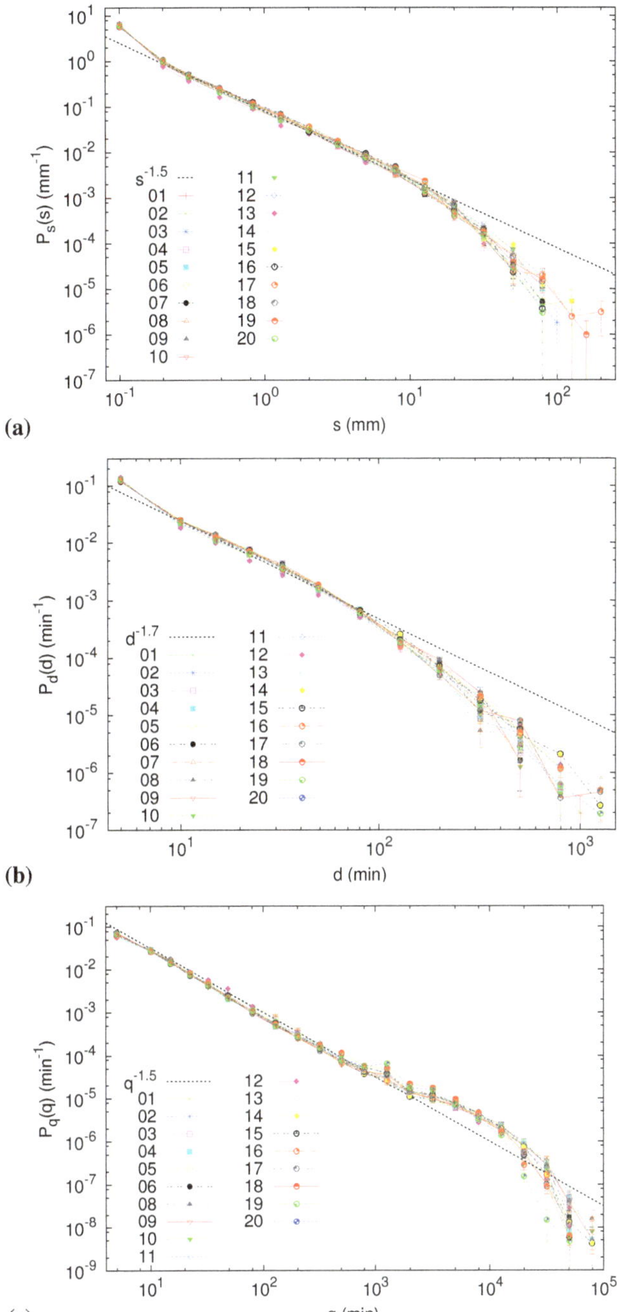

Fig. 2. Probability densities for all the sites for the whole time covered by the record of **(a)** event sizes, **(b)** event durations, and **(c)** dry spells.

As in the case in which some parameter is estimated from the data there is no closed formula to calculate the p value, we perform Monte Carlo simulations in order to compute the statistics of the Kolmogorov–Smirnov distance and from there the p value. In this way, for each $s_{\min}$ and $s_{\max}$ we get a number of data $\bar{N}_s$ in that range and, repeating the procedure many times, a p value. We look for the values of the extremes ($s_{\min}$ and $s_{\max}$) which maximize the number of data in between but with the restriction that the p value has to be greater than 10 % (this threshold is arbitrary, but the conclusions do not change if it is moved). The maximization is performed sweeping 100 values of $s_{\min}$ and 100 values of $s_{\max}$, in log-scale, in such a way that all possible ranges (within this log-resolution) are taken into account. We have to remark that, in contrast with Peters et al. (2010), we have considered always discrete probability distributions, both in the ML fit and in the simulations. Of course, it is a matter of discussion which approach (continuous or discrete) is more appropriate for discrete data that represent a continuous process. In any case, the differences in the outcomes are rather small. Notice also that the method is not based on the estimation of the probability densities shown in the previous subsections, what would be inherently more arbitrary (Clauset et al., 2009).

The results of this method are in agreement with the visual conclusions obtained in the previous subsection, as can be seen in Table 2. Starting with the size statistics, 13 out of the 20 sites yield reasonable results, with an exponent τ_s between 1.43 and 1.54 over a logarithmic range $s_{\max}/s_{\min}$ from 12 to more than 200. For the rest of the sites, the range is too short, less than one decade (a decade is understood from now as an order of magnitude). In the application of the algorithm, it has been necessary to restrict the value of $s_{\min}$ to be $s_{\min} \geq 0.2$ mm; otherwise, as the distributions have a concave shape (in log-scale) close to the origin (which means that there are many more events in that scale than at larger scales), the algorithm (which maximizes the number of data in a given range) prefers a short range with many data close to the origin than a larger range with less data away from the origin. It is possible that a variation of the algorithm in which the quantity that is maximized were different (for instance related with the range), would not need the restriction in the minimum size.

For the distribution of durations the resulting power laws turn out to be very limited in range; only 4 sites give not too short power laws, with $d_{\max}/d_{\min}$ from 6 to 12 and τ_d from 1.66 to 1.74. The other sites yield extremely short ranges for the power law be of any relevance. The situation is analogous to the case of the distribution of sizes, but the resulting ranges are much shorter here (Peters et al., 2010). Notice that the excess of events with $d = 5$ min, eliminated from the fits imposing $d_{\min} \geq 10$ min, has no counterpart in the value of the smallest rate (not shown), and therefore, we conclude that this extra number of events is due to problems in the time resolution of the data.

Considerably more satisfactory are the results for the dry spells. In all, 16 sites give consistent results, with τ_q from 1.45 to 1.55 in a range $q_{\max}/q_{\min}$ from 30 to almost 300. It is noticeable that in these cases $q_{\max}$ is always below 1 day. The removal by hand of dry spells around that value should enlarge a little the power-law range. In the rest of sites, either the range is comparatively too short (for example, for the Gaià site, the power-law behavior of $P_Q(q)$ is interrupted at around $q = 100$ min), or the algorithm has a tendency to include the bump the distributions show between the daily peak (q beyond 1000 min) and the tail. This makes the value of the exponent smaller (around 1.25). Nevertheless, the value of the exponent is much higher than the one obtained for the equivalent problem of earthquake waiting times, where the Omori law leads to values around one, or less. This points to a fundamental differences between both kind of processes (from a statistical point of view).

In summary, the power laws for the distributions of durations are too short to be relevant, and the fits for the sizes are in the limit of what is acceptable (some cases are clear and some other not). Only the distributions of dry spells give really good power laws, with $\tau_q = 1.50 \pm 0.05$, and for more than two decades in 6 sites.

4 Scaling

4.1 Non-parametric scaling

However, the fact that a power-law behavior does not exist over a broad range of values does not rule out the existence of scaling or SOC (Christensen and Moloney, 2005). In fact, the fulfillment of a power-law distribution in the form of Eq. (1) is only valid when finite-size effects are "small", which only happens for large enough systems. In general, when these effects are taken into account, scaling behavior leads to distributions of the form (Christensen and Moloney, 2005; Peters et al., 2010):

$$P_S(s) = s^{-\tau_s} \mathcal{G}_S(s/s_\xi) \text{ for } s > s_l, \tag{2}$$

where $\mathcal{G}_S(x)$ is a scaling function that is essentially constant for $x \ll 1$ and decays fast for $x \gg 1$, accounting in this way for the finite-size effects when s is above the crossover value, s_ξ; the size s_l is just a lower cutoff limiting the validity of this description. The pure power law only emerges for $s_\xi \to \infty$; nevertheless, a truncated power law holds over an appreciable range if the scales given by s_l and s_ξ are well separated (i.e., $s_l \ll s_\xi$). As s_ξ increases with system size, typically as $s_\xi \propto L^{D_s}$ (with D_s the so-called avalanche dimension, or event-size dimension), the power-law condition (1) can only be fulfilled for large enough system sizes.

Note that, in the case of a too short power-law range or a non-conclusive fit, we still could check the existence of scaling using Eq. (2) if we knew s_ξ or L. However, s_ξ is difficult to measure, needing a parameterization of the scaling function, and it is not clear what the system size L is for rainfall. It could be the vertical extension of the clouds, or the depth of the troposphere. Nevertheless, it is important to realize that the scaling ansatz (2) still can be checked from data without knowledge of L or s_ξ. First, notice that the ansatz implies that the k-order moment of s scales with L as

$$\langle s^k \rangle \propto L^{D_s(k+1-\tau_s)} \text{ for } 1 < \tau_s < k+1, \tag{3}$$

Table 2. Results of the power-law fitting and goodness-of-fit tests applied to event sizes, event durations, and dry-spell durations (in mm or in min), for the period of nine and a half years specified in the main text. The table displays the minimum of the fitting range, s_{min}, and the ratio between the maximum and the minimum of the fitting range (logarithmic range, $s_{\text{max}}/s_{\text{min}}$), total number of events, number of events in fitting range ($\bar{N}_s$, $\bar{N}_d$, and $\bar{N}_q$, for s, d, and q, respectively), and the power-law exponent with its uncertainty (one standard deviation) calculated as stated by Bauke (2007) and displayed between parenthesis as the variation of the last digit.

Site	s_{min}	$\frac{s_{\text{max}}}{s_{\text{min}}}$	N_s	$\bar{N}_s$	τ_s	d_{min}	$\frac{d_{\text{max}}}{d_{\text{min}}}$	$\bar{N}_d$	τ_d	q_{min}	$\frac{q_{\text{max}}}{q_{\text{min}}}$	N_q	$\bar{N}_q$	τ_q
1	0.2	180.5	5393	1886	1.54(2)	10	10.0	1668	1.67(4)	95	7.8	5387	743	1.75(7)
2	0.2	4.5	5236	1323	1.64(6)	10	4.0	1581	1.60(6)	5	273.0	5231	4729	1.46(1)
3	0.2	155.5	5749	2111	1.53(2)	10	6.0	1726	1.66(5)	10	80.0	5745	3207	1.53(2)
4	0.2	12.0	5108	1745	1.43(3)	10	3.5	1564	1.41(7)	5	196.0	5103	4520	1.47(1)
5	0.2	140.0	5289	2106	1.52(2)	10	3.5	1530	1.58(7)	20	47.3	5287	1706	1.45(2)
6	0.2	68.0	4924	1969	1.49(2)	10	3.5	1441	1.59(7)	20	31.3	4926	1537	1.47(3)
7	0.2	105.5	5219	2234	1.51(2)	10	3.5	1621	1.51(7)	5	256.0	5215	4734	1.51(1)
8	0.2	213.0	5112	2047	1.53(2)	10	4.0	1567	1.55(6)	15	65.0	5107	2098	1.50(2)
9	0.2	5.0	5366	1459	1.53(5)	10	4.0	1658	1.51(6)	10	90.0	5419	2889	1.55(2)
10	0.2	19.0	6691	2452	1.51(2)	10	3.5	2066	1.57(6)	25	33.0	6685	1758	1.48(3)
11	0.2	65.0	6224	2373	1.49(2)	10	5.0	1932	1.56(5)	45	468.3	6219	2005	1.24(1)
12	0.2	4.0	5967	1500	1.53(6)	10	4.0	1889	1.49(6)	5	235.0	5961	5376	1.47(1)
13	0.3	66.7	8330	1853	1.45(2)	10	12.5	2288	1.74(3)	130	158.5	8328	1501	1.27(2)
14	0.2	3.5	6525	1711	1.56(6)	10	5.0	2299	1.62(4)	5	215.0	6520	5906	1.47(1)
15	0.3	3.7	6485	1102	1.39(7)	10	5.0	2095	1.64(5)	15	49.7	6479	2560	1.51(2)
16	0.2	3.5	7491	1852	1.59(5)	10	4.0	2385	1.59(5)	5	214.0	7510	6789	1.50(1)
17	0.2	80.5	6962	2853	1.52(2)	10	3.5	2087	1.60(6)	10	68.5	6958	3719	1.52(1)
18	0.2	41.5	7511	2847	1.51(2)	10	3.5	2238	1.57(6)	20	31.0	7507	2302	1.53(2)
19	0.2	99.5	6767	2742	1.47(2)	10	3.5	1958	1.60(6)	50	21.7	6800	1378	1.26(3)
20	0.2	3.5	9012	2047	1.69(5)	10	8.5	2972	1.66(3)	15	34.3	9007	3367	1.50(2)

if $s_l \ll s_\xi$, see Christensen and Moloney (2005). Second, Eq. (2) can be written in a slightly different form, as a scaling law,

$$P_S(s) = L^{-D_s\tau_s}\mathcal{F}_S\left(s/L^{D_s}\right) \quad \text{for } s > s_l, \tag{4}$$

where the new scaling function $\mathcal{F}_S(x)$ is defined as $\mathcal{F}_S(x) \equiv x^{-\tau_s}\mathcal{G}_S(x/a)$ (a is the constant of proportionality between s_ξ and L^{D_s}). This form of $P_S(s)$ (in fact, $P_S(s,L)$), with an arbitrary $\mathcal{F}_S$, is the well-known scale-invariance condition for functions with two variables (Christensen and Moloney, 2005). Changes of scale (linear transformations) in s and L may leave the shape of the function $P_S(s,L)$ unchanged (this is what scale invariance really means, power laws are just a particular case in one dimension).

Substituting $L^{D_s} \propto \langle s^2\rangle/\langle s\rangle$ and $L^{D_s\tau_s} \propto L^{2D_s}/\langle s\rangle \propto \langle s^2\rangle^2/\langle s\rangle^3$ (from the scaling of $\langle s^k\rangle$, assuming $\tau_s < 2$) into Eq. (4) leads to

$$P_S(s) = \langle s\rangle^3\langle s^2\rangle^{-2}\tilde{\mathcal{F}}_S\left(s\langle s\rangle/\langle s^2\rangle\right), \tag{5}$$

where $\tilde{\mathcal{F}}_S(x)$ is essentially the scaling function $\mathcal{F}_S(x)$, absorbing the proportionality constants. Therefore, if scaling holds, a plot of $\langle s^2\rangle^2 P_S(s)/\langle s\rangle^3$ versus $s\langle s\rangle/\langle s^2\rangle$ for all the sites has to yield a collapse of the distributions into a single curve, which draws $\tilde{\mathcal{F}}_S(x)$ (a similar procedure is outlined in Rosso et al., 2009). In order to proceed, the mean and the quadratic mean, $\langle s\rangle$ and $\langle s^2\rangle$, can be easily estimated from data. Since no estimation of parameters is involved for this procedure, we call it non-parametric scaling.

The outcome for $P_S(s)$, $P_D(d)$, and $P_Q(q)$ is shown in Fig. 3a, b, and c, with reasonable results, especially for the distribution of dry spells. The plot suggests that the scaling function $\mathcal{G}_Q$ of the dry-spell distribution has a maximum around $x \simeq 1$, but this is not in disagreement with our approach, which only assumed a constant scaling function for small x and a fast decay for large x.

Note that the quotient $\langle s^2\rangle/\langle s\rangle$ gives the scale for the crossover value s_ξ (as $s_\xi \propto \langle s^2\rangle/\langle s\rangle$, with a constant of proportionality that depends on the scaling function $\mathcal{G}_S$ and on s_l/s_ξ), and therefore it is the ratio of the second moment to the mean and not the mean which describes the scaling behavior of the distribution. This can have important implications for extreme events: an increase in the value of the mean is not proportional to an increase of the most extreme events, represented by s_ξ. For the case of event sizes, we get values of $\langle s^2\rangle/\langle s\rangle$ between 10 and 30 mm (which is a variability much higher than that of $\langle s\rangle$), and therefore the condition $s_l \ll s_\xi$ is very well fulfilled (assuming that the moment ratio $\langle s^2\rangle/\langle s\rangle$ is of the same order as s_ξ, and with $s_l \simeq s_{\text{min}}$), which is a test for the consistency of our approach. For dry spells $\langle q^2\rangle/\langle q\rangle$ is between 5 and 13 days, which is even better for the applicability of the scaling analysis. The case of the event durations is somewhat "critical", with $\langle d^2\rangle/\langle d\rangle$ between 70

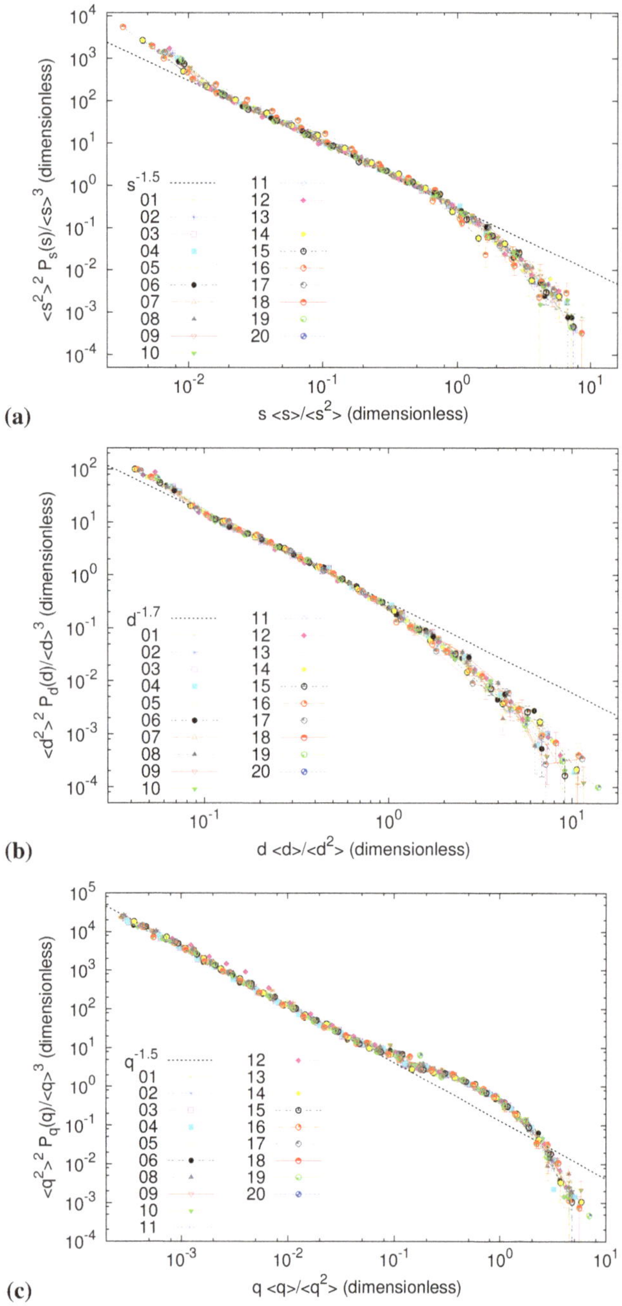

Fig. 3. Collapse of the probability densities for all the sites for the whole time covered by the record of: **(a)** event sizes, **(b)** event durations, and **(c)** dry spells. Rescaling is performed using the first and second moment of each distribution, following Eq. (5).

and 120 min, which yields d_ξ/d_1 in the range from 14 to 24. Nevertheless, we observe that the condition $s_1 \ll s_\xi$ for the power law to show up is stronger than the same condition for the scaling analysis to be valid.

4.2 Parametric scaling

Further, a scaling ansatz as Eq. (2) or (4) allows an estimation of the exponent τ_s, even in the case in which a power law cannot be fit to the data. From the scaling of the moments of s we get, taking $k=1$, $L^{D_s} \propto \langle s \rangle^{1/(2-\tau_s)}$ and $L^{D_s\tau_s} \propto \langle s \rangle^{\tau_s/(2-\tau_s)}$ (again with $\tau_s < 2$); therefore, substituting into Eq. (4),

$$P_S(s) = \langle s \rangle^{-\tau_s/(2-\tau_s)} \hat{\mathcal{F}}_S \left(s/\langle s \rangle^{1/(2-\tau_s)} \right). \tag{6}$$

One only needs to find the value of τ_s that optimizes the collapse of all the distributions, that is, that makes the previous equation valid, or at least as close to validity as possible. As the scaling depends on the parameter τ_s, we refer to this procedure as parametric scaling.

We therefore need a measurement to quantify distance between rescaled distributions. In order to do that, we have chosen to work with the cumulative distribution function, $R_S(s) \equiv \int_s^\infty P_S(s')ds'$, rather than with the density (to be rigorous, $R_S(s)$ is the complementary of the cumulative distribution function, and is called survivor function or reliability function in some contexts). Although in practice both $P_S(s)$ and $R_S(s)$ contain the same probabilistic information, the reason to work with $R_S(s)$ is double: the estimation of the cumulative distribution function does not depend of an arbitrarily selected bin width ds (Press et al., 1992), and it does not give equal weight to all scales in the representation of the function (i.e., in the number of points that constitute the function). The scaling laws (4) and (6) turn out to be, therefore,

$$R_S(s) = L^{-D_s(\tau_s-1)} \mathcal{H}_S \left(s/L^{D_s} \right) \tag{7}$$

$$R_S(s) = \langle s \rangle^{-(\tau_s-1)/(2-\tau_s)} \hat{\mathcal{H}}_S \left(s/\langle s \rangle^{1/(2-\tau_s)} \right), \tag{8}$$

where $\mathcal{H}_S(x)$ and $\hat{\mathcal{H}}_S(x)$ are the corresponding scaling functions.

The first step of the method of collapse is to merge all the pairs $\{s, R_S(s)\}_i$ into a unique rescaled function $\{x, y\}$. If $i = 1, \ldots, 20$ runs for all sites, and $j = 1, \ldots, M_s(i)$ for all the different values that the size of events takes on site i (note that $M_s(i) \le N_s(i)$),

$$x_\ell(\tau) \equiv \log \left(s_{ji}/\langle s \rangle_i^{1/(2-\tau)} \right),$$

$$y_\ell(\tau) \equiv \log \left(R_i \left(s_{ji} \right) \langle s \rangle_i^{(\tau-1)/(2-\tau)} \right),$$

where s_{ji} is the jth value of the size in site i, $\langle s \rangle_i$ the mean on s in i, $R_i(s_{ji})$ the cumulative distribution function in i, and τ a possible value of the exponent τ_s. The index ℓ labels the new function, from 1 to $\sum_{\forall i} M_s(i)$, in such a way that $x_\ell(\tau) \le x_{\ell+1}(\tau)$; this means that the pairs $x_\ell(\tau), y_\ell(\tau)$ are sorted by increasing x.

Then, we just compute

$$D(\tau) \equiv \sum_{\forall \ell} \left(\left[x_\ell(\tau) - x_{\ell+1}(\tau) \right]^2 + \left[y_\ell(\tau) - y_{\ell+1}(\tau) \right]^2 \right), \tag{9}$$

which represents the sum of all Euclidean distances between the neighboring points in a (tentative) collapse plot in logarithmic scale. The value of τ which minimizes this function is identified with the exponent τ_s in Eq. (2). We have tested the algorithm applying it to SOC models whose exponents are well known (not shown).

The results of this method applied to our data sets, not only for the size distributions but also to the distributions of d, are highly satisfactory. There is only one requirement: the removal of the first point in each distribution ($s = 0.1$ mm and $d = 5$ min), as with the ML fits. The exponents we find are $\tau_s = 1.52 \pm 0.12$ and $\tau_d = 1.69 \pm 0.01$, in agreement with the ones obtained by the power-law fitting method presented above; the corresponding rescaled plots are shown in Fig. 4. Although the visual display does not allow to evaluate properly the quality of the collapse, the reduction in the value of the function $D(\tau)$ is notable. Then, the performance of the method is noteworthy, taking into account that the mean values of the distributions show little variation in most cases. In addition, the shape of the scaling function $\mathcal{G}_S$ can be obtained by plotting, as suggested by Eq. (2), $s^{\tau_s} P_S(s)$ versus $s/\langle s\rangle^{1/(2-\tau_s)}$, and the same for the other variable, d. Figure 5 displays what is obtained for each distribution. In contrast, the application of this method to $P_Q(q)$ does not yield consistent results, as τ_q turns out to be rather small (1.24). Notice that the existence of a daily peak in the distributions is an obstacle to a data collapse, as the peak prevents a good scaling.

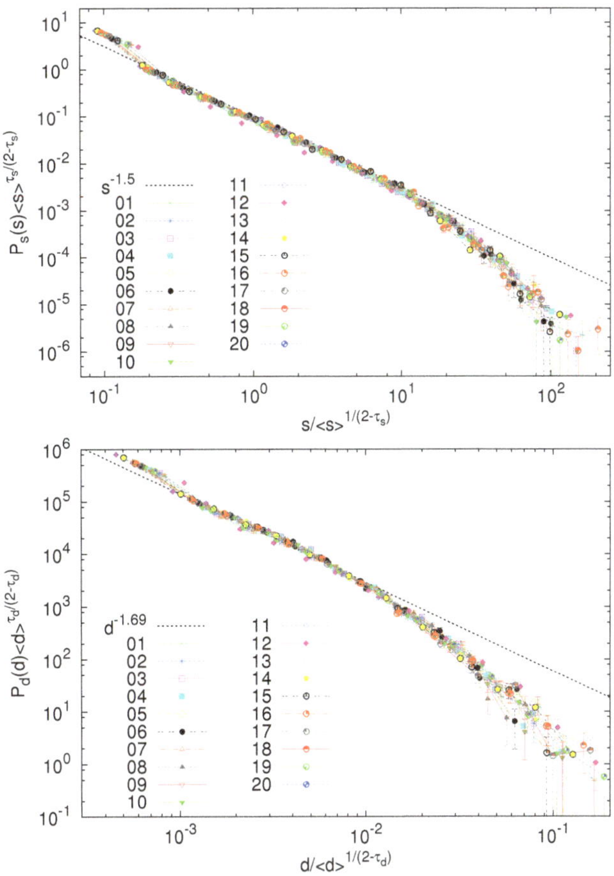

Fig. 4. Collapse of the probability densities for all the sites for the whole time covered by the record of **(a)** event sizes and **(b)** event durations; rescaled using Eq. (6) with the exponents: $\tau_s = 1.52$ and $\tau_d = 1.69$, determined minimizing the Euclidean distance between parametrically collapsed distributions. Units are mm or min to the corresponding powers appearing in the axes.

5 Discussion and conclusions

We have performed an in-depth study of the properties of rain-event observables in rainfall in the Mediterranean region in order to check if power-law distributions and scale invariance can be useful for modeling rain events and dry spells. The results support this hypothesis, which had not been checked before in this region or for this kind of data resolution. For the distributions of rain-event sizes, we get power-law exponents valid for one or two decades in the majority of sites, with exponent values $\tau_s \simeq 1.50 \pm 0.05$. For the distributions of event durations, the fitting ranges are shorter, reaching in the best case one decade, with exponents $\tau_d \simeq 1.70 \pm 0.05$. This range is expected to be shorter than for event sizes, given that these combine the event duration distribution with the rain rate (Peters et al., 2010). And finally, the dry spell distributions yield the more notable power-law fits, with exponents in the range $\tau_q \simeq 1.50 \pm 0.05$, in some cases for more than 2 decades.

These results are compatible with the ones obtained for the Baltic sea by Peters et al. (2002), which yielded $\tau_s \simeq \tau_q \simeq 1.4$ and $\tau_d \simeq 1.6$. The agreement is remarkable, taking into account the different nature of the data analyzed and the disparate fitting procedures. However, the concordance with the more recent results of Peters et al. (2010) is not very good, quantitatively. That previous study, with a minimum detection rate of $0.2\,\mathrm{mm\,h^{-1}}$ and a time resolution $\Delta t = 1$ min, found $\tau_s \simeq 1.17$ for several sites across different climates, using essentially the same statistical techniques as in the present study. For dry-spell durations the exponents were close in many cases to $\tau_q \simeq 1.3$. The difference between the value of the exponents may be due to data resolution (but then, our agreement with the values in the first paper may be fortuitous). Changes in the detection threshold have a nontrivial repercussion in the size and duration of the events and the dry spells (an increase in the threshold can split one single event into two or more separate ones but also can remove events). Further, better time resolution and lower detection threshold allow the detection of smaller events, enlarging the power-law range and reducing the weight from the part close to the crossover point (where the distribution becomes steeper); this trivially leads to smaller values of the exponent. In the dry-spell case, the power-law range in this study is enough to guarantee that our estimation of the exponents are robust, so the discrepancy with Peters et al. (2010) may

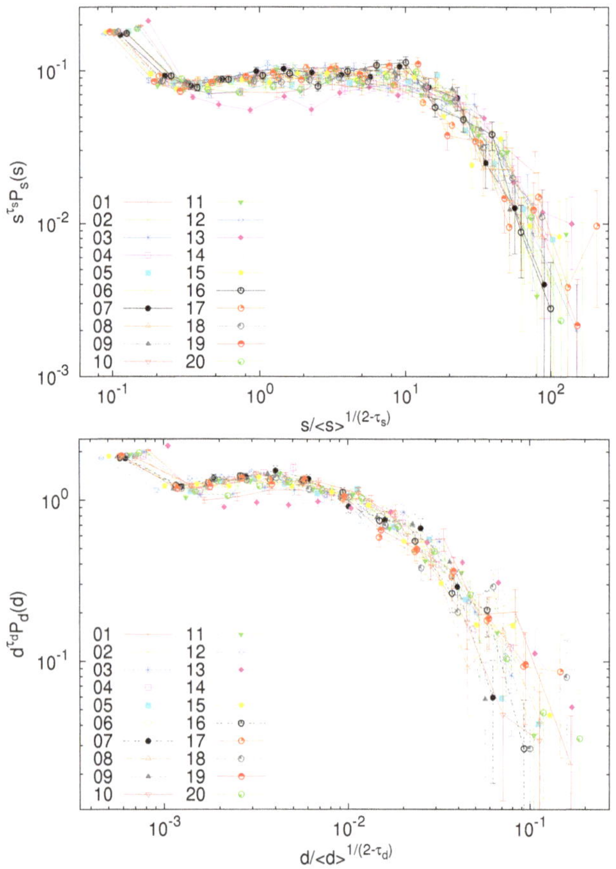

Fig. 5. Inferred scaling functions $\mathcal{G}_S$ and $\mathcal{G}_D$ (except for a proportionality factor in the x axis) corresponding to the rescaled distributions of s and d in Fig. 4, multiplied by s^{τ_s} and d^{τ_d}. Units in the abscissae are as in the previous plot, whereas in the ordinates these are mm^{τ_s-1} and min^{τ_d-1}.

be due to the non-trivial effect of the change in the detection thresholds or differences on the measurement devices.

On the other hand, finite size effects can explain the limited power-law range obtained for s and d observables, as it occurs in (self-organized and non-self-organized) critical phenomena. The finite-size analysis performed, in terms of combinations of powers of the moments of the distributions, supports this conclusion. The collapse of the distributions is a clear signature of scale invariance: different sites share a common shape of the rain-event and dry-spell distributions, with differences in the scale of those distributions, depending on system size. Then, in the ideal case of an infinite system, the power laws would lack an upper cutoff. Moreover, the collapse of the distributions allows an independent estimate of the power-law exponents, which, for event durations and sizes, are in surprising agreement with the values obtained by the maximum-likelihood fit. For dry spell durations, a daily peak in the distributions hinders their collapse.

Nevertheless, future work should consider spatially extended events. Our measurements are taken in a point of the surface which reflects information on the vertical scale, then, the results could be affected by this. Another related but different issue are the correlations between the sites, which have not been taken into account. Further, in order to establish a link with SOC or not, one would need to measure, simultaneously, other variables, such as water vapor content (Peters and Neelin, 2006). Our results show that these studies can be undertaken without high-resolution rain data.

The implications of the results for hazard assessment are also of great interest: if there is not a characteristic rain-event size, then there is neither a definite separation nor a fundamental difference between the smallest harmless rains and the most hazardous storms. Further, it is generally believed that the critical evolution of events in SOC systems implies that, at a given instant, it is equally likely that the event intensifies or weakens, which would make detailed prediction unattainable. However, this view has been recently proved wrong, as it has been reported that a critical evolution describes the dynamics of some SOC systems only on average; further, the existence of finite size effects can be used for prediction (Garber et al., 2009; Martin et al., 2010). Interestingly, in the case of rain, it has been recently shown by Molini et al. (2011) that knowledge of internal variables of the system allows some degree of prediction for the duration of the events, related also to the departure of the system from quasi-equilibrium conditions.

Finally, we urge studies which explore the effects of resolution and detection-threshold value in high-resolution rain data. A common misbelief in the SOC approach is that avalanches happen following a memoryless process, leading therefore to exponential distributions for the waiting times (Corral, 2005). This has been refuted if a threshold on the intensity is present (Paczuski et al., 2005). In this case, times between avalanches follow a power-law distribution, as we find for dry spells.

In summary, we conclude that the statistics of rainfall events in the NW Mediterranean area studied are in agreement with the SOC-paradigm expectations but also with any other mechanism for power-law genesis. This is the first time this study is realized for this region and it is a confirmation of what has been found for other places of the world, but in our case using data with lower resolution.

Acknowledgements. This work would not have been possible without the collaboration of the Agència Catalana de l'Aigua (ACA), which generously shared its rain data with us. We have benefited a lot from a long-term collaboration with Ole Peters, and are grateful also to R. Romero and A. Rosso, for addressing us towards the ACA data and an important reference (Rosso et al., 2009), and to J. E. Llebot for providing support and encouragement. We also thank Joaquim Farguell from ACA. A. Mugnai and other people of the Plinius conferences made us realize of the importance of our results for the non-linear geophysics community. A. Deluca enjoyed a Ph.D. grant of the Centre de Recerca Matemàtica. Spanish grants related to this work are FIS2009-09508, 2009SGR-164, and FIS2012-31324.

Edited by: D. Schertzer

References

ACA (Agència Catalana de l'Aigua/Catalan Water Agency), available at: http://aca-web.gencat.cat/aca (last access: April 2014), 2014.

Andrade, R. F. S., Schellnhuber, H. J., and Claussen, M.: Analysis of rainfall records: possible relation to self-organized criticality, Physica A, 254, 557–568, 1998.

Bak, P.: How Nature Works: The Science of Self-Organized Criticality, Copernicus, New York, 1996.

Bauke, H.: Parameter estimation for power-law distributions by maximum likelihood methods, Eur. Phys. J. B, 58, 167–173, 2007.

Bodenschatz, E., Malinowski, S., Shaw, R., and Stratmann, F.: Can we understand clouds without turbulence?, Science, 327, 970–971, 2010.

Cahalan, R. and Joseph, J.: Fractal statistics of cloud fields, Mon. Weather Rev., 117, 261–272, 1989.

Christensen, K. and Moloney, N. R.: Complexity and Criticality, Imperial College Press, London, 2005.

Clauset, A., Shalizi, C. R., and Newman, M. E. J.: Power-law distributions in empirical data, SIAM Rev., 51, 661–703, 2009.

Corral, Á.: Comment on "Do Earthquakes Exhibit Self-Organized Criticality?", Phys. Rev. Lett., 95, 159801, doi:10.1103/PhysRevLett.95.159801,2005.

Corral, A., Font, F., and Camacho, J.: Noncharacteristic half-lives in radioactive decay, Phys. Rev. E, 83, 066103, doi:10.1103/PhysRevE.83.066103, 2011.

Deidda, R.: Rainfall downscaling in a space-time multifractal framework, Water Resour. Res., 36, 1779–1794, doi:10.1029/2000WR900038, 2000.

Deluca, A. and Corral, A.: Fitting and goodness-of-fit test of non-truncated and truncated power-law distributions, Acta Geophys., 61, 1351–1394, 2013.

Dickman, R.: Rain, Power Laws, and Advection, Phys. Rev. Lett., 90, 108701, doi:10.1103/PhysRevLett.90.108701, 2003.

Dickman, R., Vespignani, A., and Zapperi, S.: Self-organized criticality as an absorbing-state phase transition, Phys. Rev. E, 57, 5095–5105, 1998.

Dickman, R., Muñoz, M. A., Vespignani, A., and Zapperi, S.: Paths to Self-Organized Criticality, Braz. J. Phys., 30, 27–41, 2000.

Garber, A., Hallerberg, S., and Kantz, H.: Predicting extreme avalanches in self-organized critical sandpiles, Phys. Rev. E, 80, 026124, doi:10.1103/PhysRevE.80.026124, 2009.

Hergarten, S.: Self-Organized Criticality in Earth Systems, Springer, Berlin, 2002.

Hooge, C., Lovejoy, S., Schertzer, D., Pecknold, S., Malouin, J.-F., and Schmitt, F.: Mulifractal phase transitions: the origin of self-organized criticality in earthquakes, Nonlin. Processes Geophys., 1, 191–197, doi:10.5194/npg-1-191-1994, 1994.

Jensen, H. J.: Self-Organized Criticality, Cambridge University Press, Cambridge, 1998.

Lavergnat, J. and Golé, P.: A stochastic model of raindrop release: Application to the simulation of point rain observations, J. Hydrol., 328, 8–19, 2006.

Lovejoy, S.: Area-perimeter relation for rain and cloud areas, Science, 216, 185–187, 1982.

Lovejoy, S. and Schertzer, D.: Multifractals and rain, in: New Uncertainty Concepts in Hydrology and Water Resources, edited by: Kundzewicz, Z. W., Cambridge University Press, 61–103, 1995.

Lovejoy, S. and Schertzer, D.: The weather and climate, Cambridge University Press, Cambridge, 2013.

Malamud, B. D.: Tails of natural hazards, Phys. World, 17, 31–35, 2004.

Malamud, B. D., Morein, G., and Turcotte, D. L.: Forest Fires: An Example of Self-Organized Critical Behavior, Science, 281, 1840–1842, 1998.

Martin, E., Shreim, A., and Paczuski, M.: Activity-dependent branching ratios in stocks, solar x-ray flux, and the Bak-Tang-Wiesenfeld sandpile model, Phys. Rev. E, 81, 016109, doi:10.1103/PhysRevE.81.016109, 2010.

Molini, L., Parodi, A., Rebora, N., and Craig, G. C.: Classifying severe rainfall events over Italy by hydrometeorological and dynamical criteria, Q. J. Roy. Meteor. Soc., 137, 148–154, 2011.

Olsson, J., Niemczynowicz, J., and Berndtsson, R.: Fractal analysis of high-resolution rainfall time series, J. Geophys. Res., 98, 23265–23274, 1993.

OTT (OTT Hydromet GmbH), available at: http://www.ott-hydrometry.de (last access: April 2014), 2014.

Paczuski, M., Boettcher, S., and Baiesi, M.: Interoccurrence times in the Bak-Tang-Wiesenfeld sandpile model: A comparison with the observed statistics of solar flares, Phys. Rev. Lett., 95, 181102, doi:10.1103/PhysRevLett.95.181102, 2005.

Peters, O. and Christensen, K.: Rain: Relaxations in the sky, Phys. Rev. E, 66, 036120, doi:10.1103/PhysRevE.66.036120, 2002.

Peters, O. and Christensen, K.: Rain viewed as relaxational events, J. Hydrol., 328, 46–55, 2006.

Peters, O. and Neelin, J. D.: Critical phenomena in atmospheric precipitation, Nat. Phys., 2, 393–396, 2006.

Peters, O., Hertlein, C., and Christensen, K.: A Complexity View of Rainfall, Phys. Rev. Lett., 88, 018701, doi:10.1103/PhysRevLett.88.018701, 2002.

Peters, O., Neelin, J., and Nesbitt, S.: Mesoscale convective systems and critical clusters, J. Atmos. Sci., 66, 2913–2924, 2009.

Peters, O., Deluca, A., Corral, A., Neelin, J. D., and Holloway, C. E.: Universality of rain event size distributions, J. Stat. Mech.-Theory E., P11030, doi:10.1088/1742-5468/2010/11/P11030, 2010.

Press, W. H., Teukolsky, S. A., Vetterling, W. T., and Flannery, B. P.: Numerical Recipes in FORTRAN, 2nd Edn., Cambridge University Press, Cambridge, 1992.

Rosso, A., Le Doussal, P., and Wiese, K. J.: Avalanche-size distribution at the depinning transition: A numerical test of the theory, Phys. Rev. B, 80, 144204, doi:10.1103/PhysRevB.80.144204, 2009.

Schertzer, D. and Lovejoy, S.: Hard and soft multifractal processes, Physica A, 185, 187–194, 1992.

Schertzer, D. and Lovejoy, S.: Multifractal generation of self-organized criticality, in: Proceedings of the Second IFIP Working Conference on Fractals in the Natural and Applied Sciences, edited by: Novak, M. M., Elsevier, North-Holland: Amsterdam, 325–339, 1994.

Schertzer, D. and Lovejoy, S.: Multifractals, generalized scale invariance and complexity in geophysics, Int. J. Bifurcat. Chaos, 21, 3417–3456, 2011.

Sornette, A. and Sornette, D.: Self-organized Criticality and Earthquakes, Europhys. Lett., 9, 197–202, 1989.

Sornette, D.: Critical Phenomena in Natural Sciences, 2nd Edn., Springer, Berlin, 2004.

Tang, C. and Bak, P.: Critical Exponents and Scaling Relations for Self-Organized Critical Phenomena, Phys. Rev. Lett., 60, 2347–2350, 1988.

Telesca, L., Lapenna, V., Scalcione, E., and Summa, D.: Searching for time-scaling features in rainfall sequences, Chaos Soliton. Fract., 32, 35–41, 2007.

Veneziano, D. and Lepore, C.: The scaling of temporal rainfall, Water Resour. Res., 48, W08516, doi:10.1029/2012WR012105, 2012.

von Mises, R.: Mathematical Theory of Probability and Statistics, Academic Press, New York, 1964.

Wood, R. and Field, P.: The distribution of cloud horizontal sizes, J. Climate, 24, 4800–4816, 2011.

Appendix A

Details on the estimation of the probability density

In practice, the estimation of the density from data is performed taking a value of ds large enough to guarantee statistical significance, and then compute $P_S(s)$ as $n(s)/(N_s\Delta)$, where $n(s)$ is the number of events with size in the range between s and $s+ds$, N_s the total number of events, and Δ is defined as

$$\Delta = \mathcal{R}_s\left(\lfloor (s+ds)/\mathcal{R}_s \rfloor - \lfloor s/\mathcal{R}_s \rfloor\right),$$

where $\lfloor x \rfloor$ is the integer part of x and $\mathcal{R}_s$ the resolution of s, that is, $\mathcal{R}_s = 0.1$ mm (but note that high resolution means low $\mathcal{R}_s$). Therefore, $\Delta/\mathcal{R}_s$ is the number of possible different values of the variable in the interval considered. Notice that using Δ instead of ds in the denominator of the estimation of $P_S(s)$ allows one to take into account the discreteness of s. If $\mathcal{R}_s$ tended to zero, then $\Delta \to ds$ and the discreteness effects would become irrelevant.

How large does ds have to be to guarantee the statistical significance of the estimation of $P_S(s)$? Working with long-tailed distributions (where the variable covers a broad range of scales) a very useful procedure is to take a width of the interval ds that is not the same for all s, but that is proportional to the scale, as $[s, s+ds) = [s_o, bs_o)$, $[bs_o, b^2 s_o), \ldots, [b^k s_o, b^{k+1} s_o)$, that is, $ds = (b-1)s$ (with $b > 1$). Given a value of s, the corresponding value of k that associates s with its bin is given by $k = \lfloor \log_b(s/s_o) \rfloor$. Correspondingly, the optimum choice to assign a point to the interval $[s, s+ds)$ is given by the value $\sqrt{b}s$. This procedure is referred to as logarithmic binning, because the intervals appear with fixed width in logarithmic scale (Hergarten, 2002). In this paper we have generally taken $b \simeq 1.58$, in such a way that $b^5 = 10$, providing 5 bins per order of magnitude.

As the distributions are estimated from a finite number of data, they display statistical fluctuations. The uncertainty characterizing these fluctuations is simply related to the density by

$$\frac{\sigma_{P_S}(s)}{P_S(s)} \simeq \frac{1}{\sqrt{n(s)}}, \qquad \text{(A1)}$$

where $\sigma_{P_S}(s)$ is the standard deviation of $P_S(s)$ (do not confound with the standard deviation of s). This is so because $n(s)$ can be considered a binomial variable (von Mises, 1964), and then, the ratio between its standard deviation and mean fulfills $\sigma_n(s)/\langle n(s)\rangle \simeq 1/\sqrt{n(s)}$, with $n(s)/N_s \ll 1$. As $P_S(s)$ is proportional to $n(s)$, the same relation holds for its relative uncertainty.

7

Controlling balance in an ensemble Kalman filter

G. A. Gottwald

School of Mathematics and Statistics, University of Sydney, Sydney, NSW 2006, Australia

Correspondence to: G. A. Gottwald (georg.gottwald@sydney.edu.au)

Abstract. We present a method to control unbalanced fast dynamics in an ensemble Kalman filter by introducing a weak constraint on the imbalance in a spatially sparse observational network. We show that the balance constraint produces significantly more balanced analyses than ensemble Kalman filters without balance constraints and than filters implementing incremental analysis updates (IAU). Furthermore, our filter with the weak constraint on imbalance produces good rms error statistics which outperform those of ensemble Kalman filters without balance constraints for the fast fields.

1 Introduction

In data assimilation one seeks to find the best estimation of the state of a dynamical system given a forecast model with possible model error and noisy observations at discrete observation intervals (Kalnay, 2002). This estimate is coined the *analysis*. This procedure, however, does not necessarily produce dynamically consistent analyses. In particular, the analysis may contain unbalanced gravity waves, which are absent in the true atmospheric state and which may spoil the subsequent forecast initialised with these dynamically inconsistent states. Ever since the early days of numerical weather prediction the creation of imbalance has been central to the problem of producing reliable forecasts (see for example Daley, 1993, Chapter 6, and Lynch, 2006 for a historical account). The heuristic reasoning behind the occurrence of unbalanced analyses is that there may be several states of the fast variables which are compatible with the observations of the slow state variables, most of them corresponding to unbalanced states. Furthermore, unbalanced states can be generated by the discontinuous nature of the data assimilation procedure, leading to unphysical readjustment processes of analyses by the subsequent nonlinear forecast model (Bloom et al., 1996; Ourmières et al., 2006). Examples of the creation of imbalance in variational data assimilation schemes are, for example, Bloom et al. (1996) and Lorenc (2003b). In the context of ensemble filters, unbalanced analyses are further created by the procedure of localisation which was introduced by Houtekamer and Mitchell (1998, 2001), Hamill et al. (2001), Ott et al. (2004) and Szunyogh et al. (2005) to mitigate spurious cross-correlations in the covariance matrices due to finite ensemble sizes. Localisation of any type can potentially cause imbalance in the initial conditions (Cohn et al., 1998; Lorenc, 2003a; Mitchell et al., 2002; Houtekamer and Mitchell, 2005; Oke et al., 2007; Kepert, 2009; Greybush et al., 2011).

There exist several strategies to combat undesired unbalanced analyses. These strategies can be divided into those which employ a re-balancing procedure after the data assimilation, and those which try to create balanced analyses within the data assimilation process itself. Post-processing methods include digital filtering (Lynch and Huang, 1992) and normal mode initialisation (Machenhauer, 1977; Baer and Tribbia, 1977). Within variational data assimilation algorithms balance constraints can be implemented to ensure sufficient balance (Thépaut and Courtier, 1991; Polavarapu et al., 2000; Gauthier and Thépaut, 2001; Neef et al., 2006; Watkinson et al., 2007; Cotter, 2013). To militate against the effects of intermittent discontinuous assimilations, several filtering approaches have been introduced to render the assimilation procedure more continuous. Bloom et al. (1996) introduced the method of incremental analysis updates (IAU) for 3D-Var in which the analyses increments are distributed over a fixed time window. It has since been applied to ensemble filters, see for example Polavarapu et al. (2004), and has found numerous applications in atmospheric and oceanic contact (Zhu et al., 2003; Weaver et al., 2003; Ourmières et al., 2006). Bergemann and Reich (2010a) create balanced analyses by using a continuous formulation of the Kalman analysis step (Bergemann et al., 2009; Bergemann and Reich, 2010b). Kepert (2009) modified the covariance localisation procedure so that it respects balance.

Here we will present a novel approach to generating balanced analyses within an intermittent discontinuous data assimilation procedure. We will incorporate prior information on the amount of imbalance to augment given observational information for the slow variables. This implementation of a balance constraint within the data assimilation step eliminates unwanted spurious imbalance, leading to physical analyses states and to an improved analysis skill as measured by the rms error of the fast variables.

In the next section we briefly describe the framework of variance limiting Kalman filters developed in Gottwald et al. (2011) which will form the basis of our imbalance limiting filter. In Sect. 3 we present a modified slow-fast Lorenz-96 model which incorporates balanced dynamics, introduced in Bergemann and Reich (2010a). In Sect. 4 we present results showing how controlling unbalance can produce better skill than current ensemble Kalman filters. We conclude with a summary in Sect. 5.

2 The variance limiting Kalman filter

Gottwald et al. (2011) introduced a variation of the ensemble Kalman filter, coined variance limiting Kalman filter (VLKF). This filter was designed to control overestimation of error covariances caused by finite ensemble sizes in sparse observational grids. The filter imposes weak constraints on unobserved variables and data voids using climatological information. The effect of the weak constraint was shown to drive the analysis of the unobserved variables towards their climatic mean and furthermore to limit the posterior error covariance of the unobserved variables to not exceed their climatic covariance. This yielded a remarkable increase in the skill, even in the observed variables. The filter has since been used in Mitchell and Gottwald (2012) to control noise at the grid resolution scale caused by model error.

It is our aim here to employ the VLKF to control undesirable imbalance. In general the instantaneous amount of imbalance is not available through direct observations. We assume prior knowledge of the climatological mean and of the climatological covariance of imbalance. This statistical information may be available through historical observational data or through free running simulations. We will use the weak constraint in VLKF on imbalance to drive the analysis towards balance, inhibiting excessive unphysical unbalanced fast energy.

The filter described in Gottwald et al. (2011) and Mitchell and Gottwald (2012) was formulated for large ensemble sizes, ensuring invertibility of the forecast error covariance (a situation not satisfied for data assimilation in operational numerical weather forecast centres). We recast the VLKF here in a form which allows for small ensemble sizes, and redo the derivation in a slightly different manner.

Given a D-dimensional dynamical system

$$\dot{z}_t = \mathbf{F}(z_t), \tag{1}$$

which is observed at discrete times $t_i = i\,\Delta t_{\mathrm{obs}}$, data assimilation aims at producing the best estimate of the current state given a typically chaotic, possibly inaccurate model $\dot{z} = f(z)$ and noisy observations of the true state z_t (Kalnay, 2002). We assume that we are given observations

$$\boldsymbol{y}_{\mathrm{o}}(t_i) = \mathbf{H} z_t(t_i) + \boldsymbol{r}_{\mathrm{o}},$$

where the observation operator $\mathbf{H} : \mathbb{R}^D \to \mathbb{R}^{D_{\mathrm{o}}}$ maps from the whole space into observation space, and $\boldsymbol{r}_{\mathrm{o}} \in \mathbb{R}^{D_{\mathrm{o}}}$ is assumed to be i.i.d. observational Gaussian noise with associated error covariance matrix $\mathbf{R}_{\mathrm{o}}$. Additionally we incorporate climatological information of D_{w} pseudo-observables, in particular their mean $\boldsymbol{a}_{\mathrm{clim}} \in \mathbb{R}^{D_{\mathrm{w}}}$ and their covariances $\mathbf{A}_{\mathrm{clim}} \in \mathbb{R}^{D_{\mathrm{w}} \times D_{\mathrm{w}}}$. In general, it is not advisable to incorporate simultaneously direct observations and climatological information for a variable, as this may spoil the generally more accurate information of the direct observations. In our application here the climatological information will be the mean and covariances of some measure of imbalance, but pseudo-observables may be any subset of unobserved variables or their integrated quantities such as their energy. We assume that we can determine those quantities prior to the data assimilation procedure either through historical data or through long-time numerical simulations. We remark that one may use values other than the climatic covariance to control the analysis error covariance if one interprets the variance constraint merely as a numerical tool to stabilise and regularise the filter. Furthermore, in non-equilibrium situations, when climatological information is irrelevant, such as during strong fronto-genesis in a weather forecasting context for example, we may estimate the mean and the covariance of the unobserved pseudo-observables via a running average of the analysis (this requires the analysis to be tracking).

We introduce a pseudo-observation operator $\mathbf{h} : \mathbb{R}^D \to \mathbb{R}^{D_{\mathrm{w}}}$ which maps from the whole space into the space of the pseudo-observables. The (as yet unknown) error covariance of those pseudo-observations is denoted by $\mathbf{R}_{\mathrm{w}}$. Gottwald et al. (2011) considered $D_{\mathrm{o}} + D_{\mathrm{w}} = D$, which will be relaxed here.

The Kalman filter can be formulated as a minimisation problem of the following cost function (e.g. Kalnay, 2002; Simon, 2006) with a given background z_{f} and associated error forecast covariance $\mathbf{P}_{\mathrm{f}}$ as

$$\begin{aligned} J(z) = {} & \frac{1}{2}(z - z_{\mathrm{f}})^T \mathbf{P}_{\mathrm{f}}^{-1} (z - z_{\mathrm{f}}) \\ & + \frac{1}{2}(\mathbf{H}z - \boldsymbol{y}_{\mathrm{o}})^T \mathbf{R}_{\mathrm{o}}^{-1} (\mathbf{H}z - \boldsymbol{y}_{\mathrm{o}}) \\ & + \frac{1}{2}(\mathbf{h}z - \boldsymbol{a}_{\mathrm{clim}})^T \mathbf{R}_{\mathrm{w}}^{-1} (\mathbf{h}z - \boldsymbol{a}_{\mathrm{clim}}) . \end{aligned} \tag{2}$$

The error covariance matrix $\mathbf{R}_w$ is so far undetermined. We will invoke below a constraint on the analysis error covariance, namely that the analysis error covariance projected onto the subspace spanned by the pseudo-observations equals the climatological covariance $\mathbf{A}_{\text{clim}}$. In anticipation of the analytical results below which reveal that such a constraint cannot be imposed on the whole D_w-dimensional unobserved subspace whilst simultaneously ensuring positive definiteness of $\mathbf{R}_w$, but only on a $\hat{D}_w \leq D_w$-dimensional subspace of the unobserved subspace, we introduce here a (so far undetermined) transformation matrix $\mathbf{S}_w \in \mathbb{R}^{D_w \times \hat{D}_w}$. The transformation matrix satisfies $\mathbf{S}_w^T \mathbf{S}_w = \mathbf{I}_{\hat{D}_w}$ (but not necessarily $\mathbf{S}_w \mathbf{S}_w^T = \mathbf{I}_{D_w}$). We will formulate the filter restricted to this subspace and introduce the transformed pseudo-observation operator

$$\hat{\mathbf{h}} = \mathbf{S}_w^T \mathbf{h},$$

and the transformed error covariances

$$\hat{\mathbf{R}}_w^{-1} = \mathbf{S}_w^T \mathbf{R}_w^{-1} \mathbf{S}_w$$
$$\hat{\mathbf{A}}_{\text{clim}} = \mathbf{S}_w^T \mathbf{A}_{\text{clim}} \mathbf{S}_w,$$

as well as the transformed climatological mean of the pseudo-observations $\hat{\boldsymbol{a}}_{\text{clim}} = \mathbf{S}_w^T \boldsymbol{a}_{\text{clim}}$. We now combine direct observations and pseudo-observations, and write the cost function in the more compact form

$$J(z) = \frac{1}{2}(z - z_f)^T \mathbf{P}_f^{-1} (z - z_f) + \frac{1}{2}\left(\hat{\mathbf{H}} z - \hat{y}\right)^T \hat{\mathbf{R}}^{-1} \left(\hat{\mathbf{H}} z - \hat{y}\right), \quad (3)$$

where we introduced combined observations $\hat{y}$, the observation operator $\hat{\mathbf{H}}$ and the error covariance matrix $\hat{\mathbf{R}}$ with

$$\hat{y} = \begin{pmatrix} y_o \\ \hat{a}_{\text{clim}} \end{pmatrix} \in \mathbb{R}^{D_o + \hat{D}_w},$$
$$\hat{\mathbf{H}} = \begin{pmatrix} \mathbf{H} \\ \hat{\mathbf{h}} \end{pmatrix} \in \mathbb{R}^{(D_o + \hat{D}_w) \times D},$$
$$\hat{\mathbf{R}}^{-1} = \begin{pmatrix} \mathbf{R}_o^{-1} & 0 \\ 0 & \hat{\mathbf{R}}_w^{-1} \end{pmatrix} \in \mathbb{R}^{(D_o + \hat{D}_w) \times (D_o + \hat{D}_w)}.$$

The analysis is given as the minimum of the cost function $J(z)$ and is readily calculated as

$$z_a = z_f - \hat{\mathbf{K}}\left[\hat{\mathbf{H}} z_f - \hat{y}\right], \quad (4)$$

where the Kalman gain matrix is given by

$$\hat{\mathbf{K}} = \mathbf{P}_a \hat{\mathbf{H}}^T \hat{\mathbf{R}}^{-1}, \quad (5)$$

with the error covariance matrix of the analysis given by

$$\mathbf{P}_a = \left(\mathbf{P}_f^{-1} + \hat{\mathbf{H}}^T \hat{\mathbf{R}}^{-1} \hat{\mathbf{H}}\right)^{-1}. \quad (6)$$

Using the matrix identity $(\mathbf{P}^{-1} + \mathbf{H}^T \mathbf{R}^{-1} \mathbf{H})^{-1} = \mathbf{P} - \mathbf{P}\mathbf{H}^T (\mathbf{R} + \mathbf{H}\mathbf{P}\mathbf{H}^T)^{-1} \mathbf{H}\mathbf{P}$ (see for example Simon, 2006) the analysis error covariance is recast in a form which does not involve the inverse of the forecast error covariance $\mathbf{P}_f$ as

$$\mathbf{P}_a = \left[\mathbf{I} - \hat{\mathbf{K}}\hat{\mathbf{H}}\right] \mathbf{P}_f, \quad (7)$$

and the Kalman gain matrix can be rewritten in the computationally more convenient form

$$\hat{\mathbf{K}} = \mathbf{P}_f \hat{\mathbf{H}}^T \left(\hat{\mathbf{H}} \mathbf{P}_f \hat{\mathbf{H}}^T + \hat{\mathbf{R}}\right)^{-1}, \quad (8)$$

which involves only taking the inverse of $((D_o + \hat{D}_w) \times (D_o + \hat{D}_w))$ matrices rather than of $(D \times D)$ matrices.

We remark that one can explicitly separate the updates according to the deviations from the observations and the pseudo-observations in the analysis and have

$$z_a = z_f - \mathbf{K}_o \left[\mathbf{H} z_f - y_o\right] - \hat{\mathbf{K}}_w \left[\hat{\mathbf{h}} z_f - \hat{a}_{\text{clim}}\right], \quad (9)$$

with

$$\mathbf{K}_o = \mathbf{P}_a \mathbf{H}^T \mathbf{R}_o^{-1} \quad \text{and} \quad \hat{\mathbf{K}}_w = \mathbf{P}_a \hat{\mathbf{h}}^T \hat{\mathbf{R}}_w^{-1}.$$

This shows that weighted by the error covariance of the weak constraint $\hat{\mathbf{R}}_w$ the analysis of the pseudo-observables is driven towards their climatic mean. However, due to the generically global nature of the Kalman gain matrices the inclusion of climatological information of the pseudo-observables also affects the observed degrees of freedom.

So far the error covariance $\hat{\mathbf{R}}_w$ associated with the weak constraint is undetermined. We will now determine $\hat{\mathbf{R}}_w$ and thereby control the variance of the unresolved pseudo-observables $\mathbf{h}z$ by requiring that the analysis error covariance, projected onto the pseudo-observables, equals the climatological covariance, i.e.

$$\mathbf{h} \mathbf{P}_a \mathbf{h}^T = \mathbf{A}_{\text{clim}}. \quad (10)$$

We rewrite the analysis error covariance (Eq. 6) as

$$\mathbf{P}_a = \left(\mathbf{P}^{-1} + \hat{\mathbf{h}}^T \hat{\mathbf{R}}_w^{-1} \hat{\mathbf{h}}\right)^{-1}, \quad (11)$$

where we introduced the analysis error covariance $\mathbf{P}$ for a standard Kalman filter without any weak constraint which only combines the forecast with direct observations

$$\mathbf{P} = \left(\mathbf{P}_f^{-1} + \mathbf{H}^T \mathbf{R}_o^{-1} \mathbf{H}\right)^{-1} = [\mathbf{I} - \mathbf{K}_o \mathbf{H}] \mathbf{P}_f. \quad (12)$$

Upon using the Sherman–Morrison–Woodbury formula $(\mathbf{P}^{-1} + \hat{\mathbf{h}}^T \hat{\mathbf{R}}_w^{-1} \hat{\mathbf{h}})^{-1} \hat{\mathbf{h}}^T \hat{\mathbf{R}}_w^{-1} = \mathbf{P}\hat{\mathbf{h}}^T (\hat{\mathbf{R}}_w + \hat{\mathbf{h}}\mathbf{P}\hat{\mathbf{h}}^T)^{-1}$ (see for example Simon, 2006), the error covariance matrix for the pseudo-observables $\hat{\mathbf{R}}_w$ is found from the constraint (Eq. 10) to be

$$\hat{\mathbf{R}}_w^{-1} = \hat{\mathbf{A}}_{\text{clim}}^{-1} - \left(\hat{\mathbf{h}} \mathbf{P} \hat{\mathbf{h}}^T\right)^{-1}. \quad (13)$$

Expanding Eq. (13) we obtain

$$\left(\mathbf{S}_{\mathrm{w}}^T \mathbf{R}_{\mathrm{w}} \mathbf{S}_{\mathrm{w}}\right)^{-1} = \left(\mathbf{S}_{\mathrm{w}}^T \mathbf{A}_{\mathrm{clim}} \mathbf{S}_{\mathrm{w}}\right)^{-1} - \left(\mathbf{S}_{\mathrm{w}}^T \mathbf{h} \mathbf{P} \mathbf{h}^T \mathbf{S}_{\mathrm{w}}\right)^{-1}. \quad (14)$$

This makes apparent the role of the transformation matrix $\mathbf{S}_{\mathrm{w}}$. $\mathbf{S}_{\mathrm{w}} \in \mathbb{R}^{D_{\mathrm{w}} \times \hat{D}_{\mathrm{w}}}$ can be chosen such that $\hat{\mathbf{R}}_{\mathrm{w}} = \mathbf{S}_{\mathrm{w}}^T \mathbf{R}_{\mathrm{w}} \mathbf{S}_{\mathrm{w}}$, being an error covariance, is positive definite: the transformation matrix $\mathbf{S}_{\mathrm{w}}$ projects onto the subspace of the space of pseudo-observables which in a standard Kalman filter would experience an analysis error covariance $\mathbf{h}\mathbf{P}\mathbf{h}^T$ exceeding the climatological covariance $\mathbf{A}_{\mathrm{clim}}$. All other $D_{\mathrm{w}} - \hat{D}_{\mathrm{w}}$ pseudo-observations are discarded in order to ensure a positive definite and invertible error covariance matrix $\hat{\mathbf{R}}_{\mathrm{w}} \in \mathbb{R}^{\hat{D}_{\mathrm{w}} \times \hat{D}_{\mathrm{w}}}$. In Appendix A we provide an algorithm to compute $\mathbf{S}_{\mathrm{w}}$.

We formulate the filter in the framework of ensemble Kalman filters (EnKF) (Evensen, 2006; Hamill, 2006) where an ensemble with k members z_k

$$\mathbf{Z} = [z_1, z_2, \ldots, z_k] \in \mathbb{R}^{D \times k}$$

is propagated by the model dynamics according to the model

$$\dot{\mathbf{Z}} = \boldsymbol{F}(\mathbf{Z}), \; \mathbf{F}(\mathbf{Z}) = \left[f(z_1), f(z_2), \ldots, f(z_k)\right] \in \mathbb{R}^{D \times k}.$$

The forecast ensemble is split into its mean $\overline{z}_{\mathrm{f}}$ and ensemble deviation matrix $\mathbf{Z}_{\mathrm{f}}'$. The ensemble deviation matrix $\mathbf{Z}_{\mathrm{f}}'$ can be used to provide a Monte Carlo estimate for the ensemble forecast covariance matrix via

$$\mathbf{P}_{\mathrm{f}}(t) = \frac{1}{k-1} \mathbf{Z}'(t) \left[\mathbf{Z}'(t)\right]^T \in \mathbb{R}^{D \times D}.$$

Note that $\mathbf{P}_{\mathrm{f}}(t)$ is rank-deficient for $k < D$, which is the typical situation in numerical weather prediction where N is of the order of 10^9 and k of the order of 100.

At the end of each analysis cycle an ensemble $\mathbf{Z}_{\mathrm{a}}$ is generated which must be consistent with the analysis error covariance $\mathbf{P}_{\mathrm{a}}$, and satisfies

$$\mathbf{P}_{\mathrm{a}} = \frac{1}{k-1} \mathbf{Z}_{\mathrm{a}}' \left[\mathbf{Z}_{\mathrm{a}}'\right]^T.$$

In previous work Gottwald et al. (2011) and Mitchell and Gottwald (2012) used the ensemble transform Kalman filter (ETKF) (Bishop et al., 2001; Tippett et al., 2003; Wang et al., 2004), which seeks a transformation $\mathbf{T} \in \mathbb{R}^{k \times k}$ such that the analysis deviation ensemble $\mathbf{Z}_{\mathrm{a}}'$ is given as a deterministic perturbation of the forecast ensemble $\mathbf{Z}_{\mathrm{f}}$ via $\mathbf{Z}_{\mathrm{a}}' = \mathbf{Z}_{\mathrm{f}}'\mathbf{T}$. In order to incorporate localisation needed for small ensemble sizes easily, we will implement for our VLKF here an approximate square root filter (DEnKF) proposed by Sakov and Oke (2008) where the analysis deviations are determined according to

$$\mathbf{Z}_{\mathrm{a}}' = \left(\mathbf{I} - \frac{1}{2}\mathbf{K}\mathbf{H}\right) \mathbf{Z}_{\mathrm{f}}'. \quad (15)$$

A new forecast is obtained by propagating $\mathbf{Z}_{\mathrm{a}}$ with the nonlinear forecast model to the next observation time, where a new analysis cycle will be started.

We will use here diagonal target matrices $\mathbf{A}_{\mathrm{clim}}$ where the diagonal entries are set to the mean value of the diagonal entries of the full climatic covariance. We found that otherwise the variance constraint is not "switched on" sufficiently often to drive the dynamics to the mean $\boldsymbol{a}_{\mathrm{clim}}$ (due to a lack of simultaneous diagonalisability of $\mathbf{A}_{\mathrm{clim}}$ and $\mathbf{h}\mathbf{P}\mathbf{h}^T$; cf. Appendix A). This suggests that the variance constraint is a numerical tool to regularise the filter, with the advantage however that the regularisation is performed in a dynamically consistent way, performed *within* the data assimilation procedure using only dynamical quantities such as measured imbalance.

3 The modified Lorenz-96 model

The Lorenz-96 model (Lorenz, 1996; Lorenz and Emanuel, 1998)

$$\dot{x}_j = x_{j-1}\left(x_{j+1} - x_{j-2}\right) - x_j + F \quad j = 1, \ldots, d \quad (16)$$

with periodic $x_j = x_{j+d}$ is a standard test bed for data assimilation as it is computationally manageable but still incorporates crucial ingredients of real mid-latitude atmospheric flows such as nonlinear energy conservation, advection, forcing and linear damping. Recently, Bergemann and Reich (2010a) introduced a modification of the standard Lorenz-96 model by coupling it to a purely dispersive fast wave equation mimicking the influence of fast gravity waves on slow Rossby waves in a quasi-geostrophic regime. The modified Lorenz system reads as

$$\dot{x}_j = (1-\eta) x_{j-1}\left(x_{j+1} - x_{j-2}\right) - x_j + F + \eta\left(x_{j-1} h_{j+1} - x_{j-2} h_{j-1}\right) \quad (17)$$

$$\varepsilon^2 \ddot{h}_j = -h_j + \alpha^2\left(h_{j-1} - 2h_j + h_{j+1}\right) + x_j. \quad (18)$$

The fast wave part (Eq. 18) is purely dispersive; if the dissipation and the forcing in the slow $\boldsymbol{x}$ equation (Eq. 17) is ignored the system conserves the total energy

$$H = \frac{\eta}{2} \sum_{j=1}^{d} \Big(\frac{\eta - 1}{\eta} x_j^2 + \epsilon^2 \dot{h}_j^2 + h_j^2 + \alpha^2 \left(h_{j+1} - h_{j-1}\right)^2 - 2 x_j h_j \Big) \quad (19)$$

with $0 \leq \eta \leq 1$. The modified Lorenz-96 system (Eqs. 17–18) contains an approximate slow manifold given by

$$x_j = h_j - \alpha^2\left(h_{j-1} - 2h_j + h_{j+1}\right), \quad (20)$$

which is obtained by formally setting $\varepsilon = 0$ in Eq. (18). Higher order balance relations could be derived by employing asymptotic theory.

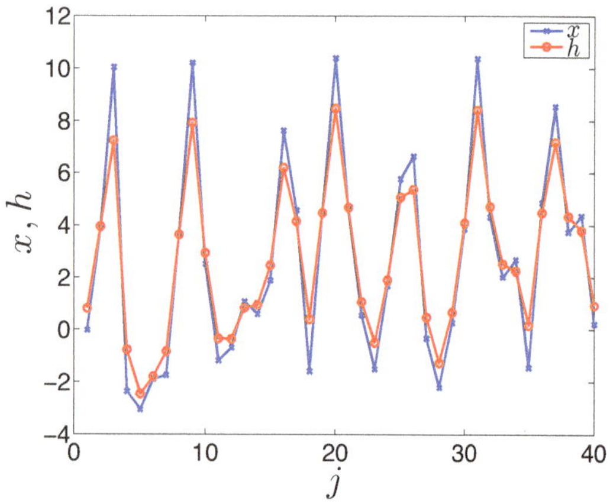

Fig. 1. Typical balanced fields $\{x_j\}$ (blue) and $\{h_j\}$ (red) for the modified slow-fast Lorenz-96 model (Eqs. 17–18).

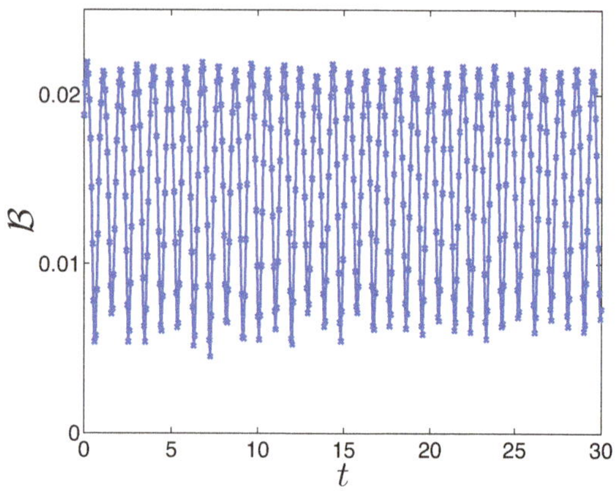

Fig. 2. Temporal evolution of the imbalance $\mathcal{B}$ for initially balanced fields of the modified slow-fast Lorenz-96 model (Eqs. 17–18).

We set the number of degrees of freedom to $d=40$ and $F=8$ for the forcing. We consider here weak coupling with $\eta=0.1$, implying sufficiently nonlinear behaviour of the slow x variables. The "Rossby number" is set to $\varepsilon=0.0025$ and the "Burgers number" is set to $\alpha^2=0.25$. In Fig. 1 we show typical initially balanced fields. Note that the balance relation (Eq. 20) implies that the balanced field $\{h_j\}=(h_1, h_2, \ldots, h_\mathrm{d})$ is smoother than $\{x_j\}=(x_1, x_1, \ldots, x_\mathrm{d})$ (h is obtained from x via the application of an inverse Helmholtz operator). Note that this is different to the situation in realistic atmospheric models where the fast variables are small scale and rapidly oscillate around the slow manifold.

We introduce the imbalance operator $\mathbf{B}$ which acts on z with $z_j=(x_j, h_j, \dot{h}_j)$ as

$$(\mathbf{B}z)_j = x_j - h_j + \alpha^2 \left(h_{j-1} - 2h_j + h_{j+1}\right), \qquad (21)$$

which according to Eq. (20) is zero to leading order if initially so. Figure 2 shows the temporal evolution of the site-averaged imbalance

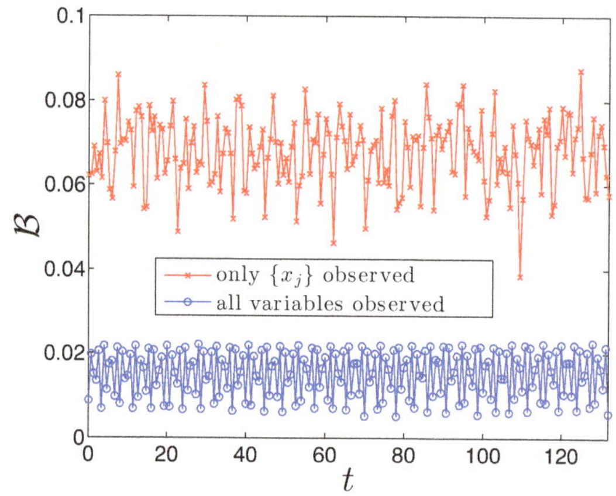

Fig. 3. Imbalance $\mathcal{B}$ of the analysis as a function of analysis cycles for ETKF with 1000 ensemble members and an observation interval of $\Delta t_\mathrm{obs}=2$ h and observational noise error variance $R_\mathrm{o}=0.84$, without covariance inflation and localisation. Results are shown for the case when all variables $\{x_j\}$, $\{h_j\}$ and $\{\dot{h}_j\}$ of the modified slow-fast Lorenz-96 model (Eqs. 17–18) are observed (blue) and for the case of a spatially sparse observations when only $\{x_j\}$ are observed (red).

$$\mathcal{B}(t) = \sqrt{\frac{1}{d}\sum_{j=1}^{\mathrm{d}} (\mathbf{B}z)_j^2} \qquad (22)$$

for initially balanced fields with a small value of $\epsilon=0.0025$. The figure clearly illustrates that balance is approximately preserved by the dynamics, provided the timescale separation is sufficiently large, i.e. ϵ sufficiently small. This justifies the terminology of Eq. (20) defining a *slow manifold*, as the initially generated imbalance does not interact with the slow variables on long timescales. The situation is very different when the dynamics is interrupted by data assimilation cycles where the data assimilation procedure introduces imbalance. In Cohn et al. (1998), Lorenc (2003a), Mitchell et al. (2002), Houtekamer and Mitchell (2005), Oke et al. (2007), Kepert (2009) and Greybush et al. (2011) the imbalance was associated with the procedure of covariance localisation. In Fig. 3 we show that ensemble filters can generate unbalanced analyses in sparse observational grids due to the intermittent discontinuous analyses updates, even without localisation. We present results for an ETKF with a large ensemble of 1000 members where only the slow $\{x_j\}$ variables are observed, and compare it to the case when all variables $\{x_j\}$, $\{h_j\}$ and $\{\dot{h}_j\}$ are observed. Whereas in the fully observed case the imbalance $\mathcal{B}$ exhibits the actual physical imbalance (cf. Fig. 2), increased imbalance is clearly seen in the sparser observational grid. We remark that this is not a finite size effect and cannot be mitigated by larger ensembles (we tested ensemble sizes of 3000), consistent with results for 3D-VAR by Bloom et al. (1996). We remark that for smaller observational noise with $R_\mathrm{o}=0.21$ the imbalance exhibits the same mean values as in Fig. 3 with $R_\mathrm{o}=0.84$. In the next section we explore how this spurious imbalance can be controlled by using the VLKF framework established in Sect. 2.

4 Numerical results

We now present results from numerical data assimilation cycles of Eqs. (17)–(18). We consider a sparse observational grid in which only every second slow $\{x_{2j}\}$ variable is observed; the variables $\{h_j\}$ and $\{\dot{h}_j\}$ are not observed. We use $D = 3\,d = 3 \times 40$, and therefore in the notation of Sect. 2 we have $D_o = 20$. We observe the system in equidistant observation intervals $\Delta t_{\rm obs}$ ranging from 1 to 6.5 h, adopting the timescales suggested by the standard Lorenz-96 system (Eq. 16), i.e. $t = 1/120$ roughly corresponds to 1 hour (see for example Lorenz and Emanuel, 1998). Observations are contaminated by Gaussian noise, with error variance $\mathbf{R}_o = (0.25\sigma_{x,{\rm clim}})^2\,\mathbf{I}_{20} = 0.84\,\mathbf{I}_{20}$ where $\sigma^2_{x,{\rm clim}} = 13.50$ is the climatic variance of $\{x_j\}$. We perform 4000 analysis cycles after a spin-up period of 1000 analysis cycles. All simulations are initialised with balanced data using Eq. (20). To generate the observations and to propagate forward the forecast model (Eqs. 17–18) we employ an implicit midpoint rule with a time step of $dt = 0.0025$ (see, for example, Leimkuhler and Reich, 2005).

Besides the variance limiting Kalman filter VLKF-$\mathcal{B}$ where we impose a climatic constraint on the imbalance $\mathbf{B}\,z$, we also employ a variance limiting Kalman filter VLKF-$\dot{h}$ where we impose a climatic constraint on the unobserved fast variables $\{\dot{h}_j\}$. Both VLKF-$\mathcal{B}$ and VLKF-$\dot{h}$ have, in the notation of Sect. 2, $D_w = 40$. With $z = (\boldsymbol{x},\, \boldsymbol{h},\, \dot{\boldsymbol{h}}) \in \mathbb{R}^{120}$, for VLKF-$\mathcal{B}$ the pseudo-observation operator $\mathbf{h} : \mathbb{R}^{120} \to \mathbb{R}^{40}$ is

$$\mathbf{h} = (\mathbf{B}\,\mathbf{0}_{40})\,,$$

with $\mathbf{B} \in \mathbb{R}^{40\times 80}$ defined in Eq. (21), and for VLKF-$\dot{h}$ the pseudo-observation operator is

$$\mathbf{h} = (\mathbf{0}_{40}\,\mathbf{0}_{40}\,\mathbf{I}_{40})\,.$$

The climatic mean and variances of $\{\dot{h}_j\}$ and those of the imbalance $\{(\mathbf{B}\,z)_j\}$ were estimated through long-time simulations of the full modified Lorenz-96 system (Eqs. 17–18) with balanced initial data as $\mathbf{B}\,z = \mathbf{0}$ and $\sigma^2_{Bz,{\rm clim}} = 8.4 \times 10^{-4}$, and $\dot{h} = -0.01$ and $\sigma_{\dot{h},{\rm clim}} = 224.35$, respectively. We set $\boldsymbol{a}_{\rm clim} = \mathbf{0}$ and $\mathbf{A}_{\rm clim} = \sigma^2_{Bz,{\rm clim}}\mathbf{I}_{40}$ for VLKF-$\mathcal{B}$, and $\boldsymbol{a}_{\rm clim} = \mathbf{-0.01}$ and $\mathbf{A}_{\rm clim} = \sigma^2_{\dot{h},{\rm clim}}\mathbf{I}_{40}$ for VLKF-$\dot{h}$, respectively. We note that both the climatic covariance of $\mathbf{B}\,z$ and of $\dot{\boldsymbol{h}}$ are concentrated near the diagonal.

For comparison with our implementations of VLKF-$\mathcal{B}$ and VLKF-$\dot{h}$ we will employ a suite of ensemble filters. In particular, we will use the EnKF with perturbed observations as in Burgers et al. (1998) and the approximate square root filter DEnKF as in Sakov and Oke (2008). Furthermore, we implement an incremental analysis update (IAU) as in Bloom et al. (1996) and Polavarapu et al. (2004), where the analysis increments are calculated by a DEnKF.

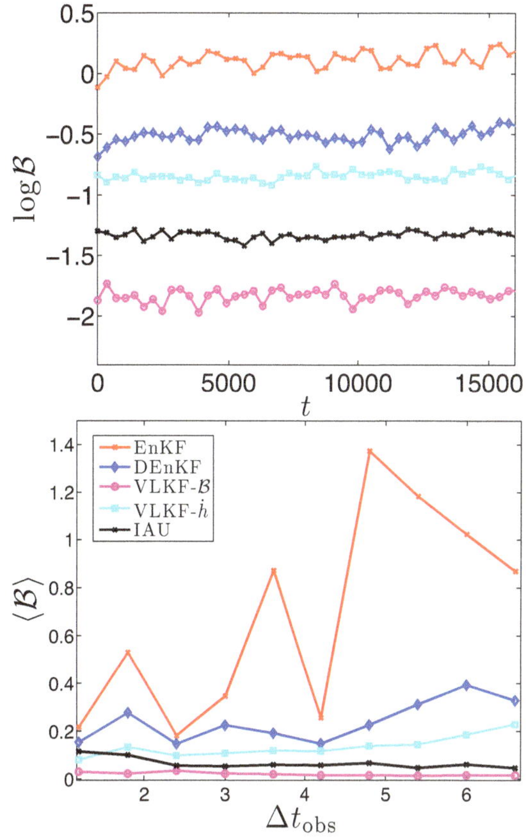

Fig. 4. Top panel: imbalance $\mathcal{B}$ of the analysis as a function of analysis cycles for $\Delta t_{\rm obs} = 5.5$ h in a log plot. In order of increasing values of $\mathcal{B}$ are VLKF-$\mathcal{B}$ (magenta open circles), IAU (black crosses) VLKF-$\dot{h}$ (cyan squares), DEnKF (blue diamonds) and EnKF (red crosses). Bottom panel: temporally averaged imbalance $\langle\mathcal{B}\rangle$ of the analysis as a function of the observation interval $\Delta t_{\rm obs}$.

All filter implementations use 10 ensemble members, which is smaller than the attractor dimension of the system (Eqs. 17–18). We employ covariance inflation whereby the prior forecast error covariance is increased by an inflation factor δ (Anderson and Anderson, 1999). Since the $\{h_j\}$ variables are not damped in the modified Lorenz system (Eq. 18), inflation of the unobserved $\{h_j\}$ variables would in general lead to an increasing growth in the associated forecast covariance. Inflation is therefore applied at each time step only to the $\{x_j\}$ components of the ensembles for DEnKF, EnKF and IAU, but to all components of the ensemble for VLKF-$\mathcal{B}$ and VLKF-$\dot{h}$ which explicitly constrain. This was found to be advantageous for all respective filters. Results are obtained for a wide range of inflation factors and only the optimal result for each particular formulation of the filter is reported here. We have optimised over 1000 equally spaced values of $\delta \in (1, 1.16)$. The small ensemble size chosen here requires localisation. We employ the localisation

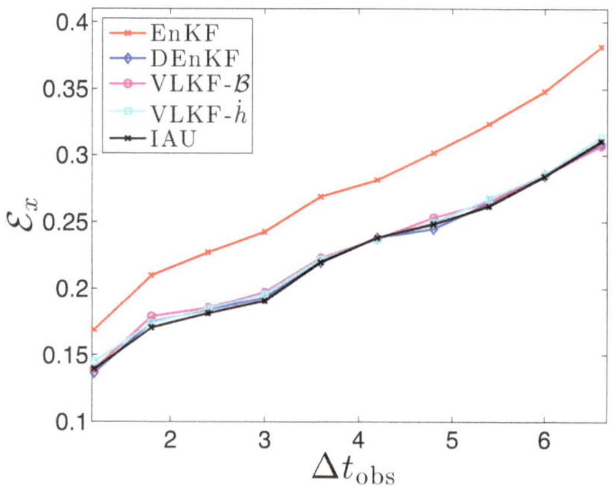

Fig. 5. Rms error of the analysis as a function of the observation interval $\Delta t_{\rm obs}$ (in hours) for the slow variables $\{x_j\}$.

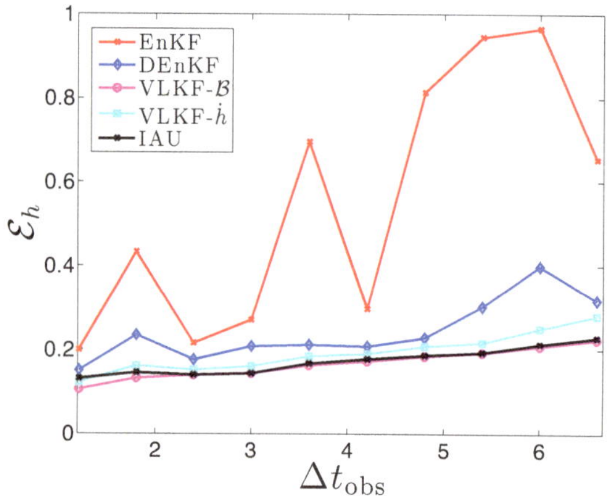

Fig. 6. Rms error of the analysis as a function of the observation interval $\Delta t_{\rm obs}$ (in hours) for the unobserved height variables $\{h_j\}$.

method along the line of Houtekamer and Mitchell (1998, 2001) and Hamill et al. (2001) whereby the forecast error covariance $\mathbf{P}_{\rm f}$ is Schur-multiplied with a localisation matrix $\mathbf{C}_{\rm loc}$. We use the compactly supported localisation function introduced by Gaspari and Cohn (1999) where correlations with distances larger than $2\rho_{\rm loc}$ are set to 0. We set the localisation radius to $\rho_{\rm loc}=8$ for all filters.

In Fig. 4 we present results of the amount of imbalance as measured by the imbalance $\mathcal{B}(t)$ and by the temporally averaged imbalance

$$\langle\mathcal{B}\rangle = \frac{1}{N}\sum_{n=1}^{N}\mathcal{B}(n\,\Delta t_{\rm obs}),$$

accrued during the data assimilation procedure for our suite of filters. EnKF and DEnKF generate a significant amount of unphysical imbalance, with values much larger than those of the actual balanced toy model with $\langle\mathcal{B}\rangle\approx 0.018$ (cf. Fig. 2). The increased imbalance in EnKF may be due to sampling errors introduced through the perturbed observations. Given the particular nature of imbalance present in the toy model (Eqs. 17–18), this may not be an issue in realistic atmospheric models. The IAU implementation strongly reduces imbalance, albeit to levels significantly larger than those expected from the actual dynamics. Our VLKF-$\mathcal{B}$ filter is able to constrain imbalance very close to the the actual physical imbalance. Note that although the pseudo-observations used in VLKF-$\mathcal{B}$ were for the imbalance of each variable $\mathbf{B}\,z$ driving the dynamics towards $\mathbf{B}\,z=\mathbf{0}$, the analysis reproduces dynamically realistic values of the integrated measure of imbalance $\langle\mathcal{B}\rangle$. VLKF-$\dot{h}$ also achieves a pronounced reduction in imbalance, but to larger values than the IAU implementation, in particular for larger observation intervals $\Delta t_{\rm obs}$. Surprisingly, for filters which only constrain the climatic variance of the height variable $\{h_j\}$, and do not impose any explicit constraints on the imbalance, i.e. $\mathbf{h}=(\mathbf{0}_{40}\,\mathbf{I}_{40}\,\mathbf{0}_{40})$, one also observes a significant reduction in imbalance (not shown).

We now investigate rms error statistics. We consider the site-averaged rms error of variables z

$$\mathcal{E} = \sqrt{\langle\frac{1}{N\,D_G}\sum_{n=1}^{N}|\overline{z}_{\rm a}(n\,\Delta t_{\rm obs}) - z_t(n\,\Delta t_{\rm obs})|^2_G\rangle} \qquad (23)$$

between the truth z_t and the ensemble mean $\overline{z}_{\rm a}$, where N is the number of analysis cycles and D_G denotes the number of variables involved. We introduce the norm $\|\mathbf{a}\|_G^2=\mathbf{a}^T\,\mathbf{G}\,\mathbf{a}$ to investigate the error over all $\{x_j\}$ variables $\mathcal{E}_x$ using $\mathbf{G}=\delta_{ij}$ for $1\le i\le 40$ and the error of the fast $\{h_j\}$ variables $\mathcal{E}_h$ using $\mathbf{G}=\delta_{ij}$ for $41\le i\le 80$. Figure 5 shows $\mathcal{E}_x$ for our suite of filters. DEnKF, IAU and our VLKF-$\mathcal{B}$ and VLKF-$\dot{h}$ exhibit very similar rms errors, with values much smaller than the observational noise with $r_{\rm o}=\sqrt{0.84}=0.91$. EnKF produces consistently worse rms errors, which again may be due to sampling errors stemming from the randomly perturbed observations.

Figure 6 shows that the correct balance statistics of VLKF-$\mathcal{B}$ manifests itself in superior rms errors for the unobserved fast height field $\{h_{\rm j}\}$ when compared to filters which do not incorporate a balance constraint. IAU and VLKF-$\mathcal{B}$ exhibit comparable rms error statistics for the height field. Furthermore, it is seen that constraining the covariance of $\{\dot{h}_j\}$, as done in VLKF-$\dot{h}$, also generates comparably good rms errors for the height field. The variations of the error in the height field $\mathcal{E}_{\rm h}$ with the observation interval $\Delta t_{\rm obs}$ mirror exactly the imbalance shown in Fig. 4.

We found that EnKF, DEnKF and IAU exhibit instances of catastrophic filter divergence whereby the forecast model develops numerical instabilities (Gottwald and Majda, 2013). Instances of this type of filter divergence were not observed in the variance constraining filters VLKF-$\mathcal{B}$ and VLKF-$\dot{h}$.

5 Conclusions

We have presented here an implementation of an ensemble filter which explicitly limits the amount of imbalance *within* the data assimilation procedure. We were able to produce balanced analyses by incorporating statistical information such as mean and variance of imbalance, available through long-time integration or historical data, as pseudo-observations. This procedure not only successfully constraints imbalance to its climatic values, but also produces very good filter performance in terms of rms errors of the fast unobserved variables.

We presented a comparison between a filter which explicitly constraints the amount of imbalance, coined VLKF-$\mathcal{B}$, a filter which constraints the statistics of the fast variables but not the imbalance, coined VLKF-$\dot{h}$, and standard implementations of EnKF, DEnKF and IAU. It was found that our balance controlling filter VLKF-$\mathcal{B}$ is able to constrain the amount of imbalance to lie within the physically observed limits. Besides improved balance of the analyses this also implied very good error statistics for the unobserved height field. We tested our method against the widely used IAU implementation and found that it generates less unphysical imbalance and has very similar rms error statistics for the observed and the unobserved variables.

The variance constraint we employ requires the determination of the overestimating subspace – the eigenspace in which $\mathbf{P}_f$ experiences covariances above the climatological – which was achieved in this work by singular vector decomposition. The extra computational cost implied has to be weighed against the cost of an additional application of the forecast model involved in IAU or of fast Fourier transforms when using digital filters, as well as whether the superior performance of VLKF-$\mathcal{B}$ in generating less unphysical imbalance is worth it.

Appendix A

Construction of the transformation matrix $\mathbf{S}_w$

We present here an algorithm of how to construct the transformation matrix $\mathbf{S}_w$. This matrix projects into the subspace of those pseudo-observable subspaces which in a standard Kalman filter would produce an analysis whose analysis error covariance matrix exceeds the prescribed (climatological) error covariance $\mathbf{A}_{\text{clim}}$. We need to find $\mathbf{S}_w$ such that Eq. (14) produces a positive error covariance matrix $\mathbf{R}_w$. For convenience we recall Eq. (14):

$$\left(\mathbf{S}_w^T \mathbf{R}_w \mathbf{S}_w\right)^{-1} = \left(\mathbf{S}_w^T \mathbf{A}_{\text{clim}} \mathbf{S}_w\right)^{-1} - \left(\mathbf{S}_w^T \mathbf{hPh}^T \mathbf{S}_w\right)^{-1}. \quad \text{(A1)}$$

Introducing $\hat{\mathbf{P}}_h = \mathbf{S}_w^T \mathbf{hPh}^T \mathbf{S}_w$, $\hat{\mathbf{A}}_{\text{clim}} = \mathbf{S}_w^T \mathbf{A}_{\text{clim}} \mathbf{S}_w$ and $\hat{\mathbf{R}}_w = \mathbf{S}_w^T \mathbf{R}_w \mathbf{S}_w$, we may rewrite this as

$$\hat{\mathbf{R}}_w^{-1} = \hat{\mathbf{A}}_{\text{clim}}^{-1} - \hat{\mathbf{P}}_h^{-1}.$$

Matrices $\mathbf{S}_w$ satisfying Eq. (A1) can be determined provided $\mathbf{A}_{\text{clim}}$ and $\mathbf{hPh}^T$ are simultaneously diagonalisable. We remark that Eq. (A1) can be readily converted into

$$\hat{\mathbf{R}}_w = \hat{\mathbf{P}}_h \left[\hat{\mathbf{P}}_h - \hat{\mathbf{A}}_{\text{clim}}\right]^{-1} \hat{\mathbf{A}}_{\text{clim}}.$$

Simultaneous diagonalisation of $\mathbf{A}_{\text{clim}}$ and $\mathbf{hPh}^T$ can be checked either by looking at the null space of the commutator $\mathbf{C} = \mathbf{A}_{\text{clim}} \mathbf{hPh}^T - \mathbf{hPh}^T \mathbf{A}_{\text{clim}}$, or by comparing the eigenspaces of the respective matrices directly. We note that if we prescribe a diagonal target covariance $\mathbf{A}_{\text{clim}} = \lambda \mathbf{I}$, simultaneous diagonalisation is automatically asserted. Let the transformation matrix onto the subspace in which $\mathbf{A}_{\text{clim}}$ and $\mathbf{hPh}^T$ are simultaneously diagonalisable be $\mathbf{M}_0^T \in \mathbb{R}^{D_w \times D_{w_0}}$ with $D_{w_0} \leq D_w$.

Consider $\mathbf{A}_{\text{clim}\,0} = \mathbf{M}_0^T \mathbf{A}_{\text{clim}} \mathbf{M}_0$ and $\mathbf{P}_0 = \mathbf{M}_0^T \mathbf{hPh}^T \mathbf{M}_0$, and simultaneously diagonalise by writing (wlog),

$$\mathbf{A}_{\text{clim}\,0} = \mathbf{S}_A \mathbf{A}_{\text{clim}\,0} \mathbf{S}_A^T = \tilde{\mathbf{S}}_A \tilde{\mathbf{S}}_A^T,$$

with $\tilde{\mathbf{S}}_A = \mathbf{S}_A \mathbf{A}_{\text{clim}\,0}^{\frac{1}{2}}$. Introducing

$$\tilde{\mathbf{P}}_0 = \tilde{\mathbf{S}}_A^{-1} \mathbf{P}_0 \tilde{\mathbf{S}}_A^{-T},$$

we transform with an orthogonal transformation $\mathbf{S}_\mathbf{P}$

$$\tilde{\mathbf{P}}_0 = \mathbf{S}_\mathbf{P} \hat{\hat{\mathbf{P}}}_0 \mathbf{S}_\mathbf{P}^T.$$

Introducing $\mathbf{Q} = \tilde{\mathbf{S}}_A \mathbf{S}_\mathbf{P}$ we may write

$$\mathbf{A}_{\text{clim}\,0} = \mathbf{Q}\mathbf{Q}^T,$$
$$\mathbf{P}_0 = \mathbf{Q} \hat{\hat{\mathbf{P}}}_0 \mathbf{Q}^T.$$

Using the diagonal matrices $\mathbf{I}$ and $\hat{\hat{\mathbf{P}}}_0$, associated with $\mathbf{A}_{\text{clim}}$ and $\mathbf{hPh}^T$, respectively, one can now readily check for overestimation of $\mathbf{R}_w$. Transforming into the subspace in which $\mathbf{A}_{\text{clim}}$ and $\mathbf{hPh}^T$ are both diagonal with the same eigenspaces, we obtain

$$\hat{\hat{\mathbf{R}}}_w^{-1} = \mathbf{Q}^T \mathbf{M}_0^T \mathbf{R}_w^{-1} \mathbf{M}_0 \mathbf{Q} = \mathbf{I} - \hat{\hat{\mathbf{P}}}_0^{-1},$$

and $\hat{\hat{\mathbf{R}}}_w$ can be calculated directly by inverting the diagonal $\hat{\hat{\mathbf{R}}}_w^{-1}$.

We have now determined a transformation $\mathbf{Q}^T \mathbf{M}_0^T$ which simultaneously diagonalises $\mathbf{A}_{\text{clim}}$ and $\mathbf{hPh}^T$. Transforming renders $\hat{\hat{\mathbf{R}}}_w \in \mathbb{R}^{D_{w_0} \times D_{w_0}}$ diagonal, but not necessarily positive definite as required. However, $\hat{\hat{\mathbf{R}}}_w$ being diagonal allows us to readily determine a transformation matrix $\mathbf{S}_{\text{red}} \in \mathbb{R}^{\hat{D}_w \times D_{w_0}}$ which projects onto the $\hat{D}_w$-dimensional overestimating subspace in which $\hat{\hat{\mathbf{R}}}_w$ is positive definite, with $\hat{D}_w \leq D_{w_0}$.

This concludes our algorithm for how to compute a positive definite invertible covariance matrix $\hat{\mathbf{R}}_w$ and also provides an expression for the transformation matrix $\mathbf{S}_{\mathrm{w}} \in \mathbb{R}^{\hat{D}_{\mathrm{w}} \times D_{\mathrm{w}}}$ as

$$\mathbf{S}_{\mathrm{w}} = \mathbf{S}_{\mathrm{red}} \mathbf{Q}^T \mathbf{M}_0^T.$$

We note that one may define formally an effective pseudo-observation operator $\hat{\mathbf{h}} \in \mathbb{R}^{\hat{D}_{\mathrm{w}} \times D_{\mathrm{w}}}$

$$\hat{\mathbf{h}} = \mathbf{S}_{\mathrm{red}} \mathbf{Q}^T \mathbf{M}_0^T \mathbf{h}.$$

Acknowledgements. I am grateful to Reiner Pope for speeding up my code and parallelising it. I am grateful to Sebastian Reich for fruitful discussions and to two anonymous referees for helpful comments. I acknowledge support from the Australian Research Council.

Edited by: G. Desroziers

References

Anderson, J. L. and Anderson, S. L.: A Monte Carlo implementation of the nonlinear filtering problem to produce ensemble assimilations and forecasts, Mon. Weather Rev., 127, 2741–2758, 1999.

Baer, F. and Tribbia, J.: On complete filtering of gravity modes through non-linear initialization, Mon. Weather Rev., 105, 1536–1539, 1977.

Bergemann, K. and Reich, S.: A mollified ensemble Kalman filter, Q. J. Roy. Meteorol. Soc., 136, 1636–1643, 2010a.

Bergemann, K. and Reich, S.: A localisation technique for ensemble Kalman filters, Q. J. Roy. Meteorol. Soc., 136, 701–707, 2010b.

Bergemann, K., Gottwald, G. A., and Reich, S.: Ensemble propagation and continuous matrix factorization algorithms, Q. J. Roy. Meteorol. Soc., 135, 1560–1572, 2009.

Bishop, C. H., Etherton, B. J., and Majumdar, S. J.: Adaptive sampling with the Ensemble Transform Kalman Filter, Part I: Theoretical aspects, Mon. Weather Rev., 129, 420–436, 2001.

Bloom, S. C., Takacs, L. L., Da Silva, A. M., and Ledvina, D.: Data assimilation using incremental analysis updates, Mon. Weather Rev., 124, 1256–1271, 1996.

Burgers, G., van Leeuwen, P. J., and Evensen, G.: Analysis scheme in the ensemble Kalman filter, Mon. Weather Rev., 126, 1719–1724, 1998.

Cohn, S. E., Da Silva, A., Guo, J., Sienkiewicz, M., and Lamich, D.: Assessing the effects of data selection with the DAO Physical-Space Statistical Analysis System, Mon. Weather Rev., 126, 2913–2926, 1998.

Cotter, C.: Data assimilation on the exponentially accurate slow manifold, Philos. T. Roy. Soc. A, 371, 20120300, doi:10.1098/rsta.2012.0300, 2013.

Daley, R.: Atmospheric Data Analysis, Cambridge University Press, Cambridge, 1993.

Evensen, G.: Data Assimilation: The Ensemble Kalman Filter, Springer, New York, 2006.

Gaspari, G. and Cohn, S. E.: Construction of correlation functions in two and three dimensions, Q. J. Roy. Meteorol. Soc, 125, 723–757, 1999.

Gauthier, P. and Thépaut, J. N.: Impact of the digital filter as a weak constraint in the pre operational 4DVAR assimilation system of Météo-France, Mon. Weather Rev., 117, 1225–1254, 2001.

Gottwald, G. A. and Majda, A. J.: A mechanism for catastrophic filter divergence in data assimilation for sparse observation networks, Nonlin. Processes Geophys., 20, 705–712, doi:10.5194/npg-20-705-2013, 2013.

Gottwald, G. A., Mitchell, L., and Reich, S.: Controlling overestimation of error covariance in ensemble Kalman filters with sparse observations: A variance limiting Kalman filter, Mon. Weather Rev., 139, 2650–2667, 2011.

Greybush, S. J., Kalnay, E., Miyoshi, T., Ide, K., and Hunt, B. R.: Balance and ensemble Kalman filter localization techniques, Mon. Weather Rev, 139, 511–522, 2011.

Hamill, T. M.: Ensemble-based atmospheric data assimilation, in: Predictability of Weather and Climate, edited by: Palmer, T. and Hagedorn, R., Cambridge University Press, Cambridge, 124–156, 2006.

Hamill, T. M., Whitaker, J. S., and Snyder, C.: Distance-dependent filtering of background covariance estimates in an ensemble Kalman filter, Mon. Weather Rev., 129, 2776–2790, 2001.

Houtekamer, P. L. and Mitchell, H. L.: Data Assimilation Using an Ensemble Kalman Filter Technique, Mon. Weather Rev., 126, 796–811, 1998.

Houtekamer, P. L. and Mitchell, H. L.: A sequential ensemble Kalman filter for atmospheric data assimilation, Mon. Weather Rev., 129, 123–136, 2001.

Houtekamer, P. L. and Mitchell, H. L.: Ensemble Kalman filtering, Q. J. Roy. Meteorol. Soc., 131, 3269–3289, 2005.

Kalnay, E.: Atmospheric Modeling, Data Assimilation and Predictability, Cambridge University Press, Cambridge, 2002.

Kepert, J. D.: Covariance localisation and balance in an ensemble Kalman filter, Q. J. Roy. Meteorol. Soc. 135, 1157–1176, 2009.

Leimkuhler, B. and Reich, S.: Simulating Hamiltonian Dynamics, Cambridge University Press, Cambridge, 2005.

Lorenc, A. C.: The potential of the ensemble Kalman filter for NWP – a comparison with 4DVAR, Q. J. Roy. Meteorol. Soc., 129, 3183–3203, 2003a.

Lorenc, A. C.: Modelling of error covariances by 4D-Var data assimilation, Q. J. Roy. Meteorol. Soc., 129, 3167–3182, 2003b.

Lorenz, E. N.: Predictability – a problem partly solved, in: Predictability, edited by: Palmer, T., European Centre for Medium-Range Weather Forecast, Shinfield Park, Reading, UK, 1996.

Lorenz, E. N. and Emanuel, K. A.: Optimal sites for supplementary weather observations: simulation with a small model, J. Atmos. Sci., 55, 399–414, 1998.

Lynch, P.: The Emergence of Numerical Weather Prediction: Richardson's Dream, Cambridge University Press, Cambridge, 2006.

Lynch, P. and Huang, X.-Y.: Initialization of the HIRLAM model using a digital filter, Mon. Weather Rev., 120, 1019–1034, 1992.

Machenhauer, B.: On the dynamics of gravity oscillations in a shallow water model with applications to normal mode initialization, Contrib. Atmos. Phys., 50, 253–271, 1977.

Mitchell, H. L., Houtekamer, P. L., and Pellerin, G.: Ensemble Size, balance, and model-error representation in an Ensemble Kalman Filter, Mon. Weather Rev., 130, 2791–2808, 2002.

Mitchell, L. and Gottwald, G. A.: Controlling model error of underdamped forecast models in sparse observational networks using a variance-limiting Kalman filter, Q. J. Roy. Meteorol. Soc., 139, 212–225, 2012.

Neef, L. J., Polavarapu, S. M., and Shepherd, T. G.: Four-Dimensional Data Assimilation and Balanced Dynamics, J. Atmos. Sci., 63, 1840–1858, 2006.

Oke, P., Sakov, P., and Corney, S.: Impact of localization in the EnKF and EnOI: Experiments with a small model, Ocean Dynam., 57, 32–45, 2007.

Ott, E., Hunt, B., Szunyogh, I., Zimin, A., Kostelich, E., Corrazza, M., Kalnay, E., and Yorke, J.: A local ensemble Kalman filter for atmospheric data assimilation, Tellus A, 56, 415–428, 2004.

Ourmières, Y., Brankart, J. M., Berline, L., Brasseur, P., and Verron, J.: Incremental analysis update implementation into a sequential ocean data assimilation system, J. Atmos. Ocean. Tech., 23, 1729–1744, 2006.

Polavarapu, S., Tanguay, M., and Fillion, L.: Four-dimensional variational data assimilation with digital filter initialisation, Mon. Weather Rev., 128, 2491–2510, 2000.

Polavarapu, S., Ren, S., Clayton, A. M., Sankey, D., and Rochon, Y.: On the relationship between incremental analysis updating and incremental digital filtering, Mon. Weather Rev., 132, 2495–2502, 2004.

Sakov, P. and Oke, P. R.: A deterministic formulation of the ensemble Kalman filter: an alternative to ensemble square root filters, Tellus A, 60, 361–371, 2008.

Simon, D. J.: Optimal State Estimation, John Wiley & Sons, Inc., New York, 2006.

Szunyogh, I., Kostelich, E., Gyarmati, G., Patil, D. J., Hunt, B., Kalnay, E., Ott, E., and Yorke, J.: Assessing a local ensemble Kalman filter: Perfect model experiments with the National Centers for Environmental Prediction global model, Tellus A, 57, 528–545, 2005.

Thépaut, J. N. and Courtier, P.: Four-dimensional variational data assimilation using the adjoint of a multilevel primitive equation model, Q. J. Roy. Meteorol. Soc., 129, 2089–2102, 1991.

Tippett, M. K., Anderson, J. L., Bishop, C. H., Hamill, T. M., and Whitaker, J. S.: Ensemble square root filters, Mon. Weather Rev., 131, 1485–1490, 2003.

Wang, X., Bishop, C. H., and Julier, S. J.: Which is better, an ensemble of positive-negative pairs or a centered spherical simplex ensemble?, Mon. Weather Rev., 132, 1590–1605, 2004.

Watkinson, L. R., Lawless, A. S., Nichols, N. K., and Roulstone, I.: Weak constraints in four-dimensional variational data assimilation, Meteorol. Z., 16, 767–776, 2007.

Weaver, A. T., Vialard, J., and Anderson, D. L. T.: Three- and four-dimensional variational assimilation with a general circulation model of the tropical Pacific Ocean, Part I: Formulation, internal diagnostics, and consistency checks, Mon. Weather Rev., 131, 1360–1378, 2003.

Zhu, Y., Todling, R., Guo, J., Cohn, S. E., Navon, I. M., and Yang, Y.: The GEOS-3 retrospective data assimilation system: The 6-hour lag case, Mon. Weather Rev., 131, 2129–2150, 2003.

8

Multiscaling and joint multiscaling description of the atmospheric wind speed and the aggregate power output from a wind farm

R. Calif[1] **and F. G. Schmitt**[2]

[1]EA 4539, LARGE laboratoire en Géosciences et Énergies, Université des Antilles et de la Guyane, 97170 P-á-P, France
[2]CNRS, UMR 8187 LOG Laboratoire d'Océanologie et de Géosciences, Université de Lille 1, 28 avenue Foch, 62930 Wimereux, France

Correspondence to: R. Calif (rcalif@univ-ag.fr)

Abstract. We consider here wind speed time series and the aggregate output wind power from a wind farm. We study their scaling statistics in the framework of fully developed turbulence and Kolmogorov's theory. We estimate their Fourier power spectra and consider their scaling properties in the physical space. We show that the atmospheric wind speed and the aggregate power output from a wind farm are intermittent and multifractal over a wide range of scales. The coupling between simultaneous data of the wind speed and aggregate power output is investigated through a joint multifractal description using the generalized correlation functions (GCFs). This multiscaling test is compatible with a linear relation between the wind speed and the aggregate power output fluctuations for timescales $T \geqslant 10^3\,\mathrm{s} \simeq 15\,\mathrm{min}$.

1 Introduction

Increasing the wind energy contribution to electrical networks requires improving the tools to forecast the electrical power produced by wind farms, in order to proportion network lines. However the wind energy is a fluctuating energy ressource, due to the high variability of the wind at all spatial and/or temporal scales. In the atmospheric boundary layer, the Reynolds number (ratio of inertial to viscous force) can be as large as $Re = 10^8$ (Burton et al., 2001). Large values of the Reynolds number lead to a huge intermittency of wind speed fluctuations at all temporal or spatial scales, ranging from large-scale variations (years) to very small scale variations (few min down to seconds) (Stull, 1988). Small-scale intermittency remains a challenging problem for the turbulence community research (Peinke et al., 2004). Several approaches can be used to consider the scaling intermittency of small-scale turbulence, the most classical one being structure functions analysis. In recent years several studies have been dedicated to the analysis of scaling laws and turbulent intermittency at small scales in the laboratory (Anselmet et al., 1984; She and Levêque, 1994) and in the atmospheric boundary layer (Schmitt et al., 1993, 1994; Katul et al., 1995; Schmitt, 2007; Böttcher et al., 2007; Morales et al., 2011; Calif and Schmitt, 2012). These studies have shown that atmospheric turbulent speed at small scales has multifractal scaling fluctuations and exhibits long-range power correlations.

However the knowledge of variations ranging from minutes to a few days – corresponding to 1–1000 km, i.e., the mesoscale range – is necessary to provide efficient tools for management and control of wind power generation. The studies concerning this scale range are fewer than those for the small-scale range, due to the possible nonuniversality of the power law slope in the mesoscale range. Recent works (Lauren et al., 1999; Muzy et al., 2010) have been dedicated to scaling and multiscaling properties of the atmospheric wind speed in the mesoscale range.

In this paper, the scaling properties of the atmospheric wind speed are provided for the small-scale and mesoscale ranges. Atmospheric wind speed data sampled at 20 and 1 Hz are analyzed through multifractal theory in order to characterize the wind speed fluctuations behavior for each scaling regime. In parallel, the aggregate power output data from a wind farm are analyzed within the same theoretical framework. The paper is organized as follows. In Sect. 2, the data sets are described. In Sect. 3, the theoretical framework, traditional spectral analysis and structure functions analysis are presented. In Sect. 4, in order to quantify the coupling of the relationship between the wind atmospheric and the aggregate power output, a joint multifractal description is presented. In Sect. 5, the Fourier spectra are estimated for the atmospheric wind speed and the aggregate power output. To provide a full characterization of experimental fluctuations, at all intensities and at all scales, multifractal and joint multifractal approaches are used.

2 Description of the data

In order to investigate the spectral and multiscaling properties of the turbulent wind speed v in the atmospheric surface layer and the aggregate power output P, a data set is collected on the wind energy production site of Petit-Canal, an island in the French West Indies. This 10 MW production site, located at 16°15′ N latitude and 60°30′ W longitude, was positioned at approximately 60 m (197 ft) a.s.l., at the top of a sea cliff. The wind speed v is measured with a three-cup anemometer (model A100L2 from Vector Instruments) black having a response time of 0.15 s and an ultrasonic anemometer (model CSAT3). Both were mounted on a 40 m (131 ft) tall mast erected 20 m (66 ft) from the cliff edge, at 38 m (125 ft) from the ground. The two experimental databases for atmospheric wind speed are used here: (i) with the three-cup anemometer, the sampling frequency and duration are respectively 1 Hz and 1 yr (January 2006–2007); (ii) with the ultrasonic anemometer, the sampling frequency and duration are respectively 20 Hz and 2 weeks (14–28 July 2005). Moreover, the aggregate power output produced by this wind farm was recorded with a sampling frequency 1 Hz, during 1 yr (January 2006–2007). Table 1 gives a description of our data set with the mean and the standard deviation values. Additionally, two examples of simultaneous wind speed and aggregate power output data sequence for different timescales, are illustrated in Fig. 1: (i) Fig. 1a presents simultaneous time series sampled at 1 Hz; (ii) Fig. 1b presents the moving averages of simultaneous time series, over a time period $T = 10^4$ s. Figure 1a shows that for very short timescales of the order of seconds, the wind speed signal exhibits high fluctuations contrary to the wind power signal, while over large scales the observed moving averages have the same trends.

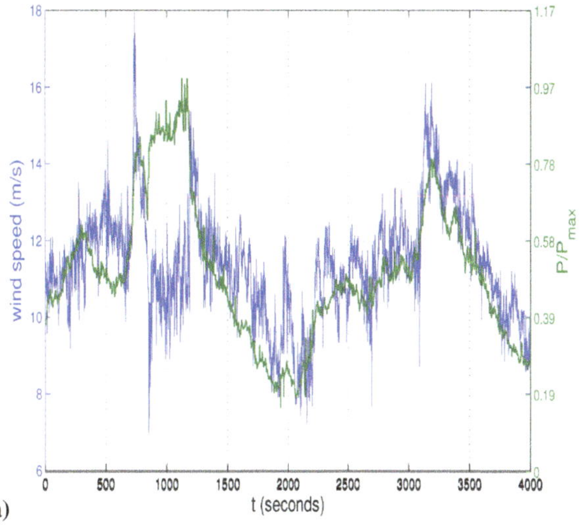

(a)

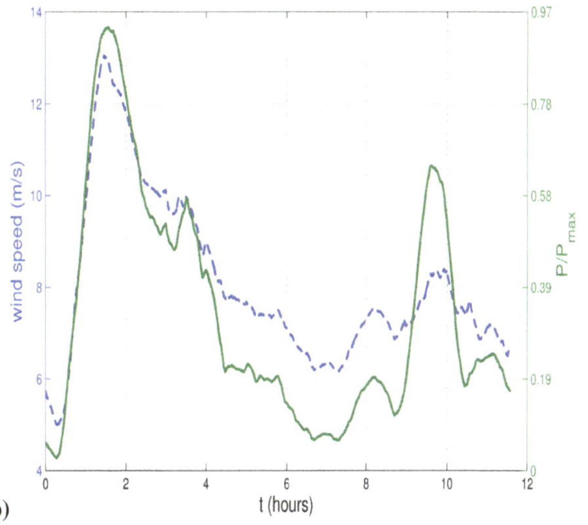

(b)

Fig. 1. Two examples of simultaneous wind power data: **(a)** simultaneous time series sampled at 1 Hz; **(b)** moving averages of simultaneous time series, over a time period $T = 10^4\,\mathrm{s} \simeq 3\,\mathrm{h}$.

3 Theoretical framework

3.1 The existence of −1 and −5/3 power laws in the atmospheric wind speed

Classically, a scale invariance can be detected by computing of power spectral density $E(f)$. For a scale-invariant process, the following power law is obtained over a range of frequency f:

$$E(f) \sim f^{-\beta}, \tag{1}$$

where β is the spectral exponent. According to some authors (Mandelbrot, 1982; Schertzer and Lovejoy, 1987; Marshak et al., 1994), it contains information about the degree of stationarity of the field:

- $\beta < 1$, the process is stationary,
- $\beta > 1$, the process is nonstationary,
- $1 < \beta < 3$, the process is nonstationary with increments stationary.

It may also be considered as characterizing the degree of correlation (Ivanov et al., 2001; Telesca et al., 2003).

In the atmospheric surface layer, it is recognized that wind speed spectra possess three spectral regions (Pope, 2000): (1) the energy-containing range black (or large-scale turbulence) in which turbulent kinetic energy is produced, (2) the inertial subrange (small-scale turbulence) in which

Table 1. Description of our database with the mean and the standard deviation values.

Data set	Sampling frequency	Total number of data points	Mean value $<.>$	Standard deviation σ
Wind speed v	20 Hz	2 886 276 (2005)	$8.39\,\mathrm{m/s^{-1}}$	$1.95\,\mathrm{m\,s^{-1}}$
Wind speed v	1 Hz	26 438 400 (2006)	$8.20\,\mathrm{m/s^{-1}}$	$3.32\,\mathrm{m\,s^{-1}}$
Aggregate power output P	1 Hz	26 438 400 (2006)	–	–

turbulent kinetic energy is transferred from large to small scales, and (3) the dissipation range (millimeters and smaller) where turbulent kinetic energy is converted to heat by the action of fluid viscosity. Figure 2 gives an illustration of the Kolmogorov quasi-equilibrium energetic cascade in the inertial range. In this zone, the Kolmogorov theory implies that the wind speed fluctuations possess a power law spectrum (Kolmogorov, 1941; Obukhov, 1941):

$$\beta = 5/3. \tag{2}$$

This relation is written here for wave numbers, but can also be used for frequencies, involving Taylor's hypothesis to relate spatial fluctuations to temporal fluctuations. This power law has been verified many times for wind atmospheric data (Schmitt et al., 1993, 1994; Katul et al., 1995; Schmitt, 2007; Morales et al., 2011; Calif and Schmitt, 2012). In the energy-containing range where the scales extend from a few minutes to a few days, the properties of turbulent wind speed depend on their strong anisotropy and their dependence on the flow domain boundary characteristics (Katul and Chu, 1998; Katul et al., 2012). However in this study, the existence of universal power laws at low wave numbers for the power spectral density of the turbulent longitudinal velocity is examined theoretically and experimentally for the near-neutral atmospheric surface layer. Indeed, several experimental studies have found a -1 power law slope at production scales for the wind speed spectrum in the atmospheric surface layer in neutral conditions (Katul and Chu, 1998; Nickels et al., 2005; Katul et al., 2012):

$$E(f) \sim f^{-1};\, f \leqslant 1\,\mathrm{Hz}. \tag{3}$$

Moreover this power law slope was predicted by three theoretical approaches. More details are given in Katul and Chu (1998). However this power law does not present the same level of universality as $\beta = 5/3$ for turbulence in the inertial range: as mentioned in Lauren et al. (1999), in the mesoscale range, the power law slope can vary with the local topography and the atmospheric conditions. Let us note also that in the wind energy community some fits of speed power spectra are often used, based on von Karman's formula of the form $E(f) = a/(b + cf^2)^{5/6}$ (Karman, 1948) or on Kaimal's formula $E(f) = A/(1 + Bf)^{5/3}$ (Kaimal et al., 1972), where a, b, c and A, B are constants. These two formulae have no theoretical ground: von Karman wanted to perform an interpolation between a low-frequency f^4 spectrum and a high-frequency Kolmogorov spectrum $f^{-5/3}$, and Kaimal et al. (1972) proposed their formula on purely empirical grounds for the low-frequency part. In both cases the main idea is to capture in a single expression the injections scales and the inertial range with Kolmogorov spectrum. Indeed, these fits are based on Kolmogorov theory (Simiu and Scalan, 1978) which defines a general form of the fluctuating wind speed shown as follows (Zhang et al., 2008). In the present paper, we mainly focus on other scaling regimes and therefore do not need to consider the fits proposed by these authors. Finally, we must note also that the power spectrum density $E(f)$ is a second-order statistic (proportional to the square of the amplitude of a given frequency fluctuation) and its slope is not sufficient to fully specify a scaling process. Multifractal analysis is a natural generalization to study the scaling behavior of a nonlinear phenomenon, using qth-order structure functions, and to obtain a full characterization of wind speed fluctuations, at all intensities and all scales; this is discussed in the next subsection.

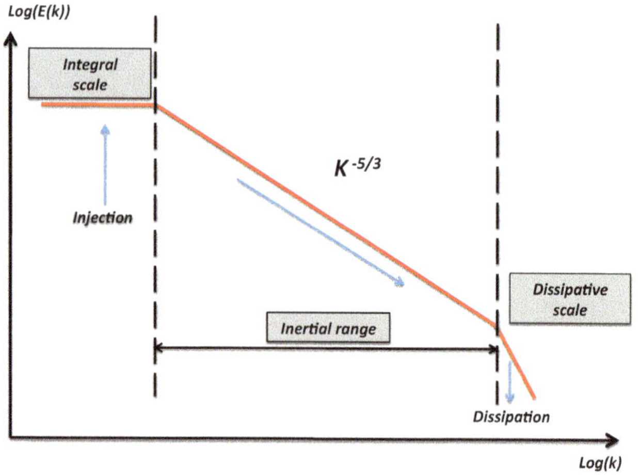

Fig. 2. An illustration of the Kolmogorov quasi-equilibrium energetic cascade in the inertial range.

3.2 Intermittency and multifractal properties

Intermittency in turbulence has been a subject of research for almost 50 yr now, following Kolmogorov's 1962 seminal works (Kolmogorov, 1962). This intermittency in the inertial range has been modeled by cascade models in various studies for velocity (Novikov and Stewart, 1964; Yaglom, 1966; Mandelbrot, 1974). These models reproduce intermittency and multiscaling in the inertial range turbulence. Here we may define intermittency as the property of having large fluctuations at all scales, with a correlated structure, or, in other words, with various orders of singularities to be distributed over fractal sets with varying dimensions (rather than a unique dimension as for monofractal process); in the case of intermittency the scaling moment function $\zeta(q)$, which is introduced below, is nonlinear. Large fluctuations are much more frequent than what would be obtained for Gaussian processes (Frisch, 1995; Schertzer et al., 1997; Vulpiani and Livi, 2004). This is typically studied considering the probability density function (PDF), or more often using the moment of order q of these fluctuations, called "structure functions of order q":

$$S_q(\Delta t) = < (\Delta X)^q > \simeq \Delta t^{\zeta(q)}, \tag{4}$$

where $\Delta X(\Delta t) = X(t+\Delta t) - X(t)$ and Δt is a time increment. We have here written the fluctuations in time, since below in this paper we deal with time series analysis. $\zeta(q)$ is the scaling exponent function. $\zeta(2) = \beta - 1$ relates the second-order moment to the power spectrum scaling exponent; the knowledge of the full $(q, \zeta(q))$ curve for integer and noninteger moments provides a full characterization of wind speed fluctuations at all scales and all intensities. The parameter $H = \zeta(1)$ is the Hurst exponent characterizing the nonconservation of the mean. Monofractal processes correspond to a linear function $\zeta(q) = qH$, with $H = 1/2$ for the Brownian motion, $0 < H < 1$ ($H \neq 1/2$) for a fractional Brownian motion (that can be defined as a fractional integration of order b ($0 < b < 1$) of a Gaussian noise, with $H = \frac{3}{2} - b$). The values of the function $\zeta(q)$ are estimated from the slope of $S_q(\Delta t)$ versus Δt in a log–log diagram for all moments q. The function $\zeta(q)$ defines the types of scaling behavior; in other words, this exponent function is useful to characterize the statistics of the random process. If $\zeta(q)$ is linear, the statistical behavior is monoscaling; if $\zeta(q)$ is nonlinear and concave, the behavior is defined as mutiscaling, corresponding to a multlfractal process. The concavity of this function is a characteristic of the intermittency: the more the curve is concave, the more the process is intermittent (Frisch, 1995; Schertzer et al., 1997; Vulpiani and Livi, 2004).

Kolmogorov's 1941 model corresponds to a linear model for the exponent function (Kolmogorov, 1941):

$$\zeta(q) = \frac{q}{3}. \tag{5}$$

For a multifractal process, $\zeta(q)$ is nonlinear. Several models have been proposed in the literature to fit the scaling exponents $\zeta(q)$. Here we consider the classical lognormal model (Kolmogorov, 1962):

$$\zeta(q) = \frac{q}{3} + \frac{\mu}{18}(3q - q^2), \tag{6}$$

with μ being the intermittency parameter. Yaglom (1966) has proposed a multiplicative cascade model compatible with Kolmogorov's ideas. Several other models have been proposed since in the literature, for example the log-stable model proposed by Schertzer and Lovejoy (1987):

$$\zeta(q) = qH - \frac{C_1}{(\alpha - 1)}\left(q^\alpha - q\right), \tag{7}$$

where $H = \zeta(1)$ is the Hurst parameter which defines the degree of smoothness or roughness of the field. The parameter C_1 is the fractal co-dimension of the set giving the dominant contribution to the mean ($q = 1$) and bounded between 0 and d (d the dimension space, here $d = 1$). It measures the mean intermittency characterizing the sparseness of the field: the larger C_1, the more the mean field is inhomogeneous. The multifractal Lévy parameter α is bounded between 0 and 2, where $\alpha = 0$ corresponds to the monofractal case and $\alpha = 2$ coresponds to the multifractal lognormal case. The parameter α measures the degree of multifractility, i.e, how fast the inhomogeneity increases with the order of the moments. But here we consider the lognormal model which provides a reasonable fit up to $q = 5$; hence here the question of the best model (among the inifinitely divisible family of models) is not the topic of the present paper, and we consider here the lognormal fit as convenient for the joint analysis done in the next section.

4 Joint analysis for multivariate data

4.1 Test for data independence using second-order correlation

In order to test independence between two random process, second-order statistics are often considered in the wind energy community (Burton et al., 2001). More precisely the cross-correlation function in the time domain and the coherence function in the frequency domain, of two processes $x(t)$ and $y(t)$, are determined for highlighting possible correlations. We recall here the expressions for the cross-correlation function C_{xy} and the coherence function H_{xy} (Papoulis and Pillai, 2002):

$$C_{xy} = \frac{< x(t+\tau)y(t) >}{\sigma_x \sigma_y}, \tag{8}$$

where $< . >$ is the statistical average, τ is a time lag, σ_x, σ_y are respectively the standard deviations of processes $x(t)$ and $y(t)$ (Papoulis and Pillai, 2002), and

$$H_{xy} = \sqrt{\frac{|E_{xy}(f)|^2}{E_x(f)E_y(f)}}. \tag{9}$$

$E_{xy}(f)$ is the Fourier co-spectrum, and $E_x(f)$ and $E_y(f)$ are the Fourier spectra of processes $x(t)$ and $y(t)$, respectively. If C_{xy} and H_{xy} are zero or close to zero, then x and y are uncorrelated. In contrast, if C_{xy} and H_{xy} are unity or close to unity, x and y are perfectly correlated; if C_{xy} and H_{xy} are close to -1, x and y are negatively correlated.

These statistical tests are based on second-order statistics. In the following section a generalization of correlation coefficient is given using the multifractal framework.

4.2 Joint multifractal description for bivariate field: generalized correlation functions (GCFs) and exponents (GCEs)

We recall here that, whereas independence implies noncorrelation, noncorrelation does not imply independence. In order to better consider the relation between two scaling time series, we apply here a testing technique proposed in Seuront and Schmitt (2005). Instead of random variables x and y, we consider here the increments of two stochastic processes $\Delta X(\Delta t) = X(t+\Delta t) - X(t)$ and $\Delta Y(\Delta t) = Y(t+\Delta t) - Y(t)$, and the normalization of the joint moments is given as (Seuront and Schmitt, 2005)

$$c(h,g) = \frac{<\Delta X^h \Delta Y^g>}{<\Delta X^h><\Delta Y^g>} \sim \Delta t^{-r(h,g)}. \qquad (10)$$

When ΔX^h and ΔY^g are independent, $r(h,g)=0$ and $c(h,g)=1$. In contrast, increasing values of $c(h,g)$ would characterize increasing dependence between ΔX^h and ΔY^g.

The generalized correlation exponent (GCE hereafter), estimated as the slope of the power law of $c(h,g)$ versus Δt in a log–log plot, is then expressed as:

$$r(h,g) = \zeta_X(h) + \zeta_Y(g) - S(h,g), \qquad (11)$$

where $\zeta_X(h)$ and $\zeta_Y(g)$ characterize the multiscaling properties of the single fluctuations $<\Delta X^h>$, and $<\Delta Y^g>$ and $S(h,g)$ characterize the multiscaling properties of the joint fluctuations $<\Delta X^h \Delta Y^g>$. Both $c(h,g)$ and $r(h,g)$ are generalizations of correlation functions. This multiscaling test for independence between two stochastic processes takes into account the multifractal character of intermittent processes. It also allows to test for the phenomenology responsible for the high-intensity (rare and unexpected) fluctuations observed in intermittent distributions, considering their potential association with both high- and low-intensity fluctuations characterized respectively by high and low orders of moment. In Seuront and Schmitt (2005), GCEs are considered in special cases:

- If X and Y are lognormal multifractal processes (Meneveau et al., 1990), then

  $$S(h,g) = a_1 h + a_2 g - a_3 h^2 - a_4 g^2 - \sigma hg \qquad (12)$$

  so that

  $$r(h,g) = S(h,0) + S(0,g) - S(h,g). \qquad (13)$$

  Consequently

  $$r(h,g) = \sigma hg. \qquad (14)$$

  In this case, it is clear that $r(1,1)$ or $r(2,2)$ are enough to estimate the only needed parameter, namely the correlation coefficient σ, so that if $r(1,1)=0$ or $r(2,2)=0$, it can be concluded that the two processes are independent.

- If X and Y are independent, $r(h,g)=0$;

- if X and Y are proportional, i.e., $X = aY$, then

  $$r(h,g) = \zeta_Y(h) + \zeta_Y(g) - \zeta_Y(h+g). \qquad (15)$$

  Additionally the shape of the obtained surface is symmetric in the h–g plane.

- If a power law exists between X and Y, i.e., $X = aY^b$, one has

  $$r(h,g) = \zeta_Y(bh) + \zeta_Y(g) - \zeta_Y(bh+g). \qquad (16)$$

  In this case $r(h,g) > 0$, but it is symmetric in the bh–g plane.

5 Spectral and multifractal analysis

In this section the Fourier power spectral densities are estimated for our database: wind speed data sampled at 20 and 1 Hz, and aggregate power output data sampled at 1 Hz. After this, the multifractal analysis is applied.

5.1 Spectral analysis

In this study, in order to estimate the power spectral densities of the wind speed and the aggregate power output from a wind farm, the discrete Fourier transform of the times series considered is computed. The expression of the power spectral density for a process $x(t)$ is recalled here. A N point-long interval used to construct the value at frequency domain point f, X_f (Bracewell, 1999):

$$X(f) = \sum_{j=0}^{N-1} x_j e^{2\pi ijf/N}, \quad f = 0,1,\ldots,N-1, \qquad (17)$$

and the power spectral density is

$$E(f) = \frac{1}{2\pi}|X(f)|^2. \qquad (18)$$

The Fourier spectra reveal two scaling ranges in the atmospheric wind speed and aggregate power output data. In Fig. 3, we have plotted the spectrum for the wind speed data sampled at 20Hz. This spectrum is obtained from an average of 2886 spectra computed with wind speed time series of length 1000 data points. This averaged spectrum

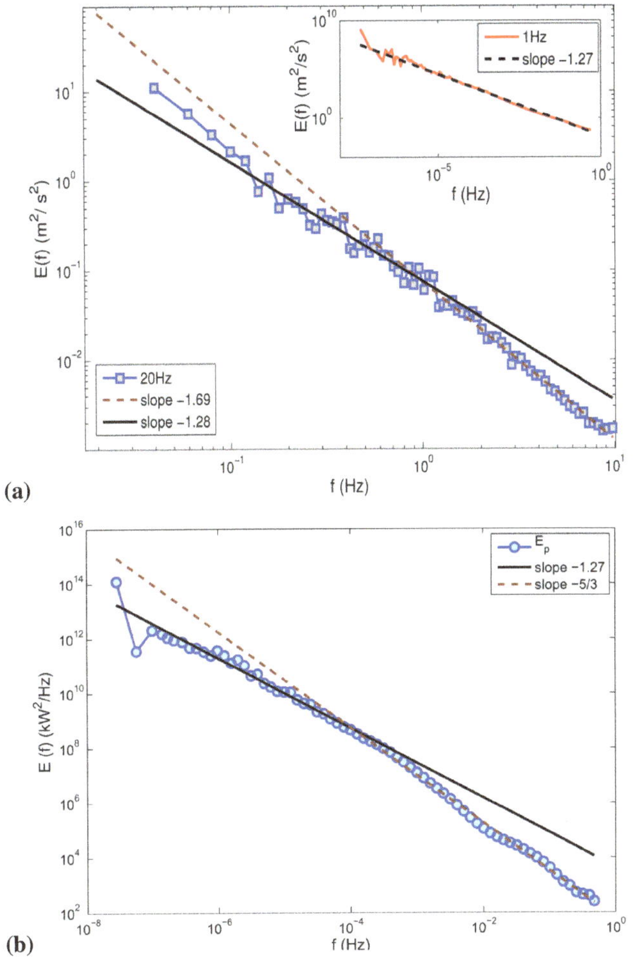

Fig. 3. The Fourier power spectra illustrating two scaling regimes: **(a)** averaged spectrum $E_v(f)$ for the atmospheric wind speed sampled at 20 Hz displaying a -1.69 scaling for $1 \leqslant f \leqslant 10$ Hz and a -1.28 scaling for $f \leqslant 1$ Hz. The inset shows that the Fourier spectra of wind atmospheric sampled at 1 Hz display a -1.27 scaling over seven decades. **(b)** Averaged spectrum $E_p(f)$ for the aggregate power output, for which a crossover between a $-5/3$ slope and -1.27 slope is also found.

highlights two spectral regimes. For the high frequencies $1 \leqslant f \leqslant 10$ Hz, corresponding to timescales $0.1 \leqslant T \leqslant 1$ s, in the inertial range, as expected the spectrum possesses a spectral slope $\beta = 1.69$ very close to 5/3 expected by Kolmogorov's theory. The slight difference is usually interpretated as coming from intermittency effects (Frisch, 1995; Schertzer et al., 1997). For the low-frequency part – i.e., $f \leqslant 1$ Hz corresponding to timescales $T \geqslant 1$ s in the energy-containing range – the spectrum follows a -1.28 spectral slope. This is consistent with Katul and Chu (1998) and recently Fitton et al. (2014). In the inset, we displayed the spectrum obtained from the wind speed data sampled at 1 Hz, exhibiting a -1.28 scaling for the low-frequency part – i.e., $10^{-7} \leqslant f \leqslant 1$ Hz corresponding to timescales $1 \leqslant T \leqslant 10^7$ s (10^7 s $\approx$ 4 months) – over seven decades. This confirms the scaling obtained for frequencies $f \leqslant 1$ Hz with the wind speed data sampled at 20 Hz. Furthermore this scaling is found also in Calif et al. (2014) using Hilbert spectral technique. Recent papers on longitudinal low-wave-number spectra provide evidence of a -1 power law in the atmospheric surface layer for neutral conditions, and a $-5/3$ power law for near-convective conditions (Katul et al., 1995; Katul and Chu, 1998; Kader and Yaglom, 1991). The power spectrum density $E_v(f)$ presented in this paper seems to show a break around 1 Hz, in agreement with the results published in Lauren et al. (1999). The models of Kader and Yaglom (1991) and Katul et al. (1995) suggest a transition frequency of 0.3 Hz, and the Kaimal model a value of 1 Hz. Figure 3a shows that our present data are comparable with the latter value.

Figure 3b illustrates the average spectrum $E_p(f)$ of the aggregate power output for the entire wind farm sampled at 1 Hz: the same slopes are observed. For frequencies $10^{-4} \leqslant f \leqslant 0.5$ Hz corresponding to timescales $2 \leqslant T \leqslant 10^4$ s, $E_p(f)$ displays a power law near the exact value 5/3, with $\beta = 1.67$. Previously published power spectra of wind generator power have shown power law regions of the power spectrum plot covering one or two decades of frequency (Sørensen et al., 2002; McNerney and Richardson, 1992). These studies have not provided a comparison between their data and the Kolmogorov spectrum. Apt (2007) has shown that the output wind power from a wind farm located follows a Kolmogorov spectrum over more than four orders of magnitude in frequency from 4.45×10^{-6} to 3.33×10^{-2} Hz, corresponding to timescales from 30 s to 2.6 days. Here, for frequencies $f \leqslant 10^{-4}$ Hz corresponding to timescales $T \geqslant 10^4$ s (approximately 3 h), a power law with $\beta = 1.27$ is observed for the first time. This highlights a scaling break approximately around 10^{-4} Hz, for the power spectrum density $E_p(f)$.

We obtain approximately the same value for the change of slope. Let us then remark here that wind speed and aggregate power output have the same type of regime with high-frequency $-5/3$ power law behavior and low-frequency -1.2 power law behavior. However the scale break for the regime change are not the same: $T_0 = 1$ s for the wind speed and $T_0 = 10^4$ s $\simeq 3$ h for the aggregate power output. We do not know whether this four-orders-of-magnitude change in the scale break is universal or is related to some spatial characteristics, such as local meteorology and spacing of wind turbine generators in the farm.

However, the power spectrum density $E(f)$ is a second-order statistic (proportional to the square of the amplitude of a given frequency fluctuation) and its slope is not sufficient to fully specify a scaling process. Multifractal analysis is a natural generalization to study the scaling behavior of a nonlinear phenomenon, using qth-order structure functions; this is discussed below.

5.2 The structure functions scaling exponent $\zeta(q)$ for the atmospheric wind speed sequences

Here, we have analyzed two databases for atmospheric wind speed v sampled at 20 and 1 Hz. The structure function analysis is performed for wind speed increments $\Delta v(\Delta t) = |v(t+\Delta t) - v(t)|$ for $0.05 \leqslant \Delta t \leqslant 500$ s of data sampled at 20 Hz and for $1 \leqslant \Delta t \leqslant 6 \times 10^5$ s of data sampled at 1 Hz and for all moments between 0.15 and 5 with increments of 0.25. Figure 4a gives the scaling of the structure functions for a sequence of length 7000 data points for $q = 1$, 1.5, 2, and 2.5 in a log–log diagram for wind speed sampled at 20 Hz. The structure functions display two scaling tendencies: for $0.05 \leqslant \Delta t \leqslant 1$ s corresponding to the inertial range and for $10 \leqslant \Delta t \leqslant 500$ s. Figure 4b gives the scaling of the structure functions for a sequence for $q = 1$, 1.5, 2, and 2.5 in a log–log diagram for wind speed v sampled at 1 Hz, for $10 \leqslant \Delta t \leqslant 5 \times 10^5$ s. The straight lines in this figure indicate that the scaling of the relationship is well respected. The two scaling tendencies observed for the wind speed v correspond to the inertial range (small scales) and the energy-containing range (large scales).

We then estimate the scaling exponent functions $\zeta_{vH}(q)$ and $\zeta_{vL}(q)$ respectively for the inertial and the mesoscale ranges. The exponent functions ζ_{vH} are estimated with 412 time series of length 7000 values sampled at 20 Hz and ζ_{vL} with 5 time series of length 5 millions of values sampled at 1 Hz. The scaling exponent function $\zeta_v(q)$ is the ensemble average obtained from all these scaling exponent functions $\zeta_i(q)$:

$$\zeta_{vH}(q) = \frac{1}{412}\sum_{i=1}^{412}\zeta_i(q); \zeta_{vL}(q) = \frac{1}{5}\sum_{i=1}^{5}\zeta_i(q). \tag{19}$$

5.2.1 Scaling exponent function ζ_{vH} for the inertial range

Figure 5 presents the empirical scaling exponent function ζ_{vH}, for the inertial range, compared to linear model K41 ($\zeta(q) = q/3$) (Kolmogorov, 1941), and the lognormal model proposed by Kolmogorov (1962) that corresponds to the following relationship:

$$\zeta(q) = \frac{q}{3} + \frac{\mu}{18}(3q - q^2), \tag{20}$$

where the parameter μ is called "intermittency parameter" and is close to 0.25 (Arneodo et al., 1996). The function $\zeta_{vH}(q)$ obtained from our database is nonlinear and concave similar to the results obtained in previous studies (Schmitt et al., 1993; Schertzer et al., 1997), with the following properties:

- $0 < q < 3$, $\zeta_{vH}(q) > q/3$ for small fluctuations,
- $q = 3$, $\zeta_{vH}(3) = 1$ in agreement with the Kolmogorov 4/5 law,
- $q > 3$, $\zeta_{vH}(q) < q/3$ for large fluctuations.

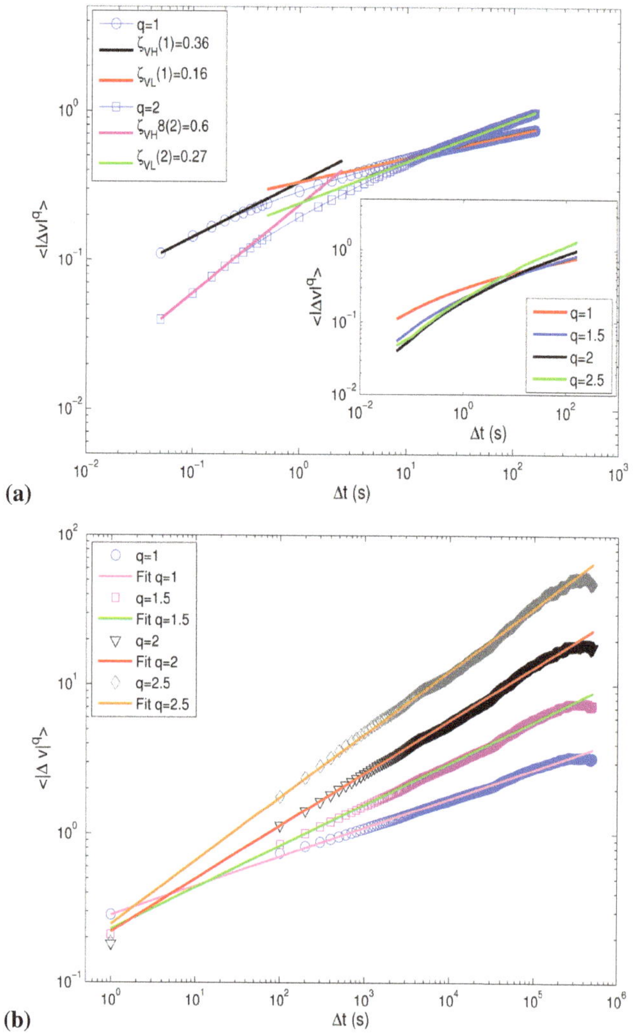

Fig. 4. The scaling of the structure functions of the wind speed $S_v(q)$ **(a)** for data sampled at 20 Hz and **(b)** for data sampled at 1 Hz. The structure functions display also two scaling regimes for $0.05 \leqslant \Delta t \leqslant 10^3$ s for data sampled at 20 Hz and a scaling regime for $1 \leqslant \Delta t \leqslant 5 \times 10^5$ s.

The lognormal model provides a reasonable fit for the empirical exponent function $\zeta_{vH}(q)$. The average of the fluctuations correspond to $q = 1$, and $H = \zeta_{vH}(1) \simeq 0.36$ is the so-called "Hurst" exponent characterizing the scaling nonconservation of the mean. The second moment $\zeta_{vH}(2) = 0.68$ is linked to the spectral exponent $\beta \simeq 1 + \zeta_{vH}(2) = 1.68$.

5.2.2 Scaling exponent function ζ_{vL} for the energy-containing range

Figure 5 illustrates the empirical scaling exponent function $\zeta_{vL}(q)$ obtained for the mesoscale range, compared to a linear model ($\zeta_{vL}(1) \simeq 1/6$) and a quadratic model

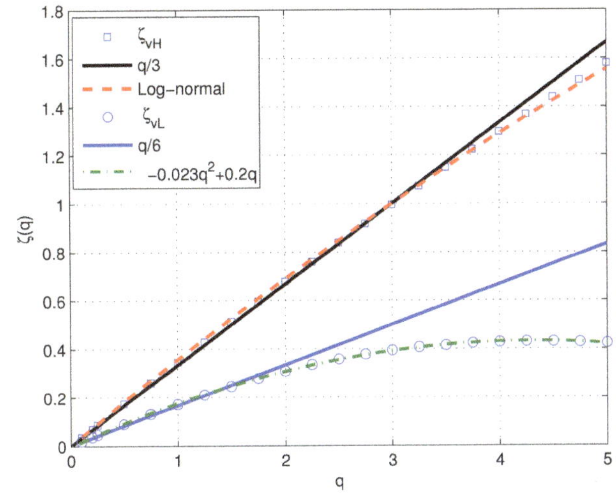

Fig. 5. The empirical scaling exponent function ζ_{vH} (□) for the inertial range compared to the linear model K41 (solid line) corresponding to homogeneous turbulence, and the lognormal model (dashed line) taking into account the intermittency effects. The empirical scaling exponent function ζ_{vL} (°) for the mesoscale range, compared to a linear function $\zeta_{vL}(1)q \simeq q/6$ (blue solid line) and a quadratic function $-0.023q^2+0.2q$ (green dashed line).

$\zeta(q) = -0.023q^2 + 0.2q$. The exponent function ζ_{vL} is nonlinear and concave, characterizing a multifractal process. We obtain the value of $H = \zeta_{vL}(1) \simeq 0.16$ and the value $\zeta_{vL}(2) \simeq 0.30$, and consequently $\beta \simeq 1.30$ compatible with the value estimated by the Fourier analysis ($\beta \simeq 1.27$). The moment function $K(q) = qH - \zeta(q)$ is often used for characterizing a type of scaling. Lauren et al. (1999) performed a multiscaling analysis for characterizing the low-wavenumber statistical properties of surface layer winds. In this study, the data were measured with a cup anemometer at a sampling rate of 1 Hz. Then, the function $K(q)$ was plotted for q ranging 0–3; for example $K(1)$ and $K(3)$ were estimated respectively to be 0 and 0.16. In our database the values of $K(1)$ and $K(3)$ were estimated respectively to be 0 and 0.13.

5.3 The structure functions scaling exponent $\zeta_P(q)$ for the aggregate power output

Here we apply the structure function analysis to aggregate power output data from a cluster of wind turbine generators, for characterizing the type (monoscaling or multiscaling) and regimes of scaling for these data. We consider the increments of output power $\Delta P(\Delta t) = |P(t+\Delta t) - P(t)|$ for $1 \leqslant \Delta t \leqslant 10^6$ s and for all moments between 0.25 and 5 with increments of 0.25. Figure 4b illustrates the scaling for the aggregate power output P for $q = 1$, 1.5, 2 and 2.5, in a log–log diagram. The structure functions display two scaling tendencics: for $1 \leqslant \Delta t \leqslant 6.5 \times 10^4$ s and for $6.5 \times 10^4 \leqslant \Delta t \leqslant 10^6$ s. Here, the straight lines black show also that the scaling of Eq. (2) is well verified. Then, we estimate the scaling ex-

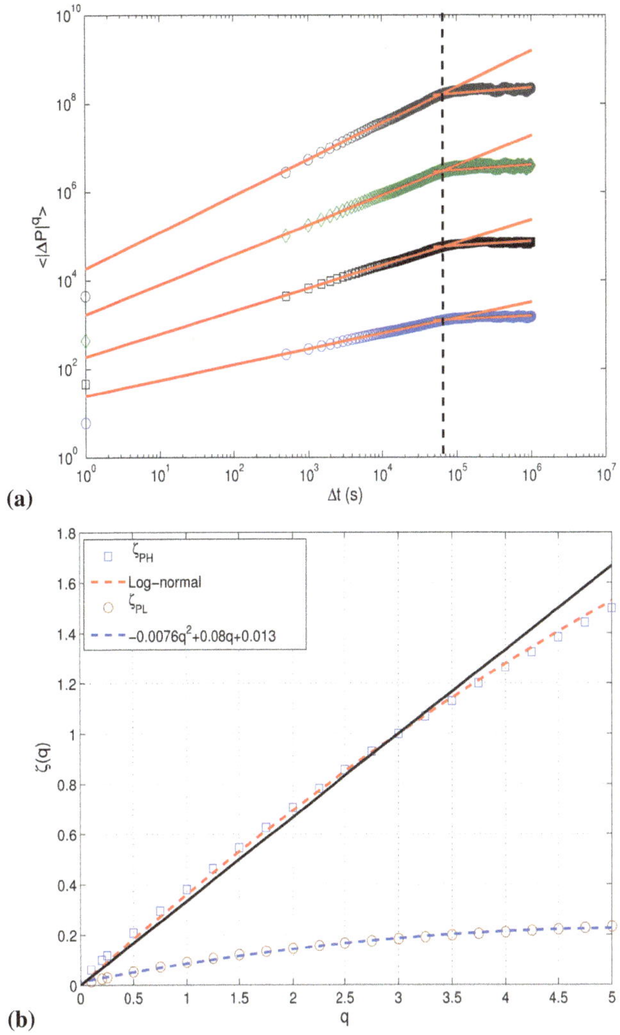

Fig. 6. (a) The scaling of the structure functions for wind power $S_P(q)$ illustrating two scaling regimes: for $1 \leqslant \Delta t \leqslant 6.5 \times 10^4$ s and for $6.5 \times 10^4 \leqslant \Delta t \leqslant 10^6$ s. **(b)** The empirical scaling exponent function $\zeta_{PH}(q)$ (□) estimated for $1 \leqslant \Delta t \leqslant 6.5 \times 10^4$ s and compared to the K41 model (black solid line) and the lognormal model (red dashed line). The empirical scaling exponent function $\zeta_{PL}(q)$ (°) compared to a quadratic fit (blue dashed line).

ponent functions $\zeta_{PH}(q)$ and $\zeta_{PL}(q)$ from the slopes of the straight lines using a linear least regression, for each scaling regime. The scaling exponent function $\zeta_{PH}(q)$ is estimated with $i = 20$ time series of length 10^6 values sampled at 1 Hz and $\zeta_{PL}(q)$ with $i = 2$ time series of length 10^7 values sampled at $1Hz$. The scaling exponent functions $\zeta_P(q)$ are the ensemble average obtained from all these exponents $\zeta_i(q)$.

The empirical curves $\zeta_{PH}(q)$ and $\zeta_{PL}(q)$ for the aggregate power output P of this wind farm are illustrated in Fig. 6. $\zeta_{PH}(q)$ is compared to the linear model K41 $(q/3)$ and the lognormal model given in Eq. (6). For the timescales $1 \leqslant \Delta t \leqslant 6.5 \times 10^4$ s (≈ 18 h), $\zeta_{PH}(q)$ is concave and nonlinear, characterizing a multifractal process. Moreover, the exponent function ζ_{PH} is close to the lognormal model. $\zeta_{PH}(q)$ presents the following properties similar to $\zeta_{vH}(q)$:

- $0 < q < 3, \zeta_{PH}(q) > q/3$ for small fluctuations,
- $q = 3, \zeta_{PH}(3) = 1$,
- $q > 3, \zeta_{PH}(q) < q/3$ for large fluctuations.

For the timescales $t \geqslant 6.5 \times 10^4$ s (~ 18 h), $\zeta_{PL}(q)$ is nonlinear and concave, characterizing a multifractal process. $\zeta_{PL}(q)$ is fitted by a quadratic model. However, a representative estimate of $\zeta_{PL}(q)$ must be performed with more data.

In short, the atmospheric wind data present two scaling regimes observed with spectral and structure function analysis. The scaling exponent functions ζ_{vH} and ζ_{vL} are fitted by an equation of type $-Aq^2 + Bq$ with A characterizing the curvature indicating the degree of intermittency. Consequently, the turbulent wind speed is less intermittent in the inertial range ($A = 0.013$) than the energy-containing range ($A = 0.023$). The aggregate power output presents also two scaling regimes. For $1 \leqslant \Delta t \leqslant 6.5 \times 10^4$ s, the wind power is a multifractal process having the same properties as the wind speed in the inertial range, and for $\Delta t \geqslant 6.5 \times 10^4$ s the wind power is also a multifractal process less intermittent than the turbulent wind atmospheric in the mesoscale range.

6 Relation between the wind speed and the aggregate power output fluctuations

6.1 A scaling test for dependence: the coherence and the cross-correlation functions

The coherence function is often used in the wind energy community (Burton et al., 2001; Sørensen et al., 2002; Nanahara et al., 2004). In this work, the coherence function H_{vP} between the measured wind speed v and the aggregate power output P from the wind farm considered is estimated from the co-spectrum $E_{vp}(f)$ for the simultaneous data and density power spectrum $E_v(f)$ and $E_p(f)$ respectively for v and P. Figure 7a illustrates the coherence function H_{vP} defined in Eq. (9), plotted for a frequency range $10^{-7} \leqslant f \leqslant 1$ Hz. Three regimes are observed: for $f \geqslant 10^{-2}$ Hz corresponding to timescales $T \leqslant 100$ s, H_{vP} shows clearly the absence of correlation. For $10^{-3} \leqslant f \leqslant 10^{-2}$ Hz, corresponding to timescales $100 \leqslant T \leqslant 1000$ s, the value of the coherence function increases from 0 to 0.7, highlighting the presence of a correlation between the wind speed v and the aggregate power output P for frequency range $f \leqslant 10^{-3}$ Hz. Figure 7b illustrates the maximum of the cross-correlation function for a temporal range $1 \leqslant T \leqslant 10^7$ s in a log–log plot. We recall here the expression of scalar M_{vP} computed for a given timescale T:

$$M_{vP}(T) = \max\left[\frac{< v_T(t+\tau)P_T(t) >}{\sigma_v \sigma_p}\right], \tag{21}$$

where v_T and P_T are respectively the moving averages of wind speed and the power output data computer over a time period T, and σ_v and σ_P are the corresponding standard deviations. Figure 7b illustrates a the maximum M_{vP} of the cross-correlation C_{vP} in a log–log plot for the simultaneous wind speed/power output data. In the inset, we give examples of cross-correlation functions C_{vP} for the simultaneous wind speed/power output data moving averages, computed for timescales $T = 10^2$ s, 10^3, 10^4, and 10^5 s. We can see that the scalar M_{vP} is increasing for small timescales and reaches a plateau for large timescales with a transition around $10^2 - 10^3$ s; it shows the presence of a correlation between the wind speed and output power for timescales $T \geqslant 10^3$ s.

6.2 A multiscaling test for dependence: generalized correlation functions (GCFs) and exponents (GCEs)

In the multiscaling framework, two stochastic processes, defined previously, $\Delta v(\Delta t)$ and $\Delta P(\Delta t)$, characterized by their hth- and gth-order structure functions, are considered. The correlation between the two processes $(\Delta v(\Delta t))^h$ and $(\Delta P(\Delta t))^g$ then becomes a function of the scale and of the statistical orders of moment h and g, expressed by the GCF.

Figure 8a illustrates the GCF $c(h,g)$ plotted in log–log versus the time increment $\Delta t \geqslant 10^3$ s, for simultaneous wind speed and output power measurements. The function $c(h,g)$ represented is estimated for a constant value of the statistical order of moment g of wind power fluctuations ($g = 3$), and different values of statistical order of moment h of wind speed fluctuations (i.e. $h = 1, 2, 3$, from bottom to top). The function $c(h,g)$ displays a scaling for timescales $10^3 < \Delta t \leqslant 2.10^6$ s. The GFEs $r(h,g)$ are estimated from the linear regression slope of $c(h,g)$, illustrated in Fig. 8b. We can observe that the correlation between the wind speed and the wind power fluctuations increases with increasing values of the statistical order. The iso-values of $r(h,g)$ follow approximately a function of the form σhg. Moreover, because of the symmetry of $r(h,g)$ in the h–g plane, we expect to obtain a relation of proportionality between $\Delta v(\Delta t)$ and $\Delta P(\Delta t)$. Figure 9 shows $S(h,0)$ versus $S(0,g)$ indicating a proportional relation between $\Delta v(\Delta t)$ and $\Delta P(\Delta t)$ of the form $\Delta P(\Delta t) = K \Delta v(\Delta t)$.

To illustrate this purpose, Fig. 10 presents the joint distribution $\mathbb{P}(\Delta v, \Delta P)$ for wind speed fluctuations Δv and output power fluctuations ΔP, for $\Delta t = 10^4\,\mathrm{s} \approx 3$ h **(a)** and for $\Delta t = 10^5\,\mathrm{s} \approx 1$ day **(b)**, with the skeleton ϕ (red dashed line), indicating that a linear relation can be considered between the wind speed and the aggregate power output fluctuations at these timescales. The skeleton ϕ obtained with the joint pdf $\mathbb{P}(\Delta v, \Delta P)$ is defined by $\phi = \mathbb{P}_{\max}(\Delta v, \Delta P) = \max_{\Delta P}(\mathbb{P}(\Delta v, \Delta P)$.

For each time increment $\Delta t \geqslant 10^4$ s, we estimate the coefficients a and b in the relation $\Delta P(\Delta t) = a\Delta v(\Delta t) + b$. We find that b is always very small and that a has a small

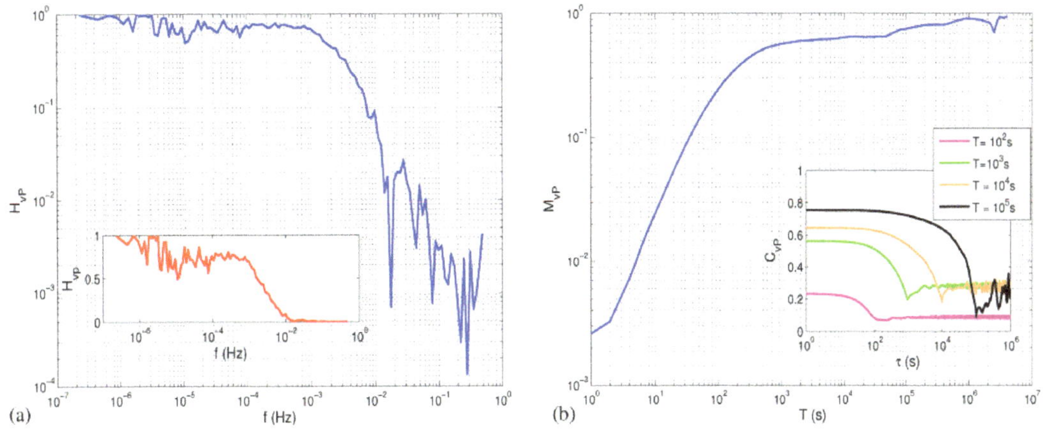

Fig. 7. The coherence function H_{vP} (frequency domain) **(a)** and the maximum M_{vP} (time domain) **(b)** of the cross-correlation function C_{vP} in log–log plot for the simultaneous wind speed/power data: the insets in **(a)** shows the functions H_{vP} in semilogx plot, and the inset in **(b)** shows the cross-correlations for the simultaneous wind speed/power output moving averages, computed for timescales $T = 10^2$, 10^3, 10^4, and 10^5 s.

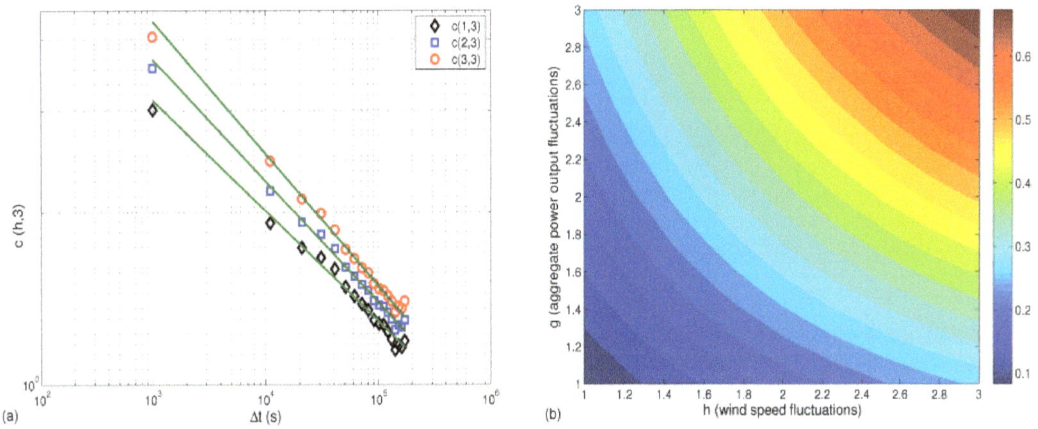

Fig. 8. (a) The generalized correlation functions (GCFs) $c(h, g)$ versus the time increments Δt in log–log plots for simultaneous wind speed and aggregate power output fluctuations. **(b)** The generalized correlation exponents (GCEs) $r(h, g)$ estimated from the linear regression of $c(h, g)$.

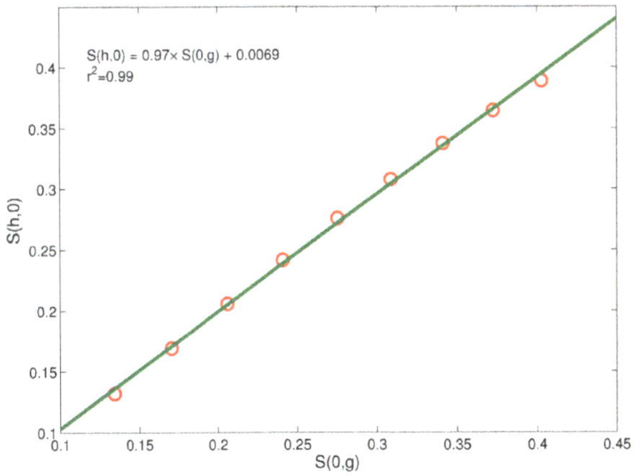

Fig. 9. Plot of the generalized correlation exponents $S(h, 0)$ versus $S(0, g)$ This confirms the symmetry of the $r(h, g)$ iso-values in the h–g plane.

Δt-dependence: $a = A\Delta t^B$, with $A = 10^{2.2}$ and $B = 0.04$. Thus, We obtain the following expression to model the skeleton ϕ for $\Delta t \geqslant 10^4$ s:

$$\Delta P(\Delta t) = A\Delta t^B \Delta v(\Delta t). \tag{22}$$

This is not incompatible with the fact that the power P typically scales cubically with the wind speed v of the form $P = cv^3$. This relation implies that $\mathrm{d}P = 3cv^2\mathrm{d}v$. Hence to a first order, since v is larger, power increments ΔP are proportional to wind speed increments Δv.

7 Discussions

The goal of this study was (i) to provide a characterization for the atmospheric wind speed fluctuations and the aggregate power output, at all intensities and at all scales, and

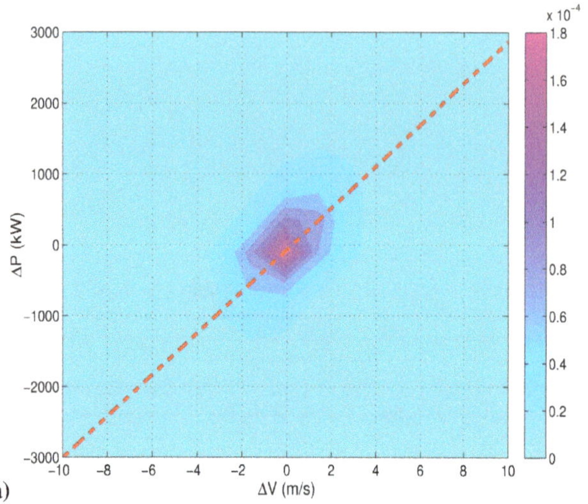

(a)

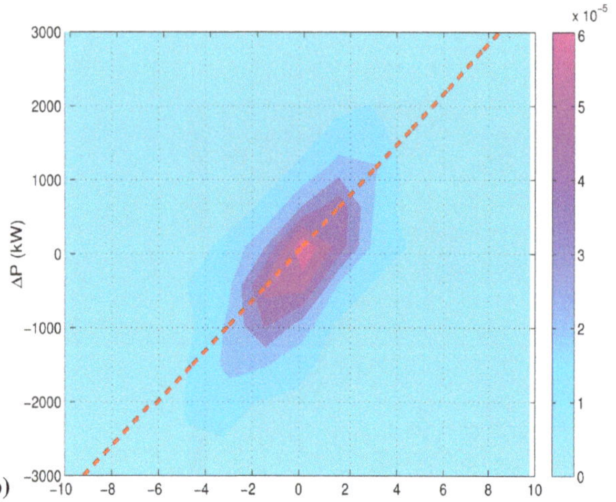

(b)

Fig. 10. The joint distribution $\mathbb{P}(\Delta v, \Delta P)$ for wind speed fluctuations Δv and output power ΔP, for **(a)** $\Delta t = 10^4$ s and **(b)** $\Delta t = 10^5$ s.

(ii) to characterize the coupling between the atmospheric wind speed and the aggregate power output measured simultaneously, in the multifractal framework.

7.1 The atmospheric wind speed

We have collected the wind speed data with a sampling rate of 20 and 1 Hz. The spectral analysis showed that the Fourier spectra $E_v(f)$ follows a power law: $E(f) \sim f^{-\beta}$ for the small-scale and the mesoscale range. Thus, for the small timescales ($0.05 \leqslant T \leqslant 1$ s) corresponding to the inertial range, the Fourier spectrum was close to the Kolmogorov spectrum, with $\beta = 1.68$, in agreement with previous studies (Schmitt et al., 1993; Katul et al., 1995; Schmitt, 2007; Böttcher et al., 2007; Morales et al., 2011; Calif and Schmitt, 2012). The structure function analysis highlighted the concavity and the nonlinearity of the exponent function $\zeta_{vH}(q)$. Furthermore, the theoretical quadratic relation for lognormal multifractals was well fitted (Böttcher et al., 2007; Morales et al., 2011; Schertzer et al., 1997). In Calif and Schmitt (2012), a lognormal continuous stochastic equation (Schmitt, 2003; Huang et al., 2008) is considered for modeling the atmospheric wind speed at these small temporal scales.

For the large timescales ($1 \leqslant T \leqslant 10^7$ s) corresponding to the mesoscale range, the Fourier analysis of the wind speed data sampled at 20 and 1 Hz showed a power law for the power spectral density $E_v(f) \sim f^{-\beta}$ with $\beta = -1.27$. Contrary to the inertial range, the turbulence effects in the mesoscale range did not possess the same degree of universality. More recent papers on longitudinal low-wave-number spectra have provided evidence of a -1 power law in the atmospheric surface layer for neutral conditions, and a $-5/3$ power law for near-convective conditions (Katul et al., 1995; Katul and Chu, 1998; Kader and Yaglom, 1991). Several theoretical and phenomenological approaches have been devoted to the explanation of such -1 power law scaling at the low frequencies. Tchen (1953) proposed a theoretical analysis of an approximate spectral budget, based on interaction between the mean flow vorticity and the fluctuating vorticity. To model the energy production and energy transfer, phemenological analogies between molecular dissipation of turbulent kinetic energy, energy extraction from large to small scales and energy removal from the mean to turbulent flows have been developed using Heisenberg's turbulent viscosity approach (Heisenberg, 1948). Tchen (1953) established that (i) a spectrum close to -1 power law strong is likely to occur close to a rough surface because of a strong interaction between the mean flow vorticity and the fluctuating vorticity, and (ii) a spectrum close to a $-5/3$ power law is likely to occur far away from the rough surface because of a weak interaction between the mean flow vorticity and the fluctuating vorticity. This was verified experimentally in Klebanoff (1954) and Katul and Chu (1998). Recently Katul et al. (2012) also proposed a phenomenological theory based on Heisenberg's eddy viscosity approach for the explanation of the existence of -1 power law scaling.

The structure functions analysis for the wind speed indicated also scaling properties for $1 \leqslant T \leqslant 10^7$ s. The exponent function $\zeta_{vL}(q)$ was nonlinear and concave, and a quadratic relation was proposed for fitting $\zeta_{vL}(q)$. However the exponent function $\zeta_{vL}(q)$ was less concave than $\zeta_{vH}(q)$: this indicated that the wind speed is more intermittent in the mesoscale range than the inertial range, as also shown in a previous study (Lauren et al., 1999).

7.2 The aggregate power output

We have also collected the aggregate output power from a wind farm with a sampling rate of 1 Hz. The spectral analysis showed two scaling regimes. For $2 \leqslant T \leqslant 9 \times 10^3$ s the slope was close to the spectral slope $-5/3$ as in a previous study (Apt, 2007). The structure functions analysis highlighted the concavity and nonlinearity of the exponent function $\zeta_{PH}(q)$. Furthermore, this function was well fitted by the lognormal model.

For timescales $T \geqslant 9 \times 10^3$ s, the Fourier analysis showed that the power spectral density $E_P(f)$ of the aggregate power output from a wind farm has a spectral slope close to 1.2. The structure function analysis confirmed the obtained scaling, indicating the concavity and nonlinearity of the scaling exponent function $\zeta_{PL}(q)$. $\zeta_{PL}(q)$ was less concave than $\zeta_{PH}(q)$, meaning a decrease of the intermittency with these timescales, as found also for the turbulent wind speed. However, for a more representative result, it is necessary to have more aggregate power output data.

7.3 The coupling between the wind speed and the output wind power fluctuations

We have collected simultaneously the wind speed v and the aggregate power output P, sampled at 1 Hz in order to highlight a possible dependence between both series. Generally, a scaling test, i.e., the coherence, is estimated in the wind energy community, for testing the dependence between the data. In this study, the coherence function H_{vP} and the scalar M_{vP} showed a coupling between the turbulent wind speed and the aggregate output wind power for timescales $T \geqslant 10^3$ s. This study involved simultaneous analysis of turbulence time series and instantaneous power data, considering the multiscale correlations between both series. For that, a multiscaling test for independence between two stochastic processes was used. The generalized correlation functions $c(h,g)$ and exponents $r(h,g)$ were estimated. For timescales $10^3 \leqslant T \leqslant 10^5$ s, the function $c(h,g)$ presented a scaling relating a coupling between both series in agreement with the coherence function H_{vP} and the scalar M_{vP}. Moreover, their GFE $r(h,g)$ indicated that larger fluctuations have stronger dependence. In fact, slow fluctuations (in the range of tens of minutes and hours) are mainly due to meteorological dynamics and are highly correlated among nearby wind farms. A smoothing effect is operated by the size of the wind farm for short timescales (or short spatial timescales). This indicates that large fluctuations in wind speed over very short timescales (wind gusts) do not give rise to similar fluctuations in the aggregate power output. There is inertia in the wind turbine generator system, corresponding to a nonlinear transfer function from turbulent wind speed to aggregate power output production. This could also be caused by the spatial repartition combined with the number of wind turbines: turbulence at different wind turbines is weakly correlated or uncorrelated (Sørensen et al., 2002). Fast fluctuations have smaller spatial correlations (Nichita et al., 2002; Petru and Thiringer, 2002) and wind gusts are smoothed in the aggregate power output from a wind farm. Furthermore, the symmetry of $r(h,g)$ in the plane $(h–g)$ is in favor of a proportional relation between the wind speed fluctuations $\Delta v(\Delta t)$, and the output power fluctuations $\Delta P(\Delta t)$ seem to be of the form $\Delta P(\Delta t) = K \Delta v(\Delta t)$ for these timescales.

8 Conclusions

This work analyzed the intermittency and the scaling properties of the wind speed v ant the aggregate power output P time series, through structure function analysis, at all scales and all intensities. The aggregate power output P possessed intermittent, multifractal properties and two scaling regimes in relation with the multifractal properties of the input wind speed v. Furthermore, we showed that the generalized correlation functions and exponents are a suitable tool for testing a dependence between two simultaneous intermittent processes. This multiscaling test allowed characterizing also the coupling of two simultaneously sampled intermittent processes by an analytical expression (power law, proportionality). Indeed, we proposed a proportionality expression for the coupling between the wind speed and the aggregate power output fluctuations. We believe that the actual relation is complex and has some nonlinear relations. We want here to understand the experimental result found about scaling relation for the power output. The relation given in relation (22) should be seen as a simplified model and not as a firm result.

We found here typical scales of 10^3–10^4 s for the aggregate power output fluctuations. The scale of $10^4\,\mathrm{s} \simeq 3$ h was found for the change of slope in aggregate power output spectral regime, and the scale of $10^3\,\mathrm{s} \simeq 15$ min was found for the correlation between both series. This scale may be due to the local topography, the local meteorology, or the influence of the spatial disposition of the different wind turbine generators composing the wind farm. A better understanding of theses scales of 10^3–10^4 s will be the subject of future works in order to consider a possible universality, or its relation with local conditions or wind farm composition.

Acknowledgements. This study is financially supported by the Regional Council of Guadeloupe, European funding (No. 1/1.4/-31614). We are thankful to the Vergnet Caraibes Company for its technical assistance during the measurement campaign. We thank the editor and the referees for useful comments.

Edited by: I. Tchiguirinskaia

References

Anselmet, F., Gagne, Y., Hopfinger, E. J. and Antonia, R. A.: High order velocity structure functions in turbulent shear flows, J. Fluid Mech., 140, 63–89, 1984.

Apt, J.: The spectrum of power from wind turbines, J. Power Sources, 169, 369–374, 2007.

Arneodo, A., Baudet C., Belin, F., Benzi, R., Castaing, B., Chabaud, B., Chavarria, R., Ciliberto and, S., Camussi, R., Chillà, F., Dubrulle, B., Gagne and, Y., Hebral, B., Herweijer, J., Marchand, M., Maurer, J., Muzy, J. F., Naert, A., Noullez, A., Peinke, J., Roux, F., Tabeling, P., van de Water, W., and Willaime, H.: Structure functions in turbulence, in various flow configurations, at Reynolds number between 30 and 5000, using extended self-similarity, EPL-Europhys. Lett., 34, 411–416, 1996.

Böttcher, F., Barth, St., and Peinke, J.: Small and large fluctuations in atmospheric wind speeds, Stoch. Env. Res. Risk A., 21, 299–308, 2007.

Bracewell, R.: The Fourier Transform and Its Applications, 3rd Edn., McGraw-Hill Science, New York, 1999.

Burton, T., Sharpe, D., Jenkins, N., and Bossanyi, E.: Wind Energy Handbook, JohnWiley & Sons, Chichester, 2001.

Calif, R. and Schmitt, F. G.: Modeling of atmospheric wind speed using a lognormal continuous stochastic equation, J. Wind Eng. Ind. Aerod., 109, 1–8, 2012.

Calif, R., Schmitt, F. G., and Huang, Y.: The scaling properties of the turbulent wind using empirical mode decomposition and arbitray-order Hilbert spectral analysis, in: Proceedings of Euromech colloqium, Springer, in press, 2014.

Fitton, G., Tchiguirinskaia, I., Schertzer, D., and Lovejoy, S.: Torque Fluctuations In the Framework Of A Multifractal 23/9-Dimensional Turbulence Model, J. Phys. Conf. Ser., in press, 2014.

Frisch, U.: Turbulence, the legacy of A. N. Kolmogorov, Cambridge University Press, Cambridge, 1995.

Huang, Y., Schmitt, F. G., Lu, Z., and Liu, Y.: An amplitude-frequency study of turbulent scaling intermittency using Hilbert spectral analysis, EPL, 84, 40010, doi:10.1209/0295-5075/84/40010, 2008.

Heisenberg, W.: Zur Statistichen Theorie der Turbulenz, Z. Phys., 124, 628–657, 1948 (in German).

Ivanov, P. C., Nunes Amaral, L. A., Havlin, S., Rosenblum, M. G., Stanley, H. E., and Struzik, Z. R.: From 1 / f noise to multifractal cascades in heartbeat dynamics, Chaos, 11, 641–652, 2001.

Kader, B. A. and Yaglom, A. M.: Turbulent Structure of an Unstable Atmospheric Layer, in: Nonlinear and Turbulent Processes in Physics, edited by: Sagdeyev, R. Z., Harwood Academic, Boston, MA, 2, 829–845, 1984.

Kader, B. A. and Yaglom, A. M.: Spectra and Correlation Functions of Surface-Layer Turbulence in Unstable Thermal Stratification, in: Turbulence and Coherent Structures, edited by: Metais, O. and Lesieur, M., Kluwer Academic Press, Dordrecht, 388–412, 1991.

Kaimal, J. C., Wyngaard, J. C., Izumi, Y., and Coté, O. R.: Spectral characteristics of surface-layer turbulence, Q. J. Roy. Meteor. Soc., 98, 563–589, 1972.

Karman, V. T.: Progress in the statistical theory of turbulence, P. Natl. Acad. Sci. USA, 34, 530–539, 1948.

Katul, G. and Chu C. R.: A Theoretical and Experimental Investigation of Energy-Containing Scales in the Dynamic Sublayer of Boundary-Layer Flows, Bound.-Lay. Meteorol., 86, 279–312, 1998.

Katul, G. G., Chu, C. R., Parlange, M. B., Albertson, J. D., and Ortenburger, T. A.: Low-wavenumber spectral characteristics of speed and temperature in the atmospheric surface layer, J. Geophys. Res., 100 14243–14255, 1995.

Katul, G. G., Porporato, A., and Nikora, V.: Existence of k^{-1} power-law scaling in the equilibrium regions of wall-bounded turbulence explained by Heisenberg's eddy viscosity, Phys. Rev. E., 86, 066311, doi:10.1103/PhysRevE.86.066311, 2012.

Klebanoff, P. S.: Characteristics of Turbulence in a Boundary Layer with Zero-Pressure Gradient, National Advisory Comiittee for Aeronautics, National Bureau of Standards, Washington D.C., Report Nr. 1247, 1–28, 1954.

Kolmogorov, A. N.: The local structure of turbulence in incompressible fluid for very large Reynolds numbers, Dokl. Akad. Nauk. SSSR, 30, 299–303, 1941.

Kolmogorov, A. N.: A refinement of previous hypotheses concerning the local structure of turbulence in a viscous incompressible fluid at high Reynolds number, J. Fluid Mech., 13, 82–85, 1962.

Lauren, M. K., Menabe, M., Seed, A. W., and Austin, G. L.: Characterisation and simulation of the multiscaling properties of the energy-containing scales of horizontal surface layer winds, Bound.-Lay. Meteorol., 90, 21–46, 1999.

McNerney, G. and Richardson, R.: The statistical smoothing of power delivered to utilities by multiples wind turbines, IEEE T. Energy Conver., 7, 644–647, 1992.

Mandelbrot, B. B.: Intermittent turbulence in self-similar cascades: divergence of high moments and dimension of the carrier, J. Fluid Mech., 62, 331–358, 1974.

Mandelbrot, B. B.: The Fractal Geometry of Nature, Freeman, New York, 1982.

Marshak, A., Davis, A., Cahalan, R., and Wiscombe, W.: Bounded cascade models as nonstationary multifractals, Phys. Rev. E, 49, 55–69, 1994.

Meneveau, Ch., Sreenivasan, K. R., Kailisnath, P., and Fan, M. S.: Joint multifractal measures: Theory and applications to turbulence, Phys. Rev. A, 41, 894–913, 1990.

Morales, A., Wächter. and Peinke, J., Characterization of wind turbulence by higher-order statistics, Wind Energy, 15, 391–406, doi:10.1002/we.478, 2011.

Muzy, J. F., Baile, R., and Poggi, P.: Intermittency of surface-layer wind velocity series in the mesoscale range, Phys. Rev. E, 81, 056308, doi:10.1103/PhysRevE.81.056308, 2010.

Nanahara, T., Asari, M., Sato, T., Yamaguchi, K., Shibata, M., and Maejima, T.: Smoothing effects of distributed wind turbines. Part 1. Coherence and smoothing effects at a wind farm, Wind Energy, 7, 61–74, doi:10.1002/we.109, 2004.

Nichita, C., Luca, D., Dakyo, B., and Ceanga, E.: Large Band Simulation of the Wind Speed for Real Time Wind Turbine Simulators, IEEE T. Energy Conver., 17, 523–529, 2002.

Nickels, T. B., Marusic, I., Hafez, S., and Chong, M. S.: Evidence of the k_1^{-1} Law in a High-Reynolds-Number Turbulent Boundary Layer, Phys. Rev. Lett., 95, 074501, doi:10.1103/PhysRevLett.95.074501, 2005.

Novikov, E. A. and Steward, R. W.: Intermittency of Turbulence and Spectrum of Fluctuations in Energy Dissipation, Izv. Akad. Nauk SSSR, Ser. Geofiz., 3, 408–413, 1964.

Obukhov, A. M.: Spectral energy distribution in a turbulent flow, Dokl. Akad. Nauk. SSSR 32, 22–24, 1941.

Papoulis, A. and Pillai, S. U.: Probability, Random variables and stochastic processes, 4th Edn., The McGraw-Hill Companies, 2002.

Peinke, J., Barth, S., Bottcher, F., Heinemann, D., and Lange, B.: Turbulence, a challenging problem for wind energy, Physica A, 338, 187–193, 2004.

Petru, T. and Thiringer, T.: Modeling of Wind Turbines for Power System Studies, IEEE T. Power Syst., 17, 1132–1139, 2002.

Pope, S. B.: Turbulent Flows Energy, Cambridge University Press, 802 pp., 2000.

Schertzer, D. and Lovejoy, S.: Physical modelling and analysis of rain and clouds by anisotropic scaling multiplicative processes, J. Geophys. Res., 92, 9693–9714, 1987.

Schertzer, D., Lovejoy, S., Schmitt, F., Chigirinskaya, Y., and Marsan, D.: Multifractal cascade dynamics and turbulent intermittency, Fractals, 5, 427–471, 1997.

Schmitt, F., Schertzer, D., Lovejoy, S., and Brunet, Y.: Estimation of universal multifractal indices for atmospheric turbulent speed fields, Fractals, 1, 568–575, 1993.

Schmitt, F., Schertzer, D., Lovejoy, S., and Brunet, Y.: Empirical study of multifractal phase transitions in atmospheric turbulence, Nonlin. Processes Geophys., 1, 95–104, doi:10.5194/npg-1-95-1994, 1994.

Schmitt, F. G.: A causal multifractal stochastic equation and its statistical properties, Eur. Phys. J. B, 34, 85–98, 2003.

Schmitt, F. G.: Gusts in intermittent wind turbulence and the dynamics of their return times, Wind Energy, in: Proceedings of the Euromech Colloquium, edited by: Peinke, J., Schaumann, P., and Barth, S., Springer, Germany, 73–79, 2007.

Seuront, L. and Schmitt, F. G.: Multiscaling statistical procedures for the exploration of biophysical couplings in intermittent turbulence. Part I. Theory, Deep-Sea Res. Pt. II, 52, 1308–1324, 2005.

She, Z. S. and Levêque, E.: Universal scaling laws in fully developed turbulence, Phys. Rev. Lett., 72, 336–339, 1994.

Simiu, E. and Scanlan, R. H.: Wind Effects on Structures: An Introduction to Wind Engineering, Wiley, New York, 1978.

Sørensen, P., Hansen, A. D., and Cavalho Rosas, P. A.: Wind models for simulation of power fluctuations from wind farms, J Wind. Eng. Ind. Aerod., 90, 1381–1402, 2002.

Stull, R. B.: An Introduction to Boundary Layer Meteorology, Springer, the Netherlands, 1988.

Tchen, C. M.: On the Spectrum of Energy in Turbulent Shear Flow, Journal of Research of the National Bureau of Standards, 50, 51–62, 1953.

Telesca, L., Colangelo, G., Lapenna, V., and Macchiato, M.: Monofractal and multifractal characterization of geoelectrical signals measured in southern Italy, Chaos Soliton. Fract., 18, 385–399, 2003.

Vulpiani, A. and Livi, R. (Eds.): The Kolmogorov legacy in physics, Springer, Berlin Heidelberg New York, 2004.

Yaglom, A. M.: The influence of fluctuations in energy dissipation on the shape of turbulent characterisitcs in the inertial interval, Soviet Physics Doklady, 2, 26–30, 1966.

Zhang, L., Lia, J., and Pengb, Y.: Dynamic response and reliability analysis of tall buildings subject to wind loading, J. Wind Eng. Ind. Aero. 96, 25–40, 2008.

Characterizing the evolution of climate networks

L. Tupikina[1,2], K. Rehfeld[1,3], N. Molkenthin[1,2], V. Stolbova[1,2], N. Marwan[1], and J. Kurths[1,2]

[1]PIK Potsdam Institute of Climate Impact Research, P.O. Box 601203, 14412 Potsdam, Germany
[2]Department of Physics, Humboldt-Universität zu Berlin, Newtonstr. 15, 12489 Berlin, Germany
[3]Alfred-Wegener-Institut, Helmholtz-Zentrum für Polar- und Meeresforschung, Telegrafenberg A43, 14473 Potsdam, Germany

Correspondence to: L. Tupikina (tupikina@pik-potsdam.de)

Abstract. Complex network theory has been successfully applied to understand the structural and functional topology of many dynamical systems from nature, society and technology. Many properties of these systems change over time, and, consequently, networks reconstructed from them will, too. However, although static and temporally changing networks have been studied extensively, methods to quantify their robustness as they evolve in time are lacking. In this paper we develop a theory to investigate how networks are changing within time based on the quantitative analysis of dissimilarities in the network structure.

Our main result is the common component evolution function (CCEF) which characterizes network development over time. To test our approach we apply it to several model systems, Erdős–Rényi networks, analytically derived flow-based networks, and transient simulations from the START model for which we control the change of single parameters over time. Then we construct annual climate networks from NCEP/NCAR reanalysis data for the Asian monsoon domain for the time period of 1970–2011 CE and use the CCEF to characterize the temporal evolution in this region. While this real-world CCEF displays a high degree of network persistence over large time lags, there are distinct time periods when common links break down. This phasing of these events coincides with years of strong El Niño/Southern Oscillation phenomena, confirming previous studies. The proposed method can be applied for any type of evolving network where the link but not the node set is changing, and may be particularly useful to characterize nonstationary evolving systems using complex networks.

1 Introduction

Networks are practical representations for complex systems with interacting components and have been used to study phenomena in sociology, engineering and natural systems (Barthélemy, 2011; Menck and Kurths, 2012; Palla et al., 2005; Holme et al., 2004). Complex network techniques, based on statistical associations between climate parameter time series at different points on Earth, have yielded new insights in the investigation of climate dynamics (Tsonis and Swanson, 2008; Donges et al., 2009; Paluš et al., 2011). Such climate networks have been used for detecting long-range correlations, or teleconnections (Martin et al., 2013; Barreiro et al., 2011), and studying such phenomena such as the El Niño/Southern Oscillation (ENSO, Gozolchiani et al., 2008; Deza et al., 2013) and the Indian Monsoon system (Rehfeld et al., 2013; Malik et al., 2011; Stolbova et al., 2014). In particular, Tsonis and Swanson (2008) found changes in the global network topology during El Niño events, with significantly fewer links and lower clustering coefficients, and inferred a lower predictability for El Niño over La Niña years. Using climate networks, Yamasaki et al. (2008) and Gozolchiani et al. (2008) also found ENSO influence on regional atmospheric processes in non-ENSO regions. Temporal and spatial variability of climate, and thus climate network structure, are of increasing interest considering ongoing environmental changes, and climate networks as evolving in time are still an open subject. The spatial-temporal developments in a given network set can be too complex to be captured by eye, and systematic approaches to quantify changes are needed. While Berezin et al. (2012) investigated the origins of the climate network stability such as the spatial embedding and physical coupling between climate in different locations using the correlation between correlation matrices, other studies describe how the network graph is changing over time to understand the behaviour of the underlying dynamical system (e.g. Rehfeld et al., 2013).

Various aspects of temporally changing networks have been considered for sociological and biological networks. Albert and Barabasi (2000) analysed random network growth and evolution in response to the addition or rewiring of links between nodes and found that the graph topology changed depending on the frequency of link changes. Fu et al. (2009) tracked node function changes using a stochastic block model for evolving networks to investigate evolutionary effects in email networks and gene regulation.

One of the most common dissimilarity measures which has been used for network comparison is Hamming distance. It was introduced by Hamming (1950) as a measure for comparing strings of symbols and was used for measuring the distance between the networks. Given the adjacency matrices $\mathbf{A}^N$ and $\mathbf{A}^M$ of two graphs N and M, their Hamming distance is determined from the sum over the number of links which are found in one, but not the other network: $H(N,M) = \sum_{i,j} |\mathbf{A}^N_{i,j} - \mathbf{A}^M_{i,j}|$. However, although Hamming distance can be generalized for directed networks with possibly differing node numbers, two networks M and N may have the same Hamming distance to the fixed network K while having different topology themselves. The Hamming distance therefore may not enough to detect topological changes.

Here we propose a common component evolution function (CCEF) based on the common set of links in pairs of networks to evaluate graph changes quantitatively in space and time. We characterize the method using Erdős–Rényi networks (Erdös and Rényi, 1959), analytically derived flow networks (Molkenthin et al., 2014a, b) and transient simulations from the START model (Rehfeld et al., 2014) for which we control changes of individual parameters over time. Then we construct annual climate networks from NCEP/NCAR reanalysis data for the Asian monsoon domain and use the CCEF to characterize the temporal evolution in the monsoon system.

2 Derivation of the common component evolution function

We consider unweighted and undirected networks, for which n nodes are joined in pairs by edges, or links. The linking structure is given in the adjacency matrix $\mathbf{A}$, a binary $n \times n$ matrix with zeros on the diagonal, as we do not allow for self-loops. An element is non-zero, $\mathbf{A}_{ij} = 1$, if and only if the vertices i and j are connected, and zero otherwise.

Let us consider a linearly ordered set of T evolving in time networks: $N_1, \ldots, N_T$. Then the *common component network* for two of these networks N_i and N_j, $\mathrm{CC}(N_i, N_j)$, is a network on the same nodes, where the set of edges is present in both original networks. If N_i and N_j have adjacency matrices $\mathbf{A}_i$ and $\mathbf{A}_j$, the number of edges in the common component network $\mathrm{CC}(N_i, N_j)$ is the number of non-zero elements above the diagonal in the binary sum of adjacency matrices $\mathbf{A}_i$ and $\mathbf{A}_j$. This common component network can be generalized for any $k+1$ networks by induction: $\mathrm{CC}(N_i, \ldots, N_{i+k+1}) = \mathrm{CC}(\mathrm{CC}(N_i, \ldots N_{i+k-1}), N_{i+k})$.

The *common component function* $\mathrm{CCF}(N_i, \ldots, N_{i+k})$ counts the number of links in a common component network of k networks: $\mathrm{CCF}(N_i, \ldots, N_{i+k}) = ||\mathrm{CC}(\mathrm{CC}(N_i, \ldots, N_{i+k-1}), N_{i+k})||$, where by $||N_i||$ we mean the number of links in the network N_i, and the common component function $\mathrm{CCF}(N_i, N_k)$ gives the number of coinciding edges in the graphs of N_i and N_k, $i, k \in \{1, 2, \ldots, T\}$. We set $\mathrm{CCF}(N_i) = \mathrm{CCF}(N_i, N_i)$. The common component function $\mathrm{CCF}(N_i)$ takes values in $[0, \max \mathrm{CCF}(N_i)]$ and is in the following normalized to $[0, 1]$ using the maximal number of links in the networks.

In analogy to covariance estimation (Chatfield, 2004) and similar to Berezin et al. (2012), we take the mean over the CCFs with the same time lags to estimate the non-normalized *common component evolution function*, CCEF*, as

$$\mathrm{CCEF}^*(\delta) = \frac{1}{T-\delta} \sum_{i=1}^{T-\delta} \mathrm{CCF}(N_i, N_{i+\delta}) \ , \tag{1}$$

where δ is the time lag between the networks, and $\delta \in [0, T-1]$. The maximum value of the CCEF* is given by $\mathrm{CCEF}^*(0)$ for zero lag, as an average number of links in the set of network

$$\mathrm{CCEF}^*(0) = \frac{1}{T} \sum_{i=1}^{T} \mathrm{CCF}(N_i) \ , \tag{2}$$

and we use it to obtain the normalized *common component evolution function*

$$\mathrm{CCEF}(\delta) = \frac{\mathrm{CCEF}^*(\delta)}{\mathrm{CCEF}^*(0)}, \tag{3}$$

which we will use exclusively in the following. As an estimation of the CCEF uncertainty we use the standard deviation over all CCEF values.

2.1 Testing the method on random networks

To test our method we generate a set of T Erdős–Rényi graphs (Erdös and Rényi, 1959) with a fixed number of n nodes and a fixed connection probability p. We artificially impose a linear ordering on the set, such that we can index them with $i \in (1, T)$. We compute the CCEF for Erdős–Rényi graphs with 100 nodes and link probabilities of 0.3, 0.5, and 0.9. The resulting functions, shown in Fig. 1, decrease from 1 to a plateau at $\mathrm{CCEF}(\delta) \approx p$ for $\delta > 0$ for each link probability p. For this example we can analytically compute the expected CCEF, since for $\delta = 0$ each network is compared with itself and therefore $\mathrm{CCEF} = 1$. For all other

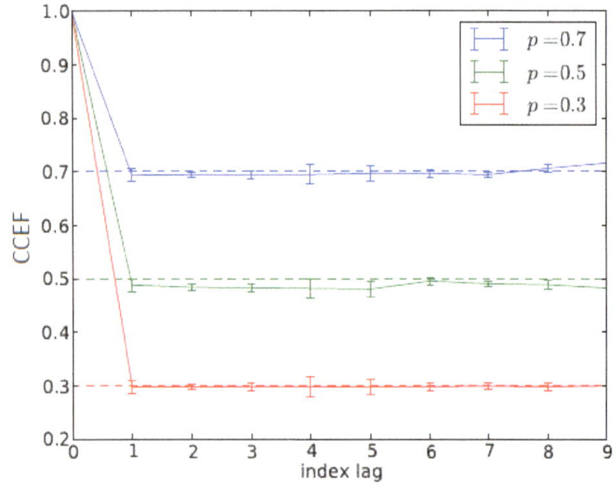

Figure 1. Mean CCEF for 10 indexed sets of random Erdős–Rényi networks with 100 nodes and different linking probabilities p. The dashed lines correspond to the analytical CCEF levels, the error bars give the 1σ standard deviation for each index lag

values of δ two random matrices with n nodes and connection probability p are compared. Then the number of totally possible links is $n(n-1)/2$, and the expectation value of the number of links in each of the networks is $pn(n-1)/2$. As the probability of each of the edges in one network to also appear in the other network is p, the total number of common links is $p^2n(n-1)/2$, which with the normalization leads to $f(p)$, the ratio of total number of common links and the expectation value of the number of links to be

$$f(p) = \frac{p^2n(n-1)}{pn(n-1)} = p \,. \tag{4}$$

The CCEF for each linking probability therefore lies close to the expected value p.

2.2 Test models

To characterize our approach further we investigate simulations of more complex, spatially embedded processes. We obtained networks (i) analytically from flow fields, as described in Molkenthin et al. (2014a) and (ii) from the **S**patio-**T**emporally **A**utoco**r**related **T**ime series model START.

2.2.1 Networks from flows

The flow networks are constructed directly from a velocity field using a correlation measure based on the temperature profiles resulting from a temperature peak via advection and diffusion (Molkenthin et al., 2014b, a). The velocity function considered here is

$$\boldsymbol{v}(x,y) = \begin{pmatrix} e^{\frac{-(y-0.5x)^2}{c}} \\ 0.5\, e^{\frac{-(y-0.5x)^2}{c}} \end{pmatrix}, \tag{5}$$

and we vary the parameter c for the flow width from 200 to 2000 in 10 steps, thus gradually changing the flow network. The positions and the number of nodes are kept constant. For each value of c we obtain a correlation matrices $\mathbf{C}_1, \ldots, \mathbf{C}_{10}$, and thresholding these matrices by different critical values we obtain set of adjacency matrices.

2.2.2 Networks from the START model

As a more complex test case we consider two transient simulations for the START model (Rehfeld et al., 2014). Networks generated from START undergo a distinct transition when the forcing parameter F is changed: for $F = -1$ the network is partitioned into two vertical connected areas; for $F = 0$ horizontal cross-links have appeared and link the two sections. At maximal forcing, for $F = 1$, there is one large, horizontally oriented component. We performed two transient simulations with a 6×7 sampling grid (Rehfeld et al., 2014) for 20 000 timesteps and 100 ensemble members each. In the first run the forcing parameter was increased linearly from the start to the end of the simulation. In the second run we periodically changed the forcing parameter $F(t) = \sin(t/2000)$. Networks were constructed based on the 20 % strongest links in the correlation matrices obtained for each 100-step-long time window. Due to the stochastic component, networks constructed for different ensemble members, but for the same time period, may be quite different, and networks for different periods of the same ensemble member may be quite similar.

2.2.3 Asian monsoon data

In a real-world application we used daily NCEP/NCAR reanalysis temperature anomaly data (NOAA, 2014) for the Asian monsoon domain for the years 1970–2010 CE. The spatial resolution was $2.5° \times 2.5°$, covering the area between 2.5° S–42.5° N and 57.5–122.5° E, resulting in time series for 468 nodes. Networks were constructed using Pearson correlation in windows for each full year and by thresholding the correlation matrix such that we obtain a link density of 5 %. The same data set and time period was used in Molkenthin et al. (2014b) to investigate the influence of changing node topologies in space on the estimates of node degree and betweenness.

3 Results

3.1 Flow networks

We computed the CCEF for flow networks with linearly increasing flow-width parameter c. As Fig. 2 shows, the common component size decreases monotonically with the width parameter difference of the networks. The higher the threshold of the correlation matrix is, the faster the CCEF decays but the general shape does not change.

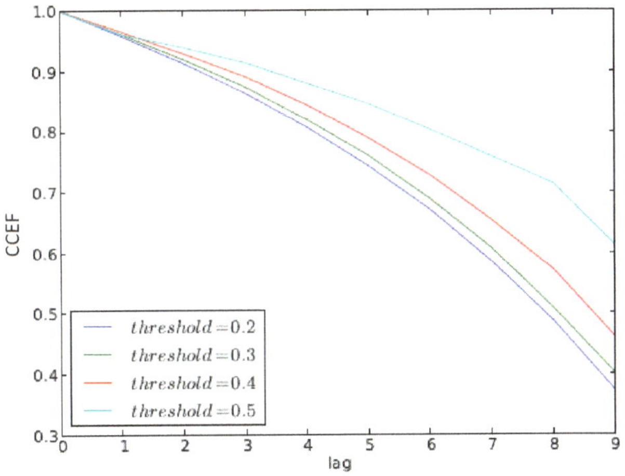

Figure 2. CCEF for the indexed flow networks with increasing flow-width parameter c, and for different threshold values of adjacency matrices of the flow networks.

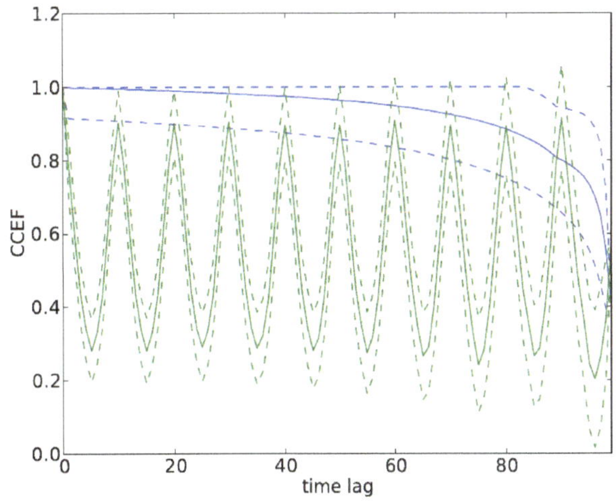

Figure 3. CCEF for networks from the START model with periodic (green) and linearly increasing (blue) forcing parameter F. The error bars indicate the standard deviation of the CC size estimates.

3.2 START-model networks

The START model undergoes a more distinct transition from a network with two distinct parts through a connected stage with three regions to one single component (Rehfeld et al., 2014) in response to a single forcing parameter F. To characterize the CCEF response to different network evolution patterns we use two test cases, in which we vary F from its minimum to its maximum. In the first example, the forcing parameter is varied linearly along time. The CCEF response is a slow decline from its maximum CCEF(0) = 1 to a minimum value CCEF(99) ≈ 0.4, as shown in Fig. 3. In the second test the forcing parameter was varied periodically as a function of time, $F = \sin\left(\frac{2\pi}{P}t\right)$, with $P = 10$. In response to the sinusoidal forcing, periodic behaviour is also observable in the CCEF and with the same period as the forcing parameter.

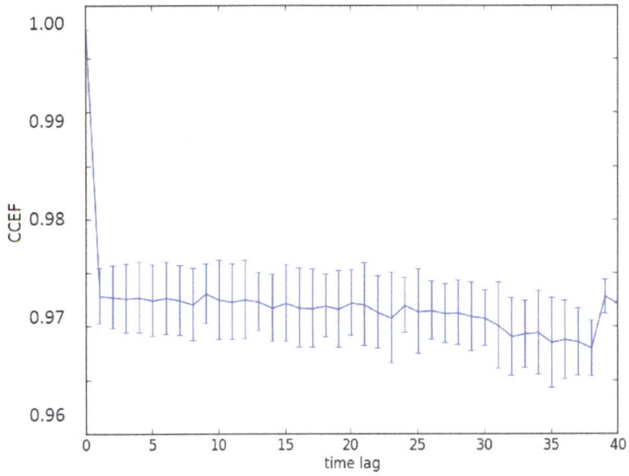

Figure 4. CCEF of annual climate networks for the time period 1970–2011 CE. Error bars are presented as CCEF standard deviation of the respective time lag in years.

3.3 Application to the Asian Monsoon domain

Finally, we used the CCEF to investigate the evolution of climate networks from observations. The networks were constructed using a link density of an annual basis for 41 years, 1970–2011 CE. The obtained CCEF in Fig. 4 is reminiscent of the Erdős–Rényi networks in Sect. 2.1, with an initial quick decline followed by a plateau.

However, while in the case of Erdős–Rényi networks (Fig. 1) the baseline is equal to the set link density, it is significantly higher than the link density here. We thus conclude that a high degree of persistence and a low amount of spatio-temporal variance can be found in climate networks from the Asian Monsoon domain at annual timescale.

The spatial domain selected for this study is the host of very distinct seasonal dynamics during the Asian Monsoon seasons, see e.g. Wang (2006). At inter-annual timescales, however, teleconnections such as that to the ENSO phenomenon plays a significant role (Turner and Annamalai, 2012; Clarke, 2008). In order to identify reasons for the variability of climate networks in this region we compared the variation of common component functions CCF(N_i, N_j) for different $i, j \in [1970, 2011]$, where i is kept fixed. in this way, we obtained a common-links-recurrence diagram, illustrated in Fig. 5a, with maximum values on the diagonal. Each pair (i, j) for $i, j \in [1970, 2011]$ corresponds to the value of the common component function CCF(N_i, N_j), as in Eq. (3).

Figure 5a shows rows and columns with distinct lower values for the CCF. In these lines the overall sum, $S_i = \sum_j \text{CCF}(N_i, N_j)$, takes smaller values in the years $i \in (1971-1973, 1975, 1984, 1989, 1993, 1999)$. We compared this sequence with a list with strong El Niño phenomena

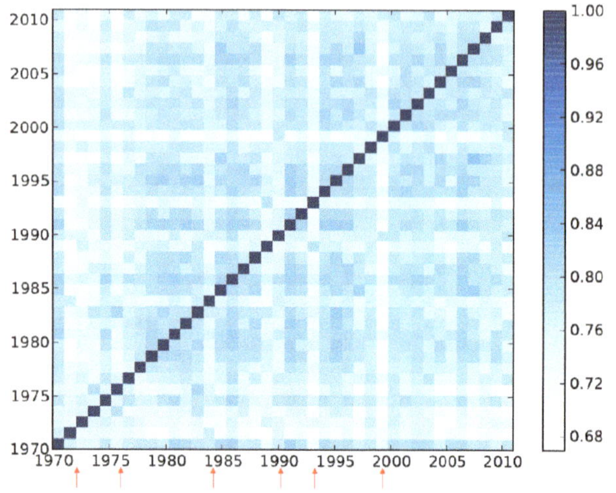

(a) Common link recurrence

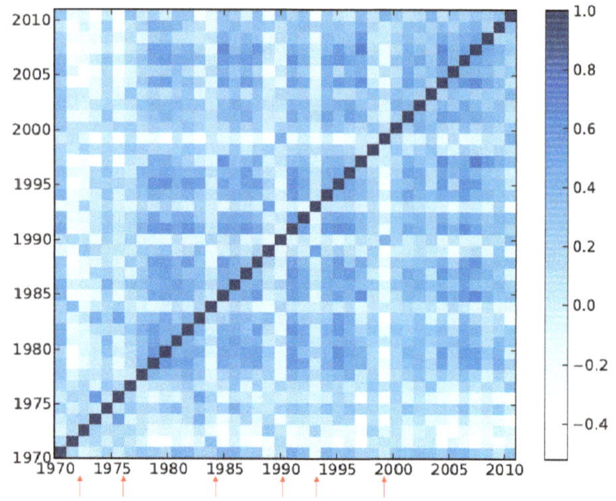

(b) Correlation of the common link evolution

Figure 5. Common-links-recurrence diagram **(a)** and correlation matrix of the common link evolution **(b)**. Each point (i, j) in diagram **(b)** corresponds to the value of the correlation coefficient $\mathrm{corr}(i, j)$ between the common component functions $\mathrm{CCF}(N_i, N_k)$ and $\mathrm{CCF}(N_j, N_k)$. Lines with low values (marked at the bottom with arrows) are observable around strong ENSO years.

according to the El Niño 3.4 index (Trenberth, 1997), and observed that 1972, 1982, 1988, 1992, and 1997 were the strongest ENSO event years in this time period. At the same time, the correlation between the CCF functions of these years and all others, given in Fig. 5b, also takes on very low values. Around stronger El Niño years the surface temperature networks have fewer common links, and the correlation of their CCF with all others is considerably lower. ENSO events occur during the Northern Hemisphere winter season and thus the main effect of the link breakdown in our networks occurs in the year after the event started.

4 Discussion

The CCEF enables us to investigate the evolution of linearly ordered, or evolving, network sets quantitatively. We tested its response to three different types of model networks and find that the responses enable us to characterize their evolution.

Unlike random networks, the flow-network CCEF level is, within limits, not related to the threshold value but displays a deterministic decrease of network similarity: two flow networks separated by bigger index lag have fewer links in the intersection, hence the common component function $\mathrm{CCF}(N_i, N_{i+\delta})$ decreases with the growth of δ, and the intersection of two flow networks decreases with the difference in the width parameter. Furthermore, we find that the links in a network set with higher threshold are more persistent.

The START model examples, on the other hand, illustrate the distinct difference between slow, linear changes of the processes generating the networks over time – and periodic, rapid transitions. While in the first case the CCEF decreases slowly, and only considerably for large time difference, in case of periodic and rapid transitions the CCEF response is also periodic over the time lag. In this case it is particularly important that the time window a single network corresponds to is sufficiently small compared to the ongoing evolution to avoid aliasing effects which would occur in the case of window width as a multiple of the forcing period and to be able to detect the changes at all.

The year-long daily temperature anomaly networks of the Asian Monsoon domain show a high degree of spatio-temporal persistence. This is consistent with the results of Berezin et al. (2012), who found similarly high values over large regions of the South Atlantic and the equatorial Pacific. While the general shape of the CCEF in Fig. 4 agrees with that for the Erdős–Rényi random networks (Fig. 1), the CCEF fluctuates around a much higher level compared to the link density.

This points towards a highly non-random, deterministic general structure in the network on which the inter-annual variability is imprinted. Links in this network are comparatively stable but lose some of their stability when the external disturbance of an El Niño event is added. This agrees well with the findings of Gozolchiani et al. (2008) and Tsonis and Swanson (2008), who showed that, for global networks, fluctuations, or "blinking" of links could be related to the global signature of ENSO variability. Therefore, despite the large persistence in the monsoon network, the monsoon–ENSO teleconnection is also visible in the common link recurrence (Fig. 5b).

To check whether the main changes in the climate networks investigated here occurred due to changes in the degrees of "supernodes" (nodes with higher degree), we plotted the variability of the degree for each node in Fig. 6a) and find that, indeed, degree variability is high (low) where node degree is high (low). Computing the correlation between time

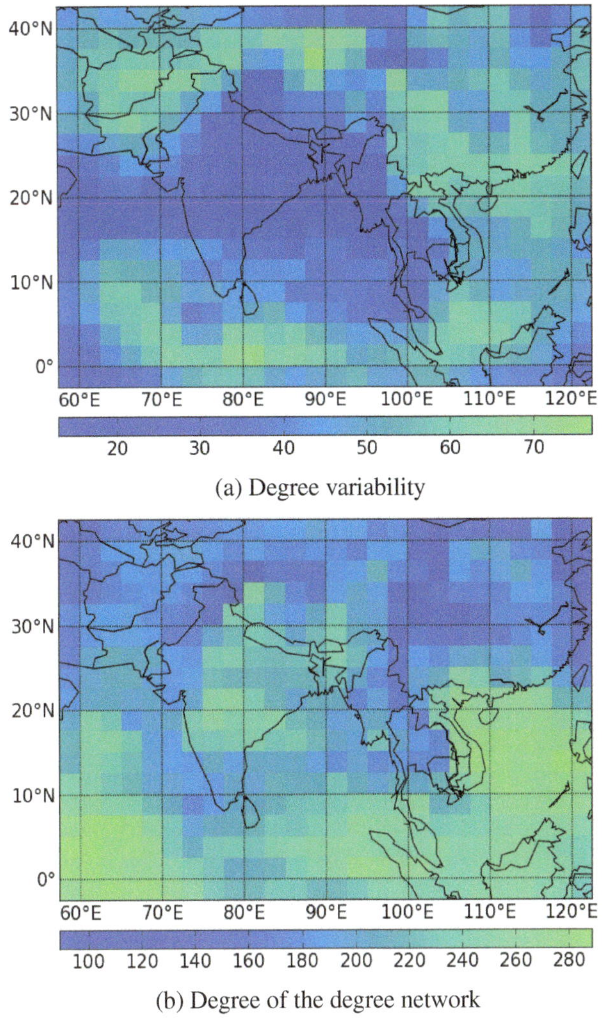

Figure 6. Mean degree variability **(a)** and degree of the degree variability **(b)** in the annual temperature anomaly networks (1970–2011).

series of node degrees, we obtain a network of degree variability. Figure 6b shows the degree of the resulting graph. Where the degree variability is low, over mainland India, we find high degrees in the degree network, which means that links in this region are mostly persistent. Where the degree variability is high, over the adjacent Indian Ocean and the South China Sea, we also observe high degree values in the network of the degrees, suggesting that degree changes here are large, but synchronized. In the northern part of the Asian monsoon domain considered here, spanning from Afghanistan through Pakistan, the Himalayas to China, we find a higher degree variability with less synchronized degree changes. Desynchonization in this region may occur due to the additional effect of the continental Westerlies and the large altitudinal gradients.

5 Conclusions

We have presented a generic approach to characterize the evolution of networks. With model tests we established that it is possible to use it to distinguish random, deterministic and periodic evolution behaviours in a set of networks. The new quantity to measure variability and persistence in networks is suitable for different network types. For example, the network set may be linearly ordered by time – or by parameter difference. The method can be extended in a straightforward manner, but it currently requires that the node structure and link density remain constant. Applying the CCEF analysis to data from the Asian Monsoon domain we found that El Niño years are accompanied by a distinct network imprint, leading to small common components with non-ENSO years and high agreement with ENSO years. In the future, the CCEF could be a particularly useful tool in the investigation of change points in network evolution.

Acknowledgements. We thank Jakob Runge for helpful discussions. The authors acknowledge funding by the German Science Foundation (DFG Graduate School 1539), the German Federal Ministry for Education and Research (BMBF project PROGRESS) and the Marie-Curie ITN LINC (P7-PEOPLE-2011-ITN, grant no. 289447) and via Helmholtz grant VG-900NH.

Edited by: J. Davidsen

References

Albert, R. and Barabasi, A.: Topology of evolving networks: local events and universality, Phys. Rev. Lett., 85, 5234–5237, 2000.

Barreiro, M., Marti, A., and Masoller, C.: Inferring long memory processes in the climate network via ordinal pattern analysis, Chaos, 21, 013101, doi:10.1063/1.3545273, 2011.

Barthélemy, M.: Spatial networks, Phys. Rep., 499, 1–100, doi:10.1016/j.physrep.2010.11.002, 2011.

Berezin, Y., Gozolchiani, A., Guez, O., and Havlin, S.: Stability of climate networks with time, Scientific Reports, 2, 666, doi:10.1038/srep00666, 2012.

Chatfield, C.: The analysis of time series: an introduction, CRC Press, Florida, USA, 6th Edn., 2004.

Clarke, A.: An introduction to the dynamics of El Niño & the Southern Oscillation, Academic Press, 2008.

Deza, J., Barreiro, M., and Masoller, C.: Inferring interdependencies in climate networks constructed at inter-annual, intra-season and longer time scales, Eur. Phys. J.-Spec. Top., 222, 511–523, doi:10.1140/epjst/e2013-01856-5, 2013.

Donges, J., Zou, Y., Marwan, N., and Kurths, J.: Complex networks in the climate system, Eur. Phys. J.-Spec. Top., 174, 157–179, doi:10.1140/epjst/e2009-01098-2, 2009.

Erdös, P. and Rényi, A.: On random graphs I., Publicationes Mathematicae (Debrecen), 6, 290–297, 1959.

Fu, W., Song, L., and Xing, E. P.: Dynamic mixed membership blockmodel for evolving networks, in: Proceedings of the 26th

Annual International Conference on Machine Learning – ICML '09, 1–8, doi:10.1145/1553374.1553416, 2009.

Gozolchiani, A., Yamasaki, K., Gazit, O., and Havlin, S.: Pattern of climate network blinking links follows El Niño events, EPL, 83, 28005, doi:10.1209/0295-5075/83/28005, 2008.

Hamming, R. W.: Error detecting and error correcting codes, The Bell System Technical Journal, 29, 147–160, doi:10.1002/j.1538-7305.1950.tb00463.x, 1950.

Holme, P., Edling, C. R., and Liljeros, F.: Structure and time evolution of an Internet dating community, Soc. Networks, 26, 155–174, doi:10.1016/j.socnet.2004.01.007, 2004.

Malik, N., Bookhagen, B., Marwan, N., and Kurths, J.: Analysis of spatial and temporal extreme monsoonal rainfall over South Asia using complex networks, Clim. Dynam., 39, 971–987, doi:10.1007/s00382-011-1156-4, 2011.

Martin, E. A., Paczuski, M., and Davidsen, J.: Interpretation of link fluctuations in climate networks during El Niño periods, EPL-Europhys. Lett., 102, 48003, doi:10.1209/0295-5075/102/48003, 2013.

Menck, P. and Kurths, J.: Topological Identification of Weak Points in Power Grids, Nonlinear Dynamics of Electronic Systems, 144–147, 2012.

Molkenthin, N., Rehfeld, K., Marwan, N., and Kurths, J.: Networks from Flows – From Dynamics to Topology, Scientific Reports, 4, 4119, doi:10.1038/srep04119, 2014a.

Molkenthin, N., Rehfeld, K., Stolbova, V., Tupikina, L., and Kurths, J.: On the influence of spatial sampling on climate networks, Nonlin. Processes Geophys., 21, 651–657, doi:10.5194/npg-21-651-2014, 2014b.

NOAA: NCEP/NCAR, http://www.esrl.noaa.gov/psd/data/gridded/data.ncep.reanalysis.html (last access: February 2014), 2014.

Palla, G., Derényi, I., Farkas, I., and Vicsek, T.: Uncovering the overlapping community structure of complex networks in nature and society, Nature, 435, 814–818, doi:10.1038/nature03607, 2005.

Paluš, M., Hartman, D., Hlinka, J., and Vejmelka, M.: Discerning connectivity from dynamics in climate networks, Nonlin. Processes Geophys., 18, 751–763, doi:10.5194/npg-18-751-2011, 2011.

Rehfeld, K., Marwan, N., Breitenbach, S. F. M., and Kurths, J.: Late Holocene Asian Summer Monsoon dynamics from small but complex networks of palaeoclimate data, Clim. Dynam., 41, 3–19, doi:10.1007/s00382-012-1448-3, 2013.

Rehfeld, K., Molkenthin, N., and Kurths, J.: Testing the detectability of spatio-temporal climate transitions from paleoclimate networks with the START model, Nonlin. Processes Geophys., 21, 691–703, doi:10.5194/npg-21-691-2014, 2014.

Stolbova, V., Martin, P., Bookhagen, B., Marwan, N., and Kurths, J.: Topology and seasonal evolution of the network of extreme precipitation over the Indian subcontinent and Sri Lanka, Nonlin. Processes Geophys., in review, 2014.

Trenberth, K. E.: The Definition of El Niño, B. Am. Meteorol. Soc., 78, 2771–2777, 1997.

Tsonis, A. and Swanson, K. L.: Topology and Predictability of El Niño and La Niña Networks, Phys. Rev. Lett., 100, 228502, doi:10.1103/PhysRevLett.100.228502, 2008.

Turner, A. and Annamalai, H.: Climate change and the South Asian summer monsoon, Nature Climate Change, 2, 587–595, doi:10.1038/nclimate1495, 2012.

Wang, B.: The Asian Monsoon, Springer-Verlag Berlin Heidelberg New York, ISBN 3-540-40610-7, 2006

Yamasaki, K., Gozolchiani, A., and Havlin, S.: Climate Networks around the Globe are Significantly Affected by El Niño, Phys. Rev. Lett., 100, 228501, doi:10.1103/PhysRevLett.100.228501, 2008.

10

Trend analysis using non-stationary time series clustering based on the finite element method

M. Gorji Sefidmazgi[1], M. Sayemuzzaman[2], A. Homaifar[1], M. K. Jha[3], and S. Liess[4]

[1]North Carolina A&T State University, Dept. of Electrical Engineering, Greensboro, USA
[2]North Carolina A&T State University, Dept. of Energy and Environmental Systems, Greensboro, USA
[3]North Carolina A&T State University, Dept. of Civil, Architectural and Environmental Engineering, Greensboro, USA
[4]University of Minnesota, Department of Soil, Water and Climate, St. Paul, USA

Correspondence to: A. Homaifar (homaifar@ncat.edu)

Abstract. In order to analyze low-frequency variability of climate, it is useful to model the climatic time series with multiple linear trends and locate the times of significant changes. In this paper, we have used non-stationary time series clustering to find change points in the trends. Clustering in a multi-dimensional non-stationary time series is challenging, since the problem is mathematically ill-posed. Clustering based on the finite element method (FEM) is one of the methods that can analyze multidimensional time series. One important attribute of this method is that it is not dependent on any statistical assumption and does not need local stationarity in the time series. In this paper, it is shown how the FEM-clustering method can be used to locate change points in the trend of temperature time series from in situ observations. This method is applied to the temperature time series of North Carolina (NC) and the results represent region-specific climate variability despite higher frequency harmonics in climatic time series. Next, we investigated the relationship between the climatic indices with the clusters/trends detected based on this clustering method. It appears that the natural variability of climate change in NC during 1950–2009 can be explained mostly by AMO and solar activity.

1 Introduction

In recent years, new methods have been developed in the area of machine learning and computational statistics to analyze massive data. In the climate science area, it is important to find connections between results of such data-based analysis and physical phenomena. In this paper, we address the problem of climate variability and its relation to physical phenomena. The trend analysis is useful for a better understanding of climate change and variability. Furthermore, interpreting the results of trend detection is useful for an understanding of the physical relations behind both warming and cooling on inter-annual timescales and climate indices.

The linear trend is the most straightforward assessment of the long-term behavior of a time series (Wilks, 2011). During the last decade, numerous papers have discussed long-term linear tendencies of climate parameters such as precipitation, temperature and the North Atlantic Oscillation (NAO) index (Tabari and Hosseinzadeh Talaee, 2011; Sonali and Nagesh Kumar, 2013). Due to changes in the trends, real-world time series do not generally fit in a straight line. As outlined by the IPCC Fifth Assessment Report (IPCC, 2013), in the Northern Hemisphere, 1983–2012 was likely the warmest 30-year period of the last 1400 years and has a larger temperature trend. Also, a visual inspection of global temperature time series for the period 1880–1997 shows that the mean warming obtained by fitting a straight line did not occur in a consistent manner, except in two continuous periods, one beginning around 1910 and the other starting in the mid-1970s (Tomé and Miranda, 2004). It has been statistically shown that a linear trend does not adequately describe low-frequency behavior of temperature time series (Karl et al., 2000; Lyubchich et al., 2013). Seidel and Lanzante (2004) analyzed global surface temperature anomalies and proved using the Schwarz Bayesian information criterion (SIC) that the piecewise linear model is better than a single linear trend. Other analysis methods, such as Fourier analysis, cannot find change points that result from heterogeneity in the data and many of the features become hidden. Therefore, it is useful to find local trends and detect the breakpoints in the time series.

Mudelsee (2010) indicated some parametric and nonparametric methods to find models for trend change. In parametric methods, a cost function is defined using ordinary least squares (OLS), and a brute-force search is performed over all candidate points to estimate the change points. Also, an algorithm is presented in Tomé and Miranda (2005) for fitting a continuous regression model with several break points to data and then it was applied in order to study changes in Azores temperature, the NAO index and Lisbon winter precipitation. In addition, Liu et al. (2010) used the same method to find partial trends of wind variability in the mesosphere and the lower thermosphere. In nonparametric smoothing, the points inside a sliding window of length L are averaged using a kernel function and the window runs along the time axis (Godtliebsen et al., 2012; Scherrer et al., 2013).

Methods based on statistical tests are also used in the literature to find changes in the linear trends. They consist of performing several kinds of tests to confirm or reject the hypothesis of a change occuring. Mann–Kendall is used to detect change points in trends of rainfall and stream flow (Modarres and Sarhadi, 2009; Ehsanzadeh et al., 2011). Toreti and Desiato (2008) analyzed mean temperature trends in Italy using two separated lines detected by the progressive Mann–Kendall method. Furthermore, Martínez et al. (2010) used four homogeneity tests to find one break point in the temperature trend of Spain and then related these changes to volcanic eruptions. The Bai–Perron test is also used in change point analysis of time series (Toreti et al., 2010a; Verbesselt et al., 2010). The former paper connected temperature trend change over Italy to NAO. Beaulieu et al. (2012a, b) assumed that the most likely position for a shift in model parameters of time series is one that minimizes the SIC and found best change points in linear trends of atmospheric CO_2. Kiely (1999) found change points in precipitation in Ireland using the Pettitt–Mann–Whitney method after smoothing using a 10 year moving average window.

Bayesian and Markovian methods are also used in the literature to find change points in climatic time series. In Bayesian methods, times of the change points and other model parameters are assumed to be random variables and their statistical distribution is estimated. Markov methods assume that change points in the time series are generated by a Markov process. Dose and Menzel (2004) and Rutishauser et al. (2007) analyzed phenological time series to find piecewise linear sections using a Bayes estimator and Monte Carlo and assessed the impacts of global change. Seidou and Ouarda (2007) proposed a Bayesian method to detect multiple change points in the trend of river stream flows of Quebec. Ruggieri (2013) used a combination of Bayesian analysis and dynamic programming to find change points in the trend of the $\delta^{18}O$ record of the Plio-Pleistocene. Kehagias and Fortin (2006) identified annual discharge phases in the Senegal River during 1903–1982 using the Hidden Markov Model.

It is very important that the mean-shift change points caused by changes in observational procedures, instrumentation or station relocation are not discarded (Martínez et al., 2010). Many of these change point times are undocumented in station histories. Gradual changes may also be caused by trees growing taller. Different methods such as Hidden Markov Modeling (Toreti et al., 2012), minimum description length (Lu et al., 2010), higher-order moments (Della-Marta and Wanner, 2006; Toreti et al., 2010b), wavelet power spectrum (Li et al., 2013) and Bayesian analysis (Hannart and Naveau, 2009) are proposed to detect artificial changes in climatic time series. Finding these types of change points in the time series, which also have long-term trends, is very difficult (Gallagher et al., 2013).

All the aforementioned methods are applied only to one-dimensional time series. In addition, the methods based on brute-force are not suitable for longer time series because the search space becomes very large. In fact, time series containing N data points have approximately N^k distinct placements of k change points. Bayesian and Markov methods are based on some statistical assumptions that are not true in general. The precision of kernel methods depends highly on the length of the defined window. In practice, it is very difficult to discriminate between natural (climatic) and artificial (non-climatic) variations in climatic time series. Conrad (1944) defined *relative homogeneity*, indicating that the variations in weather have similar tendencies over rather large regions. Therefore, neighboring stations should have the same pattern of temperature change. It is very unlikely that artificial change occurred at the same time for most of the sensors (Martínez et al., 2010). As a result, an algorithm to find breakpoints in climatic time series that processes several stations *in parallel* and also imposes minimal assumptions on the observed data is attracting ever-increasing attention in modern environmental studies.

In this paper, we study the change detection problem from the point of view of time series clustering. An important characteristic of climatic time series is nonstationarity where statistical properties of time series change under time translations. Assuming that the time series has a mathematical model with time-dependent parameters, a cluster (regime, phase or segment) is defined as a period such that in each of these periods, the model parameters are constant. The problem of finding these clusters and change points between them is called *non-stationary time series clustering* (Horenko, 2010a).

Time series clustering based on the finite element method (FEM) was recently introduced in Horenko (2010a), which can find the clusters without any statistical assumptions. Through this method, a multidimensional time series is processed to find clusters such that the change points between clusters are present in all dimensions at the same time. In FEM clustering, an optimization problem for clustering is defined. Using Tikhonov regularization and FEM, this optimization is converted into a familiar *linear quadratic programming* (LQP) and a least mean square problem. Iteratively, these two sets of optimizations are solved. Assuming a different linear trend (Horenko, 2010b) and also a differential equation in each cluster (Horenko, 2010c), FEM clustering was used to analyze climate dynamics in ERA-40 reanalysis time series by finding fast switched clusters.

In this paper, FEM clustering is used to find long-term trends in sensors data in the form of time series. Section 2 describes basic definitions and the procedure of the FEM-clustering algorithm in detail. For the special case of trend detection, necessary equations are derived and it is explained how the optimal number of clusters and length of clusters can be chosen. In Sect. 3, the algorithm is tested for a generated time series. In Sect. 4, the method is applied to mean temperature time series of North Carolina (NC) consisting of 249 stations. The results represent a spatio–temporal pattern of change in the area of study. In Sect. 5 the relationship between simulation results and AMO and solar activity is presented, which shows the relationship between detected clusters and these indices.

2 Finite element method for time series clustering

Modeling is the process of finding an appropriate parametric model to explain the observed phenomenon accurately. Let x_t be an observed m-dimensional time series defined over period of time $[0,T]$. To describe this time series with partially linear trends, we assume a model in each cluster as $x_t = c_{0k} + c_{1k} \times t +$ noise, where k is the index of cluster and $k = 1,\ldots,K$. Now define the distance from the observation x_t and the linear trend model.

$$d(x_t, c_k) = \|x_t - (c_{0k} + c_{1k} \times t)\|^2 \tag{1}$$

The problem of model approximation is treated as the optimization for c_k. Assuming the time series is available on $t = 1,\ldots,T$, the optimization problem can be rewritten as

$$\sum_{t=0}^{T}\sum_{k=1}^{K} \mu_k(t).\mathrm{d}(x_t, c_k) \xrightarrow{M(t),C} \min \tag{2}$$

where $C = [c_1,\ldots,c_K]$ are the set of cluster parameters means the linear trend coefficients. $M(t) = [\mu_1(\mathrm{t}),\ldots,\mu_K(\mathrm{t})]$ are the cluster membership functions that show how much the data in time t belong to cluster k, which fulfills these constraints (Horenko, 2010b):

$$\mu_k(t) \geq 0; \qquad \sum_{k=1}^{K} \mu_k(t) = 1 \tag{3}$$

Solving the optimization in Eq. (2) is not possible, since the number of unknowns is more than the number of known variables. Different values of initial parameters may lead to different local minimums. Hence the problem is *ill-posed* in the sense of *Hadamard* and thus requires *regularization*. For example in Tikhonov regularization, the additional assumption (called the regularization term) is included in the minimization problem (Aster and Thurber, 2012). In FEM clustering, it is assumed that the cluster membership functions $\mu_k(t)$ are smooth and that their derivatives are bounded. For a given time series, this constraint limits the total number of switchings between the clusters and adds *persistency* to $\mu_k(t)$ (Horenko, 2010a). Since $\mu_k(t)$ is defined for all the times in $[0,T]$, it belongs to infinite-dimensional *Hilbert space*. Thus the smoothness assumption in Hilbert space is (Poznyak, 2008)

$$\left\|\frac{\partial \mu_k(t)}{\partial t}\right\|_{\mathbf{H}(0,T)} = \int_0^T \left(\frac{\partial \mu_k(t)}{\partial t}\right)^2 \mathrm{d}t < +\infty. \tag{4}$$

The new optimization problem using Tikhonov regularization can be defined as in Eq. (5). Tikhonov's regularized problem can be used as a trade-off among the objective function (first term) and the penalty function (second term), and this trade-off is controlled by $\delta > 0$. Here δ is the *regularization factor* that controls the persistency of the clusters (Horenko, 2010a).

$$\sum_{k=1}^{K}\sum_{t=0}^{T}\left[\mu_k(t).\mathrm{d}(x_t, c_k) + \delta\left(\frac{\partial \mu_k(t)}{\partial t}\right)^2\right] \xrightarrow{M(t),C} \min \tag{5}$$

Unfortunately there is no analytical solution for Eq. (5), but it can be solved using the subspace iteration procedure. The optimization can be solved iteratively with respect to two sets of unknown parameters ($M(t)$ and C) by solving one and substituting in the other one. In order to solve the optimization numerically with respect to $M(t)$, it should be converted from a continuous-time to a discrete-time domain. For this reason, in FEM clustering, the Galerkin method is utilized for discretization. Galerkin methods can convert a continuous operator problem (such as a differential equation) to a discrete problem. They are widely used in the FEM literature to solve differential equations (Zienkiewicz et al., 2013). Assuming $\mu_k(t)$ is defined over $[0,T]$, we define a set of N continuous functions (α_n) called FEM-basis over the support of $\mu_k(t)$. The FEM-basis functions are linearly independent in Hilbert space and have local support on $[0,T]$. Then mapping equations for $\mu_k(t)$ using FEM-basis functions can be written as Eqs. (6) and (7). In fact, $\mu_k(t)$ is mapped to a space spanned by the FEM-basis, where $\tilde{\mu}_k^{(n)}$ are scalars called *Galerkin coefficients*. In our problem, the function $\mu_k(t)$ is unknown (like in differential equations, etc.). By determining Galerkin coefficients and using FEM-basis functions, one can build an approximated version of functions $\mu_k(t)$. The FEM-basis functions here are defined in the form of N triangular functions each one called a *hat function* as in Eq. (8)

and Fig. 1 (Horenko, 2010a):

$$\mu_k(t) = \tilde{\mu}_k(t) + \text{error} = \frac{1}{\Delta}\sum_{n=1}^{N} \tilde{\mu}_k^{(n)}.\alpha_n(t) + \text{error} \tag{6}$$

$$\tilde{\mu}_k^{(n)} = \sum_{t=0}^{T} \mu_k(t)\cdot\alpha_n(t); \quad k=1,\dots,K; \quad n=1,\dots,N \tag{7}$$

$$\alpha_n(t) = \begin{cases} \frac{t_2-t}{\Delta} & n=1, t\in[t_1,t_2] \\ \frac{t-t_{n-1}}{\Delta} & 2\le n\le N-1, t\in[t_{n-1},t_n] \\ \frac{t_{n+1}-t}{\Delta} & 2\le n\le N-1, t\in[t_n,t_{n+1}] \\ \frac{t-t_{N-1}}{\Delta}, & n=N, t\in[t_{N-1},t_N] \end{cases} \tag{8}$$

Applying Eqs. (6), (7) and (8) in Eq. (5) and using the locality of FEM-basis functions support, one can find an optimization after some mathematical simplification in the form of Eqs. (9), (10), (11) and (12). **H** is a tri-diagonal matrix called *stiffness matrix*. Constraints on the cluster membership functions are converted to a set of constraints on Galerkin coefficients as Eq. (13) (Horenko, 2010a):

$$\tilde{L}^{\varepsilon} = \sum_{k=1}^{K} [\beta(c_k)^T \overline{\mu}_k + \delta\ \overline{\mu}_k^T\ \mathbf{H}\ \overline{\mu}_k] \xrightarrow{\overline{\mu}_k, C} \min \tag{9}$$

$$\beta(c_k) = [\sum_{t_1}^{t_2}\alpha_1(t).\text{d}(x_t,c_k)\dots\sum_{t_{n-1}}^{t_{n+1}}\alpha_n(t).\text{d}(x_t,c_k) \tag{10}$$

$$\dots\sum_{t_{N-1}}^{t_N}\alpha_N(t).\text{d}(x_t,c_k)]$$

$$\overline{\mu}_k = [\tilde{\mu}_k^{(1)},\dots\tilde{\mu}_k^{(n)},\dots,\tilde{\mu}_k^{(N)}] \tag{11}$$

$$\mathbf{H} = \begin{bmatrix} \frac{1}{\Delta} & \frac{-1}{\Delta} & 0 & \dots & 0 & 0 \\ \frac{-1}{\Delta} & \frac{2}{\Delta} & \ddots & 0 & 0 & 0 \\ 0 & \ddots & \ddots & \frac{-1}{\Delta} & 0 & 0 \\ 0 & 0 & \frac{-1}{\Delta} & \frac{2}{\Delta} & \ddots & 0 \\ \vdots & 0 & 0 & \ddots & \ddots & \frac{-1}{\Delta} \\ 0 & 0 & 0 & 0 & \frac{-1}{\Delta} & \frac{1}{\Delta} \end{bmatrix} \tag{12}$$

$$\sum_{k=1}^{K}\tilde{\mu}_k^{(n)} = 1; \qquad \tilde{\mu}_k^{(n)} \ge 0; \tag{13}$$

$$\forall n = 1,\dots,N; \qquad k=1,\dots,K$$

The minimization problem of Eq. (9) is in the form of a summation of K linear quadratic programming (LQP) optimizations. To solve this optimization, all $\overline{\mu}_k$ are augmented in a vector λ to make one optimization in the form of

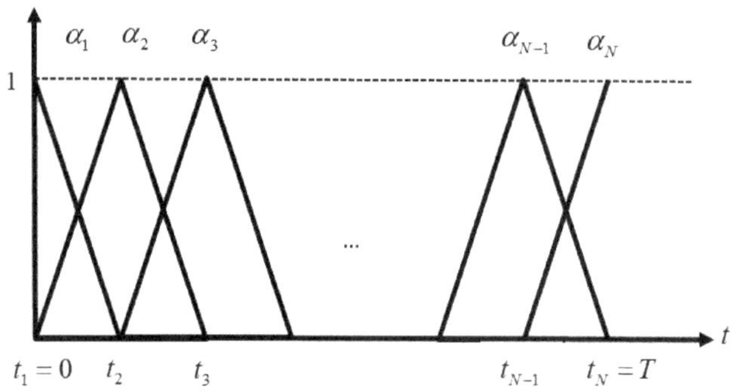

Figure 1. Finite element basis functions are in the form of an N hat function. Increasing N makes the algorithm more computationally expensive, but helps to find shorter clusters.

$$\frac{1}{2}\lambda^T G\lambda + B^T\lambda \xrightarrow{\lambda} \min \tag{14}$$

where

$$\lambda = [\overline{\mu}_1,\dots,\overline{\mu}_k,\dots\overline{\mu}_K]^T \tag{15}$$

$$B = [\beta(c_1),\dots,\beta(c_k),\dots\beta(c_K)] \tag{16}$$

$$G = 2\times\delta \begin{bmatrix} \mathbf{H} & 0 & \dots & 0 \\ 0 & \mathbf{H} & \ddots & \vdots \\ \vdots & \ddots & \ddots & 0 \\ 0 & \dots & 0 & \mathbf{H} \end{bmatrix} \tag{17}$$

As a result, the optimization constraints are converted to

$$\lambda_i \ge 0 \text{ for } i = 1,\dots,N\times K \tag{18}$$

$$F.\lambda = Q \tag{19}$$

$$F = \underbrace{[\ I_{N\times N}\quad I_{N\times N}\quad \dots\quad I_{N\times N}\ ]}_{K \text{ times}} \tag{20}$$

$$Q = [\ 1\quad 1\quad \dots\quad 1\ ]_{N\times 1}^T \tag{21}$$

The resulting optimization problem is a large sparse quadratic programming, which includes constraints in the form of equalities and inequalities. In the past, different techniques such as the *interior point* method and the *active set* method have been applied to solve the LQP (Boyd and Vandenberghe, 2004). The large sparse problem in this study is solved by the toolbox developed in Gurobi (2014). This toolbox is specially designed to solve large sparse optimization problems and has options to choose among several solvers.

Assuming that $\mu_k(t)$ is known, one can solve the optimization with respect to C analytically using the least mean square. Note that the second term in Eq. (5) is not a function of C and thus can be eliminated in this step. Based on the distance functional defined in Eq. (1),

$$c_{0k} = \frac{\begin{vmatrix} \sum_{t=0}^{T} x_t.\mu_k(t) & \sum_{t=0}^{T} t.x_t.\mu_k(t) \\ \sum_{t=0}^{T} t.x_t.\mu_k(t) & \sum_{t=0}^{T} x_t^2.\mu_k(t) \end{vmatrix}}{\begin{vmatrix} \sum_{t=0}^{T} \mu_k(t) & \sum_{t=0}^{T} t.\mu_k(t) \\ \sum_{t=0}^{T} t.\mu_k(t) & \sum_{t=0}^{T} t^2.\mu_k(t) \end{vmatrix}}; \tag{22}$$

$$c_{1k} = \frac{\begin{vmatrix} \sum_{t=0}^{T} \mu_k(t) & \sum_{t=0}^{T} x_t.\mu_k(t) \\ \sum_{t=0}^{T} t.\mu_k(t) & \sum_{t=0}^{T} t.x_t.\mu_k(t) \end{vmatrix}}{\begin{vmatrix} \sum_{t=0}^{T} \mu_k(t) & \sum_{t=0}^{T} t.\mu_k(t) \\ \sum_{t=0}^{T} t.\mu_k(t) & \sum_{t=0}^{T} t^2.\mu_k(t) \end{vmatrix}}$$

The general procedure for solving optimization problems is shown in Table 1. K, δ, and Δ are the numbers of clusters, the regularization factor, and finite discretization of time interval (width of the hat functions), respectively. It should be set at the beginning of the procedure. Changing Δ has an effect on the accuracy of clustering and the computational processing. To cluster a time series with K clusters and using N hat functions, the size of the vector to be minimized (λ) in Eq. (14) will be $N \times K$. Decreasing Δ will increase the number of FEM basis functions, and this leads to an increase in the size of LQP and, consequently, the volume of the computations. On the other hand, the accuracy of the algorithm will increase and help detect very short clusters. A more challenging problem arises when choosing K. A trivial solution exists when all the lines connecting every two points are in separate clusters. By increasing K, the value of the cost function always decreases and when $K =$ *length of time series – 1*, this value approaches zero. As a result it is not possible to find the *optimal* number of clusters. Therefore, we assume an upper bound on the number of clusters. Trial and error and human judgment are used to select K subjectively. The criterion for choosing K is based on the assumption that clusters are deterministic. This means that each datum in the time series belongs to only one of the clusters at each time. As a result, the values of $\mu_k(t)$ should be about 1 or 0. When the $\mu_k(t)$ is 1, it means that the datum in time t completely belongs to cluster k. When it is 0, it means that this datum does not belong to cluster k. One important issue in the time series clustering problem is the length of the clusters. For a fixed number of clusters, switching may occur many times or just a few times between the clusters. If the number of switchings between clusters increases, the duration in which the system stays in each of the clusters decreases. To investigate low-frequency variability in the system, small switching numbers are desirable. In contrast, for high-frequency variability we are interested in many transitions between clusters. It is shown that by increasing δ, clusters will become more *metastable* and the number of transitions between the clusters will be decreased. In fact, the observed process will stay for longer times in the identified metastable cluster before making a transition to the next identified cluster (Horenko, 2010a, b). Horenko (2010b) used the value of $\delta = 0$ to find less-metastable regimes in analyzing climate dynamics.

In summary, to assign clusters number K, we first assign a large $K \ll$ length of the time series and find clusters and $\mu_k(t)$ for different values of δ. If any of the cluster membership functions has values between 0 and 1, we decrease K and repeat the procedure until the cluster membership functions for all of the clusters becomes either about 0 or 1. We continue this procedure to find the largest K with a value of δ such that all of the clusters are deterministic.

Table 1. This procedure should be followed to solve clustering optimization and find two sets of known parameters: cluster membership functions and linear trend parameters in each cluster.

Step	Length
1	Set K the number of clusters, δ the regularization factor, Δ the width of hat functions.
2	Iteration number $L = 1$
3	Choose random initial λ^{initial} satisfying related constraints.
4	Solve the minimization problem with respect to C for fixed λ^{initial} to find C^{initial}.
5	Solve the minimization problem with respect to λ for fixed $C^{[L]}$ to find $\lambda^{[L+1]}$.
6	Solve the minimization problem with respect to C for a fixed $\lambda^{[L+1]}$ to find $C^{[L+1]}$.
7	Go to step 5 and continue until the pre-defined number of iterations.

3 Synthetic data testing

In order to test the algorithm and observe the effects of mean shift and outliers, we generated a two-dimensional time series with three clusters. Noise with mean value 0 and variance 4 is added to the time series. The models for the time series

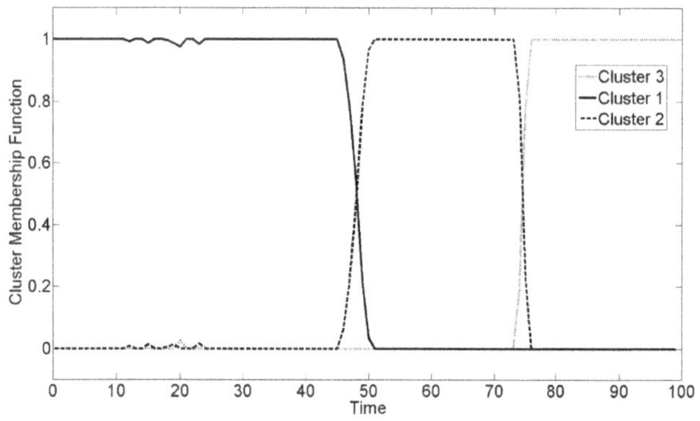

Figure 2. Cluster membership functions for synthetic data when $K = 3$ and $\delta = 1$. It can be seen that the clusters are found correctly.

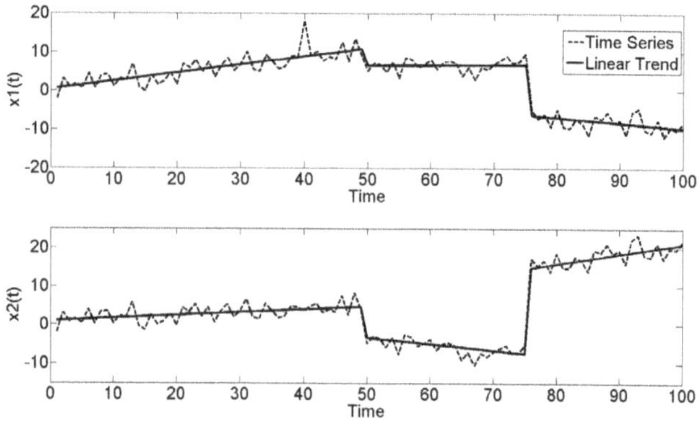

Figure 3. Detected clusters and linear trends for synthetic data when $K = 3$ and $\delta = 1$. Results show good clustering.

are

$$x_1(t) = \begin{cases} 0.2t + 0.1 & t \in [0, 50) \\ 0.1t - 0.1 & t \in [50, 75) \\ -0.1t + 0.2 & t \in [75, 100] \end{cases}$$

$$x_2(t) = \begin{cases} 0.3t + 0.4 & t \in [0, 50) \\ 0.2t + 0.1 & t \in [50, 75) \\ -0.1t - 0.4 & t \in [75, 100]. \end{cases} \quad (23)$$

Level shifts equal to 0.5 and 0.6 are added to $x_1(t)$ at $t = 20$ in the first cluster and to $x_2(t)$ in the third cluster at $t = 75$, respectively. In order to simulate an outlier the value of $x_1(t)$ at $t = 40$ is set to 18. The cluster membership function is shown in Fig. 2, where $\delta = 4$ and $K = 3$. Here, the change points are found correctly in the presence of a level shift and outlier since mean-shift changes and outliers are not at the same time in the two dimensions. Figure 3 shows detected trends in two dimensions of the original time series.

Now assume a case in which we set $K = 4$. The new values of $\mu_k(t)$ are shown in Fig. 4. In this figure, cluster membership functions for the first and fourth clusters are about either 0 or 1. Thus, these two clusters are found correctly, but the cluster membership functions for the second and third clusters are about 0.5 in the interval [50, 75] and 0 elsewhere. This means that clusters 2 and 3 are not *separable* and that K must be decreased from 4 to 3.

To observe the effect of the regularization factor δ on the clustering result, set $K = 3$ and increase the value of δ from 1 to larger numbers. The FEM clustering does not produce the deterministic clusters. It means that the values of $\mu_k(t)$ are always between 0 and 1. On the other hand, if we decrease the number of clusters and set $K = 2$, an acceptable result can be found when $\delta = 7$, as shown in Fig. 5. It is clear that the clusters [0, 50] and [50, 75] are merged together to create a new cluster. The second cluster is between [75, 100], which is found correctly. Thus, new clusters are found with a small number of switchings between them and with longer duration. In conclusion, in FEM clustering of time series, it is necessary to repeat clustering procedures with different values of K and δ until acceptable results are found.

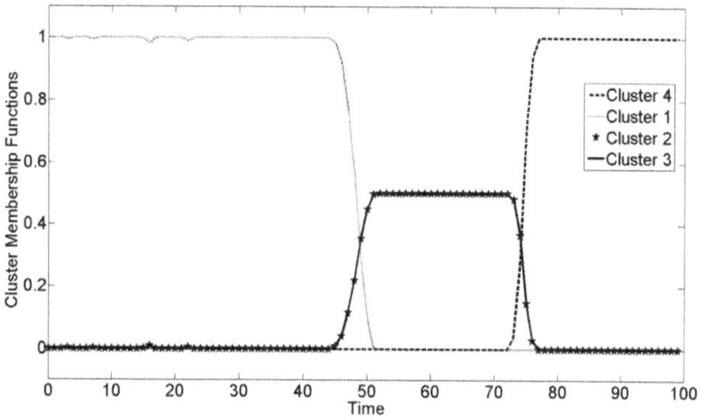

Figure 4. Cluster membership functions for synthetic data when $K = 4$; $\mu(t)$ for the first and fourth clusters are about 0 or 1 and thus these clusters are found correctly, but $\mu(t)$ for the second and third clusters are equal to about 0.5 in time domain [50, 75] and 0 elsewhere. This means that clusters two and three are not separable and that K must be decreased from 4 to 3.

4 Application: NC temperature trend analysis

In this paper, we study climate clusters/trends in NC. Boyles and Raman (2003) and Bililign et al. (2012) analyzed temperature trends in NC and showed the effect of global warming. The state of NC is located in the southeastern United States (75°30′–84°15′ W, 34°–36°21′ N). The study area covers approximately 52 664 square miles (136 399 km^2), encompassing 100 counties. A data set of average temperatures in 249 stations across NC is analyzed for the period beginning in 1950 until the end of October 2009. Figure 6 shows the locations of measuring stations. Daily temperatures are collected from the United States Department of Agriculture-Agriculture Research Service. This data set was facilitated by the National Oceanic and Atmospheric Administration (NOAA). These data contain quality control information from the Cooperative Observer Program (COOP) network and the Weather Bureau Army Navy (WBAN) and NOAA agencies.

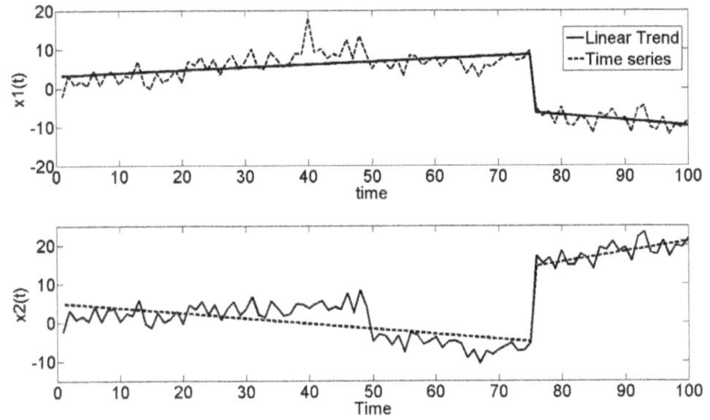

Figure 5. Result of clustering for synthetic data when $K = 2$ and $\delta = 7$. The first and second clusters are merged together. Increasing the δ leads to an answer with a smaller number of switching.

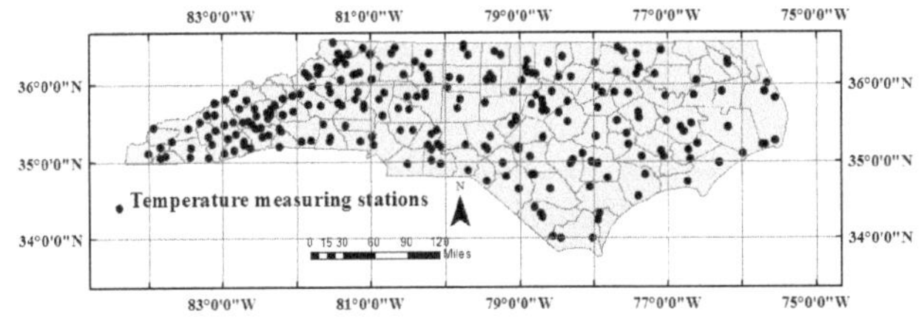

Figure 6. The state of NC is located in the southeastern United States. The study area covers an area of approximately 136 399 km^2. A data set of average temperatures at 249 stations across NC is analyzed for the period including the beginning of 1950 till the end of October 2009.

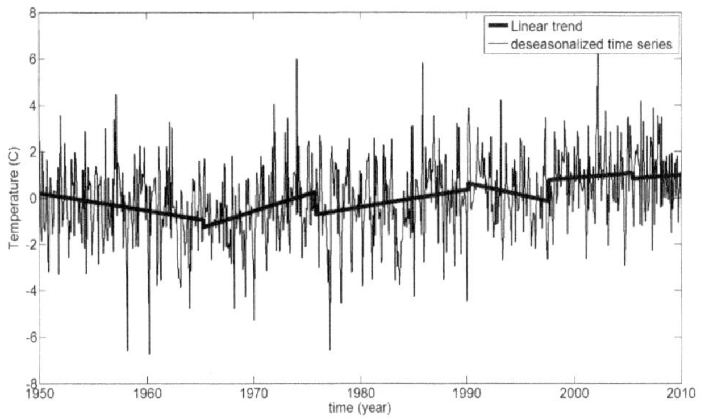

Figure 7. Time series at one station and its clusters/trends are shown. Thin lines show deseasonalized temperature time series and thick lines are linear trends found by the FEM algorithm. All 249 dimensions of the time series are analyzed at the same time, but the figure only shows the result at one of these stations.

Table 2. By increasing δ, clusters will become more metastable and the number of transitions between the clusters will be decreased. In fact, the observed process will stay for longer and longer times in the identified metastable cluster before making a transition to the next identified cluster.

Regularization factor	0.01	1	10	25	60	80	100
Number of switching	50	40	24	18	6	5	4

The data are examined for consistency and missing information. About 0.1 % of the data is missing and filled with cubic interpolation. Daily data are averaged to derive monthly data to eliminate high-frequency harmonics and also reduce the complexity of calculations. Monthly average is generally preferred over daily data due to the fact that this time frame is longer than the period of most large-scale synoptic waves in the troposphere (Nigam, 2003). The result is a 249×718 dimension time series. In order to find the trends, the annual cycle has been removed by subtracting the multi-year monthly means.

In applying FEM clustering to the time series, the value of K is initially assumed to be 10 and the algorithm is executed for different values of the regularization parameter δ. The regularization parameter is a real value larger than 0. For each δ, the algorithm ran several times to avoid convergence to local minimums in the optimization problem. When we reach the largest K such that all the $\mu(t)$ are about 0 or 1, an optimal solution is found. The optimal solution for this particular time series is achieved when $K = 6$, $\delta = 80$ and $\Delta = 4$. This result represents six clusters with different trends varying in magnitude and direction (positive or negative) with different length in NC. Table 2 shows the effect of δ on the number of switchings between clusters. As expected, the number of switchings decreases by decreasing the δ, while the length of the clusters increases. Once the linear trends are obtained, the *Mann–Kendall* trend approach can be used to test the statistical significance of trends (Modarres and Sarhadi, 2009).

Figure 7 shows the linear trends detected for the deseasonalized monthly time series at one of the stations. Note that all of the 249 dimensions of the time series are analyzed in batch by applying the FEM-clustering algorithm, and this figure only shows results at one of these stations. Figure 8 represents the linear trends for all 249 stations. Table 3 shows the average temperature, the average change in temperature and the corresponding percentage change in temperature over 60 years. The average trend is calculated by averaging between trends at all 249 stations. The average temperature change is found by multiplying the average trend by the length of the cluster. Percentage change is calculated by average temperature change divided by the average temperature during that cluster. Figure 9a–f shows the map corresponding to the spatial distribution of linear trends in NC. Twelve colors are used in *ArcGIS 10.0* to sketch the spatial distribution of trends in maps; contours are generated using an inverse distance weighted (IDW) algorithm with a power of 2.0, 12 grid points and variable radius location (Tewolde et al., 2010).

A closer observation of annual linear trends of clusters at all of the stations (Fig. 8), the spatial interpolated annual

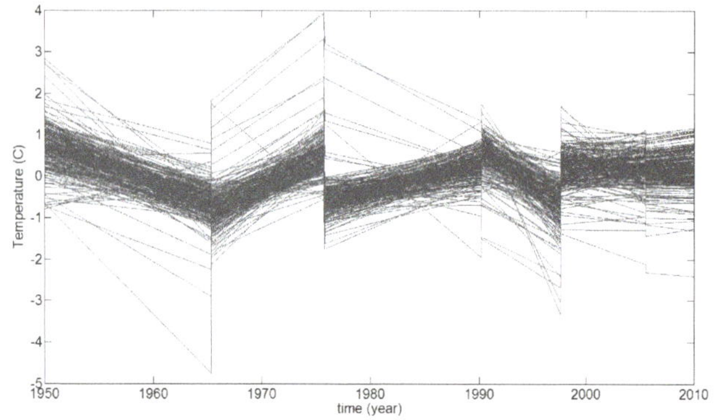

Figure 8. Linear trends and clusters at 249 stations during 60 years are shown. There are two notable decreasing trends between 1950–1964 and 1990–1998 and also a remarkable increase in 1964–1976. Other values of trends are smaller.

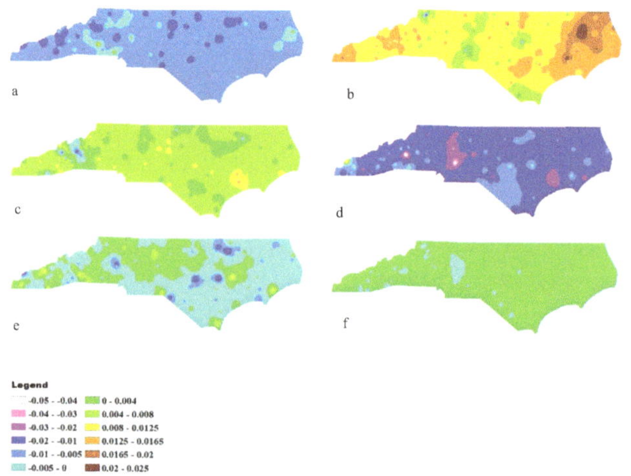

Figure 9. Six clusters with linear trends found in NC. **(a)** A notable negative trend between 1950 and 1965. **(b)** A positive trend after 1965. The coastal zone shows the highest warming trend. **(c)** Smaller positive trend all over the state between 1976 and 1990. **(d)** Negative trend in the 1990s with about 8 % decrease in temperature. **(e)** Positive trend in Piedmont and negative trend in other parts. **(f)** Most areas of the state have mild positive trends.

Table 3. Here are shown patterns of climate change in 60 years. They consist of the average temperature, the average change in temperature and the corresponding percentage change in temperature over 60 years. Average trend is calculated by averaging between trends at all 249 stations. Average temperature change is found by multiplying average trend by the length of the cluster. Percentage change is calculated by average temperature change divided by the average temperature during that cluster.

Clusters	Length	Average trend (°C per month)	Average temperature change (°C)	Percent change
Cluster 1	1950–1965	−0.0077	−1.42	−9.8
Cluster 2	1965–1976	0.0107	1.34	9.1
Cluster 3	1976–1990	0.0047	0.83	5.8
Cluster 4	1990–1998	−0.0133	−1.20	−8.0
Cluster 5	1998–2005	−0.0005	−0.05	−0.3
Cluster 6	2005–2009	0.0011	0.06	0.4

trends of NC (Fig. 9) and the corresponding Table 3 clearly shows two notable decreasing trends during 1965–1964 and 1990–1998, and a remarkable increase in temperature during 1964–1976. The longest and shortest periods of 15 and 5 years were found in cluster 1 and cluster 6, respectively. Also, clusters 2 and 4 show the warmest and coolest trends, respectively. Figure 9a corresponds to cluster 1, with the coolest trend statewide. After 1965, temperature in NC trends shifted to a positive trend (i.e., warmer). The coastal zone shows the highest warming trend in cluster 2 during the period 1950–1976, which indicates the greatest warming trends in 60 years (Fig. 9b). Cluster 3 (Fig. 9c) has a positive trend almost equally in all of the regions. Cluster 4 (Fig. 9d) has a negative trend in all areas except Piedmont, in which the amount of cooling is more significant. Cluster 5 (Fig. 9e) has a positive trend in the Piedmont area and a negative trend in the coastal and mountainous area. Cluster 6 shows the very mild warming and cooling trends statewide (Fig. 9f).

5 Comparison with climatic indices

In this section, the relationship between some climate indices and the results based on our proposed clustering method are discussed. Among the many available indices, we particularly analyzed the Atlantic Multi-decadal Oscillation (AMO) and the sunspot cycle. Investigation of these indices reveals that there exists some relation between temperature in NC and these indices.

The AMO is a pattern of multi-decadal variability in surface temperature centered in the North Atlantic Ocean. The index shows persistent warming and cooling phases that typically last a few decades. AMO has been linked with the occurrence of Sahel drought, variability in northeastern Brazilian rainfall, North American climate, river flows, and the frequency of Atlantic hurricanes (Knight et al., 2005). Figure 10 shows the AMO index time series (Schlesinger and Ramankutty, 1994; Enfield et al., 2001). FEM clustering was applied to the AMO time series using $\delta = 1$ and found four clusters and linear trends for the period spanning 1950–2010, which are shown in Fig. 10. The sunspot number is the sum of the number of individual sunspots plus ten times the number of sunspot groups. The number of sunspots visible on the sun, with waxes and wanes, has an approximate 11-year cycle (Frame and Gray, 2010). The total solar irradiance is measured from dark sunspots and bright faculae (Lean, 2009). Figure 11 shows the change of solar irradiance for the period of 1950–2010 (Lean, 2009; Hathaway, 2010). The annual climate trends in NC could be related to the AMO/solar activity.

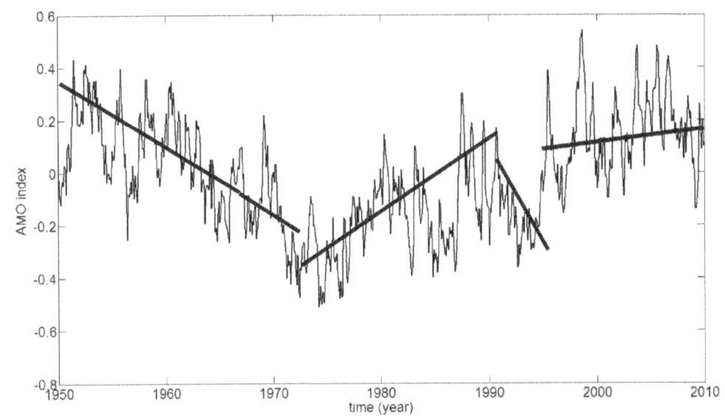

Figure 10. AMO is a pattern of multi-decadal variability in surface temperature centered in the North Atlantic Ocean. The index shows persistent warm and cool phases typically lasting a few decades. The figure shows the AMO index for the period of 1950–2010 and also its linear trends (thick lines).

- There is a cooling trend during 1950–1965, which coincides with the AMO dropping period in Fig. 10. As shown in Table 3, the ensemble average of trends in NC temperature is equal to -0.007 (°C month^{-1}) and trends in the AMO index -0.001 (°C month^{-1}).
- During 1965–1970 (cluster 2), the trend is positive. However, during this period, the AMO index continued to decrease, which is in contradiction with the positive trend in cluster 2. Nevertheless, comparing the trend in cluster 2 with solar activity, as in Fig. 11, shows that during this period solar activity increases, which coincides with the temperature increase in cluster 2. This comparison shows that the temperature increase could be due to an increase in solar activity. The ensemble average trend in NC temperature is 0.0107 (°C month^{-1}) and the trend in AMO is -0.007 (°C month^{-1}). After 1970, the AMO index has increased and at the same time the NC climate in cluster 2 has risen, too.
- During 1976–1990, NC climate is in cluster 3, with a smaller trend compared to cluster 2. We can observe that during the same period, the AMO index is negative but increasing with a positive trend. The AMO trend is 0.001 (°C month^{-1}) and the ensemble average trend in NC temperature is 0.0047 (°C month^{-1}).
- There is a negative trend in cluster 4 during 1990–1998. During this period the AMO index and solar activity decreases. Trends in AMO, solar irradiance and the NC temperature are -0.007 (°C month^{-1}), -0.0012 (watt $\text{m}^{-2} \times$ month) and -0.0133 (°C month^{-1}), respectively. Thus, the negative trend in temperature may be due to the negative trends in the AMO and the solar irradiation.

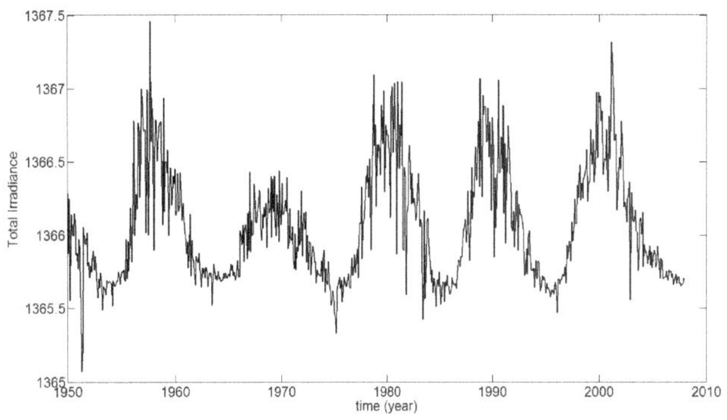

Figure 11. The total solar irradiance is measured from dark sunspots and bright faculae. This figure shows the change in solar irradiance for the period of 1950–2010 in Watts m^{-2}.

- Clusters 5 and 6 in NC have a very small negative and positive trend during 1998–2009. The AMO index oscillates around 0.2 (°C month^{-1}) in this period and does not have a notable trend.

6 Conclusions

In this paper, multiple change points in the trend of multi-dimensional climatic time series are found using FEM-based clustering. Since the time series of all stations are processed at the same time, the result of clustering in this study only shows climate changes due to natural phenomena. FEM clustering is applied to the time series with different linear trends in each cluster to find change points. After setting up an appropriate cost function, FEM clustering utilized Tikhonov regularization and Galerkin discretization to convert the optimization problem into a linear quadratic and a least mean square problem. By trial and error, the optimal number and length of clusters can be determined in FEM clustering. A climatic time series of NC is analyzed by this method. The results represent a spatio–temporal pattern of climate change corresponding to the areas of study. Then, some relations between clusters and climatic indices were discussed. It appears that the natural variability of climate change in NC during 1950–2009 can be explained mostly by AMO and solar activity.

Acknowledgements. This work is partially supported by the Expeditions in Computing by the National Science Foundation under award number CCF-1029731.

Edited by: A. Ganguly

References

Aster, R. C. and Thurber, C. H.: Parameter estimation and inverse problems, Academic Press, Waltham, MA, 2012.

Beaulieu, C., Chen, J., and Sarmiento, J. L.: Change-point analysis as a tool to detect abrupt climate variations, Philos. Trans. R. Soc. Math. Phys. Eng. Sci., 370, 1228–1249, 2012a.

Beaulieu, C., Sarmiento, J. L., Mikaloff Fletcher, S. E., Chen, J., and Medvigy, D.: Identification and characterization of abrupt changes in the land uptake of carbon, Glob. Biogeochem. Cy., 26, GB1007, doi:10.1029/2010GB004024, 2012b.

Bililign, S., Lin, Y.-L., Davis, R., Ilias, S., Kurkalova, L., Kyei, Y., Rastigejev, Y., Uzochukwu, G., and Bae, S.: Effects of global warming on North Carolina, Int. J. Clim. Change Impacts Responses, 3, 53–70, 2012.

Boyd, S. P. and Vandenberghe, L.: Convex optimization, Cambridge University Press, Cambridge, UK, 2004.

Boyles, R. P. and Raman, S.: Analysis of climate trends in North Carolina (1949–1998), Environ. Int., 29, 263–275, 2003.

Conrad, V.: Methods in climatology, Harvard University Press, Cambridge, MA, 1944.

Della-Marta, P. M. and Wanner, H.: A method of homogenizing the extremes and mean of daily temperature measurements, J. Climate, 19, 4179–4197, 2006.

Dose, V. and Menzel, A.: Bayesian analysis of climate change impacts in phenology, Global Change Biol., 10, 259–272, 2004.

Ehsanzadeh, E., Ouarda, T. B. M. J., and Saley, H. M.: A simultaneous analysis of gradual and abrupt changes in Canadian low streamflows, Hydrol. Process, 25, 727–739, 2011.

Enfield, D. B., Mestas-Nuñez, A. M., and Trimble, P. J.: The Atlantic Multidecadal Oscillation and its relation to rainfall and river flows in the continental US, Geophys. Res. Lett., 28, 2077–2080, 2001.

Frame, T. H. A. and Gray, L. J.: The 11-Yr solar cycle in ERA-40 data: An update to 2008, J. Climate, 23, 2213–2222, 2010.

Gallagher, C., Lund, R., and Robbins, M.: Changepoint detection in climate time series with long-term trends, J. Climate, 26, 4994–5006, 2013.

Godtliebsen, F., Holmström, L., Miettinen, A., Erästö, P., Divine, D. V., and Koc, N.: Pairwise scale space comparison of time series with application to climate research, J. Geophys. Res.-Ocean., 117, C03046, doi:10.1029/2011JC007546, 2012.

Gurobi Optimization: Gurobi optimizer reference manual, available at: http://www.gurobi.com (last access: 15 January 2014), 2014.

Hannart, A. and Naveau, P.: Bayesian multiple change points and segmentation: Application to homogenization of climatic series, Water Resour. Res., 45, W10444, doi:10.1029/2008WR007689, 2009.

Hathaway, D. H.: The solar cycle, Living Rev. Sol. Phys., 7, doi:10.12942/lrsp-2010-1, 2010.

Horenko, I.: Finite element approach to clustering of multidimensional time series, SIAM J. Sci. Comput., 32, 62–83, 2010a.

Horenko, I.: On clustering of non-stationary meteorological time series, Dynam. Atmos.-Ocean., 49, 164–187, 2010b.

Horenko, I.: On the Identification of Nonstationary Factor Models and Their Application to Atmospheric Data Analysis, J. Atmos. Sci., 67, 1559–1574, 2010c.

IPCC: Climate Change 2013 – The Physical Science Basis: Working Group I Contribution to the Fifth Assessment Report of the Intergovernmental Panel on Climate Change, 2013.

Karl, T. R., Knight, R. W., and Baker, B.: The record breaking global temperatures of 1997 and 1998: Evidence for an increase in the rate of global warming?, Geophys. Res. Lett., 27, 719–722, 2000.

Kehagias, Ath. and Fortin, V.: Time series segmentation with shifting means hidden markov models, Nonlin. Processes Geophys., 13, 339–352, doi:10.5194/npg-13-339-2006, 2006.

Kiely, G.: Climate change in Ireland from precipitation and streamflow observations, Adv. Water Resour., 23, 141–151, 1999.

Knight, J. R., Allan, R. J., Folland, C. K., Vellinga, M., and Mann, M. E.: A signature of persistent natural thermohaline circulation cycles in observed climate, Geophys. Res. Lett., 32, L20708, doi:10.1029/2005GL024233, 2005.

Lean, J.: Calculations of solar irradiance: monthly means from 1882 to 2008, available at: http://www.geo.fu-berlin.de/en/met/ag/strat/forschung/SOLARIS/Input_data/Calculations_of_Solar_Irradiance.pdf, 2009.

Li, Z., Yan, Z., Cao, L., and Jones, P.: Adjusting inhomogeneous daily temperature variability using wavelet analysis, Int. J. Climatol., 34, 1196–1207, doi:10.1002/joc.3756, 2013.

Liu, R. Q., Jacobi, C., Hoffmann, P., Stober, G., and Merzlyakov, E. G.: A piecewise linear model for detecting climatic trends and their structural changes with application to mesosphere/lower thermosphere winds over Collm, Germany, J. Geophys. Res.-Atmos., 115, D22105, doi:10.1029/2010JD014080, 2010.

Lu, Q., Lund, R., and Lee, T. C. M.: An MDL approach to the climate segmentation problem, Ann. Appl. Stat., 4, 299–319, 2010.

Lyubchich, V., Gel, Y. R., and El-Shaarawi, A.: On detecting non-monotonic trends in environmental time series: a fusion of local regression and bootstrap, Environmetrics, 24, 209–226, 2013.

Martínez, M. D., Serra, C., Burgueño, A., and Lana, X.: Time trends of daily maximum and minimum temperatures in Catalonia (ne Spain) for the period 1975–2004, Int. J. Climatol., 30, 267–290, 2010.

Modarres, R. and Sarhadi, A.: Rainfall trends analysis of Iran in the last half of the twentieth century, J. Geophys. Res.-Atmos., 114, D03101, doi:10.1029/2008JD010707, 2009.

Mudelsee, M.: Climate time series analysis classical statistical and bootstrap methods, Springer, Dordrecht, 2010.

Nigam, S.: Teleconnections, in: Encyclopedia of Atmospheric Sciences, 2243–2269, 2003.

Poznyak, A. S.: Advanced mathematical tools for automatic control engineers, Vol. 1, Elsevier, Amsterdam, 2008.

Ruggieri, E.: A Bayesian approach to detecting change points in climatic records, Int. J. Climatol., 33, 520–528, 2013.

Rutishauser, T., Luterbacher, J., Jeanneret, F., Pfister, C., and Wanner, H.: A phenology-based reconstruction of interannual changes in past spring seasons, J. Geophys. Res.-Biogeosci., 112, G04016, doi:10.1029/2006JG000382, 2007.

Scherrer, S. C., Wüthrich, C., Croci-Maspoli, M., Weingartner, R., and Appenzeller, C.: Snow variability in the Swiss Alps 1864–2009, Int. J. Climatol., 3162–3173, 2013.

Schlesinger, M. E. and Ramankutty, N.: An oscillation in the global climate system of period 65–70 years, Nature 367, 723–726, 1994.

Seidel, D. J. and Lanzante, J. R.: An assessment of three alternatives to linear trends for characterizing global atmospheric temperature changes, J. Geophys. Res.-Atmos., 109, D14108, doi:10.1029/2003JD004414, 2004.

Seidou, O. and Ouarda, T. B. M. J.: Recursion-based multiple changepoint detection in multiple linear regression and application to river streamflows, Water Resour. Res., 43, W07404, doi:10.1029/2006WR005021, 2007.

Sonali, P. and Nagesh Kumar, D.: Review of trend detection methods and their application to detect temperature changes in India, J. Hydrol., 476, 212–227, 2013.

Tabari, H. and Hosseinzadeh Talaee, P.: Analysis of trends in temperature data in arid and semi-arid regions of Iran, Global Planet. Change, 79, 1–10, 2011.

Tewolde, M. G., Beza, T. A., Costa, A. C., and Painho, M.: Comparison of different interpolation techniques to map temperature in the southern region of Eritrea, in: 13th AGILE International Conference on Geographic Information Science, 1–5, 2010.

Tomé, A. R. and Miranda, P. M. A.: Piecewise linear fitting and trend changing points of climate parameters, Geophys. Res. Lett., 31, L02207, doi:10.1029/2003GL019100, 2004.

Tomé, A. R. and Miranda, P. M. A.: Continuous partial trends and low-frequency oscillations of time series, Nonlin. Processes Geophys., 12, 451–460, doi:10.5194/npg-12-451-2005, 2005.

Toreti, A. and Desiato, F.: Temperature trend over Italy from 1961 to 2004, Theor. Appl. Climatol., 91, 51–58, 2008.

Toreti, A., Desiato, F., Fioravanti, G., and Perconti, W.: Seasonal temperatures over Italy and their relationship with low-frequency atmospheric circulation patterns, Clim. Change, 99, 211–227, 2010a.

Toreti, A., Kuglitsch, F. G., Xoplaki, E., Luterbacher, J., and Wanner, H.: A novel method for the homogenization of daily temperature series and its relevance for climate change analysis, J. Climate, 23, 5325–5331, 2010b.

Toreti, A., Kuglitsch, F. G., Xoplaki, E., and Luterbacher, J.: A novel approach for the detection of inhomogeneities affecting climate time series, J. Appl. Meteorol. Climatol., 51, 317–326, 2012.

Verbesselt, J., Hyndman, R., Newnham, G., and Culvenor, D.: Detecting trend and seasonal changes in satellite image time series, Remote Sens. Environ., 114, 106–115, 2010.

Wilks, D. S.: Statistical methods in the atmospheric sciences, Academic Press, Waltham, MA, 2011.

Zienkiewicz, O. C., Taylor, R. L., and Zhu, J. Z.: Finite Element Method: its basis and fundamentals, Butterworth-Heinemann, Amsterdam, 2013.

11

Lagrangian study of surface transport in the Kuroshio Extension area based on simulation of propagation of Fukushima-derived radionuclides

S. V. Prants, M. V. Budyansky, and M. Yu. Uleysky

Laboratory of Nonlinear Dynamical Systems, Pacific Oceanological Institute of the Russian Academy of Sciences, 43 Baltiyskaya st., 690041 Vladivostok, Russia

Correspondence to: S. V. Prants (prants@poi.dvo.ru)

Abstract. Lagrangian approach is applied to study near-surface large-scale transport in the Kuroshio Extension area using a simulation with synthetic particles advected by AVISO altimetric velocity field. A material line technique is proposed and applied to find out the origin of water masses in cold-core cyclonic rings pinched off from the jet in summer 2011. Tracking and Lagrangian maps provide the evidence of cross-jet transport. Fukushima-derived caesium isotopes are used as Lagrangian tracers to study transport and mixing in the area a few months after the 11 March 2011 tsunami that caused heavy damage of the Fukushima Nuclear Power Plant (FNPP). Tracking maps are computed to trace the origin of water parcels with measured levels of ^{134}Cs and ^{137}Cs concentrations collected during two research vessel (R/V) cruises in June and July 2011 in the large area of the northwest Pacific (Kaeriyama et al., 2013; Buesseler et al., 2012). It is shown that Lagrangian simulations are useful for finding the surface areas that are potentially dangerous due to the risk of radioactive contamination. The results of simulation are supported by tracks of the surface drifters that were deployed in the area.

1 Introduction

The Kuroshio Extension (KE) prolongs the Kuroshio Current, a western boundary current in the northwest Pacific, where the latter separates from the continental shelf of the Japanese island Honshu at Cape Inubō at about 35°42′ N. It flows eastward from this point as a strong unstable meandering jet constituting a front separating the warm subtropical and cold subpolar waters of the North Pacific Ocean. There are cyclonic and anticyclonic recirculation gyres on the northern and southern flanks of the jet. The main features of the KE are described by Qiu and Chen (2005) and Itoh and Yasuda (2010). The Kuroshio and the KE transport a large amount of heat and release that to the atmosphere, strongly affecting climate. It is a region with one of the most intense air–sea heat exchanges and the highest eddy kinetic energy level. It is also a region with commercial fishing grounds of Pacific saury, tuna, squid, Japanese sardine and other species.

The Kuroshio–Oyashio frontal zone contains various types of mesoscale and submesoscale eddies that transfer heat, salt, nutrients, carbon, pollutants and other tracers across the ocean. They originate, besides from the KE, from the Tsugaru Warm Current, flowing between the Honshu and Hokkaido islands, and from the cold Oyashio Current flowing out of the Arctic along the Kamchatka Peninsula and the Kuril Islands. Those eddies may persist for the periods ranging from a few weeks to a couple of years and have a strong influence on the local climate, hydrography and fishery.

A study of the role of KE rings and their interaction with the mean flow is important for many reasons. They act to transfer energy to the mean currents, influence on the KE jet dynamics and drive the recirculation gyres. They transport for a long distance water masses with biophysical properties different from ambient waters that may have a great impact on living organisms. The strongest mesoscale eddies of both polarities are generated along the KE. The warm-core anticyclonic rings (ACRs) are pinched off from the meandering KE mainly to the north whereas the cold-core cyclonic ones (CR) to the south of it. The occurrence, distribution and behavior of the KE rings, moving generally westward due to the planetary β effect, have been studied in a number of papers via hydrographic observations, infrared imaging and altimetry data (Tomosada, 1986; Ebuchi and Hanawa, 2001; Waseda, 2003; Itoh and Yasuda, 2010). However, the process of their separation from the parent jet is not fully understood.

Lagrangian tools have been successfully used to obtain a detailed description of different advective transport phenomena in the ocean and atmosphere. There is a vast literature on this topic (for a review see Mancho et al., 2004; Wiggins, 2005; Koshel' and Prants, 2006; Prants, 2013, and references therein). As for the problem of eddy separation from strong jet currents and a cross-jet transport, there are papers on Lagrangian approach to the loop current eddy separation in the Gulf of Mexico (Kuznetsov et al., 2002; Andrade-Canto et al., 2013) and on Lagrangian description of cross-jet transport in the Kuroshio Current (Mendoza et al., 2010; Mendoza and Mancho, 2012). Near-surface velocity fields from numerical models of circulation in the Gulf of Mexico have been used to study the eddy separation process by computing effective invariant manifolds (Kuznetsov et al., 2002) and finite-time Lyapunov exponents (Andrade-Canto et al., 2013). It has been shown that the Lagrangian methods are a useful supplement to traditional approaches as they reveal flow details not easily extracted from Eulerian point of view.

In Mendoza et al. (2010) and Mendoza and Mancho (2012), a special lobe technique from dynamical system theory (Wiggins, 1992) and a method of distinguished hyperbolic trajectories (Ide et al., 2002) have been applied to find a geometrical skeleton of some transport processes in an altimetric velocity field in the Kuroshio region including surface cross-jet transport. In Samelson and Wiggins (2006), Uleysky et al. (2007), Uleysky et al. (2010a) and Uleysky et al. (2010b), a dynamical system-theory-based analysis, revealing mechanisms of chaotic zonal and cross-jet transport, has been carried out for a few kinematic and dynamical analytic models of meandering jets.

In this paper we study numerically the process of interaction of cold-core CRs with the KE main current, the events of their separation from the parent jet and their role in near-surface cross-jet transport. The special aim is to know whether it was possible for Fukushima-derived radionuclides to cross the KE jet, which is supposed to be an impenetrable barrier. Simulation is based on solving advection equations for synthetic particles in the AVISO velocity fields. The results are plotted as (1) backward-in-time Lagrangian latitudinal maps, where colors code the latitudes from which particles in a given region came to their final positions on the map, and (2) tracking maps showing where they were walking and how frequently they visited different places in the region for a given period of time. In Sect. 3.1 we compute both types of the maps to trace origin of water masses inside two CRs pinched off from the KE jet in summer 2011 and to document the surface cross-jet transport. The Fukushima-derived Cs isotopes are used as Lagrangian tracers to study transport and mixing processes. We apply the proposed material line technique in Sect. 3.2 to trace the origin of water parcels with measured levels of concentrations of Fukushima-derived Cs isotopes collected during two R/V cruises in June and July 2011 in the large area of the northwest Pacific (Kaeriyama et al., 2013; Buesseler et al., 2012). The results of our simulation are supported by tracks of the surface drifters that were deployed in the area.

2 Data and methodology

Geostrophic velocities were obtained from the AVISO database (http://www.aviso.oceanobs.com). The data are gridded on a $1/3° \times 1/3°$ Mercator grid. It has been justified by Mancho et al. (2006) that a bicubical spatial interpolation and third-order Lagrangian polynomials in time are able to provide accurate numerical results. Geostrophic velocities, derived from altimetry, are a good approximation to the surface fields. We do not expect essentially different results with respect to the present analysis, taking into account ageostrophic corrections such as an Ekman correction from wind data, because we are interested in transport phenomena for a comparatively long period of time, more than a few weeks. In fact, we study transport of particles and Fukushima-derived radionuclides for 3 months or more. It is reasonably to expect that wind variations are averaged out for such a long period of time.

Lagrangian trajectories are computed by integrating the advection equations with a fourth-order Runge–Kutta scheme with a fixed time step of 0.001th part of a day. Our results are based not on individual trajectories but on statistics for thousands and millions of trajectories. Most of them have been found to be chaotic with positive values of the maximal finite-time Lyapunov exponent. We cannot, of course, guarantee that we compute "true" trajectories for individual particles. We can only guarantee with our tracking and other Lagrangian maps that particles have visited some areas more frequently than the other ones. Typical chaotic systems are highly robust against a high-frequency noise mimicking small-scale diffusion. Thus, an additional small noise in the advection equations is not expected to change our results based on statistics for a large number of particles.

Satellite altimetry demonstrates clearly that the KE alternates between two dominant states: one with two quasistationary meanders and another one when the meanders are not especially prominent (Qiu and Chen, 2005; Sugimoto and Hanawa, 2012). During the meandering state, steep troughs develop stimulating ring pinch-off events on both flanks of the jet. In this paper we will focus on that state with increased eddy activity. A sketch of the meandering KE state in Fig. 1 shows the eastward jet current with two crests near $x = 143°$ E and $x = 151°$ E, which are anticyclonic side of the jet and two troughs near $x = 147°$ E and $x = 153°$ E with cyclonic rotations. In reality the KE jet is highly unstable; the

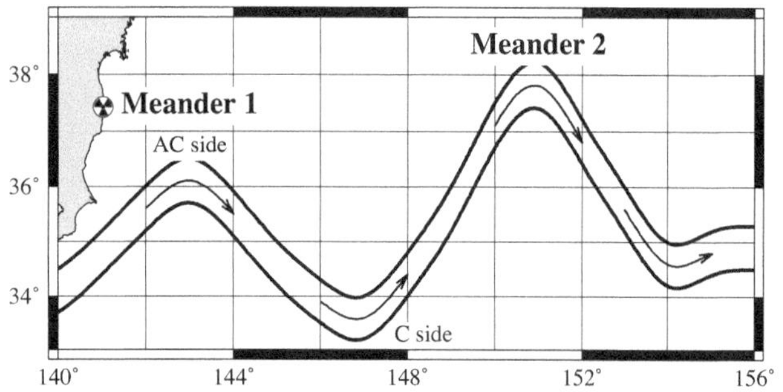

Fig. 1. Schematic view of the Kuroshio Extension state with two quasistationary meanders. Location of the Fukushima Nuclear Power Plant is shown.

meander's amplitude may change in the course of time, and locations of the crests and troughs may fluctuate strongly both in meridional and zonal directions.

Motion of a fluid particle in a two-dimensional flow is the trajectory of a dynamical system with given initial conditions governed by the velocity field. We solve advection equations on the Earth sphere in the form

$$\dot{x} = u(x, y, t), \qquad \dot{y} = v(x, y, t), \tag{1}$$

where time t is in days. Coordinates x and y of a passive particle are related with its latitude ϕ and longitude λ in degrees as follows:

$$\lambda = \frac{x}{60}, \quad \phi = \frac{180}{\pi} \arcsin \tanh \left(\frac{\pi}{180} \left(\frac{y}{60} + y_0 \right) \right),$$
$$y_0 = \frac{180}{\pi} \operatorname{artanh} \sin \left(\frac{\pi}{180} \phi_0 \right), \quad \phi_0 = -82. \tag{2}$$

We use the transformation (Eq. 2) because the AVISO grid is homogeneous in those coordinates. The velocities u and v in Eq. (1) are expressed through the the latitudinal U_ϕ and longitudinal U_λ components of the linear velocity U in cm s^{-1} as follows:

$$u = \frac{10\,800}{\pi R_E \cos\phi} \frac{86\,400}{100\,000} U_\lambda \approx \frac{0.466}{\cos\phi} U_\lambda,$$
$$v = \frac{10\,800}{\pi R_E \cos\phi} \frac{86\,400}{100\,000} U_\phi \approx \frac{0.466}{\cos\phi} U_\phi, \tag{3}$$

where R_E is the Earth radius in km.

The Lagrangian technique we use is based on calculation of particle motion in an altimetric velocity field forward and backward in time. When integrating advection equations (Eq. 1) forward in time, we compute particle trajectories to know their fate; when integrating them backward in time, we know origin of particles and their history. A graphic view of transport and mixing is provided by Lagrangian synoptic maps that are plots of one of the Lagrangian indicators versus particle's initial positions (Prants et al., 2011a, 2013; Prants, 2013). The region under study is seeded by a large number of synthetic particles whose trajectories are computed forward or backward in time for a given period of time. There are a number of Lagrangian indicators (or descriptors in terminology of Mendoza and Mancho, 2012) that may be used to characterize transport and mixing processes in the ocean and atmosphere (for a recent review see Mancho et al., 2013). Among them are zonal and absolute displacements of water parcels, the number of their cyclonic and anticyclonic rotations, vorticity, their residence time in a given area, the number of times particles visit different places in the region and others. The so-called M-function, which measures the Euclidean arc length of the curves outlined by trajectories for a finite-time interval (Madrid and Mancho, 2009), can be also used to plot the corresponding Lagrangian maps. Sometimes it is useful to compute composite maps with two or more Lagrangian indicators plotted together.

To compute the Lagrangian maps, we apply the methodology elaborated in our recent papers and applied to study transport and mixing processes in different basins, from marine bays (Prants et al., 2013) and seas (Prants et al., 2011a) to the ocean scale (Prants et al., 2011b; Prants, 2013). In the first part of Sect. 3.1 we calculate for a large number of particles the latitudes, ϕ, from which each particle came to its final position and plot Lagrangian latitudinal maps in order to document a cross-jet transport. Similar latitudinal maps have been used in another context by Abraham and Bowen (2002) and d'Ovidio et al. (2010).

The material line technique developed in this paper is a tool to trace origin, history and fate of water masses. A large number of particles (markers) are placed on a material line, crossing a feature under study, and evolve backward in time. Different kinds of outputs can be obtained with the help of that technique. In the context of this study, it is useful to get as an output tracking maps showing by density plots where the corresponding markers were walking for a given period of time. It has been done in the second part of Sect. 3.1 with markers placed on material lines crossing two Kuroshio CRs. This method is especially useful if one places material lines along the transects with in situ measurements. In Sect. 3.2 we carry out such simulations with material lines placed along the transects in the northwest Pacific where concentrations of radioactive Cs isotopes have been measured in surface seawater during two R/V cruises in 2011 after the Fukushima Nuclear Power Plant (FNPP) accident. Our simulations are complemented by tracks of available surface drifters (http://www.aoml.noaa.gov/phod/dac/index.php).

3 Results

3.1 Lagrangian study of origin of Kuroshio Extension rings and possibility of surface cross-jet transport

Documenting separation of rings from jets, merging with jets and tracking their propagation are long-standing problems in oceanography. An altimeter velocity field along with drifter

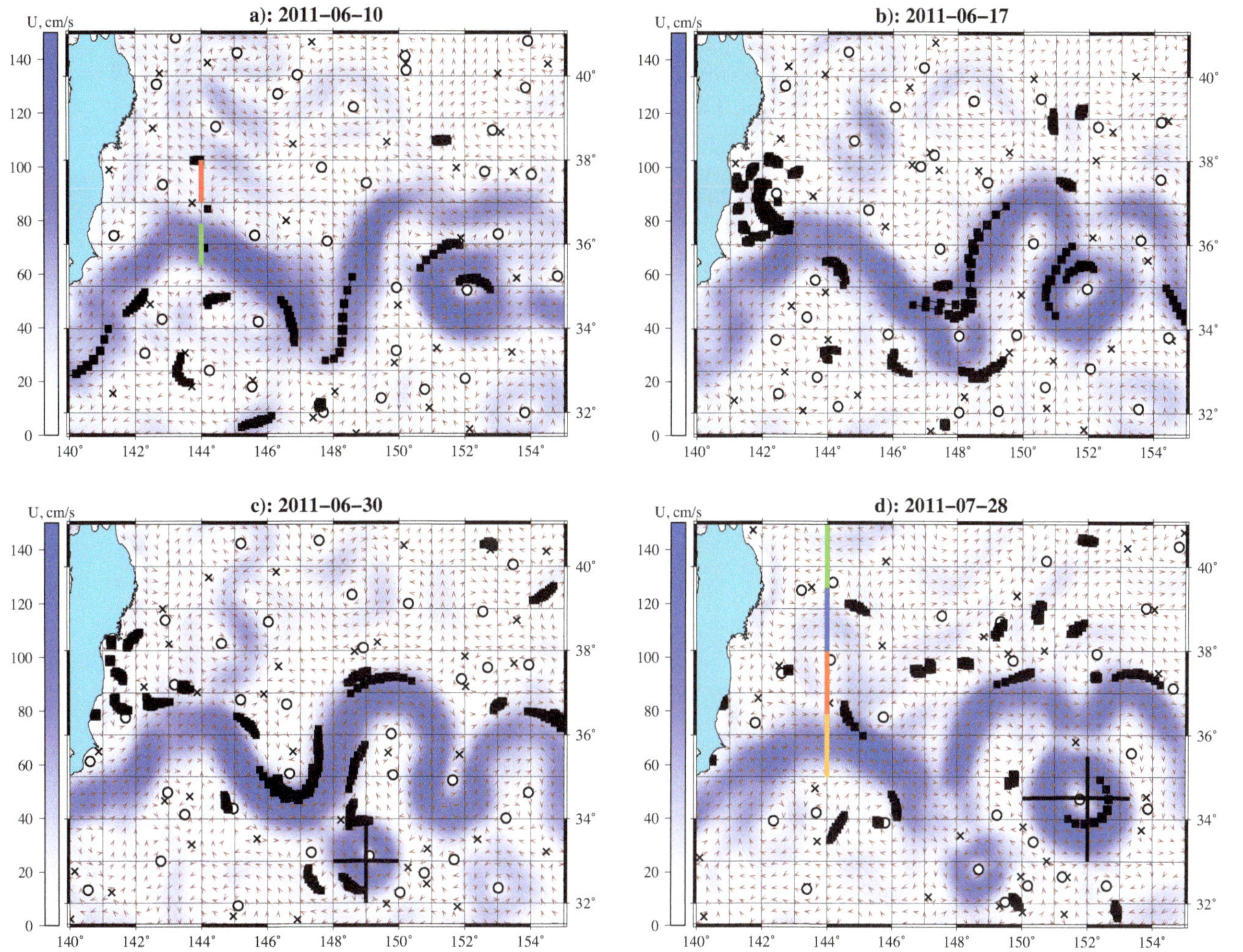

Fig. 2. **(a)–(c)** Metamorphoses of the velocity field in the process of formation of the CR1 on 10 June, 17 June and 30 June 2011, respectively, with tracks of drifters imposed. **(d)** Velocity field on 28 July 2011 with the CR2 separated from the jet. Nuances of blue color measure the modulus of the linear velocity U in cm s^{-1}. Crosses and open circles mark "instantaneous" hyperbolic and elliptic stagnation points, respectively. The straights are material lines that have been evolved backward in time to trace origin of the corresponding markers. Perpendicular black lines cross the CR1 and CR2. Colored lines in **(a)** and **(d)** are placed along the transects where seawater samples have been collected on 10 and 11 June 2011 during the R/V *Ka'imikai-O-Kanaloa* cruise (Buesseler et al., 2012) and on 26–29 July 2011 during the R/V *Kaiun maru* cruise (Kaeriyama et al., 2013), respectively.

observations may be used with these aims. Near-surface velocity fields on fixed days are shown in Fig. 2, where crosses and open circles mark "instantaneous" hyperbolic and elliptic stagnation points, respectively. Color codes the value of the modulus of the linear velocity $U = \sqrt{U_\phi^2 + U_\lambda^2}$ in cm s^{-1} Tracks of some surface drifters for 3 days are plotted by black squares.

In Fig. 2 we illustrate the process of formation of a CR (which we denote as CR1) from the KE meander in June 2011 when the KE was in the state with two prominent quasistationary meanders. In the beginning of June, the trough of the first meander starts to steepen with its edges becoming day by day closer and closer to each other (Fig. 2a). The edges merge eventually closing a volume of water with a cyclonic rotation to be connected with the parent jet by an arch (Fig. 2b). An elliptic point in its center and a hyperbolic point in the neck of the meander appear. Approximately 5 days later, the velocity field bifurcates, the CR1 is separated from the jet and the meander amplitude decreases correspondingly (Fig. 2c). Tracks of two drifters, encircling partly the ring, are shown in Fig. 2c.

There are a number of eddies around the KE. In this study we focus on the CR1, the CR2 (Fig. 2d), which is a ring separated from the jet on 28 July 2011, and the Tōhoku ACR, visible in all the panels of Fig. 2, with the center on 28 July 2011 to be at 144° E, 38° N. Here we apply the material line technique to trace the origin and history of water masses inside the CR1 and CR2 crossing them by black perpendicular lines shown in Fig. 2c and d, respectively. In the next section we apply the same technique to trace the origin and history of markers placed on colored segments along the transects where seawater samples have been collected on 10 and 11 June 2011 during the R/V *Ka'imikai-O-Kanaloa* cruise (Buesseler et al., 2012) (Fig. 2a) and on 26–29 July 2011 during the R/V *Kaiun maru* cruise (Kaeriyama et al., 2013) (Fig. 2d).

Transport of water masses across strong jet currents, like the Gulf Stream and the Kuroshio, is important because they separate waters with distinct bio-physico-chemical properties. It may cause heating and freshening of waters with a great impact on the weather and living organisms. As for transport across the KE, one should mention the paper by Mendoza et al. (2010) in which a turnstile mechanism (Wiggins, 1992) has been supposed to provide KE cross-jet transport in altimetric data sets. Mendoza et al. (2010) have computed stable and unstable manifolds of relevant distinguished hyperbolic trajectories and found a lobe transport across a Lagrangian barrier defined from pieces of those manifolds. It is a mechanism of chaotic advection well studied with analytical model flows (Wiggins, 1992; Koshel' and Prants, 2006). However, it is not direct evidence of cross-jet transport because of difficulties and inevitable errors in computing the manifolds and relevant trajectories in altimetric data sets. Moreover, the cross-jet transport has been found by Mendoza et al. (2010) in a far downstream region between 155 and 165° E where the KE jet is highly unstable and may even bifurcate.

The question of whether the much more stable upstream KE jet between 141 and 153° E is an impenetrable barrier for cross-jet transport is still open. The new aspect of that problem arose suddenly after the Fukushima accident on 11 March 2011. As common opinion (Buesseler et al., 2012), it is difficult to expect observation of Fukushima-derived radionuclides on the southern side of the KE jet. Could contaminant waters from the Fukushima area cross the KE jet and appear on the southern side of the jet or not?

In order to document directly the cross-jet transport of Fukushima-derived radionuclides, we compute a special kind of Lagrangian maps by integrating the advection equations (Eq. 1) backward in time from a fixed day till the day of the accident for a large number of particles distributed in the study area. The latitude, ϕ, from which each particle came to its final position is coded by color. The latitudinal map on 30 June 2011 in Fig. 3a demonstrates that "red" waters crossed the latitude 31° N from the south, "green" waters crossed the latitude of the FNPP (141°05′ E, 37°25′ N) from the north, whereas nuances of the grey color code the particles that originated from the latitudes between 31 and 37° N. Zoom in Fig. 3b shows the CR1 with "green" water in its core (originated from the latitudes > 37° N) that may contain increased concentration of Fukushima-derived radionuclides.

The simulation results are confirmed by tracks of two surface drifters that were trapped by the CR1 (Fig. 3b). The

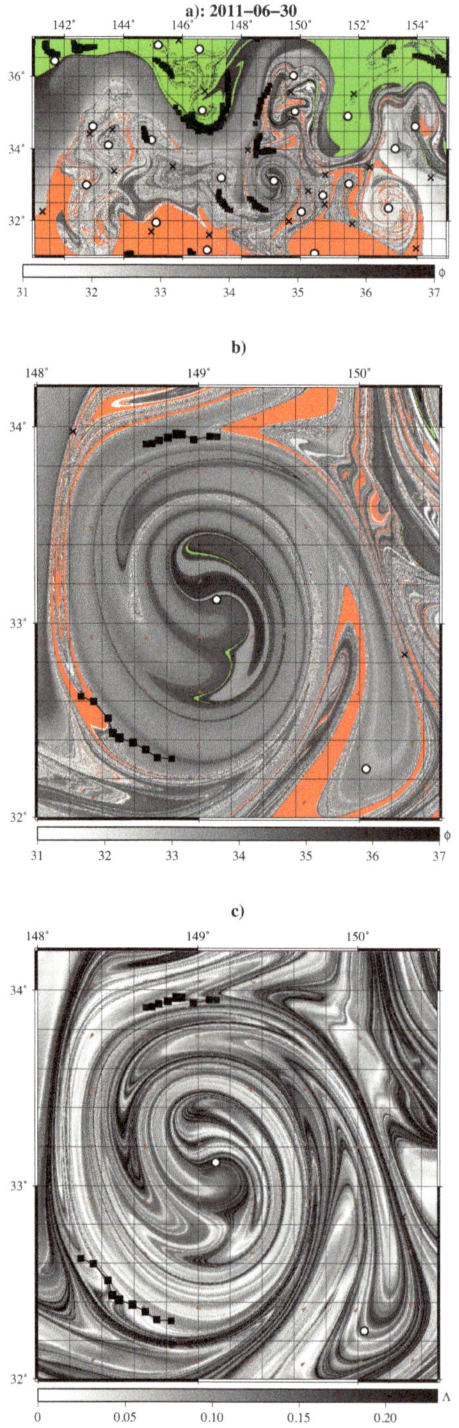

Fig. 3. **(a)** Lagrangian latitudinal map on 30 June 2011 computed from that date to the day of the accident, 11 March 2011, with "red" and "green" particles that originated from the latitudes < 31 and > 37° N, respectively. Nuances of the grey color in geographic degrees code the particles that originated from the latitudes between 31 and 37° N. **(b)** Zoom of the area with the CR1 demonstrates transport across the KE jet ("green" particles). Tracks of two drifters to be trapped by the CR1 are shown as black squares. Crosses and open circles mark "instantaneous" hyperbolic and elliptic stagnation points, respectively. **(c)** Lyapunov map of the same area as in **(b)**.

southern track belongs to the drifter no. 98875 released on December 2010 at the point 126.784° E, 15.778° N (http://www.aoml.noaa.gov/phod/dac/index.php). It was transported by the Kuroshio Current from the southwest. The northern track belongs to the drifter no. 36473 released on 11 June during the R/V *Ka'imikai-O-Kanaloa* cruise (Buesseler et al., 2012) at the point 144.087° E, 35.901° N. It crossed the KE jet and was trapped by the CR1. It is a "green" drifter in Fig. 2 in Buesseler et al. (2012). Some of their "red" drifters in that Fig. 2 have been trapped by the meander trough that formed the CR2 in July 2011.

The map in Fig. 3b is clear evidence of transport of water across the KE jet. To check that finding we initialized a material line crossing the CR1 core along 31°06′ N (the horizontal black line in Fig. 2c) and evolved it backward in time till the day of the accident. It has been found that the fragments of that line, containing "green" particles, really came from the area near the location of the FNPP, whereas the other particles of that line came from the west, mainly along the KE jet. However, the amount of potentially dangerous Fukushima waters in the core of the CR1 is comparatively small.

To quantify chaos in irregular flows, it is useful to compute the maximal Lyapunov exponent that characterizes quantitatively mixing along with directions of maximal stretching and contracting (see, e.g., Koshel' and Prants, 2006). We show in Fig. 3c the map of the finite-time Lyapunov exponent, Λ, which is the finite-time average of the maximal separation rate for a pair of neighboring advected particles. That quantity has been computed backward in time from 30 June to 30 May by the method of the singular-value decomposition of an evolution matrix for the linearized advection equations (Prants et al., 2011a). The Lyapunov map demonstrates the same large-scale structures as the latitude map. It provides also a typical fine-scale pattern of mixing in the eddy CR1 with alternating black and white bands. The black bands are locus of those particles that separated from each other for 1 month in the past at a maximal rate. The boundaries between particles of different origin in Fig. 3b correspond to the black bands with high values of Λ in Fig. 3c.

Now we apply the material line technique to trace the origin of water masses in the CR1 and the CR2 that was born after a separation of the trough of the second KE meander from the parent jet in July 2011. In June–July 2011 it was deforming strongly and eventually produced a ring-like structure with a diameter of about 300 km that has been detached to the south of the parent jet and then reabsorbed in a short time. That ring-like structure with the center near $x = 152°$ E and $y = 34°30'$ N is seen in Fig. 2d in the altimetric velocity field on 28 July 2011.

Starting on 30 June, we evolve backward in time until the day of the accident two perpendicular black material lines with a large number of markers, crossing the CR1 (Fig. 2c). The other two black material lines were chosen to cross the CR2 on 28 July (Fig. 2d), when a hyperbolic point appeared between the ring and the KE jet. We checked that the results did not change essentially when integrating not crossed segments but all the ring's interiors except for a larger density of points in some areas on the corresponding tracking maps.

Coming back to the question of whether the KE jet is an impenetrable barrier for Fukushima-derived radionuclides, we compute tracking maps. The region under study, 130–170° E and 25–50° N, is divided into 500 × 500 cells. We advect particles of the corresponding material lines backward in time from 30 June (CR1) and from 28 July (CR2) until the day of the accident, 11 March 2011. Then one fixes each day how many times markers visited each cell during the month after the accident, from 11 March to 10 April, when the maximal leakage of radionuclides from the FNPP directly into the ocean and their atmospheric fallout on the ocean surface were registered. Summarizing those numbers, we get a total number of visits in each cell to be proportional to the density of marker's traces, ν, which is coded by color in the logarithmic scale. The probability that the CR1 contains contaminated water is comparatively low because the density of points in the region around the FNPP is small (Fig. 4a). This map confirms the result of direct calculation of the cross-jet transport in Fig. 3b, where only a small amount of potentially contaminated "green" water, originated from the latitudes > 37° N, is visible. It is clear from Fig. 4a that the CR1 consists mainly of Kuroshio water.

The method we proposed above provides "strict" proof of a cross-jet transport different from forward-in-time simulation of a patch of particles initially located around the Fukushima area. Previous modeling by Buesseler et al. (2012) (see Fig. S3 in a supplementary material of that paper) and Kaeriyama et al. (2013) (see their Fig. 2) has found some particles from the initial patch on the southern KE flank. However, day-by-day inspection of simulated propagation has shown that most of those particles, in fact, did not cross the jet. They propagated to the east along the northern KE side, and after reaching the longitude about 155° E, where the KE becomes unstable and even may break down for a while, part of them were advected to the south and then to the west propagating along the southern KE flank. It is hardly to name "cross-jet transport". In order to document real cross-jet transport, it is necessary to select candidate particles on the southern KE flank and compute the corresponding longitudinal and tracking maps.

As to CR2 markers, the density of their traces in the area, which is supposed to be contaminated, is much higher as compared to the CR1 case (Fig. 4b). It means that the probability of observing higher concentrations of Fukushima-derived radionuclides in surface waters of the CR2 is expected to be comparatively high. The CR2 contains water parcels that moved during the month after the accident around the mesoscale eddies present to the north and east of the FNPP location. On the day of the accident, there was the eddy system with the large Tōhoku ACR with the center around 144° E and 39° N, a small anticyclonic eddy to the north of it and a medium cyclonic eddy at the traverse of

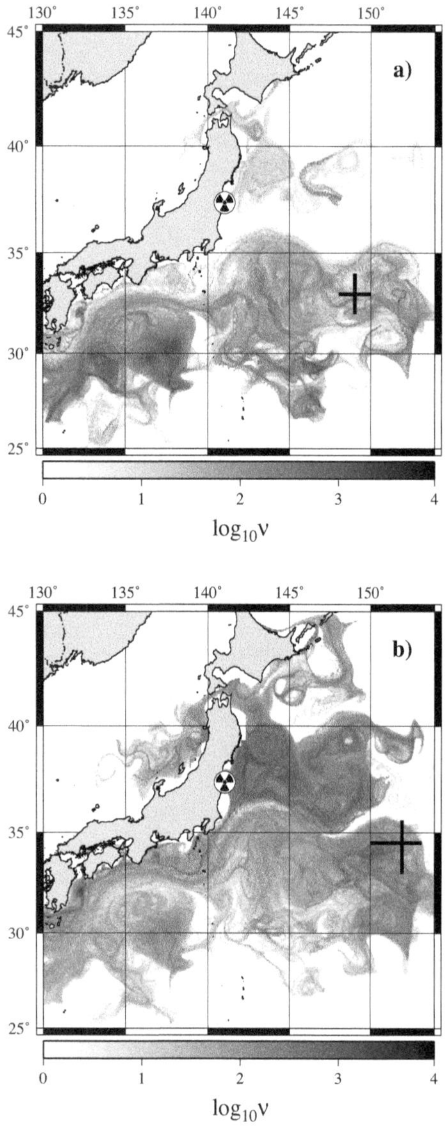

Fig. 4. Tracking maps for the markers placed on perpendicular material lines crossing **(a)** CR1 (Fig. 2c) and **(b)** CR2 (Fig. 2d). The maps show where the corresponding markers were walking after the Fukushima Nuclear Power Plant accident, from 11 March to 10 April 2011. The density of traces is in the logarithmic scale.

the Tsugaru Strait. It has been shown in our paper (Prants et al., 2011b) that namely this eddy system has governed mixing and transport of radioactive water, part of which has been trapped by the Tōhoku ACR and advected around the adjacent eddies to the north. The concentration of radionuclides around those eddies might be significantly greater than in other places because they are able to trap ambient contaminated waters (Prants et al., 2011b). The influence of the Tōhoku ACR is evident on both the tracking maps as a patch with increased density of traces at its place.

We conclude this section by emphasizing that the material line technique may be useful for finding the surface areas in the North Pacific that are potentially dangerous due to the risk of radioactive contamination. Before choosing the track of a planned R/V cruise, it is instructive to make a simulation by initializing backward-in-time evolution of material lines, crossing eddies in the region visible on Lagrangian maps. The corresponding tracking maps computed would help to know where one could expect higher or lower concentrations of Fukushima-derived radionuclides in this or that eddy.

3.2 Fukushima-derived radionuclides as Lagrangian tracers

In this section we apply the material line technique to trace the origin of water parcels with measured levels of ^{134}Cs and ^{137}Cs concentrations collected during two R/V cruises in June and July 2011 (Kaeriyama et al., 2013; Buesseler et al., 2012). Starting from the dates of sampling, we evolve backward in time material lines placed along the transects, where stations with collected surface water samples were located. Results of direct observation of radioactive Cs in surface seawater collected from R/V *Kaiun maru* in a broad area in the western and central North Pacific in July and October 2011 and July 2012 have been reported by Kaeriyama et al. (2013). In this study, we focus on the results of measurements to be carried out at their stations C43–C55 from 26 to 29 July 2011 along the 144° E meridian from 35 to 41° N. That transect is shown in Fig. 2d. The measured ^{137}Cs concentrations at the stations C43–C55 have been found to be in the range from the background level 1.9 ± 0.4 mBq kg^{-1} (station C52) to 153 ± 6.8 mBq kg^{-1} (station C47). The ^{134}Cs / ^{137}Cs ratio was close to 1. The level of the concentration of the caesium isotopes in the sea surface waters in the North Pacific before the accident did not exceed 2–3 mBq kg^{-1}.

We placed a material line along that transect and divided it into four colored segments: (1) the orange segment, 35°–36°30′ N, crossing the first meander's crest with initialization on 26 July; (2) the red segment, 36°30′–38° N, crossing the southern part of the Tōhoku ACR with initialization on 27 July; (3) the blue segment, 38°–39°30′ N, crossing its northern part with initialization on 28 July; and (4) the green segment, 39°30′–41° N with initialization on 29 July.

The tracking maps in Fig. 5, colored as the corresponding segments, show where markers were walking from 11 March to 10 April 2011. Markers from the orange segment, as expected, were advected to their places at the end of July mainly by the Kuroshio Current from the southwest and did not cross the latitude of the segment's northern end, 36°30′ N (Fig. 5a). The risk of their radioactive contamination is small. It is confirmed by the measured ^{137}Cs concentrations at the stations C52–C55 of that segment to be 2–5 mBq kg $^{-1}$ (Kaeriyama et al., 2013), which is slightly higher than the background level. Markers from the red segment have been found walking mainly in the area to the north of the latitude 36°30′ N (Fig. 5b). Part of that segment crossed the Tōhoku ACR existing in the area from the day of the accident to the

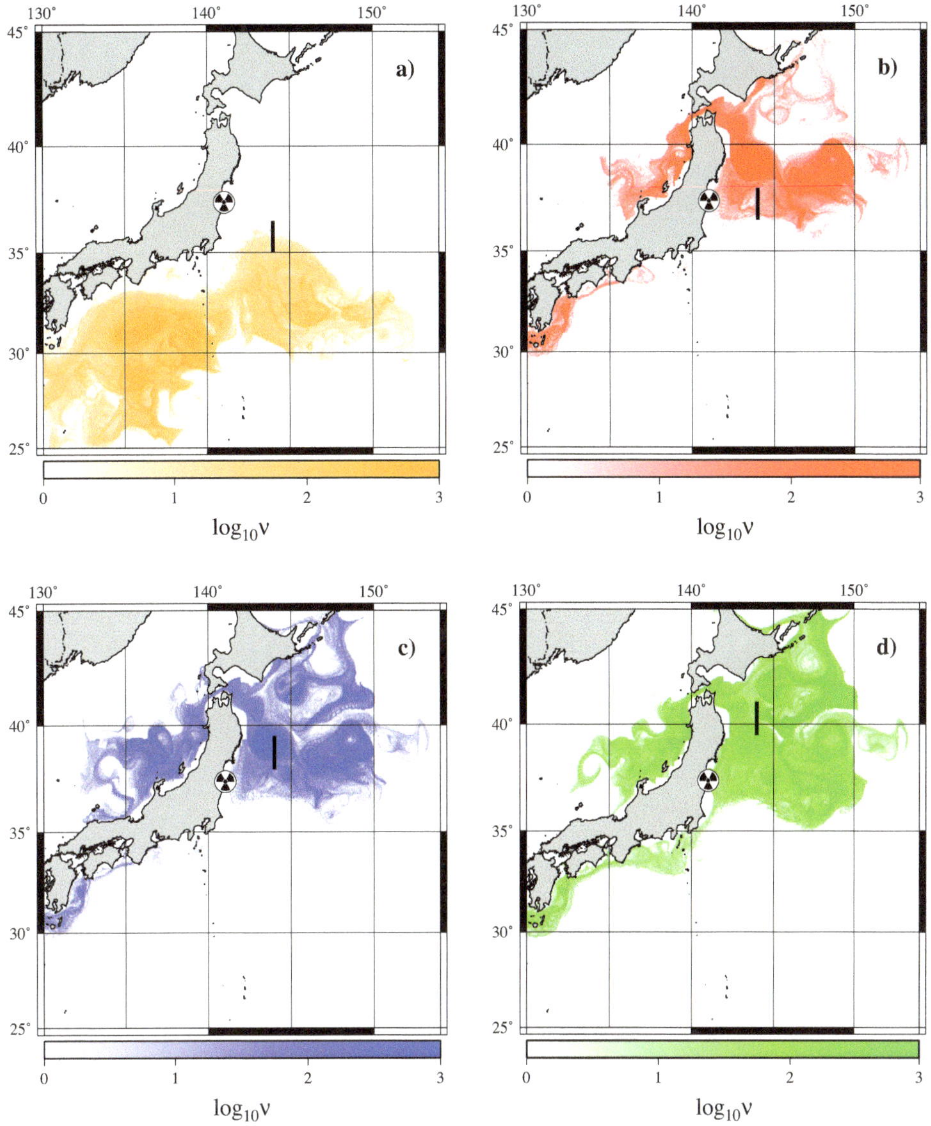

Fig. 5. Tracking maps for the markers placed on four material line segments along the transect where seawater samples have been collected on 26–29 July 2011 (Kaeriyama et al., 2013). The maps are colored as the corresponding segments in Fig. 2d and show where markers were walking from 11 March to 10 April 2011. Markers were initially placed along the 144° E meridian at the following segments: **(a)** 35°–36°30′ N, **(b)** 36°30′–38° N, **(c)** 38°–39°30′ N and **(d)** 39°30′–41° N.

days of in situ measurements. That is why we see increased density of points in the area where it has been located in the range from 11 March to 10 April. A comparatively high level of Fukushima-derived caesium isotopes is expected in the corresponding water samples. It is really the case. The ^{137}Cs concentrations at the stations C49 and C50 of that segment were measured to be 36 ± 3.3 and $50 \pm 3.6\,\mathrm{mBq\,kg^{-1}}$ (Kaeriyama et al., 2013). The high caesium concentration levels up to $153 \pm 6.8\,\mathrm{mBq\,kg^{-1}}$ (Kaeriyama et al., 2013) have been measured at the stations C46, C47 and C48 (blue segment) and C43, C44 and C45 (green segment). The tracking maps in Fig. 5c and d show clearly an increased density of traces of the corresponding markers in the area where the maximal leakage of radionuclides from the FNPP directly into the ocean and their atmospheric fallout on the ocean surface were registered from 11 March to 10 April 2011. Those maps also demonstrate strong mixing in the Kuroshio–Oyashio frontal zone. Traces of the markers, initialized at the end of July at a comparatively compact material line, have been found in the range 11 March–10 April in an extended area, including some parts of the Sea of Japan and the Okhotsk Sea.

Fukushima-derived ^{134}Cs and ^{137}Cs were measured in surface and subsurface waters, as well as in zooplankton and fish, at 50 stations on 4–18 June 2011 during the R/V *Ka'imikai-O-Kanaloa* cruise (Buesseler et al., 2012). We initialize a material line as shown in Fig. 2a where ^{137}Cs concentrations have been measured on 10 and 11 June at 25 stations in the range from 1.4 ± 0.2 mBq/kg (station 13) to 173.6 ± 9.9 mBq kg^{-1} (station 10). The ^{134}Cs / ^{137}Cs ratio was close to 1. Markers are placed on the green segment, 35°30′–36°30′ N, crossing the first meander's crest (11 June), and the red one, 37–38° N, to the north of the meander's crest (10 June). Traces of the markers of the green segment are found on both sides of the KE jet, whereas traces of the markers of the red one are on the northern side of the jet only. The southern part of the green segment crosses the jet itself, but its northern part is outside of it (Fig. 2a). That is why the tracking map in Fig. 6a consists of two disconnected domains: one is to the south of the jet and another one is to the north. The measured ^{137}Cs concentrations at the stations 13 and 14, situated in the green segment, were at the background level, in the range 1.4–3.6 mBq kg^{-1} (Buesseler et al., 2012), because the corresponding markers were advected by the KE current (Fig. 6a).

Density of traces of the markers from the red segment is comparatively high in the area around the FNPP (Fig. 6b). This finding is confirmed by measurements at stations 10, 11 and 12 (Buesseler et al., 2012) where the concentrations of Fukushima-derived ^{134}Cs and ^{137}Cs were in the range 21.9–173.6 mBq kg^{-1}. The red segment crosses partly the Tōhoku ACR, which is visible in Fig. 2a, and the traces of its markers are dense at the place of that ring. Some markers were advected to their places on the initial segment by the Tsushima Current from the Sea of Japan to the Pacific Ocean through the Tsugaru Strait.

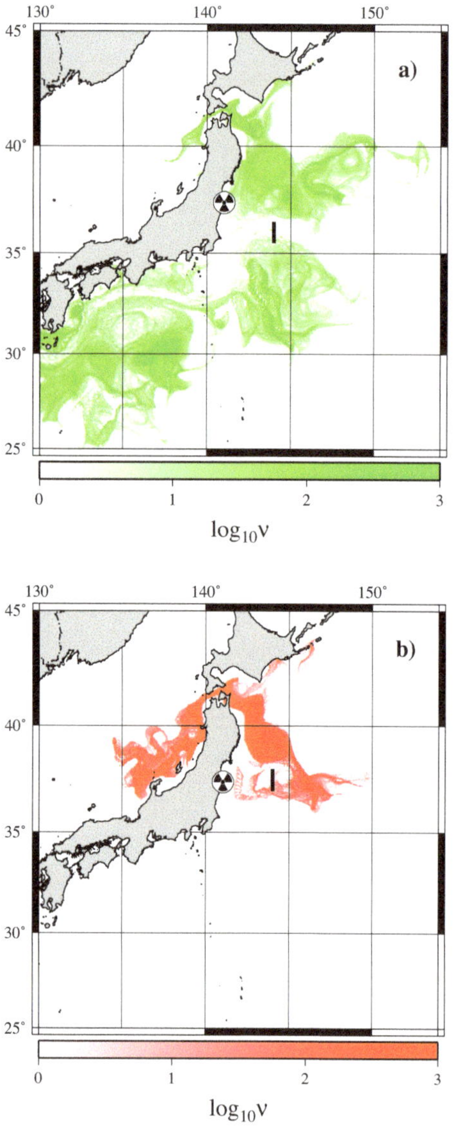

Fig. 6. Tracking maps for the markers placed on two material line segments along the transect where seawater samples have been collected on 10 and 11 June 2011 (Buesseler et al., 2012). The maps are colored as the corresponding segments in Fig. 2a and show where markers were walking from 11 March to 10 April 2011. Markers were initially placed along the 144° E meridian at the following segments: **(a)** 35°30′–36°30′ N and **(b)** 37°–38° N.

4 Summary

In this research, we used numerical simulations to study near-surface large-scale transport in the KE area based on AVISO altimetric velocity field. After solving advection equations for passive particles backward in time, we have computed Lagrangian maps for their displacements and tracking maps for the number of times markers visited different places in the region.

Two KE cold-core rings, CR1 and CR2, were chosen to illustrate the process of pinching-off from the main jet in summer 2011. The tracking and Lagrangian maps were computed to trace the origin of water masses in the cores of those rings. They revealed near-surface cross-jet transport. This conclusion is supported by tracks of the surface drifters that were deployed in the area. Water masses, constituting the CR1, were advected mainly from the southwest by the Kuroshio and the KE, and only a small amount of its water was originated from the area to the north of the KE jet. Traces of the CR2 markers have been found in a large area to the south and the north of the jet, including the area around the FNPP location.

We used Fukushima-derived caesium isotopes as Lagrangian tracers comparing the results of our simulations with in situ observations of ^{134}Cs and ^{137}Cs concentrations in water samples collected during two R/V cruises in June and July 2011 (Kaeriyama et al., 2013; Buesseler et al., 2012). Evolving backward in time material lines along the transects where measurements have been carried out during those cruises, we computed the corresponding tracking maps. It is shown that the water parcels with caesium concentration, exceeding greatly the background level, were walking during

1 month after the accident in the area where the leakage of radionuclides from the FNPP directly into the ocean and their atmosphere fallout on the ocean surface were registered to be maximal. The density of traces of markers with low caesium concentration in that area was found to be comparatively small.

We would like to emphasize that the tracking technique elaborated in this paper may be useful for planning R/V cruises in the ocean. Before choosing the track of a planned R/V cruise, it is instructive to make a simulation by initializing backward-in-time evolution of material lines, crossing potentially interesting coherent structures in the region visible on Lagrangian maps. The corresponding tracking maps would help to know where one could expect higher or lower concentrations of radionuclides, pollutants or other Lagrangian tracers.

In spite of inevitable and well-known errors in altimetry-derived geostrophic velocity fields, the tracking technique seems to be the perspective in finding out the history and origin of water masses in coherent structures. The results obtained with that technique have been confirmed in this paper by a comparison with a large number of in situ measurements of concentration of Fukushima-derived radionuclides and drifter tracks. Measuring future concentrations of Fukushima-derived radionuclides and other pollutants and comparing those with the corresponding tracking maps, one gets an opportunity to estimate the temporal limits of the tracking technique.

Acknowledgements. This work was supported by the ICMAT Severo Ochoa project SEV-2011-0087 and the Russian Foundation for Basic Research (projects 11-01-12057, 12-05-00452, 13-05-00099 and 13-01-12404). The altimeter products were distributed by AVISO with support from CNES.

Edited by: A. Turiel

References

Abraham, E. R. and Bowen, M. M.: Chaotic stirring by a mesoscale surface-ocean flow, Chaos, 12, 373–381, doi:10.1063/1.1481615, 2002.

Andrade-Canto, F., Sheinbaum, J., and Zavala Sansón, L.: A Lagrangian approach to the Loop Current eddy separation, Nonlin. Processes Geophys., 20, 85–96, doi:10.5194/npg-20-85-2013, 2013.

Buesseler, K. O., Jayne, S. R., Fisher, N. S., Rypina, I. I., Baumann, H., Baumann, Z., Breier, C. F., Douglass, E. M., George, J., Macdonald, A. M., Miyamoto, H., Nishikawa, J., Pike, S. M., and Yoshida, S.: Fukushima-derived radionuclides in the ocean and biota off Japan, P. Natl. Acad. Sci., 109, 5984–5988, doi:10.1073/pnas.1120794109, 2012.

d'Ovidio, F., De Monte, S., Alvain, S., Dandonneau, Y., and Levy, M.: Fluid dynamical niches of phytoplankton types, P. Natl. Acad. Sci., 107, 18366–18370, doi:10.1073/pnas.1004620107, 2010.

Ebuchi, N. and Hanawa, K.: Trajectory of Mesoscale Eddies in the Kuroshio Recirculation Region, J. Oceanogr., 57, 471–480, doi:10.1023/A:1021293822277, 2001.

Ide, K., Small, D., and Wiggins, S.: Distinguished hyperbolic trajectories in time-dependent fluid flows: analytical and computational approach for velocity fields defined as data sets, Nonlin. Processes Geophys., 9, 237–263, doi:10.5194/npg-9-237-2002, 2002.

Itoh, S. and Yasuda, I.: Characteristics of Mesoscale Eddies in the Kuroshio–Oyashio Extension Region Detected from the Distribution of the Sea Surface Height Anomaly, J. Phys. Oceanogr., 40, 1018–1034, doi:10.1175/2009JPO4265.1, 2010.

Kaeriyama, H., Ambe, D., Shimizu, Y., Fujimoto, K., Ono, T., Yonezaki, S., Kato, Y., Matsunaga, H., Minami, H., Nakatsuka, S., and Watanabe, T.: Direct observation of ^{134}Cs and ^{137}Cs in surface seawater in the western and central North Pacific after the Fukushima Dai-ichi nuclear power plant accident, Biogeosciences, 10, 4287–4295, doi:10.5194/bg-10-4287-2013, 2013.

Koshel', K. V. and Prants, S. V.: Chaotic advection in the ocean, Physics-Uspekhi, 49, 1151–1178, doi:10.1070/PU2006v049n11ABEH006066, 2006.

Kuznetsov, L., Toner, M., Kirwan Jr., A., Jones, C., Kantha, L., and Choi, J.: The Loop Current and adjacent rings delineated by Lagrangian analysis of the near-surface flow, J. Mar. Res., 60, 405–429, 2002.

Madrid, J. A. J. and Mancho, A. M.: Distinguished trajectories in time dependent vector fields, Chaos: An Interdisciplinary Journal of Nonlinear Science, 19, 013 111, doi:10.1063/1.3056050, 2009.

Mancho, A. M., Small, D., and Wiggins, S.: Computation of hyperbolic trajectories and their stable and unstable manifolds for oceanographic flows represented as data sets, Nonlin. Processes Geophys., 11, 17–33, doi:10.5194/npg-11-17-2004, 2004.

Mancho, A. M., Small, D., and Wiggins, S.: A comparison of methods for interpolating chaotic flows from discrete velocity data, Comput. Fluids, 35, 416–428, doi:10.1016/j.compfluid.2005.02.003, 2006.

Mancho, A. M., Wiggins, S., Curbelo, J., and Mendoza, C.: Lagrangian descriptors: A method for revealing phase space structures of general time dependent dynamical systems, Commun. Nonlin. Sci. Num. Simul., 18, 3530–3557, doi:10.1016/j.cnsns.2013.05.002, 2013.

Mendoza, C. and Mancho, A. M.: Review Article: "The Lagrangian description of aperiodic flows: a case study of the Kuroshio Current", Nonlin. Processes Geophys., 19, 449—72, doi:10.5194/npg-19-449-2012, 2012.

Mendoza, C., Mancho, A. M., and Rio, M.-H.: The turnstile mechanism across the Kuroshio current: analysis of dynamics in altimeter velocity fields, Nonlin. Processes Geophys., 17, 103–111, doi:10.5194/npg-17-103-2010, 2010.

Prants, S. V.: Dynamical systems theory methods to study mixing and transport in the ocean, Physica Scripta, 87, 038115, doi:10.1088/0031-8949/87/03/038115, 2013.

Prants, S. V., Budyansky, M. V., Ponomarev, V. I., and Uleysky, M. Yu.: Lagrangian study of transport and mixing in a mesoscale eddy street, Ocean Model., 38, 114–125, doi:10.1016/j.ocemod.2011.02.008, 2011a.

Prants, S. V., Uleysky, M. Y., and Budyansky, M. V.: Numerical simulation of propagation of radioactive pollution in the ocean from the Fukushima Dai-ichi nuclear power plant, Dokl. Earth Sci., 439, 1179–1182, doi:10.1134/S1028334X11080277, 2011b.

Prants, S. V., Ponomarev, V. I., Budyansky, M. V., Uleysky, M. Y., and Fayman, P. A.: Lagrangian analysis of mixing and transport of water masses in the marine bays, Izvest. Atmos. Ocean. Phys., 49, 82–96, doi:10.1134/S0001433813010088, 2013.

Qiu, B. and Chen, S.: Variability of the Kuroshio Extension Jet, Recirculation Gyre, and Mesoscale Eddies on Decadal Time Scales, J. Phys. Oceanogr., 35, 2090–2103, doi:10.1175/JPO2807.1, 2005.

Samelson, R. M. and Wiggins, S.: Lagrangian Transport in Geophysical Jets and Waves: The Dynamical Systems Approach (Interdisciplinary Applied Mathematics), Springer, New York, 2006.

Sugimoto, S. and Hanawa, K.: Relationship between the path of the Kuroshio in the south of Japan and the path of the Kuroshio Extension in the east, J. Oceanogr., 68, 219–225, doi:10.1007/s10872-011-0089-1, 2012.

Tomosada, A.: Generation and decay of Kuroshio warm-core rings, Deep-Sea Res. Pt. I, 33, 1475–1486, doi:10.1016/0198-0149(86)90063-4, 1986.

Uleysky, M. Y., Budyansky, M. V., and Prants, S. V.: Effect of dynamical traps on chaotic transport in a meandering jet flow, Chaos, 17, 043105, doi:10.1063/1.2783258, 2007.

Uleysky, M. Y., Budyansky, M. V., and Prants, S. V.: Chaotic transport across two-dimensional jet streams, J. Exp. Theor. Phys., 111, 1039–1049, doi:10.1134/S1063776110120174, 2010a.

Uleysky, M. Y., Budyansky, M. V., and Prants, S. V.: Mechanism of destruction of transport barriers in geophysical jets with Rossby waves, Phys. Rev. E, 81, 017202, doi:10.1103/PhysRevE.81.017202, 2010b.

Waseda, T., Mitsudera, H., Taguchi, B., and Yoshikawa, Y.: On the eddy-Kuroshio interaction: Meander formation process, J. Geophys. Res., 108, 3220, doi:10.1029/2002JC001583, 2003.

Wiggins, S.: Chaotic Transport in Dynamical Systems, in: Interdisciplinary applied mathematics, Springer-Verlag, New York, 1992.

Wiggins, S.: The dynamical systems approach to Lagrangian transport in oceanic flows, Annu. Rev. Fluid Mech., 37, 295–328, doi:10.1146/annurev.fluid.37.061903.175815, 2005.

12

Four-dimensional ensemble variational (4D-En-Var) data assimilation for the HIgh Resolution Limited Area Model (HIRLAM)

N. Gustafsson[1] **and J. Bojarova**[2]

[1]Swedish Meteorological and Hydrological Institute, 60176 Norrköping, Sweden
[2]Norwegian Meteorological Institute, P.O. Box 43 Blindern, 0313 Oslo, Norway

Correspondence to: N. Gustafsson (nils.gustafsson@smhi.se)

Abstract. A four-dimensional ensemble variational (4D-En-Var) data assimilation has been developed for a limited area model. The integration of tangent linear and adjoint models, as applied in standard 4D-Var, is replaced with the use of an ensemble of non-linear model states to estimate four-dimensional background error covariances over the assimilation time window. The computational costs for 4D-En-Var are therefore significantly reduced in comparison with standard 4D-Var and the scalability of the algorithm is improved.

The flow dependency of 4D-En-Var assimilation increments is demonstrated in single simulated observation experiments and compared with corresponding increments from standard 4D-Var and Hybrid 4D-Var ensemble assimilation experiments. Real observation data assimilation experiments carried out over a 6-week period show that 4D-En-Var outperforms standard 4D-Var as well as Hybrid 4D-Var ensemble data assimilation with regard to forecast quality measured by forecast verification scores.

1 Introduction

Data assimilation for numerical weather prediction (NWP) is the process of using observations to create initial conditions for NWP models. The number of observed values at any particular moment of time is generally much smaller than the number of model state variables of the NWP model. For this reason, a priori information has to be taken into account. In most data assimilation schemes, a short-range forecast from the NWP model, valid at the time of the observations, is used as a *background* field for the data assimilation and, in addition, statistical information about the uncertainty of this background field may be utilized. This was realized already by Bergthorsson and Döös (1955), who applied a successive correction spatial interpolation to a quasi-geostrophic barotropic forecast model. The successive correction interpolation was essentially based on weights given to the observation-minus-background deviations being proportional to the distances between the observation positions and the grid points. During the 1960s and 1970s more advanced spatial interpolation techniques, for example the statistical interpolation or OI (optimum interpolation, Gandin, 1963), were introduced and brought into operational use in three-dimensional versions, including also the idea of balancing between mass and wind field information (Rutherford, 1976; Gustafsson, 1981; Lorenc, 1981).

Variational data assimilation (Sasaki, 1958; Le Dimet and Talagrand, 1986; Lewis and Derber, 1985) provides an efficient framework for data assimilation with NWP models, since the possibility to apply dynamical and physical constraints for the data assimilation is introduced. In three-dimensional variational data assimilation (3D-Var) a cost function measuring the distance to the observations and the distance to the background field is minimized. In its simplest form, 3D-Var is equivalent to OI, with the exception that there are no requirements for data selection, while OI generally has to rely on local data selection schemes to find the observed values to influence each grid point of the model. One of the most important reasons of 3D-Var replacing OI is that 3D-Var can easily use non-linear observation operators such that satellite radiance data can be directly assimilated. 3D-Var is applied to global NWP models (Parrish and Derber, 1992; Courtier et al., 1998) and also to regional NWP models (Gustafsson et al., 2001; Lindskog et al., 2001). In four-dimensional variational data assimilation (4D-Var) the model initial state is found by minimizing a cost function measuring the distance to observations from an assimilation time window (e.g. 6 h) including also the forecast model as a strong constraint for the model development over the assimilation time window. The concept of incremental 4D-Var (Courtier et al., 1994) made it possible to practically solve the computational problem of 4D-Var by introducing simplification and linearization of the forecast model and the observation operators and by introducing a reduced spatial resolution for the 4D-Var minimization. Incremental 4D-Var schemes have been successfully introduced operationally for global NWP models (Rabier et al., 2000) as well as for regional high-resolution models (Kawabata et al., 2007; Gustafsson et al., 2012).

The application of 4D-Var for operational NWP has been very successful, but there are a few problems associated with its further development. Firstly, the application of a static background error constraint, similar to those applied in 3D-Var, at the start of the assimilation window limits the possibilities to describe flow-dependent background error structures, for example those associated with baroclinicity. This can be alleviated by introduction of longer 4D-Var assimilation time windows, allowing the implicit assimilation structure functions to develop these flow dependencies. With longer time windows it will, however, be necessary to treat model errors more efficiently, for example by introducing the weak constraint 4D-Var (Trémolet, 2006). Secondly, the incremental 4D-Var necessitates the development of the tangent linear (TL) and adjoint (AD) versions of the NWP forecast models. This requires substantial development and maintenance efforts and, furthermore, the computational cost of incremental 4D-Var is dominated by the costs for the TL and AD model integrations.

The Kalman filter (KF, Kalman, 1960) provides a framework for estimating the uncertainty of forecast model states for linear forecast models, and the extended Kalman filter (EKF, Kalman and Bucy, 1961) extends the Kalman filter theory to weakly non-linear models and observation operators. The KF and the EKF cannot be applied directly to large-scale NWP models, due to the huge size of the corresponding model state covariance matrix to be handled. The ensemble Kalman filter (EnKF, Evensen, 1994) is essentially a Monte Carlo approximation to the Kalman filter. The model uncertainty is represented through an ensemble of model states, from which the required forecast error covariances are estimated. The implementation of EnKF can roughly be subdivided into three different approaches: the EnKF with perturbed observations, which samples different sources of uncertainty (Evensen, 1994; Houtekamer and Mitchell, 2001), the square-root EnKF, which directly handles covariances involved (Tippett et al., 2003) and the reduced-rank KF (Heemink et al., 2001; Cohn and Todling, 1996), which focuses on a low-dimensional approximation of the model state covariances. One main advantage of the EnKF is the possibility to describe flow-dependent uncertainties of the background model state. Another advantage of EnKF, as compared with 4D-Var, is that it is no longer necessary to develop and maintain TL and AD versions of the NWP forecast models. These advantages have the price of a low rank of the background error covariance for full-scale NWP data assimilation, due to the necessity to apply a limited number ($\sim$ 10–100) of ensemble members. The low rank of the background error covariance matrix can be handled to some extent by, for example, application of localization to avoid spurious correlations at large distance separations, caused by sampling errors. Additive and multiplicative inflation techniques can also be used to compensate for the effects of the low rank.

The advantages and disadvantages of EnKF, as compared to 4D-Var, were discussed by Lorenc (2003), Kalnay et al. (2007), and Gustafsson (2007). A synthesis of this discussion is the recognition of the need to develop combinations of the robust and full rank 4D-Var algorithms and the ensemble-based methods that are able to describe flow dependencies. A first step in this direction is the development of hybrid variational ensemble data assimilation methods (Hamil and Snyder, 2000; Lorenc, 2003; Wang et al., 2008a, b; Clayton et al., 2013; Gustafsson et al., 2014) that incorporate, for example, ensemble-based error covariances within the framework of variational data assimilation. A further step is to replace the TL and AD model integrations in 4D-Var with the application of ensembles of non-linear model trajectories over the data assimilation window (4D-En-Var, Liu et al., 2008, 2009; Liu and Xiao, 2013; Buehner et al., 2010a, b; Desroziers et al., 2014). The subject of the present paper is to present the 4D-En-Var developed for the High Resolution Limited Area Model (HIRLAM, Undén et al., 2002). A more detailed review of 4D-En-Var developments is provided in Sect. 2. Sections 3 and 4 focus on the HIRLAM implementation of 4D-En-Var. Section 5 presents results of experiments, both with single simulated observations to demonstrate the abilities of the algorithm, and with real observations to show the performance of 4D-En-Var. Section 6 discusses the computational performance and challenges with the scheme and the concluding remarks are presented in Sect. 7.

2 Review of 4D-En-Var developments

The following cost function is minimized with respect to the assimilation increment $\boldsymbol{\delta x}$ in incremental 4D-Var:

$$J = J_\mathrm{b} + J_\mathrm{o} = \frac{1}{2}(\boldsymbol{\delta x})^T \mathbf{B}^{-1} \boldsymbol{\delta x} + \frac{1}{2}\sum_{t_k=t_0}^{t_K} (\mathbf{H}_k \mathbf{M}_k \boldsymbol{\delta x} - \boldsymbol{d}_k)^T \mathbf{R}_k^{-1} (\mathbf{H}_k \mathbf{M}_k \boldsymbol{\delta x} - \boldsymbol{d}_k), \quad (1)$$

where $\mathbf{B}$ is the background error covariance, $t_k = t_0, \ldots, t_K$ the data assimilation time window, $\boldsymbol{d}_k = \boldsymbol{y}_k - H_k(M_k(\boldsymbol{x}_\mathrm{b}))$

the innovations with y_k being the vector of observations at time t_k, x_b the model background state valid at time t_0, $M_k(.)$ denotes integration of the non-linear model from time t_0 until time t_k and $\mathbf{M}_k$ the corresponding tangent linear model integration. $H_k(.)$ is the non-linear observation operator, $\mathbf{R}_k$ is the observation error covariance and $\mathbf{H}_k$ the linearized observation operator, all valid at time t_k.

The background error covariance matrix $\mathbf{B}$ has such a large dimension that the inverse $\mathbf{B}^{-1}$ cannot be obtained directly by matrix inversion techniques. Therefore we introduce a pre-conditioning matrix $\mathbf{U}$ such that $\mathbf{B} = \mathbf{U}\mathbf{U}^T$, and $\delta x = \mathbf{U}\chi$ with χ the assimilation control variable. The transformation matrix $\mathbf{U}$ is not formulated explicitly but given, for example, as a series of simpler transform operators. The cost function to be minimized and its gradient with respect to the assimilation control variable χ are given by

$$J = J_b + J_o = \frac{1}{2}\chi^T\chi + \frac{1}{2}\sum_{t_k=t_0}^{t_K}(\mathbf{H}_k\mathbf{M}_k\mathbf{U}\chi - d_k)^T\mathbf{R}_k^{-1}(\mathbf{H}_k\mathbf{M}_k\mathbf{U}\chi - d_k) \quad (2)$$

and

$$\nabla_\chi J = \chi + \sum_{t_k=t_0}^{t_K}\mathbf{U}^T\mathbf{M}_k^T\mathbf{H}_k^T\mathbf{R}_k^{-1}(\mathbf{H}_k\mathbf{M}_k\mathbf{U}\chi - d_k). \quad (3)$$

The gradient calculation (3) is applied for every iteration during the 4D-Var minimization. A forward integration of the tangent linear model $\mathbf{M}_k$ and a backward integration of the adjoint model $\mathbf{M}_k^T$ are required for each such iteration.

Following Liu et al. (2008) we replace the static error covariance $\mathbf{B}$ with a flow-dependent error covariance $\mathbf{B} \approx X'_b(X'_b)^T$ estimated from an ensemble of background model states. X'_b is a matrix whose columns are the normalized deviations of the ensemble background states from their mean:

$$X'_b = \frac{1}{\sqrt{N-1}}(x_{b1} - \overline{x_b}, \ldots, x_{bN} - \overline{x_b}), \quad (4)$$

where N is the number of ensemble members.

Also following Liu et al. (2008) we may apply X'_b for the pre-conditioning $\delta x = X'_b\chi$. Here the control vector will have the dimension of the number of ensemble members and we may notice that the assimilation increment is just a linear combination of the ensemble perturbations. The cost function and its gradient will have a similar form as above (X'_b will replace $\mathbf{U}$). We have achieved one of our aims, to utilize a flow-dependent background error covariance at the start of the 4D-Var assimilation window, but we still need to integrate the tangent linear model forward in time and the adjoint model backward in time over the data assimilation window. To avoid this, Liu et al. (2008) used an ensemble of non-linear model integrations over the data assimilation window to replace the forward integration of the tangent linear model with pre-calculated values derived from these non-linear model trajectories in observation space.

The background error covariance of this 4D-En-Var formulation has the maximum rank equal to the number of ensemble members minus 1 ($N-1$) and cannot directly be applied to a full-scale NWP data assimilation problem. For demonstration purposes Liu et al. (2008) applied the method to a low-dimensional shallow-water model, with the degrees of freedom of the same order as the number of ensemble members. They showed that the performance of this 4D-En-Var for the low-order problem was very similar to the performance of 4D-Var and other assimilation algorithms like EnKF and 3D-Var. An even more important message is that the computational cost of 4D-En-Var was much lower than the corresponding cost of 4D-Var, since the integrations of the TL and AD models over the assimilation window were avoided.

In a second study Liu et al. (2009) applied 4D-En-Var to observing system simulation experiments (OSSE) with the WRF (weather research and forecasting) model (Skamarock et al., 2008). The number of ensemble members is in this case several orders of magnitude smaller than the model dimension, leading to a severe rank deficiency of the ensemble-based background error covariance matrix. The standard techniques in EnKF to counteract this rank deficiency are to apply covariance localization and/or to apply the EnKF algorithm locally, for example by solving separate filtering problems for each grid point of the model domain. The application of a local EnKF brings us back to the need for local data selection and will not be further discussed here. The most common algorithm for covariance localization is to apply an element-by-element multiplication (Schur product) of the original low-rank ensemble-based covariance matrix $\mathbf{B}_{ens} = X'_b(X'_b)^T$ with a localization correlation matrix $\mathbf{C}$, constructed such that the final covariances $\mathbf{B} = \mathbf{C} \circ \mathbf{B}_{ens}$ applied in the EnKF will be eliminated at long distances (Houtekamer and Mitchell, 2001). Liu et al. (2009) applied the correspondence to this covariance localization within the 4D-En-Var formulation with a technique similar to the one described by Buehner (2005). The pre-conditioning was done with the matrix $\mathbf{P}'_b$ given by

$$\mathbf{P}'_b = (\mathbf{C}' \circ X'_{b1}, \mathbf{C}' \circ X'_{b2}, \ldots, \mathbf{C}' \circ X'_{bN}), \quad (5)$$

where $\mathbf{C}'\mathbf{C}'^T = \mathbf{C}$, $\mathbf{C}'$ is an $n \times r$ matrix with $r \le n$ and X'_{bl} is an r column matrix with every column being equal to the lth column in X'_b. N is the number of ensemble members and n is the model space dimension. The columns of the transformation matrix $\mathbf{C}'$ consist of eigenvectors of the correlation matrix $\mathbf{C}$ multiplied by the square root of the eigenvalues. In order to make the eigenvector decomposition possible, the eigenvectors are evaluated on a coarser resolution grid, are possibly truncated and are interpolated to the assimilation grid. It is straightforward to show that $\mathbf{P}'_b(\mathbf{P}'_b)^T = \mathbf{C} \circ \mathbf{B}_{ens}$.

Liu and Xiao (2013) applied 4D-En-Var including covariance localization developed for WRF to a full-scale data assimilation experiment over Antarctica. They managed to demonstrate overall superiority and robust performance in comparison with 3D-Var and 3D-Var FGAT (first guess at appropriate time).

Buehner et al. (2010a, b) implemented a 4D-En-Var similar to the 4D-En-Var of Liu et al. (2009) into the Canadian operational global NWP system. The covariance localization was the same as in Buehner (2005). 4D-En-Var was compared with 4D-Var and 3D-Var FGAT, both applied with static background error covariances as well as with ensemble-based background error covariances, and was also compared with EnKF (Houtekamer and Mitchell, 2001). The ensemble-based error covariances were all based on the ensemble of short-range forecasts produced by the EnKF. Comparisons were made for single simulated observation impact experiments (Buehner et al., 2010a) and for full-scale data assimilation and forecasts over a 1 month period (Buehner et al., 2010b). The 4D-En-Var performed as well as 4D-Var with a static background error covariance, but slightly worse than 4D-Var Hybrid with an ensemble-based background error covariance in addition.

Desroziers et al. (2014) compared different 4D-En-Var formulations, including pre-conditioning based on the **B** matrix, the full background error covariance, rather than $\sqrt{\mathbf{B}}$ matrix as suggested by Liu et al. (2009) and applied in the present work with the HIRLAM 4D-En-Var. It is argued by Desroziers et al. (2014) that this alternative pre-conditioning might provide larger flexibility to implement the localization needed in 4D-En-Var. The choice of pre-conditioning method can also have an influence on the dimension of the assimilation control variable and on the characteristics of the minimization problem.

3 Formulation of the HIRLAM 4D-En-Var Hybrid

The 4D-En-Var for the HIgh Resolution Limited Area Model (HIRLAM, Undén et al., 2002) is an extension of the HIRLAM Hybrid variational ensemble data assimilation scheme (Gustafsson et al., 2014). It builds on the idea of an augmentation of the control vector by Lorenc (2003) and use of the $\mathbf{B}^{\frac{1}{2}}$ for pre-conditioning. Furthermore, the HIRLAM 4D-En-Var is a hybrid between 4D-En-Var and 3D-Var FGAT (first guess at appropriate time), such that the assimilation increment $\boldsymbol{\delta x}(t_k)$ at time t_k within the assimilation time window $t_0 \le t_k \le t_K$ is formed as a linear combination of a 3D-Var FGAT increment $\boldsymbol{\delta x}_{\text{3dvar}}\left(\frac{t_0+t_K}{2}\right)$ and a 4D-En-Var increment $\boldsymbol{\delta x}_{\text{ens}}(t_k)$:

$$\boldsymbol{\delta x}\left(t_k\right) = \boldsymbol{\delta x}_{\text{3dvar}}\left(\frac{t_0+t_K}{2}\right) + \boldsymbol{\delta x}_{\text{ens}}\left(t_k\right). \quad (6)$$

According to HIRLAM 3D-Var FGAT experiences, in order to utilize time-averaging effects, the assimilation increment related to the static part of the background error covariance should preferably be applied for non-linear model initialization at time $\frac{t_0+t_K}{2}$ in the middle of the data assimilation window.

Pre-conditioning for the two components of the assimilation increment is done separately. The pre-conditioning for the 3D-Var FGAT component $\boldsymbol{\delta x}_{\text{3dvar}} = \mathbf{U}\boldsymbol{\chi}_{\text{3dvar}}$ with $\mathbf{B}_{\text{3dvar}} = \mathbf{U}\mathbf{U}^T$ is done as in the HIRLAM 3D-Var (Berre, 2000; Gustafsson et al., 2001). The transform operator **U** includes vertical transforms of the spectral control vector $\boldsymbol{\chi}_{\text{3dvar}}$ from vertical eigenvector space to model levels, inversion of vertical balance operators and projection by Fourier transforms from spectral space to grid point space. An area extension is applied to the regional model domain in order to make application of fast Fourier transform possible (Haugen and Machenhauer, 1993). For the 4D-En-Var component $\boldsymbol{\delta x}_{\text{ens}}$ we follow Liu et al. (2009), see also Eq. (5):

$$\begin{aligned}\boldsymbol{\delta x}_{\text{ens}} &= \mathbf{P}'_{\text{b}}\boldsymbol{\chi}_{\text{ens}} = \left(\mathbf{C}' \circ \boldsymbol{X}'_{\text{b1}}, \mathbf{C}' \circ \boldsymbol{X}'_{\text{b2}}, \ldots, \mathbf{C}' \circ \boldsymbol{X}'_{\text{bN}}\right)\boldsymbol{\chi}_{\text{ens}} \\ &= \sum_{l=1}^{N} \mathbf{C}' \circ \boldsymbol{X}'_{\text{bl}}\left(\boldsymbol{\chi}_{\text{ens}}\right)_l,\end{aligned}$$

where $\boldsymbol{\chi}_{\text{ens}}$ is the control vector of dimension $N \cdot r$ and $(\boldsymbol{\chi}_{\text{ens}})_l$ the lth component of the control vector, corresponding to ensemble member l. If we introduce a vector $\boldsymbol{\alpha}_l = \mathbf{C}'(\boldsymbol{\chi}_{\text{ens}})_l$ we can show by substitution into component form that

$$\boldsymbol{\delta x}_{\text{ens}} = \sum_{l=1}^{N} \boldsymbol{\alpha}_l \circ \left(\boldsymbol{X}'_{\text{b}}\right)_l \quad (7)$$

and this is 4D-En-Var in the form of the control vector augmentation method (the α-method) suggested by Lorenc (2003). The $\boldsymbol{\alpha}_l$ can be considered as a localized weight field for the ensemble perturbations of ensemble member l. In the most general case, and without any truncation of the localization correlation function, $\boldsymbol{\alpha}_l$ will be a vector of dimension $K \times n$. In the first version of HIRLAM 4D-En-Var Hybrid, we will assume the same localized ensemble weights for each observation time window, corresponding to a strong constraint 4D-Var, and also for each model level, corresponding to application of horizontal localization only. The impact of an additional vertical localization is studied separately, see Sect. 5.2.3. In the HIRLAM 4D-En-Var Hybrid the transform matrix $\mathbf{C}'$ is derived from the localization correlation matrix **C** in a similar way as for the static background error covariance matrix. The localization correlation matrix **C** is thus generated in the extended domain with bi-periodic variations, without any loss of generality we may assume homogeneity and isotropy with regard to the horizontal localization correlation function and then we may obtain a correlation spectrum representing the localization correlation in spectral space. Its square root, the transform matrix $\mathbf{C}'$, consists of horizontal inverse fast Fourier transforms to grid point space. The transforms are carried out to obtain the weights $\boldsymbol{\alpha}_l$ for the different ensemble members. By construction these weights include the implicit localization of ensemble perturbations $(\boldsymbol{X}'_{\mathrm{b}})_l$. The same procedure is also applied in the HIRLAM hybrid variational ensemble assimilation algorithm and is discussed in Gustafsson et al. (2014).

For the localization in the horizontal direction we apply the power law correlation function with a length scale c of 500 km:

$$\rho(c,r) = \frac{1}{1+\frac{1}{2}(\frac{r}{c})^2}, \tag{8}$$

where r is the distance. In order to limit the negative localization effects on model balances, we perform localization on vorticity, divergence, temperature, specific humidity and surface pressure perturbations. A similar localization of stream function and velocity potential was discussed and applied by Kepert (2009).

The cost function expression for the HIRLAM 4D-En-Var Hybrid is given by

$$J(\chi_{\mathrm{3dvar}}, \chi_{\mathrm{ens}}) = \beta_{\mathrm{3dvar}} J_{\mathrm{b,3dvar}}(\chi_{\mathrm{3dvar}}) + \beta_{\mathrm{ens}} J_{\mathrm{b,ens}}(\chi_{\mathrm{ens}}) + J_{\mathrm{o}}, \tag{9}$$

where the weights β_{3dvar} and β_{ens} for the two different components of the background error constraint should fulfil

$$\frac{1}{\beta_{\mathrm{3dvar}}} + \frac{1}{\beta_{\mathrm{ens}}} = 1. \tag{10}$$

Slightly different from the 4D-En-Var implementation by Liu et al. (2009) we apply explicitly the linearized observation operator $\mathbf{H}_k$ in the observation constraint part J_{o} of the cost function to be minimized:

$$J_{\mathrm{o}} = \frac{1}{2}\sum_{t_k=t_0}^{t_K} (\mathbf{H}_k \boldsymbol{\delta x}(t_k) - \boldsymbol{d}_k)^T \mathbf{R}_k^{-1} (\mathbf{H}_k \boldsymbol{\delta x}(t_k) - \boldsymbol{d}_k), \tag{11}$$

where $\boldsymbol{\delta x}(t_k)$ given by Eq. (6) can be expressed in the assimilation control variables as

$$\begin{aligned}\boldsymbol{\delta x}(t_k) &= \mathbf{U}\chi_{\mathrm{3dvar}} + \kappa \sum_{l=1}^{N} \boldsymbol{\alpha}_l \circ \left(\boldsymbol{X}'_{\mathrm{b}}\right)_l (t_k) \\ &= \mathbf{U}\chi_{\mathrm{3dvar}} + \kappa \sum_{l=1}^{N} \left(\mathbf{C}'(\chi_{\mathrm{ens}})_l\right) \circ \left(\boldsymbol{X}'_{\mathrm{b}}\right)_l (t_k),\end{aligned} \tag{12}$$

where κ is a tuning factor.

There are several parameters of the HIRLAM 4D-En-Var Hybrid that can be considered as tuning parameters – for example, the form of the horizontal localization correlation function and its length scale, the weights β_{3dvar}, β_{ens} associated with the two different components of the background error constraint and the tuning factor κ to scale the ensemble perturbations for estimation of the flow-dependent forecast error covariance. Here we will make experiments with $\beta_{\mathrm{3dvar}} = \beta_{\mathrm{ens}} = 2$, thus with equal weights given to the two components of the background error constraint, and with almost full weight given to the ensemble-based background error constraint. We will also compare experiments without any scaling of the ensemble perturbations ($\kappa = 1$) with experiments with such a scaling ($\kappa = 4$). For sensitivity of the 4D-Var Hybrid scheme to these tuning parameters, see Gustafsson et al. (2014).

4 ETKF re-scaling

The 4D-En-Var Hybrid as implemented in HIRLAM provides an initial state for the ensemble control only. For the generation of the perturbed ensemble members we apply an ensemble transform KF (ETKF) based re-scaling scheme (Bishop et al., 2001). Essentially this is a low-rank estimation of the analysis error variances, preserving also dynamical structures. Further details on the HIRLAM ETKF re-scaling are described and discussed in Bojarova et al. (2010) and Gustafsson et al. (2014).

5 Results

Data assimilation experiments to validate the performance of the HIRLAM 4D-En-Var Hybrid and to compare it with HIRLAM 4D-Var, as well as with the HIRLAM 4D-Var Hybrid data assimilation, were done over the period 17 January–28 February 2008. Single simulated observation impact experiments were carried out in order to illustrate the flow-dependent assimilation increments due to the use of ensemble-based background error covariances. The average effects on forecast quality are illustrated through a data assimilation and forecast experiment, using real observations, over the whole data period (43 days). The model domain applied during the experiments is shown in Fig. 1; the horizontal grid resolution was 11 km and the number of vertical levels of the forecast model was 40.

The forecast model used in the experiments was the HIRLAM grid point forecast model. It is hydrostatic and it utilizes a semi-implicit, semi-Lagrangian two time level integration scheme (Undén et al., 2002). The physical parameterizations were the CBR turbulence scheme (Cuxart et al., 2000), the Kain–Fritsch convection scheme (Kain, 2004), the Rasch–Kristjánsson cloud water scheme (Rasch and Kristjánsson, 1998), the Savijärvi (1990) radiation and the ISBA surface scheme (Noilhan and Mahfouf, 1996).

5.1 Single observation experiments with 4D-Var, 4D-Var Hybrid, and 4D-En-Var Hybrid

To illustrate the influence of ensemble based error covariances on flow dependency, we selected a case with a strong baroclinic development over the Northern Atlantic south of Iceland on 7 February 2008. The 6 h data assimilation window selected was from 6 February 2008, 21:00 UTC,

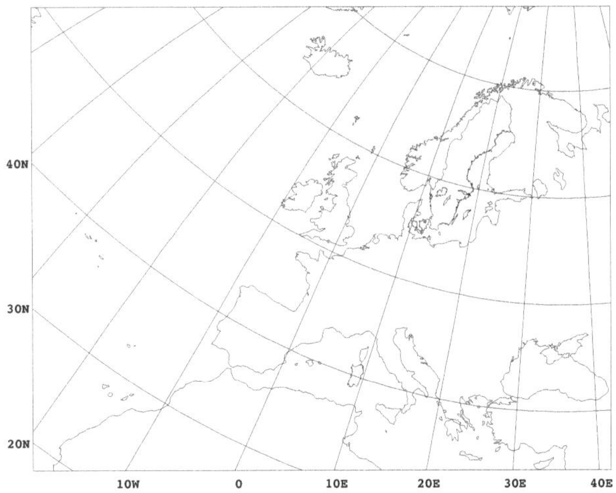

Figure 1. Model domain for the data assimilation and forecast experiments.

Table 1. Description of experiments carried out with the HIRLAM 4D-Var, the HIRLAM Hybrid 4D-Var ensemble data assimilation and with the HIRLAM 4D-En-Var for the winter period 17 January–28 February 2008.

Experiment name	Description
4DVar	4D-Var, no ensemble constraint, two outer loops, 66 and 44 km increment resolution.
4DVar-Hyb25-Sc	4D-Var Hybrid, $\beta_{\text{var}} = 4$, $\beta_{\text{ens}} = 4/3$ (25 % ensemble, 75 % static cov.), two outer loops, 66 and 44 km increment resolution. Ensemble perturbations scaled by a factor 4 in the covariance estimation.
4DEnVar-Hyb50-Sc	4D-En-Var, $\beta_{\text{var}} = \beta_{\text{ens}} = 2$ (50 % ensemble, 50 % 3D-Var FGAT cov.), one outer loop, 33 km increment resolution. Ensemble perturbations scaled by a factor 4 in the covariance estimation.
4DEnVar-Hyb50-NoSc	As 4DEnVar-Hyb50-Sc but without any scaling of the ensemble perturbations in the covariance calculation.
4DEnVar-Hyb90-NoSc	As 4DEnVar-Hyb50-NoSc with the exception that $\beta_{\text{var}} = 11$, $\beta_{\text{ens}} = 11/10$ ($\approx$ 90 % ensemble, $\approx$ 10 % 3D-Var FGAT cov.).
VertLoc	As 4DEnVar-Hyb50-Sc but with a vertical localization of ensemble-based covariances added

until 7 February, 03:00 UTC. Analysis fields for 7 February 2008, 03:00 UTC, taken from a 4D-En-Var Hybrid assimilation experiment (experiment name 4DEnVar-Hyb50-Sc, see Table 1), are shown for mean sea level pressure and 700 hPa temperature in Fig. 2a and for 500 hPa geopotential height and wind in Fig. 2b. Note the strong small-scale surface pressure development south of Iceland, the sharp frontal zones in the 700 hPa temperature field and the associated small-scale trough in the 500 hPa geopotential height field.

To investigate how the standard 4D-Var develops implicit flow-dependent assimilation structure functions, a simulated wind observation was inserted into the assimilation process at 7 February 2008, 02:00 UTC, 5 h into the assimilation window. The horizontal position of the simulated observation was 58° N, 15° W and the vertical level was 500 hPa. The corresponding observation increment was approximately southwesterly 12 m s^{-1} (10 m s^{-1} in the model geometry u component and 5 m s^{-1} in the v component).

500 hPa geopotential height and wind assimilation increments induced by the single simulated observation on 7 February 2008, 02:00 UTC, are presented in Fig. 3 for the following three assimilation configurations: (1) standard 4D-Var with a single outer loop; (2) hybrid 4D-Var ensemble assimilation with a single outer loop, with equal weights given to the static background error covariance and the ensemble-based background error covariance and with the tuning factor $\kappa = 4.0$; (3) 4D-En-Var Hybrid with a single outer loop, with equal weights given to the 3D-Var FGAT and the four-dimensional ensemble parts of the background error covariance and with the tuning factor $\kappa = 4.0$ as in the 4D-Var Hybrid. All experiments applied a horizontal grid resolution of 33 km for the assimilation increments. The 20 ensemble members from a 4D-En-Var Hybrid experiment (experiment name 4DEnVar-Hyb50-Sc) were utilized in experiments (2) and (3). It should also be emphasized that the HIRLAM implementation of the 4D-Var, 4D-Var Hybrid, and 4D-En-Var Hybrid algorithms are completely consistent and based on the same computer code.

The single simulated observation impact assimilation increments clearly illustrate the utilization of flow-dependent structure functions in all three experiments (4D-Var, 4D-Var Hybrid, and 4D-En-Var Hybrid), in particular with regard to the sharpening of the small-scale trough in the 500 hPa geopotential height and wind fields, needed in order satisfy the simulated single observation. The flow dependency is more accentuated for the 4D-Var Hybrid and 4D-En-Var Hybrid experiments than for the standard 4D-Var experiment. This result is consistent with application of a static background error covariance at the start of the assimilation window in the standard 4D-Var experiment. In contrast, there are contributions from an ensemble of forecast perturbations to this background error covariance in the 4D-Var Hybrid and 4D-En-Var Hybrid experiments. Furthermore, in the 4D-En-Var Hybrid experiment there is an additional contribution from the ensemble of background error perturbations valid

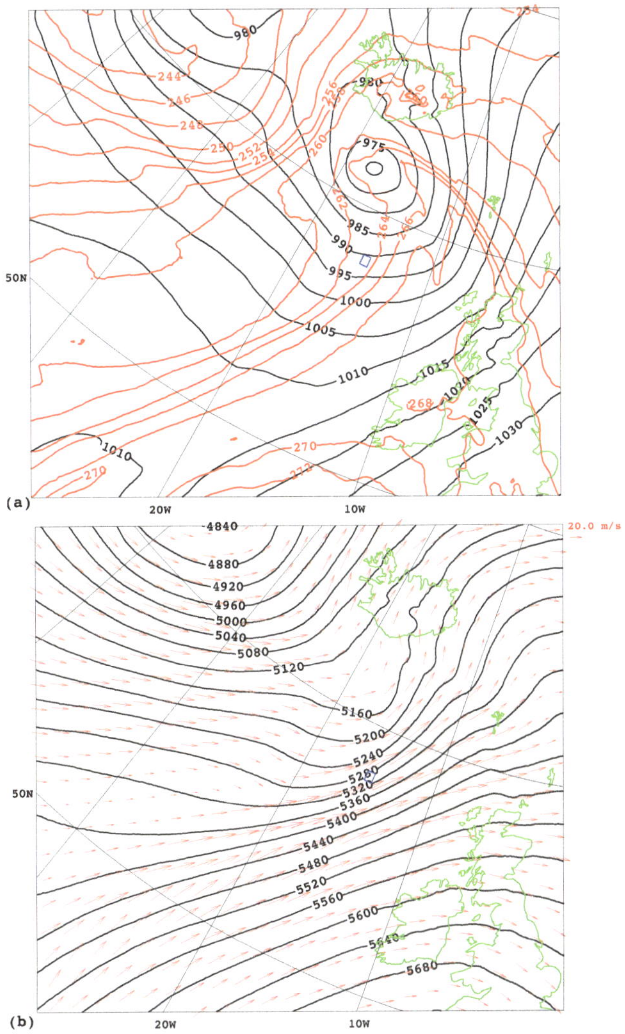

Figure 2. Mean sea level pressure and 700 hPa temperature **(a)**, 500 hPa geopotential height and wind **(b)**, 7 February 2008, 03:00 UTC. Analyses fields are from one of the 4D-En-Var Hybrid experiments (experiment name 4DEnVar-Hyb50-Sc). The position of the single simulated observation at 58° N, 15° W is indicated by a blue square. Isoline spacing is 5 hPa, 2 °K and 40 m for mean sea level pressure, 700 hPa temperature and 500 hPa geopotential height, respectively. A reference wind arrow is provided to the right of the 500 hPa map.

at the time of the simulated observation (7 February 2008, 02:00 UTC). It is satisfactory to notice that the assimilation increments from 4D-Var Hybrid and the 4D-En-Var Hybrid are quite similar, taking into account that the derivations of the two sets of increments are quite different. A tangent linear (TL) model integration of the increment over 5 h from the start of the assimilation window is applied in 4D-Var Hybrid, while in the case of the 4D-En-Var Hybrid there is no such TL model involved. The time evolution of the increments is provided by the four-dimensional background error covariance estimated from an ensemble on non-linear model integrations over the assimilation window.

In order to illustrate the effects on cross-correlations between different model variables, assimilation increments of surface pressure and 700 hPa temperature due to the single 500 hPa simulated wind observation are presented in Fig. 4. First of all we may notice that the assimilation increments are more pronounced (amplified) in 4D-En-Var Hybrid experiment. The 4D-Var increments are much weaker and the 4D-Var Hybrid increments are somewhere in between the 4D-Var and the 4D-En-Var Hybrid increments. The shape of the surface pressure increments indicates that the low-pressure system is forced to move faster when the simulated 500 hPa wind is assimilated and, in particular in the case of the 4D-En-Var Hybrid, there is also a stronger deepening of the low. From the 700 hPa temperature assimilation increment fields we can notice a consistent faster propagation of the atmospheric frontal zones.

5.2 Real observation assimilation experiments

Real observation data assimilation and forecast experiments were carried out for the period 17 January–28 February 2008 and for the six data assimilation configurations described in Table 1. The same conventional observations (TEMP, PILOT, SHIP, AIREP, DRIBU, SYNOP) and satellite radiance data (AMSU-A) were used in all experiments. For each experiment, forecasts up to +30 h were produced and verified against radiosonde and surface observations. The first three days of the experiment period were excluded from the verification in order to allow a spin-up of the LAM ensemble perturbations. We will first describe and discuss forecast verification scores based on the three data assimilation schemes 4D-Var, 4D-Var Hybrid and 4D-En-Var Hybrid. Secondly, we will describe, compare and discuss forecast verification scores for three different versions of 4D-En-Var Hybrid with the aim of checking the sensitivity to two tuning parameters of the HIRLAM 4D-En-Var Hybrid. These tuning parameters are the magnitude of the scaling of the ensemble perturbations for estimation of the flow-dependent background error covariances (tuning factor κ) and the relative weights given to the ensemble and the 3D-Var FGAT components of the background error covariance. Finally, we have also carried out a small impact study on the sensitivity of forecast verification scores to introduction of a very simple vertical localization of the ensemble-based background error covariance matrix in the HIRLAM 4D-En-Var Hybrid scheme.

5.2.1 Verification of forecasts based on 4D-Var, 4D-Var Hybrid, and 4D-En-Var Hybrid

Standard deviation (STDV) and mean (bias) verification scores for forecasts of vertical profiles of +12 and +24 h wind speed and relative humidity valid at 00:00 UTC are presented in Fig. 5 for the 4D-Var, 4D-Var Hybrid and

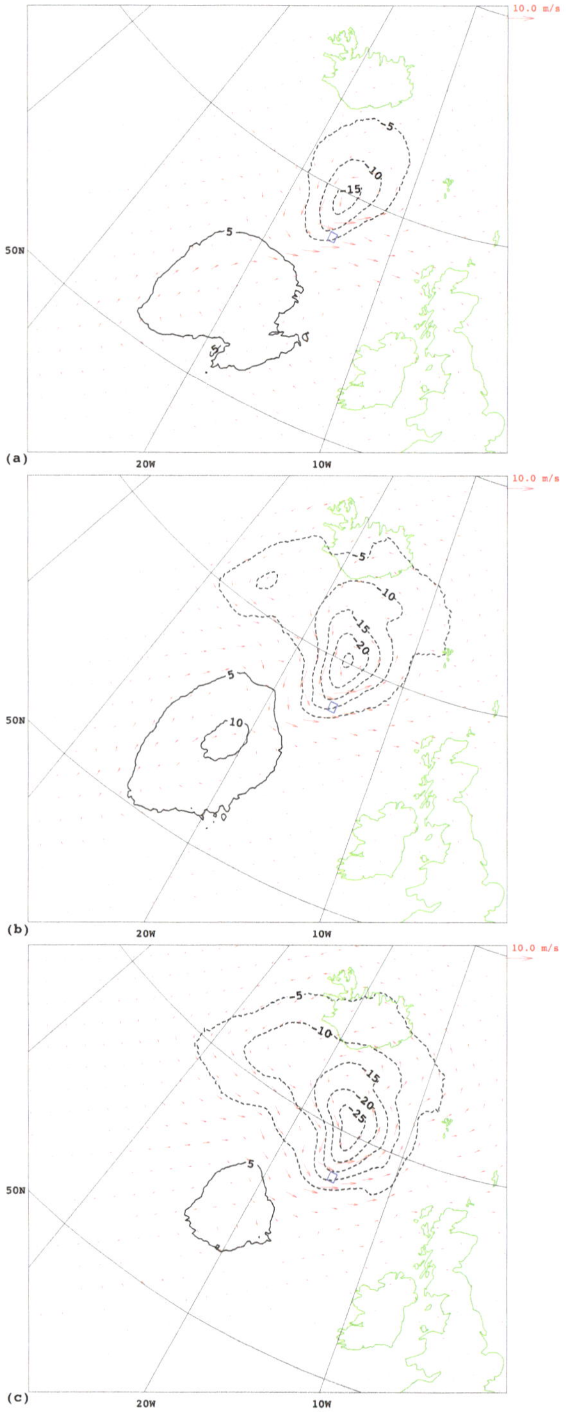

Figure 3. 500 hPa geopotential height and wind analysis increments for 7 February 2008, 02:00 UTC, 5 h into the assimilation window between 6 February 2008, 21:00 UTC, and 7 February 2008, 03:00 UTC. 4D-Var (**a**), 4D-Var Hybrid (**b**), and 4D-En-Var Hybrid (**c**). Simulated single 500 Pa wind observation, with a 10 m s^{-1} observation increment at 58° N, 15° W (marked with a blue square) in the model geometry u component and a 5 m s^{-1} observation increment in the v component. Isoline spacing is 5 m for the geopotential height and a reference wind error is provided to the right of the map.

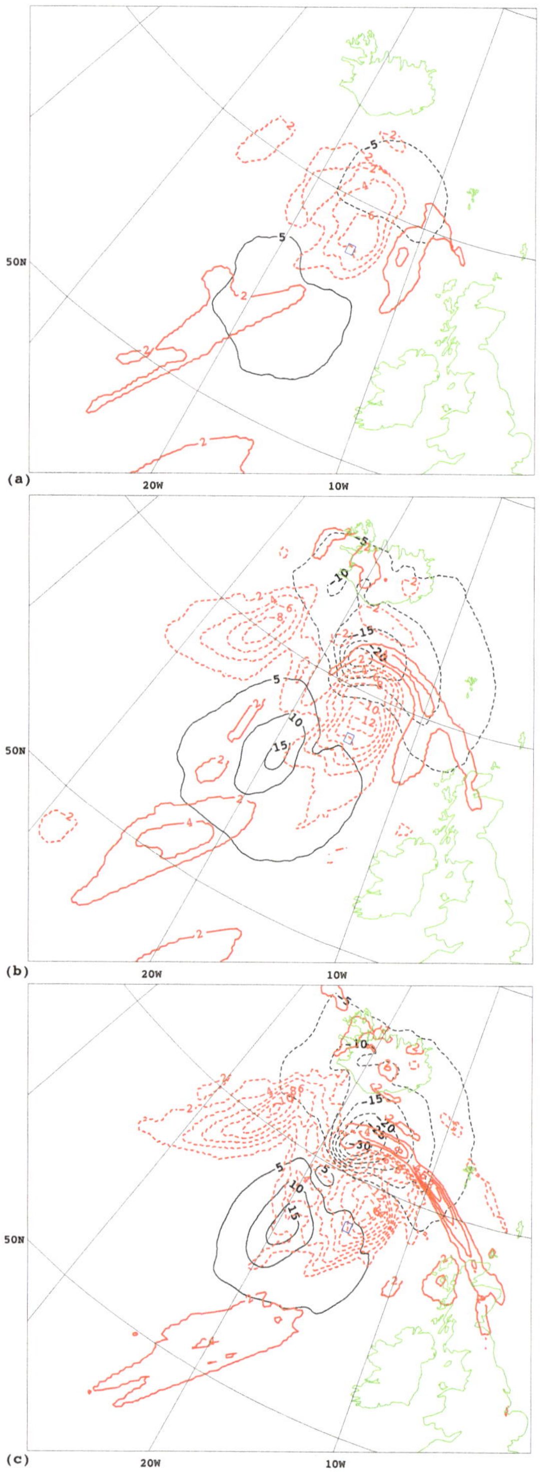

Figure 4. Surface pressure and 700 hPa temperature analysis increments for 7 February 2008, 02:00 UTC, 5 h into the assimilation window between 6 February 2008, 21:00 UTC, and 7 February 2008, 03:00 UTC. 4D-Var (**a**), 4D-Var Hybrid (**b**), and 4D-En-Var Hybrid (**c**). Simulated single 500 hPa wind observation, with a 10 m s^{-1} observation increment at 58° N, 15° W (marked with a blue square) in the model geometry u component and a 5 m s^{-1} observation increment in the v component. Isoline spacing is 5 hPa for the surface pressure and 2 °K for the 700 hPa temperature.

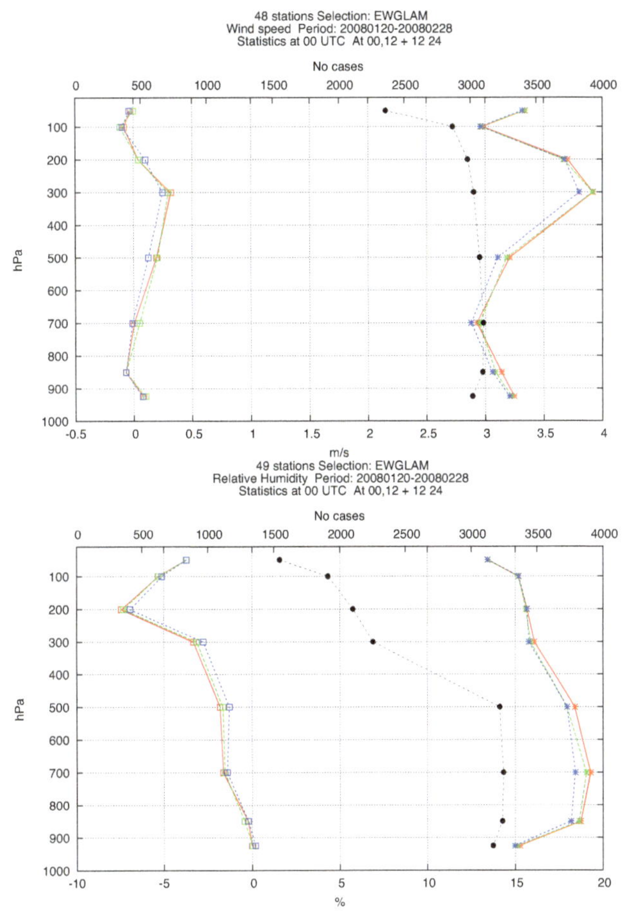

Figure 5. Verification of vertical profiles of wind speed and relative humidity forecasts against radiosonde data from stations in a European network. Standard deviation (STDV) and mean (bias) verification scores. Average verification scores for +12 and +24 h forecasts valid at 00:00 UTC over the period 20 January–28 February 2008; 4D-Var (experiment name 4DVar, red lines), 4D-Var Hybrid (experiment name 4DVar-Hyb25-Sc, green lines), and 4D-En-Var Hybrid (experiment name 4DEnVar-Hyb50-Sc, blue lines). The black dotted line indicates the number of verification comparisons.

4D-En-Var Hybrid experiments. With regard to the standard deviation verification scores we can notice a small but consistent improvement of the scores for 4D-Var Hybrid in comparison with the standard 4D-Var, and a further larger improvement of the scores for 4D-En-Var Hybrid in comparison with 4D-Var Hybrid. These improvements are most significant for tropospheric winds and for relative humidity in the mid-troposphere (500–700 hPa). Vertical profiles of temperature forecasts were also verified against radiosonde data. The temperature verification scores show an almost neutral impact from the choice of assimilation method and for the current ensemble generation methods (results are not shown).

The improvements introduced by 4D-Var Hybrid in comparison with standard 4D-Var are certainly related to the use of an ensemble-based and flow-dependent component of the background error covariance at the start of the assimilation window – for a more detailed discussion see Gustafsson et al. (2014). The reasons for the clear improvement of forecast scores produced by 4D-En-Var Hybrid in comparison with standard 4D-Var and 4D-Var Hybrid are less obvious, and may be even a bit surprising. The main difference is the replacement of the tangent linear and adjoint models in 4D-Var and 4D-Var Hybrid with the use of a four-dimensional ensemble of non-linear model integrations (trajectories) over the data assimilation window. From a computing efficiency point of view, this result is highly satisfactory, since the computing time is significantly reduced when the integration of the TL and AD models is avoided (see below for details). The particularly improved forecast verification scores for relative humidity with 4D-En-Var Hybrid may possibly be explained by the too simplified TL and AD models for moist processes in HIRLAM 4D-Var (no condensation is included, for example).

Standard deviation (STDV) and mean (bias) verification scores for mean sea level pressure (MSLP) forecasts verified against surface observations are presented in Fig. 6 for the 4D-Var, 4D-Var Hybrid and 4D-En-Var Hybrid experiments. There are two features in these MSLP verification scores that need to be discussed and possibly also further investigated in order to arrive at plausible explanations. On one hand, there is a reduced bias in the MSLP forecasts based on 4D-En-Var Hybrid compared to the MSLP forecasts based on 4D-Var and 4D-Var Hybrid. On the other hand, the 4D-En-Var Hybrid forecasts are associated with larger STDV of the forecast errors for the shortest timescales +6 and +12 h, while the STDV of the scores are very similar for longer timescales, +18, +24 and +30 h.

Here we will only present an educated guess as to the behaviour of the verification scores for MSLP; further evidence from complementary investigations will be elaborated on below.

It is a general experience from the NWP practice that MSLP forecast biases are associated with physical inconsistencies causing, for example, adjustment processes in the form of mass fluxes between land and sea surfaces or, in the case of limited area models, mass fluxes over the lateral boundaries. One may thus speculate as to whether the time series of non-linear model perturbations, applied as assimilation basis functions in 4D-En-Var Hybrid, improves consistency with the non-linear model compared with assimilation increments derived from time integrations with a highly simplified tangent linear model.

There may be several reasons behind the degraded MSLP STDV verification scores for 4D-En-Var Hybrid for the shortest forecast timescales (+6 and +12 h). The HIRLAM 4D-Var and 4D-Var Hybrid assimilation algorithms include a weak digital filter constraint for control of high-frequency (gravity wave) oscillations. Such a constraint is not yet a part of the 4D-En-Var Hybrid algorithm with increased high frequency oscillations (noise) as a possible effect. This

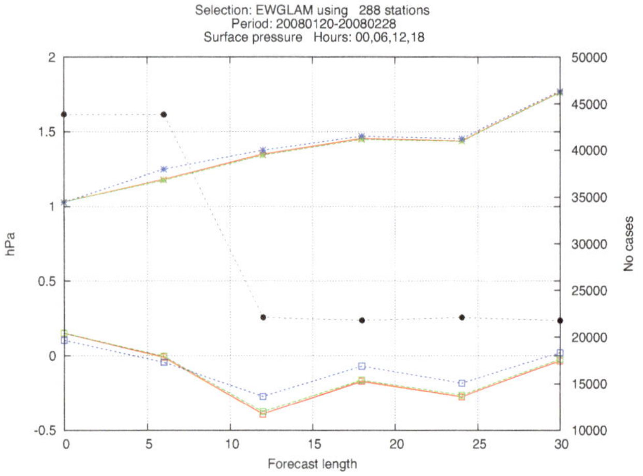

Figure 6. Verification of mean sea level pressure (MSLP) forecasts against surface (SYNOP) data from stations in a European network. Standard deviation (STDV) and mean (bias) verification scores. Average verification scores over the period 20 January–28 February 2008 as a function of forecast length. 4D-Var (experiment name 4DVar, red lines), 4D-Var Hybrid (experiment name 4DVar-Hyb25-Sc, green lines) and 4D-En-Var Hybrid (experiment name 4DEnVar-Hyb50-Sc, blue lines). The black dotted line indicates the number of verification comparisons.

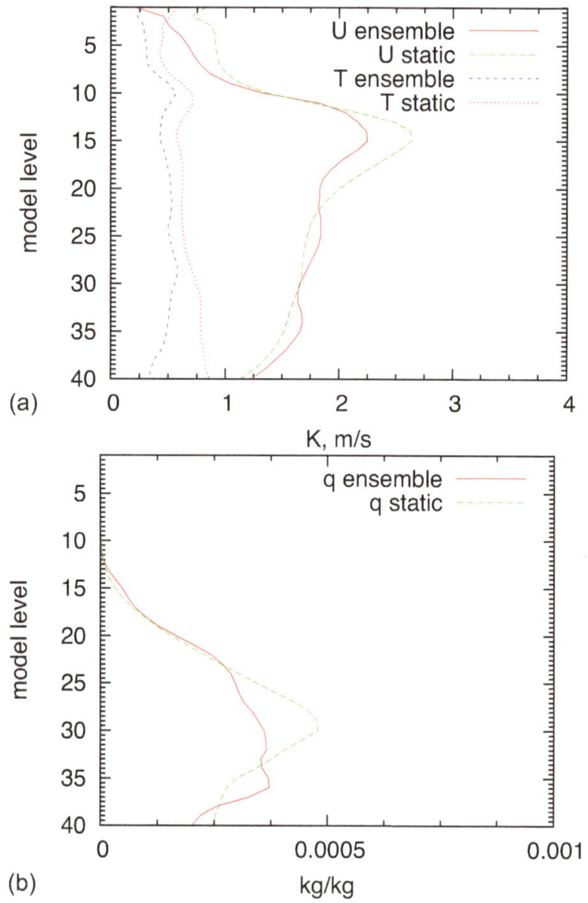

Figure 7. Vertical profiles of climatological (static) and area-averaged ensemble-based background error standard deviations for one randomly selected case 22 January 2008, 12:00 UTC + 6 h. **(a)** U component of wind (ensemble red curve, static green curve) and temperature (ensemble blue curve, static pink curve). **(b)** Specific humidity (ensemble red curve, static green curve).

possibility is further explored below. Furthermore, the rather primitive ETKF re-scaling scheme may also provide noisy surface pressure perturbations that may degrade their usefulness for assimilation purposes. Another explanation to the degraded MSLP STDV verification scores for the shortest timescales may be that the relatively small ensemble of background states (20 members) does not provide enough variability to fit the shortest timescale surface pressure variations equally well as with the tangent linear model in 4D-Var. Investigations of the fit between hourly surface pressure observations and the assimilation model states in the three different experiments seem to give some evidence for this hypothesis (see below).

5.2.2 Sensitivity of forecast verification scores to tuning parameters of the 4D-En-Var Hybrid scheme

The HIRLAM 4D-En-Var Hybrid scheme includes several tuning parameters. Two of the most important ones are (1) the relative weights given to the ensemble and the static 3D-Var FGAT components of background error covariance (parameters β_{ens} and β_{var}) and (2) the scaling factor κ applied to the ensemble perturbations that are used to estimate the flow-dependent (ensemble) component of the background error covariance. Data assimilation and forecast experiments were carried out over the time period 17 January–28 February 2008 to test the sensitivity of the forecast verification scores to different values of these 4D-En-Var Hybrid tuning parameters. It needs to be mentioned that the two tuning parameters are not completely unrelated. A scaling of the ensemble perturbations is efficiently equivalent to changing the weight given to the ensemble part of the background error covariance while keeping the weight for the static part unchanged (neglecting Eq. 10).

In order to understand better the results of these tuning experiments, it is necessary to examine the relative magnitudes of the contributing components of the background error covariance in the 4D-En-Var Hybrid experiments. Figure 7 shows vertical profiles of climatological (static) background error standard deviations and ensemble-based background error standard deviations, obtained by horizontal averaging of variances for one randomly selected case, 22 January 2008, 12:00 UTC, +6 h. It appears that the vertical variations and the magnitudes of these background error standard deviations are quite similar. It is encouraging that the simple tuning of the inflation factor in the ETKF re-scaling scheme, based on innovations (Bojarova et al., 2010), provides similar results as a more elaborate tuning of the climatological

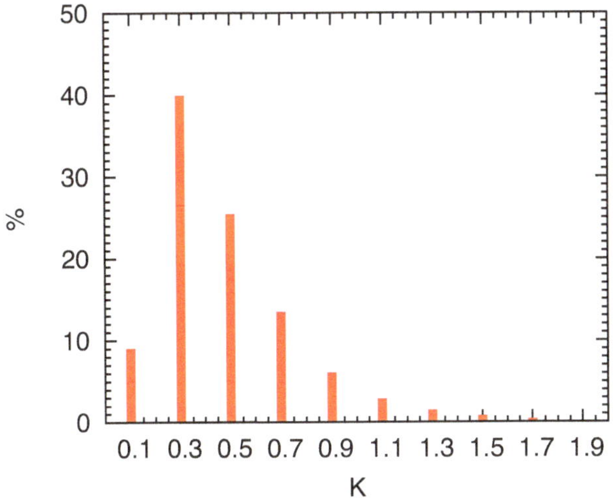

Figure 8. Histogram of ensemble-based background error standard deviations for temperature on model level 30 (≈ 800 hPa) for one randomly selected case, 22 January 2008, 12:00 UTC + 6 h.

background error standard deviations (also based on innovations). One must also keep in mind that the ensemble based background error standard deviations in Fig. 7 represent horizontal averages – these standard deviations have significant spatial variations (Gustafsson et al., 2014). To further illustrate this, we present a histogram of individual grid point background error standard deviations for temperature at model level 30 valid on 22 January 2008, 12:00 UTC, +6 h in Fig. 8. The standard deviation calculated by horizontal averaging is 0.56 K. The frequency is highest for the interval 0.2–0.4 K, but contributions from much larger standard deviations in fewer grid points are significant. This result is consistent with the experience that large forecast uncertainty is associated with quite narrow zones of dynamical activity, in the present case frontal zones in the low tropospheric temperature field (see Gustafsson et al., 2014).

Figure 9 illustrates the sensitivity of verification scores of vertical forecast profiles of wind speed and relative humidity to the tuning of the scaling of the ensemble perturbations by comparing the verification results of experiment 4DEnVar-Hyb50-NoSc (no scaling) with the verification results of experiment 4DEnVar-Hyb50-Sc (scaling with a factor 4). The results indicate that the forecast quality is more or less insensitive to the scaling of ensemble perturbations used to estimate the forecast error covariance, see also the verification scores for MSLP in Fig. 10. One may argue that the HIRLAM ETKF rescaling scheme includes a multiplicative inflation scheme and that the background forecast perturbations are representative of background errors (Bojarova et al., 2010). Any further scaling is not needed. Furthermore, it was shown in experiments with the HIRLAM 4D-Var Hybrid assimilation (Gustafsson et al., 2014) that the *richness* in spatial structures of the ensemble is more important than the scaling of the amplitude of the ensemble perturbations.

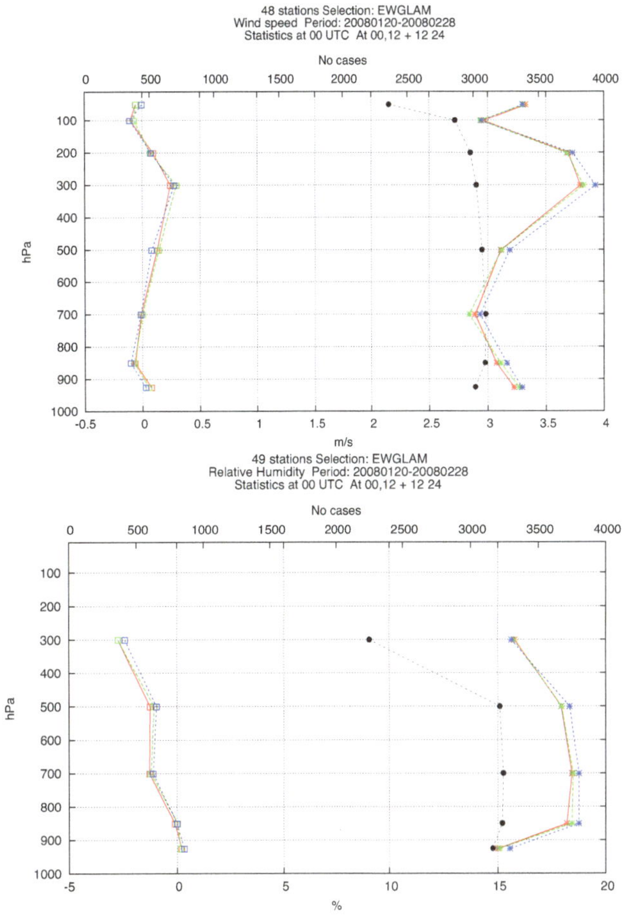

Figure 9. Verification of vertical profiles of wind speed and relative humidity forecasts against radiosonde data from stations in a European network. Standard deviation (STDV) and mean (bias) verification scores. Average verification scores for +12 and +24 h forecasts valid at 00:00 UTC over the period 20 January–28 February 2008. 4D-En-Var Hybrid with scaling of ensemble perturbations and with equal weights given to the ensemble and 3D-Var FGAT components of the background error covariance (experiment name 4DEnVar-Hyb50-Sc, red lines), 4D-En-Var Hybrid without scaling of ensemble perturbations and with equal weights given to the ensemble and 3D-Var FGAT components of the background error covariance (experiment name 4DEnVar-Hyb50-NoSc, green lines). 4D-En-Var Hybrid without scaling of ensemble perturbations and with ≈ 90 % weight given to the ensemble component of background error covariance (experiment name 4DEnVar-Hyb90-NoSc, blue lines). The black dotted line indicates the number of verification comparisons.

Figures 9 and 10 also illustrate the sensitivity of the verification scores to the relative weighing between the ensemble and the 3D-Var FGAT contributions to the background error covariance by comparing the verification results of experiment 4DEnVar-Hyb50-NoSc (equal contributions of the two components) and experiment 4DEnVar-Hyb90-NoSc (≈ 90 % contributions of the ensemble component and ≈ 10 % of the static 3D-Var FGAT component). The results show

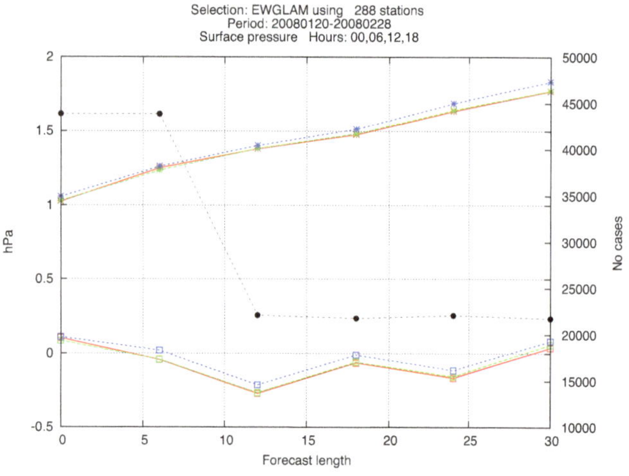

Figure 10. Verification of mean sea level pressure (MSLP) forecasts against surface (SYNOP) data from stations in a European network. Standard deviation (STDV) and mean (bias) verification scores. Average verification scores over the period 20 January–28 February 2008 as a function of forecast length. 4D-En-Var Hybrid with scaling of ensemble perturbations and with equal weights given to the ensemble and 3D-Var FGAT components of the background error covariance (experiment name 4DEnVar-Hyb50-Sc, red lines), 4D-En-Var Hybrid without scaling of ensemble perturbations and with equal weights given to the ensemble and 3D-Var FGAT components of the background error covariance (experiment name 4DEnVar-Hyb50-NoSc, green lines). 4D-En-Var Hybrid without scaling of ensemble perturbations and with ≈ 90 % weight given to the ensemble component of background error covariance (experiment name 4DEnVar-Hyb90-NoSc, blue lines). The black dotted line indicates the number of verification comparisons.

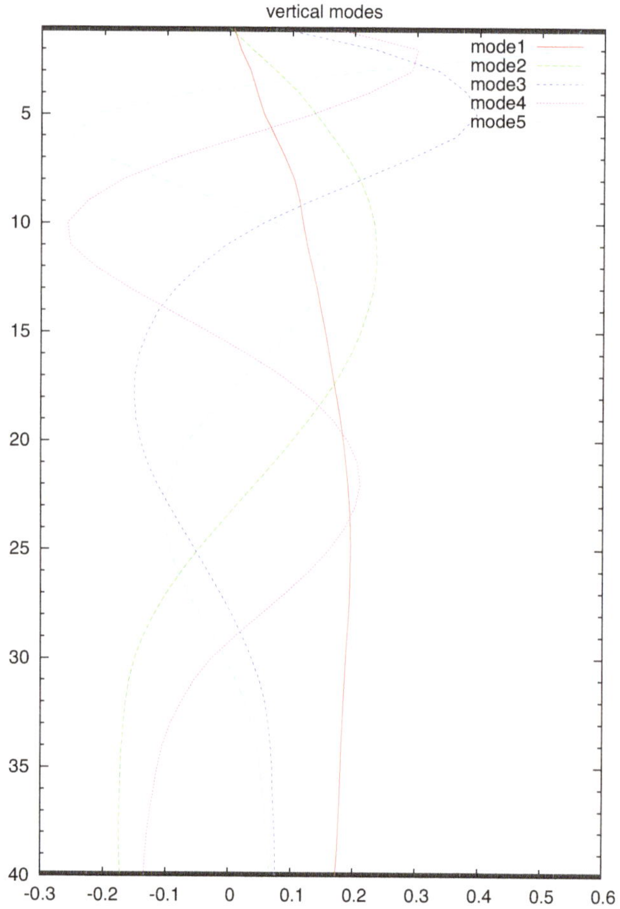

Figure 11. First five eigenvectors (vertical modes) of the vertical correlation matrix for large horizontal scale vorticity (wave numbers 1–5) background errors, as applied in the HIRLAM variational data assimilation: mode 1 (red curve), mode 2 (green curve), mode 3 (dark blue curve), mode 4 (pink curve) and mode 5 (light blue curve).

that the forecast quality, as given by the forecast verification scores, is consistently degraded when the static component of the background error covariance is reduced to 10 % only. Our interpretation is that an ensemble of background model states with 20 members only does not provide variability enough (or, in other words, richness in structures enough) for an accurate estimation of the background error covariance.

5.2.3 The impact of a very simple vertical covariance localization on forecast verification scores

All results presented so far in this paper have been based on 4D-En-Var Hybrid with a horizontal localization only of the ensemble-based covariance matrix. In order to avoid spurious influence of, for example, near-surface observations in the stratosphere, and vice versa, it is necessary to introduce also a vertical localization, see Clayton et al. (2013).

A first trial to introduce a vertical localization has been carried out also for the HIRLAM 4D-En-Var Hybrid. We simply started from the climatological background error vertical correlation matrix for vorticity for the largest horizontal scales (horizontal wave numbers 1–5). The eigenvectors of this vertical correlation matrix are presented in Fig. 11. The first eigenvector (mode number 1) represents an average over the total atmospheric column, the second eigenvector represents a separation between the stratosphere and the low troposphere, while the third eigenvector has a maximum in the middle troposphere. A vertical localization correlation matrix, see Fig. 12, was constructed from these three vertical modes by setting eigenvalues of the remaining vertical modes (modes 4–40 in the present case with 40 vertical model levels) to zero and by re-normalizing the vertical correlations based on these three first vertical modes.

One real observation data assimilation experiment (4DEnVar-Hyb50-NoSc) was re-run for the whole data period with vertical localization of the ensemble-based covariance added (experiment name Vertloc). Most forecast verification scores were slightly but consistently improved by this vertical localization, see for example verification scores for mean sea level pressure (MSLP) over a Scandinavian domain in Fig. 13. One exception to this positive impact of vertical

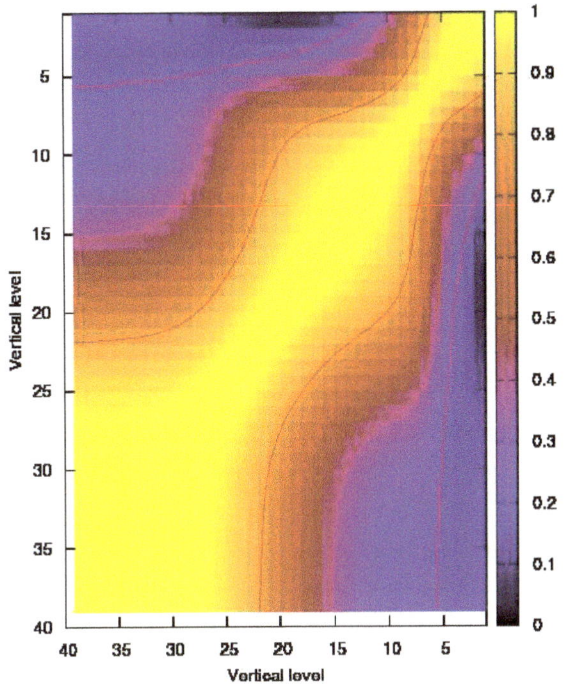

Figure 12. Vertical localization correlation matrix applied to ensemble-based background error covariances in the Vertloc assimilation experiment. The vertical localization matrix is based on the vertical correlation matrix for vorticity in the HIRLAM variational data assimilation, truncated to be valid for horizontal wave numbers 1–5 and to the three first vertical modes.

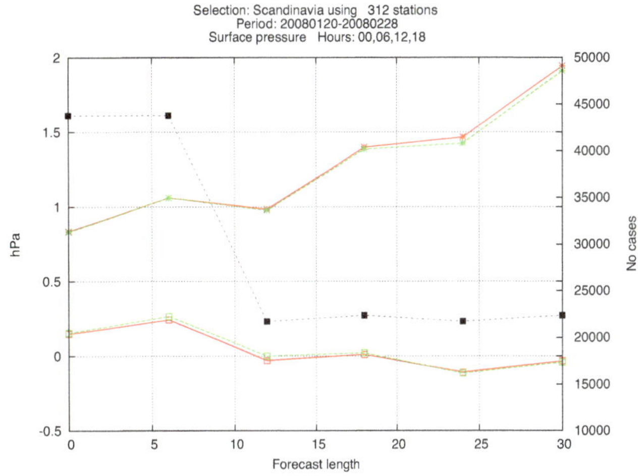

Figure 13. Verification of mean sea level pressure (MSLP) forecasts against surface (SYNOP) data from stations in a Scandinavian network. Standard deviation (STDV) and mean (bias) verification scores. Average verification scores over the period 17 January–28 February 2008 as a function of forecast length. 4D-En-Var Hybrid without vertical localization of the ensemble component of the background error covariance (experiment name 4DEnVar-Hyb50-NoSc, red lines) and 4D-En-Var with vertical localization of the ensemble component of the background error covariance (experiment name Vertloc, green lines). The black dotted line indicates the number of verification comparisons.

covariance localization on forecast verification scores was noticed for 700 hPa relative humidity forecasts (results are not shown). One likely explanation is that the vertical localization correlation matrix, derived by truncation from vorticity background error vertical correlations, does not fit very well the vertical variability of humidity background errors. Specific humidity background error vertical correlation patterns indicate two main vertical domains of activity, the planetary boundary layer and the tropospheric cloud layers above, and these are positioned vertically out of phase with the vertical correlation patterns seen in Fig. 12.

It need to be emphasized that this should be considered as an ad hoc trial to test the sensitivity of forecast verification scores to vertical localization of the ensemble-based background error covariances. Research is certainly needed to bring vertical localization, and localization in general, to a more scientifically founded level of development.

5.2.4 Noise characteristics

A weak digital filter constraint is included in the HIRLAM 4D-Var to control the level of high-frequency (gravity wave) oscillations during integration of the tangent linear and adjoint models over the data assimilation window. Although applied during the coarse-resolution tangent linear and adjoint model integrations, the application of this weak constraint during the data assimilation also efficiently prevents the presence of high-frequency (gravity wave) oscillations induced by the non-linear model during integrations from initial data created by the assimilation (Gustafsson et al., 2012). It is less straightforward to apply a similar digital filter constraint within the framework of 4D-En-Var since there is no model integration applied during the data assimilation process. The time resolution of the derived assimilation increments may not be sufficient for application of a digital filter. For this reason it is of interest to investigate the level of high-frequency oscillations in non-linear model integrations from initial model states produced by 4D-En-Var.

Figure 14 presents time series of the Sundqvist noise parameter, i.e. the horizontal average of the absolute value of the surface pressure tendency, for every time step during the first 12 h of the non-linear model integration, for the assimilation control (ensemble member 0) and for 4D-Var Hybrid and 4D-En-Var Hybrid. Figure 14 confirms that 4D-Var Hybrid, including a weak digital filter constraint, provides a quite noise-free non-linear model integration. It is also seen that 4D-En-Var Hybrid provides an initial state that causes a slightly increased noise level (from $\approx 2.0\,\mathrm{hPa}/3\,\mathrm{h}$ for 4D-Var Hybrid to $2.5\,\mathrm{hPa}/3\,\mathrm{h}$ for 4D-En-Var Hybrid) during the first 1–2 h of model integration. We may consider that such a modest increase of the noise level hardly can harm the data assimilation process.

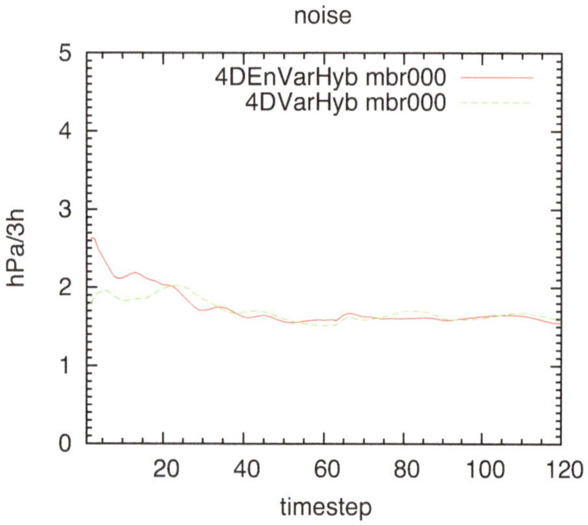

Figure 14. Time variation of the Sundqvist noise parameter, horizontal average of the absolute value of the surface pressure tendency (hPa/3 h), for each time step over the first 12 h of the non-linear model integration (model time step is 6 min). Ensemble control (mbr000). 4D-En-Var Hybrid (experiment name 4DEnVar-Hyb50-Sc, red curve) and 4D-Var Hybrid (experiment name 4DVar-Hyb25-Sc, green curve).

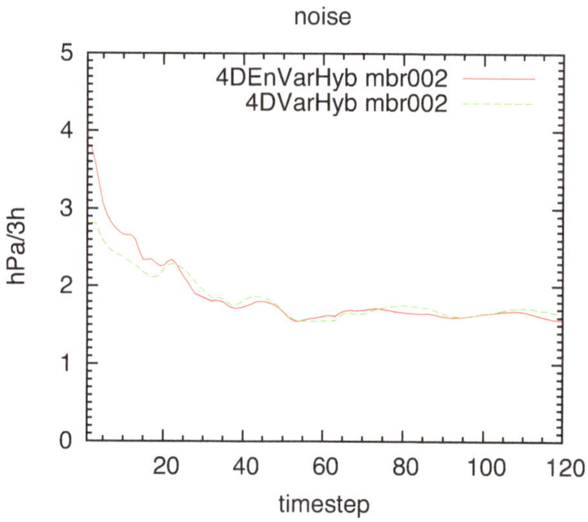

Figure 15. Time variation of the Sundqvist noise parameter, horizontal average of the absolute value of the surface pressure tendency (hPa/3 h), for each time step over the first 12 h of the non-linear model integration (model time step is 6 min). Ensemble member 2 (mbr002). 4D-En-Var Hybrid (experiment name 4DEnVar-Hyb50-Sc, red curve), and 4D-Var Hybrid (experiment name 4DVar-Hyb25-Sc, green curve).

For the ensemble members other than the ensemble control we do not apply data assimilation but rather an ETKF rescaling of the forecast background perturbations to analysis perturbations representing analysis errors. The ETKF re-scaling may influence the balances of the forecast background model states and increase the noise level in non-linear model integration for these ensemble members. Figure 15 shows one example of the Sundqvist noise parameter from the time integration of ensemble member 2 (mbr002) and for both 4D-Var Hybrid and 4D-En-Var Hybrid. Indeed, the noise level, as measured by the Sundqvist parameter, is increased from ≈ 2 hPa/3 h for the ensemble control to ≈ 3 hPa/3 h for ensemble member 2 during the first few hours of non-linear model integration in the 4D-Var Hybrid experiment. Furthermore, it can be seen that the effects in the form of noise add up due to the imbalances caused by ETKF re-scaling and the 4D-En-Var Hybrid assimilation algorithm. With regard to possible negative effects of the combined imbalances created by the ETKF re-scaling and the 4D-En-Var Hybrid algorithms, one must keep in mind that the ensemble perturbations are only applied for estimation of background error covariances at the start of the next data assimilation cycle, thus at +3 h. At this forecast lead time model balances appear to have been stabilized through the adjustment processes of the high-frequency oscillations.

5.2.5 Observation fit statistics

It was noticed that MSLP forecasts for the shortest time range (+6 h) had slightly degraded Standard Deviation (STDV) verification scores for the 4D-En-Var Hybrid experiment as compared to the 4D-Var and 4D-Var Hybrid experiments. Furthermore, it was also noticed that the level of high-frequency oscillations as measured by the surface pressure tendency was increased for the 4D-En-Var Hybrid, also compared with the 4D-Var and 4D-Var Hybrid experiments. In order to investigate further the use of surface pressure information in the 4D-En-Var Hybrid assimilation, we checked the behaviour of the surface pressure contribution to the observation part of the cost function during the minimization for one assimilation cycle, 22 February 2008, 12:00 UTC (the behaviour is similar for every assimilation cycle).

Figure 16 shows the SYNOP surface pressure contribution to the observation cost function, normalized with the number of surface pressure observations, as a function of observation time window within the 6 h data assimilation window for the 4D-Var, 4D-Var Hybrid and the 4D-En-Var Hybrid experiments. Separate curves are plotted before (O − BG, observation minus background) and after (O − A, observation minus analysis) the minimization. We may notice that the behaviour of these observation fit statistics are different for the 4D-En-Var Hybrid experiment as compared to the 4D-Var and 4D-Var Hybrid experiments.

The (O − BG) statistics indicate that the surface pressure forecasts for the observation windows 1–4 (corresponding background forecast length in the range 3–6 h) have larger errors for the 4D-En-Var Hybrid experiment than for the 4D-Var and 4D-Var Hybrid experiments. This confirms what we

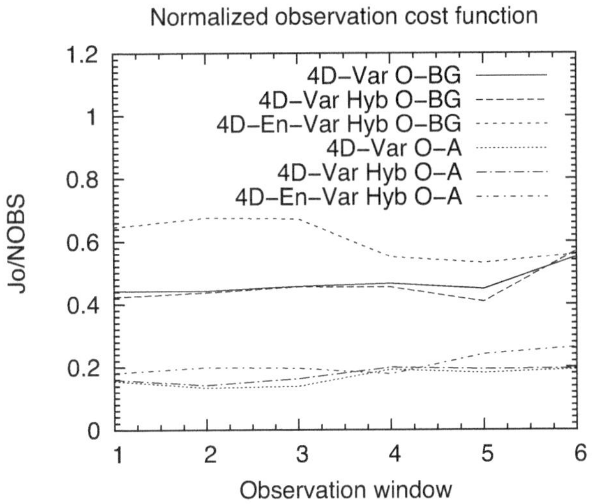

Figure 16. The contribution of SYNOP surface pressure observations to the observation cost function (J_O), normalized with the number of contributing observations (NOBS), as a function of the observation time window (−3, −2, −1, 0, +1, and +2 h). Assimilation cycle 22 February 2008, 12:00 UTC. Before minimization (O − BG) and after minimization (O − A). 4D-Var (experiment name 4DVar), 4D-Var Hybrid (experiment name 4DVar-Hyb25-Sc), and 4D-En-Var (experiment name 4DEnVar-Hyb50-Sc).

already have noticed with the forecast verification scores (see above).

The fit of the analysis to the surface pressure observations at the main observation hour (12:00 UTC, observation window 4) is closer for the 4D-En-Var Hybrid experiment as compared to the 4D-Var and 4D-Var hybrid experiments, while the fit is significantly worse for the 4D-En-Var Hybrid experiment for the other observation windows.

The cause of the different behaviour of surface pressure variations at short timescale in the different experiments is not obvious. At first sight, the increased level of high-frequency oscillations (measured by surface pressure tendencies) could possibly explain the worse surface pressure scores for the 4D-En-Var Hybrid experiment at short time range. However, this effect of imbalances in the initial 4D-En-Var Hybrid mainly affects forecasts in the range 1–2 h, while forecasts in the range 3–8 h, when the initial adjustments have already occurred, are applied as background states in the 4D-En-Var Hybrid.

Another possible explanation may be that the ensemble of 20 non-linear model trajectories in 4D-En-Var Hybrid does not provide the needed temporal variations to describe surface pressure variations with a timescale of 1 h, as provided by the observations, while the tangent linear model in 4D-Var is able to provide solutions that fit the observations more closely. This is not necessarily an advantage for the TL and AD model based HIRLAM 4D-Var. With a lack of a proper handling of biases and other representativity errors, the HIRLAM 4D-Var may over-fit the surface pressure data when observations from the same stations are available for every hour. The increased surface pressure biases in longer forecasts for the 4D-Var based experiments and the recovery of quality of the 4D-En-Var Hybrid forecasts for the longer forecast may give some support to this hypothesis. Further investigations are needed to establish a better understanding.

6 Computational issues

The computing cost in 4D-En-Var Hybrid is quite different from the computing cost in 4D-Var. 4D-Var is dominated by heavy CPU utilization in the TL and AD models, while 4D-En-Var Hybrid is more dominated by input and handling of the ensemble of model trajectories needed for the estimation of the background error covariances. This can be seen in Table 2, which shows the measured wall clock computing times from a single assimilation cycle (minimization only) of 4D-Var, 4D-Var Hybrid and 4D-En-Var Hybrid. The computing times were measured from assimilation runs with 32 MPI tasks on a single compute node in an IBM Power 6-575 system. Since 4D-En-Var Hybrid is utilizing background error covariances based on non-linear model trajectories, 4D-En-Var Hybrid is using a single outer loop, while 4D-Var and 4D-Var Hybrid are using two outer loops in order to improve the linearization for the TL and AD models. Re-linearization for the observation operators could be motivated also for 4D-En-Var Hybrid, but the possibility for this has not yet been introduced in HIRLAM 4D-En-Var Hybrid.

The measured wall clock computing times of different components of 4D-En-Var Hybrid in Table 2 show that half of the wall clock time is spent in reading the ensemble of non-linear model trajectories over the data assimilation window (full model states for 20 ensemble members and for six observation time windows). In order to fully utilize the reduction in computing time that comes with avoiding the tangent linear and adjoint models in 4D-En-Var Hybrid, the reading of the model trajectories needs to be improved by making it parallel. This has not yet been implemented in the present version of HIRLAM 4D-En-Var Hybrid. Furthermore, the storing in memory of the ensemble of non-linear model trajectories has to be reduced by packing in fewer computer memory bits than occupied by double precision floating point numbers, taking the needs of higher model resolutions into account, both with regard to resolution in space and with regard to an increased number of observation time windows.

The longer computing time spent in reading and writing of observations, as well as in writing of field data, in 4D-Var and 4D-Var Hybrid, as compared to 4D-En-Var Hybrid, is explained by the two outer loops in 4D-Var and 4D-Var Hybrid, while a single outer loop is applied in 4D-En-Var Hybrid. The total number of inner loop minimization iterations is the same in all runs, as can be seen in the computing time for the observation operators.

Table 2. Wall clock computing time of different components of 4D-Var, 4D-Var Hybrid and 4D-En-Var Hybrid with one assimilation cycle, 22 February 2008, 12:00 UTC. 4D-Var and 4D-Var Hybrid are run with two outer loops, while 4D-En-Var Hybrid is run with a single outer loop. The computing times were measured from assimilation runs with 32 MPI tasks on a single compute node in an IBM Power 6-575 system (OpenMP was not utilized).

Assimilation component	4D-Var	4D-Var Hybrid	4D-En-Var Hybrid
Total	693 s	846 s	418 s
Read field data	46 s	56 s	216 s
Read obs. data	61 s	71 s	23 s
Minimization, total	373 s	339 s	91 s
TL model	141 s	109 s	–
AD model	191 s	146 s	–
Observation operators	20 s	20 s	20 s
Write field data	118 s	107 s	53 s
Write obs. data	44 s	30 s	15 s

7 Discussion and concluding remarks

A four-dimensional ensemble variational (4D-En-Var) data assimilation scheme has been developed for the HIRLAM forecasting system. The scheme is a hybrid between 4D-En-Var, with a flow-dependent background error covariance component estimated from an ensemble of non-linear model trajectories over the data assimilation window, and 3D-Var FGAT (first guess at appropriate time) with a static background error covariance component applied in the middle of the data assimilation window. The HIRLAM 4D-En-Var Hybrid is applied to the ensemble control only; initial states for other ensemble members are obtained by a basic ETKF (ensemble transform Kalman filter) re-scaling scheme. In order to prevent negative effects of the low rank of the ensemble-based background error covariance matrix, a localization of the ensemble covariance matrix based on a Schur product with a localization correlation matrix is applied.

Single simulated observation data assimilation experiments have been carried out to demonstrate the effects of flow dependency of the ensemble-based background error covariances. These experiments show an increased level of realistic flow dependency, both for 4D-Var Hybrid in comparison with standard 4D-Var and for 4D-En-Var Hybrid in comparison with 4D-Var Hybrid. The experiments also give support to the hypothesis that the ETKF re-scaling algorithm provides initial perturbations with a rich and physically relevant variability.

Real observation experiments over a period of 6 weeks to demonstrate the effects of the assimilation on forecast quality have also been carried out. Comparison is made with standard HIRLAM 4D-Var and with HIRLAM Hybrid 4D-Var ensemble data assimilation, where the ensemble based background error covariance is applied at the start of the assimilation window only. The real observation data assimilation experiments indicate that the 4D-En-Var Hybrid outperforms standard 4D-Var as well as 4D-Var Hybrid with regard to verification scores for forecasts based on initial data from the respective assimilation method. In particular, relative humidity verification scores are significantly improved by 4D-En-Var Hybrid, and this is likely to be linked to the too simplified tangent linear model applied in the HIRLAM 4D-Var.

In contrast to 4D-Var and 4D-Var Hybrid, no time integration of tangent linear and adjoint models is needed in 4D-En-Var Hybrid. This results in a significant reduction of computing time, and possibilities for better scalability, for 4D-En-Var Hybrid. The handling of the ensemble of non-linear model trajectories over the assimilation window needs to be designed carefully, however.

The successful implementation of HIRLAM 4D-En-Var Hybrid for synoptic-scale numerical weather prediction has encouraged us to develop a 4D-En-Var Hybrid also for numerical weather prediction at the convection permitting scales with model grid resolutions of a few km. The increased importance of moist processes and non-linearities at these smaller scales should be an advantage for 4D-En-Var since the assimilation increments are formed as localized combinations of non-linear model forecast perturbations rather than derived from time integrations with a simplified tangent linear model.

There are certainly many scientific uncertainties associated with the development of 4D-En-Var Hybrid, and data assimilation in general, for the convection permitting scales. Which spatial scales are described by present and future observing systems, and which spatial scales are meaningful to assimilate for convection permitting models? For synoptic-scale 4D-Var data assimilation, the application of incremental data assimilation with tangent linear model grid resolutions increased by a factor of 3–5 as compared to the non-linear model grid resolution, has been quite successful (Gustafsson et al., 2012). Does this also apply to the data assimilation at convection permitting scales? In 4D-En-Var for example, do we need an ensemble generated from full resolution non-linear model integrations in order to derive the appropriate background error covariances for data assimilation at convection-permitting scales? Since such a full resolution convection-permitting ensemble of model integrations would be computational demanding, it might be tempting to derive the ensemble needed for 4D-En-Var with a slightly reduced non-linear model resolution.

The 4D-En-Var method described here can easily be generalized to a simple method for nowcasting by extending the same weighting of ensemble background forecast perturbations beyond the data assimilation window.

Most results presented here were based on HIRLAM 4D-En-Var Hybrid using horizontal localization of ensemble-based background error covariances only. The results from an ad hoc trial with a simple vertical localization have encouraged us to continue research on horizontal as well as vertical localization and, in order to permit a weak constraint formulation, also on temporal localization.

Acknowledgements. We would like to thank Dale Barker, Andrew Lorenc and Neil Bowler at the UK Met Office for information and discussions that helped us to develop the HIRLAM 4D-En-Var. Lars Axell inspired us with his pioneering work to develop 4D-En-Var for oceanographic models and discussions with Tomas Landelius and Magnus Lindskog were also helpful. Two anonymous reviewers helped to improve this paper. This work was supported by the Swedish National Space Board (Contract Dnr 105/08:3). The HIRLAM (HIgh Resolution Limited Area Model) project is a joint effort among the weather services in Denmark, Estonia, Finland, France, Iceland, Ireland, Lithuania, the Netherlands, Norway, Spain and Sweden for development of short-range numerical weather prediction techniques.

Edited by: O. Talagrand

References

Bergthorsson, P. and Döös, B.: Numerical weather map analysis, Tellus, 7, 329–340, 1955.

Berre, L.: Estimation of synoptic and mesoscale forecast error covariances in a limited area model, Mon. Weather Rev., 128, 644–667, 2000.

Bishop, C. H., Etherton, B. J., and Majumdar, S. J.: Adaptive Sampling with the Ensemble Transform Kalman Filter. Part I: Theoretical Aspects, Mon. Weather Rev., 129, 420–436, 2001.

Bojarova, J., Gustafsson, N., Johansson, A., and Vignes, O.: The ETKF rescaling scheme in HIRLAM, Tellus, 63A, 385–401, 2010.

Buehner, M.: Ensemble-derived stationary and flow-dependent background-error covariances: Evaluation in a quasi-operational nwp setting, Q. J. Roy. Meteor. Soc., 131, 1013–1043, 2005.

Buehner, M., Houtekamer, P., Charette, C., Mitchell, H., and He, B.: Intercomparison of variational data assimilation and the ensemble Kalman filter for global deterministic NWP. Part I: Description and single-observation experiments, Mon. Weather Rev., 138, 1550–1566, 2010a.

Buehner, M., Houtekamer, P., Charette, C., Mitchell, H., and He, B.: Intercomparison of variational data assimilation and the ensemble Kalman filter for global deterministic NWP. Part II: One-month experiments with real observations, Mon. Weather Rev., 138, 1567–1586, 2010b.

Clayton, A. M., Lorenc, A. C., and Barker, D. M.: Operational implementation of a hybrid ensemble/4D-Var global data assimilation at the Met Office., Q. J. Roy. Meteor. Soc., 139, 1445–1461, 2013.

Cohn, S. E. and Todling, R.: Approximate data assimilation scheme for stable and unstable dynamics, J. Met. Soc. Jpn., 74, 63–75, 1996.

Courtier, P., Thepaut, J.-N., and Hollingsworth, A.: A strategy for operational implementation of 4D-Var, using an incremental approach, Q. J. Roy. Meteorol. Soc., 120, 1367–1387, 1994.

Courtier, P., Andersson, E., Heckley, W., Pailleux, J., Vasiljevic, D., Hamrud, M., Hollingsworth, A., Rabier, F., and Fisher, M.: The ECMWF implementation of three dimensional variational assimilation (3D-Var). Part I: Formulation, Q. J. Roy. Meteorol. Soc., 124, 1783–1808, 1998.

Cuxart, J., Bougeault, P., and Redelsperger, J.-L.: A turbulence scheme allowing for mesoscale and large-eddy simulations, Q. J. Roy. Meteorol. Soc., 126, 1–30, 2000.

Desroziers, G., Camino, J.-T., and Berre, L.: 4DEnVar: link with 4D state formulation of variational assimilation and different possible implementations, Q. J. Roy. Meteorol. Soc., doi:10.1002/qj.2325, in press, 2014.

Evensen, G.: Sequential data assimilation with a nonlinear QG model using Monte Carlo methods to forecast error statistics, J. Geophys Res., 99, 10143–10162, 1994.

Gandin, L. S.: Objective analysis of meteorological fields, Leningrad Hydromet. Press, 1963.

Gustafsson, N.: A review of methods for objective analysis, in: Dynamic Medeorology: Data Assimilation Methods, edited by: Bengtsson, L., Ghil, M., and Källén, E., Springer Verlag, New York, 17–76, 1981.

Gustafsson, N.: Discussion on "4D-Var or EnKF", Tellus A, 59, 774–777, 2007.

Gustafsson, N., Berre, L., Hörnquist, S., Huang, X.-Y., Lindskog, M., Navascués, B., Mogensen, K. S., and Thorsteinsson, S.: Three-dimensional variational data assimilation for a limited area model. Part I: General formulation and the background error constraint, Tellus A, 53, 425–446, 2001.

Gustafsson, N., Huang, X.-Y., Yang, X., Mogensen, K., Lindskog, M., Vignes, O., Wilhelmsson, T., and Thorsteinsson, S.: Four-dimensional variational data assimilation for a limited area model, Tellus A, 64, 14985, doi:10.3402/tellusa.v64i0.14985, 2012

Gustafsson, N., Bojarova, J., and Vignes, O.: A hybrid variational ensemble data assimilation for the HIgh Resolution Limited Area Model (HIRLAM), Nonlin. Processes Geophys., 21, 303–323, doi:10.5194/npg-21-303-2014, 2014.

Hamill, T. M. and Snyder, C.: A hybrid ensemble Kalman filter-3D variational analysis scheme, Mon. Weather Rev., 128, 2905–2919, 2000.

Haugen, J. E. and Machenhauer, B.: A spectral limited-area model formulation with time-dependent boundary conditions applied to the shallow water equations, Mon. Weather Rev., 121, 2618–2630, 1993.

Heemink, A. W., Verlaan, M., and Segers, A. J.: Variance reduced ensemble Kalman filtering, Mon. Weather Rev., 129, 1718–1728, 2001.

Houtekamer, P. and Mitchell, H.: A sequential ensemble Kalman filter for atmospheric data assimilation, Mon. Weather Rev., 129, 123–137, 2001.

Kain, J. S.: The Kain-Fritsch convective parameterization: an update, J. Appl. Meteorol., 43, 170–181, 2004.

Kalman, R. E.: A new approach to linear filter and prediction theory, J. Basic. Eng.-T. ASCE, 82, 35–45, 1960.

Kalman, R. E. and Bucy, R.: New results in linear filtering and prediction theory, J. Basic Eng.-T. ASCE, 83D, 95–108, 1961.

Kalnay, E., Li, H., Miyoshi, T., Yang, S., and Ballabrera-Poy, J.: 4D-Var or ensemble Kalman filter?, Tellus A, 59, 758–773, 2007.

Kawabata, T., Seko, H., Saito, K., Kuroda, T., Tamiya, K., Tsuyuki, T., Honda, Y., and Wakazuki, Y.: An assimilation and forecasting experiment of the Nerima heavy rainfall with a cloud-resolving nonhydrostatic 4-dimensional variational data assimilation system, J. Met. Soc. Jpn., 85, 255–276, 2007.

Kepert, J. D.: Covariance localisation and balance in an ensemble Kalman filter, Q. J. Roy. Meteorol. Soc., 135, 1157–1176, 2009.

Le Dimet, F. X. and Talagrand, O.: Variational algorithms for analysis and assimilation of meteorological observations. Theoretical aspects, Tellus A, 37, 309–322, 1986.

Lewis, J. M. and Derber, J. C.: The use of adjoint equations to solve a variational adjustment problem with advective constraints, Tellus A, 37, 309–322, 1985.

Lindskog, M., Gustafsson, N., Navascués, B., Mogensen, K. S., Huang, X.-Y., Yang, X., Andrae, U., Berre, L., Thorsteinsson, S., and Rantakokko, J.: Three-dimensional variational data assimilation for a limited area model. Part II: Observation handling and assimilation experiments, Tellus A, 53, 447–468, 2001.

Liu, C. and Xiao, Q.: An ensemble-based four-dimensional variational data assimilation scheme. Part III: Antarctic applications with advanced WRF using real data, Mon. Weather Rev., 141, 2721–2739, 2013.

Liu, C., Xiao, Q., and Wang, B.: An ensemble-based four-dimensional varia- tional data assimilation scheme. Part I: Technical formulation and prelim- inary test, Mon. Weather Rev., 136, 3363–3373, 2008.

Liu, C., Xiao, Q., and Wang, B.: An ensemble-based four-dimensional vari- ational data assimilation scheme. Part II: Observing system simulation experiments with advanced research WRF (ARW), Mon. Weather Rev., 137, 1687–1704, 2009.

Lorenc, A.: A global three-dimensional multivariate statistical interpolation scheme, Mon. Weather Rev., 109, 701–721, 1981.

Lorenc, A.: The potential of the ensemble Kalman filter for NWP – a comparison with 4D-Var, Q. J. Roy. Meteorol. Soc., 129, 3183–3203, 2003.

Noilhan, J. and Mahfouf, J.-F.: The ISBA land surface parameterization scheme, Global Planet. Change, 13, 145–159, 1996.

Parrish, D. F. and Derber, J. C.: The National Meteorological Centre's spectral statistical interpolation analysis system, Mon. Weather Rev., 120, 1747–1763, 1992.

Rabier, F., Järvinen, H., Klinker, E., Mahfouf, J. F., and Simmons, A.: The ECMWF operational implementation of fourdimensional variational assimilation. I: Experimental results with simplified physics, Q. J. Roy. Meteorol. Soc., 126, 1143–1170, 2000.

Rasch, P. J. and Kristjánsson, J. E.: A comparison of the CCM3 model climate using diagnosed and predicted condensate parameterizations, J. Climate, 11, 1587–1614, 1998.

Rutherford, I. D.: An operational three-dimensional multivariate statistical objective analysis scheme, GARP WGNE Report No. 11, 98–121, 1976.

Sasaki, Y.: An objective analysis based on the variational method, J. Met. Soc. Jpn., 36, 77–88, 1958.

Savijärvi, H.: Fast radiation parameterization schemes for mesoscale and short-range forecast models, J. Appl. Meterol., 29, 437–447, 1990.

Skamarock, W. C., Klemp, J. B., Dudhia, J., Gill, D. O., Barker, D. M., Duda, M. G., Huang, X.-Y., Wang, W., and Powers, J. G.: A description of the Advanced Research WRF version3, NCAR Tech. Note NCAR/TN-4751 STR, 125 pp., 2008.

Tippett, M. K, Anderson, J. L., Bishop, C. H, Hamill, T. M., and Whitaker, J. S.: Ensemble square root filters, Mon. Weather Rev., 131, 1485–1491, 2003.

Trémolet, Y.: Accounting for an imperfect model in 4D-Var, Q. J. Roy. Meteorol. Soc., 132, 2483–2504, 2006.

Undén, P., Rontu, L., Järvinen, H., Lynch, P., Calvo, J., Cats, G., Cuxart, J., Eerola, K., Fortelius, C., Garcia-Moya, J. A., Jones, C., Lenderlink, G., McDonald, A., McGrath, R., Navascues, B., Woetman Nielsen, N., Odegaard, V., Rodriguez, E., Rummukainen, M., Rööm, R., Sattler, K., Hansen Sass, B., Savijärvi, H., Wichers Schreur, B., Sigg, R., The, H., and Tijm, A.: HIRLAM-5 Scientific Documentation, HIRLAM-5 project, c/o Per Undén, SMHI, SE-60176 Norrköping, Sweden, 2002.

Wang, X., Barker, D. M., Snyder, C., and Hamill, T. M.: A hybrid ETKF-3DVar data assimilation scheme for the WRF model. Part i: Observing system simulation experiment, Mon. Weather Rev., 136, 5116–5131, 2008a.

Wang, X., Barker, D. M., Snyder, C., and Hamill, T. M.: A hybrid ETKF-3DVar data assimilation scheme for the WRF model. Part II: Real observation experiments, Mon. Weather Rev., 136, 5132–5147, 2008b.

13

Lagrangian descriptors and the assessment of the predictive capacity of oceanic data sets

C. Mendoza[1], A. M. Mancho[2], and S. Wiggins[3]

[1]ETSI Navales, Universidad Politécnica de Madrid, Av. Arco de la Victoria 4, 28040 Madrid, Spain
[2]Instituto de Ciencias Matemáticas, CSIC-UAM-UC3M-UCM, C/ Nicolás Cabrera 15, Campus Cantoblanco UAM, 28049 Madrid, Spain
[3]School of Mathematics, University of Bristol, Bristol BS8 1TW, UK

Correspondence to: A. M. Mancho (a.m.mancho@icmat.es)

Abstract. We use a recently developed Lagrangian transport tool, *Lagrangian descriptors*, to compare the transport properties of data distributed by AVISO and numerical simulations obtained from the HYCOM model in the Yucatán–Florida current system. Our data correspond to the months from June through August 2010. Structures obtained from HYCOM are noisier than those from AVISO; however, both AVISO and HYCOM succeed in identifying Lagrangian structures that influence the paths of drifters, such as eddies, currents, lobes, etc. We find evidence in which AVISO gives the positions of important hyperbolic trajectories in a manner that is inconsistent with the trajectories of the drifters, while for the same examples HYCOM succeeds to this end.

1 Introduction

This research is concerned with the use of theoretical and computational techniques coming from dynamical systems theory that are used to assess the performance of gridded velocity fields originating from realistic ocean models and data sets. The development of the Lagrangian aspects of fluid transport in the context of oceanic flows has received a great deal of motivation from the popularity and from the successes of the dynamical systems approach to fluid transport in the 1980s and 1990s. Background and reviews can be found in Jones and Winkler (2002), Mancho et al. (2006b), Wiggins (2005), and Samelson and Wiggins (2006). Hyperbolic trajectories and their stable and unstable manifolds provide a generalization to the time-dependent setting of saddle-type stagnation points and their separatrices. In general, invariant manifolds are material curves (or surfaces in three dimensions; see for instance Bettencort et al., 2012; Froyland et al., 2012) that form a *Lagrangian skeleton* that governs the spatiotemporal evolution of the transport of fluid particles. Methods for computing hyperbolic trajectories and their stable and unstable manifolds for velocity fields defined as finite time data sets can be found in Miller et al. (1997), Malhotra and Wiggins (1998), Haller (2000), Mancho et al. (2003), Mancho et al. (2004), and Branicki and Wiggins (2009). Finite time Lyapunov exponents (FTLEs) and finite size Lyapunov exponents (FSLEs) have proved useful for determining proxies for invariant manifolds (see, e.g., Pierrehumbert, 1991; Haller and Yuan, 2000; Haller, 2001a; Haller, 2001b; Shadden et al., 2005; Lekien et al., 2007; and d'Ovidio et al., 2004). In this paper we use a recently developed Lagrangian tool, referred to as a *Lagrangian descriptor* (Mancho et al., 2013), that is described in detail in Sect. 5.

Nowadays, a huge amount of "realistic" ocean data supplied by direct observations is available. Surface currents are measured either through high-frequency coastal radars and satellite altimeters or, alternatively, through measurements with in situ devices (i.e., acoustic Doppler current profilers, ADCPs, mounted on vessels, fixed moorings, gliders, drifters). Many of these data are assimilated into numerical oceanic models, which in turn provide outputs of velocity and other scalar fields in full 3-D domains. Obtaining accurate representations of oceanic currents is of tremendous importance not only for a better management of catastrophic events, such as, for instance, contaminant dispersion, but also for their direct impact on marine life and fishing banks, for optimizing shipping routes, for search and rescue operations, etc.

Lagrangian techniques have been applied to different oceanic data sets for purposes related to the goals of this work. For instance, the correlation between Lagrangian structures obtained from radar data in different ocean areas is discussed by Shadden et al. (2009) and Haza et al. (2010). Kuznetsov et al. (2002) have examined the role played by Lagrangian structures in the Gulf of Mexico. In particular, they have analyzed data from the Colorado University Princeton Ocean Model and have compared the geometry of the dynamics with simultaneous trajectories of drifters measured in an independent way. Indirectly the agreement between the drifter evolution and the Lagrangian patterns serves as a confirmation of the model quality. The ability of coastal models to capture the true Lagrangian structure of ocean flows has been discussed also in Haza et al. (2007) by using FSLEs, and they concluded that this model-derived Lagrangian metric was capable of predicting the general fate of observed drifter pairs. Griffa et al. (2013) point out that accuracy of Lagrangian diagnosis is directly tied to the accuracy of the Eulerian velocity and perform experiments that prove that the Lagrangian diagnosis relying on Eulerian fields from accurate operational coastal models can successfully guide real-time drifter launch strategies. In turn hyperbolic trajectories detected by Lagrangian techniques have guided drifter launch strategies for a better Lagrangian assimilation (Molcard et al., 2006).

Lagrangian techniques have been used to provide feedback that improve numerical models. For instance Titaud et al. (2011) implement an assimilation method that uses correspondence between advected passive scalars and FTLEs and vectors. Nencioli et al. (2011) propose an iterative method that uses satellite data, ADCP and drifters to improve the detection of Lagrangian coherent structures (LCSs). Nencioli et al. (2013) use FTLE to estimate horizontal eddy diffusivity, a parameter required by numerical simulations.

Lagrangian techniques have also been used in the past to compare predictions based on different data sets. Beron-Vera et al. (2010) evaluate the surface ocean mixing predicted by two different altimetry AVISO products, REF and UPD, constructed with, respectively, two and four satellite altimeters. Their analysis is based on a mixing characterization obtained from statistics based on FTLE. They also compare the quality of the data based on the agreement between a drifter trajectory and the Lagrangian skeleton supplied by the FTLE. With their analysis they improve the previous purely Eulerian analysis of the data (Pascual et al., 2006); however, a weakness of their approach is that FTLEs develop spurious structures that have no Lagrangian interpretation (Branicki and Wiggins, 2010; Mancho et al., 2013), and this could have biased their mixing diagnosis.

In this article we analyze the correlation between drifter trajectories and Lagrangian structures in the Gulf of Mexico obtained from two data sets: one comes from a numerical simulation and the second from satellite altimetry. The ocean model that we have considered is the HYCOM model (Bleck, 2002; Chassignet et al., 2007); as for altimeter data, we consider the AVISO product UPD, which is reconstructed from four satellites. In order to perform our analysis, we use a recently developed Lagrangian tool (Mendoza and Mancho, 2010; Mendoza et al., 2010; Mancho et al., 2013), the so-called M function or Lagrangian descriptor. We expect that the sharp outputs obtained from this tool as confirmed in de la Cámara et al. (2012) and Mancho et al. (2013) will help us to obtain accurate conclusions from our analysis. The agreement between the drifters and the obtained geometry serves as a benchmark for the predictive capacity of the data sets.

This paper is organized as follows. Section 2 describes the data sets used to perform the analysis, which includes the details of the drifters that we have considered. Section 3 provides a description of the ocean area under study. Section 4 describes the construction of a dynamical system from the data set. Section 5 describes the Lagrangian descriptors used to reveal the geometry of the dynamics. Section 6 describes the results, and finally Sect. 7 discusses the conclusions.

2 Data sets

In this section we give a brief description of the types of data used to study the flow in the particular regions that we consider.

2.1 HYCOM

The Hybrid Coordinate Ocean Model (HYCOM) (Bleck, 2002; Chassignet et al., 2007) is part of the US Global Ocean Data Assimilation Experiment (GODAE) within the National Ocean Partnership Program (NOPP). This model uses the Navy Coupled Ocean Data Assimilation (NCODA) system: satellite sea-surface height and temperature data are assimilated, and the model is forced by surface winds and air–sea fluxes. Specific details can be found at http://hycom.org/dataserver/goml0pt04. The output of this model has been previously used to study transport processes in the Gulf of Mexico by Olascoaga (2010), Mezic et al. (2010), and Sulman et al. (2013).

In particular, we have used experiment 31.0 (http://hycom.org/dataserver/goml0pt04/expt-31pt0), which provides data from 2009 to the present. The experiment has 1/25 degree equatorial resolution and latitudinal resolution of 1/25°cos (lat). This represents approximately a grid size at mid-latitudes of 3.5 km. The experiment uses 20 layers along the vertical coordinate and is carried out in a domain in which latitudes range from 18.09 to 31.96° and the longitude ranges from 98.0 to 76.4° W. The data are provided hourly.

Our calculations are performed in the fourth layer, which corresponds to a depth of 30 m. This is an appropriate choice, as it mitigates excess noise of the data in the surface layer.

A similar depth choice for this model is used, for instance, in Sulman et al. (2013) or for CUPOM (Colorado University Princeton Ocean Model) in Branicki and Kirwan (2010). In the latter work the authors show how the resulting Lagrangian structure is close to 2-D surfaces that extend nearly vertically into the water column, thus confirming that the choice of a deeper layer just affects the noise on the observed structures. We also note that continent shapes at 30 m depth are not the same as those at 0 m depth, for which we have more experience.

2.2 AVISO

Surface velocity fields are obtained from satellite altimetry and produced by Ssalto/Duacs and distributed by AVISO, with support from Cnes (http://www.aviso.oceanobs.com/duacs/).

The data are given on a spatial grid of 1080×915 points (longitude/latitude). The latitude ranges from 82° S to 81.9746° N using a Mercator projection, and the longitude ranges from 0 to 359.667° using a uniform grid. The spatial precision is thus 1/3° at the Equator and is provided with daily frequency.

AVISO data are nowadays a standard source for analysis of transport processes in the ocean, not only in the area of interest in our study in the Gulf of Mexico (Andrade-Canto et al., 2013) but also in many other areas all around the world (Prants et al., 2014).

2.3 Drifters

Surface drifter tracks are obtained from data distributed by the Global Drifters Program (GDP), from the National Oceanic and Atmospheric Administration (NOAA) and Atlantic Oceanographic and Meteorological Laboratory (AOML) (http://www.aoml.noaa.gov/phod/dac/index.php) for the time period and regions corresponding to our simulations that we carry out with HYCOM and AVISO data. Each drifter has an identification number having five digits, which allows for the identification of different drifters in our simulations. The sampling of the drifter position is once every 6 hours.

3 The Gulf of Mexico

Our focus is the study of the Gulf of Mexico, in the area between latitudes 22–28° N and longitudes 80–91° W, from 11 June through July to 7 August in the year 2010. This choice corresponds to the months just after the Deepwater Horizon oil spill in the year 2010, although this work does not aim to achieve any conclusion regarding those events.

The Yucatán–Florida current system is in this region, also known as the Loop Current (Forristall et al., 1992; Andrade-Canto et al., 2013; Sulman et al., 2013). This current enters the gulf through the Yucatán channel, then circulates clockwise and exits the gulf through the Straits of Florida, flowing eventually into the Gulf Stream. During the period under consideration, our data show that this current is pinched off forming first a large anticyclonic loop, which survives until 24 June. After this date, the big loop breaks into several eddies, which become weaker and weaker by the end of the period of interest.

Our study attempts to understand the consistency between the evolution of drifters during this period and the Lagrangian features provided by the two data sets. In particular, agreement in the following aspects will be analyzed: the role played by manifolds that represent barriers to transport, drifter dispersion produced by hyperbolic trajectories, drifter confinement produced by eddies, and material ejection through lobe dynamics.

4 The dynamical system

We are interested in the study of transport due solely to advection. In this case a particle evolves according to

$$\frac{\mathrm{d}\boldsymbol{x}}{\mathrm{d}t} = \boldsymbol{v}(\boldsymbol{x},t), \boldsymbol{x} \in \mathbb{R}^n, t \in \mathbb{R}. \tag{1}$$

In our geophysical applications, we consider two-dimensional motion of particles, i.e., $n = 2$. We assume that $\boldsymbol{v}(\boldsymbol{x},t)$ is C^r ($r \geq 1$) in $\boldsymbol{x}$ and continuous in t. In our study, the velocity field $\boldsymbol{v}$ given in Eq. (1) is defined as a data set, provided by observations or simulations and has been described in the previous section.

The equations of motion that describe the horizontal evolution of particle trajectories on a sphere are

$$\frac{\mathrm{d}\phi}{\mathrm{d}t} = \frac{u(\phi,\lambda,t)}{R\cos(\lambda)}, \tag{2}$$

$$\frac{\mathrm{d}\lambda}{\mathrm{d}t} = \frac{v(\phi,\lambda,t)}{R}. \tag{3}$$

Here the variables (ϕ, λ) are longitude and latitude; u and v, respectively, represent the eastward and northward components of the velocity field. The particle trajectories must be integrated using Eqs. (2)–(3); since information is provided solely on a discrete space–time grid, the first issue to deal with is that of interpolation. We have daily or hourly maps of the velocity field, and these are too coarse of a time grid to provide a time step for the integration of particle trajectories. However, this frequency sampling is adequate in the sense that changes of the velocity field below that resolution are smooth enough to be adequately captured by an interpolator. Days are a typical timescale for the system (Eqs. 2–3), and this is the unit of time in which results are reported. A recent paper by Mancho et al. (2006a) compares different interpolation techniques for tracking particle trajectories. Bicubic interpolation in space (Press et al., 1999) and third-order Lagrange polynomials in time are shown to provide a computationally efficient and accurate method. In our calculations we replace the bicubic spatial interpolation in space by the spline interpolation supplied by MATLAB as we have verified that it provides similar accuracy.

5 Lagrangian descriptors

Solutions of dynamical systems are qualitatively described according to Poincaré's idea to seek geometrical structures in the phase space that can be used to organize trajectories schematically into distinct regions corresponding to qualitatively different types of trajectories.

In time-independent systems, i.e., those for which Eq. (1) does not depend explicitly on time, fixed points are essential for describing the solutions geometrically. This is so because stable and unstable manifolds of hyperbolic fixed points act as separatrices that divide the phase portrait into regions in which particles have different dynamical fates. To achieve this geometrical representation in time-dependent aperiodic dynamical systems, special hyperbolic trajectories have been distinguished, which have been called distinguished hyperbolic trajectories (Ide et al., 2002; Madrid and Mancho, 2009), and the stable and unstable manifolds of such trajectories have been successfully used to describe transport in geophysical flows (Mancho et al., 2008; Branicki et al., 2011; Mendoza et al., 2010).

In this work our geometrical representation is built with Lagrangian descriptors (LDs). LDs are based on a function referred to as M that was proposed in Madrid and Mancho (2009) as a building block of the definition of *distinguished trajectories*. Later in Mendoza and Mancho (2010), this function was introduced as a Lagrangian descriptor, i.e., a function that provides a global dynamical picture of the geometrical structures governing transport in arbitrary time-dependent flows. Lagrangian descriptors were extended and theoretically justified in simple examples in Mancho et al. (2013) and Mendoza and Mancho (2012). In this article we will focus on the first approach to LDs given by a function M that measures the Euclidean arc length of the curve outlined by a trajectory passing through $\boldsymbol{x}^*$ at time t^*. The trajectory is integrated from $t^*-\tau$ to $t^*+\tau$. This is mathematically expressed as follows: for all initial conditions $\boldsymbol{x}^*$ in an open set $\mathcal{B} \in \mathbb{R}^n$, at a given time t^*, the Lagrangian descriptor is a function $M(\boldsymbol{x}^*, t^*)_{v,\tau}\colon (\mathcal{B}, t) \rightarrow \mathbb{R}$ given by

$$M(\boldsymbol{x}^*, t^*)_{v,\tau} = \int_{t^*-\tau}^{t^*+\tau} \sqrt{\sum_{i=1}^{n} \left(\frac{\mathrm{d}x_i(t)}{\mathrm{d}t}\right)^2}\, \mathrm{d}t. \tag{4}$$

Here $x_1(t), x_2(t), \ldots, x_n(t)$ are the components in $\mathbb{R}^n$ of a trajectory $\boldsymbol{x}(t)$. The function M is defined for dynamical systems in arbitrary dimension n, but for our current application we will take $n = 2$. It depends on τ and also, of course, on the velocity field $\boldsymbol{v}$.

For small τ the output provided by M is smooth, almost without structure and resembling that of Eulerian currents. This is confirmed from Fig. 1a, which shows the evaluation of M on AVISO in the Loop Current region in the Gulf of Mexico for $\tau = 3$ days. Figure 1b and c confirm that for increasing τ the function M loses regularity along lines, which, from the numerical point of view, have been systematically confirmed to be aligned with the stable and unstable manifolds of hyperbolic trajectories (Mendoza and Mancho, 2010; Mendoza et al., 2010; de la Cámara et al., 2012, 2013; Mancho et al., 2013). In Fig. 1b the lines aligned with the stable and unstable manifolds of a hyperbolic point are clearly seen, and further details are distinguished in Fig. 1c. The Lagrangian features are not provided by the specific values taken by M (M has length dimension and the units are those of the plane on which the integration of the trajectory is performed – see Eqs. 2–3), but the Lagrangian information is contained *in the structure* of M; for this reason the color scale in the figures is not included. To make an interpretation of M, it is sufficient to know that the red color expresses the maximum arc-length values, obtained for particles moving rapidly, as they are placed in jets, and blue color are minimum arc-length values obtained in regions where particles move calmly.

There is no a priori knowledge on how long the time integration interval τ should be taken in order to highlight Lagrangian structures. It has been observed that it depends on the data set under study. On the other hand, the analytical arguments given for a simple example in Mancho et al. (2013) suggest that, in hyperbolic regions with large expansion or contraction rates, the formation of these singular features could be achieved at shorter τ. Figure 1 suggests $\tau = 15$ days as an appropriate choice for the AVISO data.

The increasing complexity of M versus τ is expected from the nature of M, since it is reflecting the history of initial conditions on open sets, and in highly chaotic systems this history is expected to be more tangled for longer time intervals. Thus τ values larger than 15 days could have been taken, which would have made visible many details of the manifold structure, as reported for instance in Mendoza and Mancho (2010) and Mancho et al. (2013). This is consistent with what is obtained from the direct computation of manifolds (Mancho et al., 2003, 2004) for which the longer the integration time is, the larger the pieces of manifolds are. At the moment there exist no formal proof that guarantees that these singular features are truly advected by the flow, although some analytical arguments are discussed in this respect in Mancho et al. (2013).

Figure 2 shows the output on the same region for the same τ selection on the HYCOM data. Again it is observed that for increasing τ more and more Lagrangian structure is observed. The position of the hyperbolic point here is shifted with respect to that observed in the AVISO data. Also the manifold structure for a given τ is much richer than in the AVISO data. In fact for $\tau = 15$ the mesoscale structure is

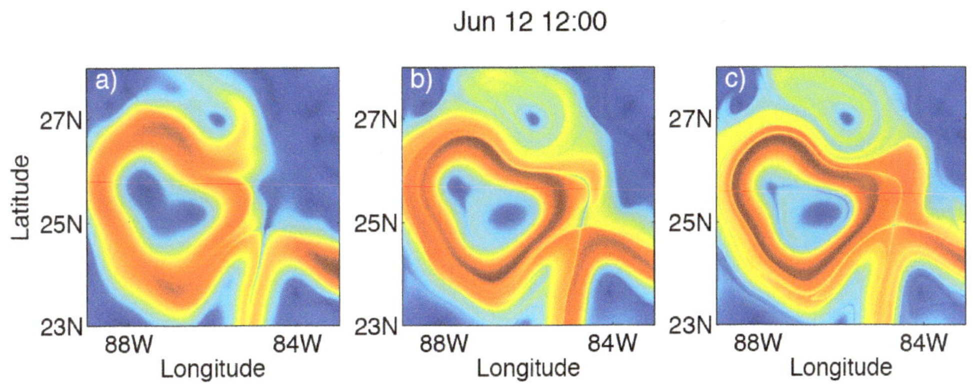

Figure 1. Evaluation of the M function in the Gulf of Mexico for AVISO data. **(a)** $\tau = 3$ days; **(b)** $\tau = 9$ days; **(c)** $\tau = 15$ days. The position of the manifolds is highlighted in this figure.

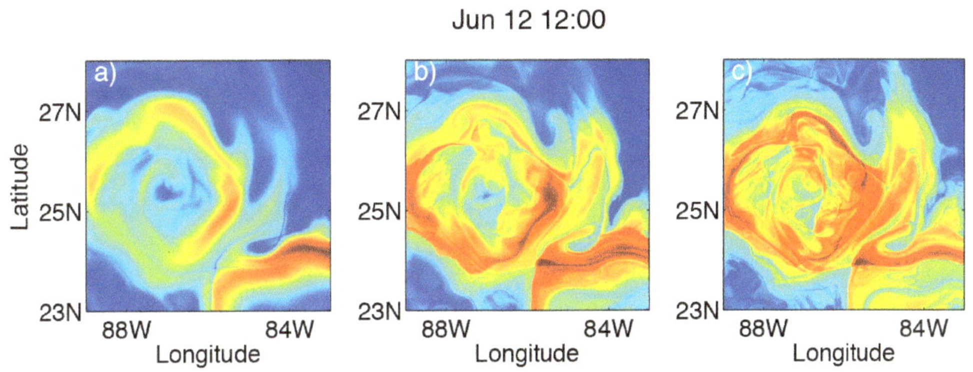

Figure 2. Evaluation of the M function in the Gulf of Mexico for HYCOM data. **(a)** $\tau = 3$ days; **(b)** $\tau = 9$ days; **(c)** $\tau = 15$ days.

rather blurred when compared to $\tau = 9$. This confirms that the appropriate τ, which allows the manifold visualization, depends on the data. In the following sections we will produce M maps with $\tau = 9$ days for the HYCOM data as that is enough to highlight the mesoscale structure.

The absence of regularity of M at large τ highlights the position of invariant manifolds. These contrasts correspond to transport barriers as confirmed in Fig. 3. The direct computation of the stable and unstable manifolds is displayed side by side with the structure of M, and the coincidence is evident. This calculation is done following the method developed by Mancho et al. (2003, 2004, 2006b) and Mendoza and Mancho (2012), and as explained this is done by integrating particle trajectories. In particular, stable and unstable manifolds are, respectively, composed of particle trajectories that at a given time are placed in a small segment along the stable or unstable subspaces of a given distinguished hyperbolic trajectory (DHT). Manifold structures displayed in Fig. 3c have been obtained after integrating backwards and forwards in time using AVISO data for a period of 15 days, and those in Fig. 3d have been obtained after integrating backwards and forwards in time with HYCOM data for a period of 9 days. The direct computation of manifolds displays a less complex appearance than those obtained from the M function because, as noted in Mendoza and Mancho (2012), it is done for selected DHTs that may leave out many other DHTs in the neighborhood, while M accounts for all all stable and unstable manifolds from all possible DHTs in the vicinity.

The presence of a rich manifold structure for HYCOM data when compared to AVISO is due to the presence of small "noisy" structures in the velocity fields themselves for the HYCOM output, which is absent in AVISO. "Noisy" structures in the HYCOM velocity field are possible due to the smaller grid size used in this case. However, below the resolution grid these "noisy" structures are assumed to follow the bicubic interpolation, which is smooth within each cell and C^1 at boundaries of the grid. Figure 4 confirms this point. In this figure we show the eastwards u and northwards v velocity components along the longitude 86° W on 12 June 2010. The AVISO velocity (in blue) is smoother than the HYCOM velocity (in black).

Hernández-Carrasco et al. (2011) have discussed the reliability and dependence on the grid size of a Lagrangian diagnosis based on FSLE. In particular they have considered two types of grid sizes. One of the sizes is related to the grid in which the velocity field is given (called the V-grid). This grid is also present in our case. Changing the size here is related to introducing slight modifications in the underlying velocity

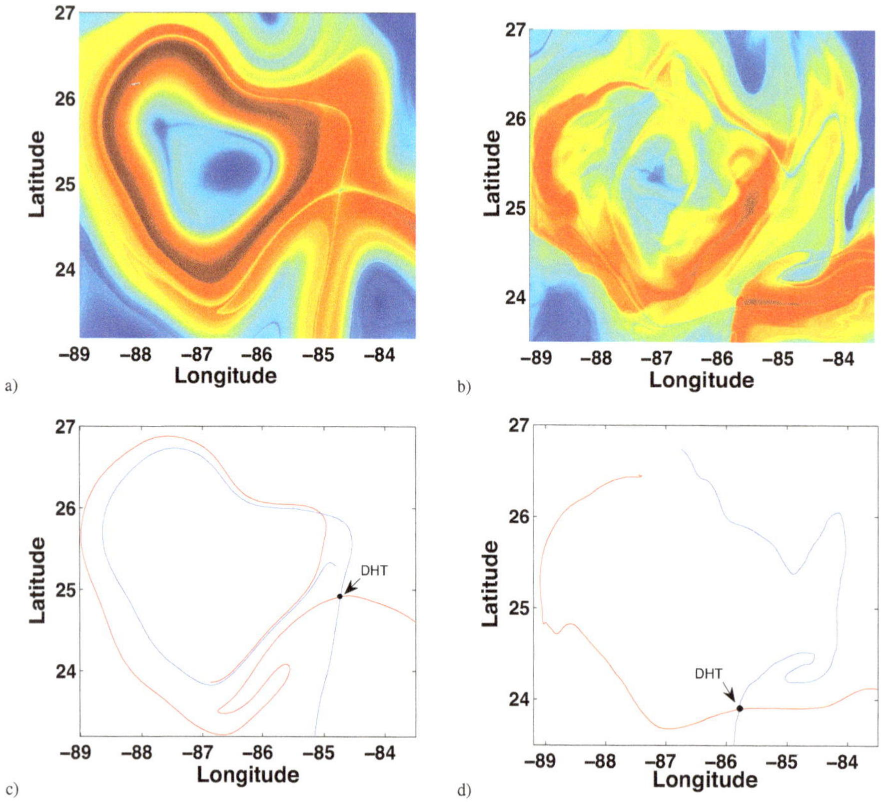

Figure 3. Evaluation of the M function and visualization of the manifold structure on 12 June 2010 at 12:00 UTC. **(a)** Evaluation of M with AVISO for $\tau = 15$ days; **(b)** M on HYCOM for $\tau = 9$ days; **(c)** manifolds with AVISO computed for 15 days; **(d)** manifolds with HYCOM computed for 9 days.

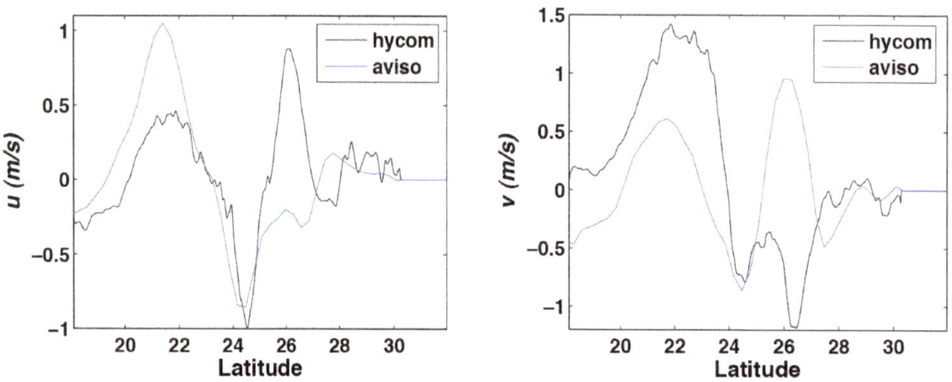

Figure 4. Velocity components along the longitude 86° W for HYCOM (black) and AVISO (blue) data on 12 June 2010: **(a)** eastwards velocity component and **(b)** northwards velocity component.

field defining the system (Eqs. 2–3), as this continuous dynamical system is produced from an interpolation that is not identical for each grid choice. Lagrangian structures are expected to be robust versus slight perturbations to the velocity field (see for instance Haller, 2002) given that chaotic saddles are structurally stable objects (although this has not been proven for aperiodically time-dependent velocity fields, it is a reasonable conjecture). Consistently Lagrangian descriptors have been shown to produce similar outputs for slightly modified velocity fields (Mendoza and Mancho, 2010, 2012). A second grid size effect discussed by Hernández-Carrasco et al. (2011) – the F-grid – is related to the adjustment of a parameter, δ, that measures the initial separation of trajectories on which FSLEs depend. However, this does not influence our results as LDs do not depend on such a parameter.

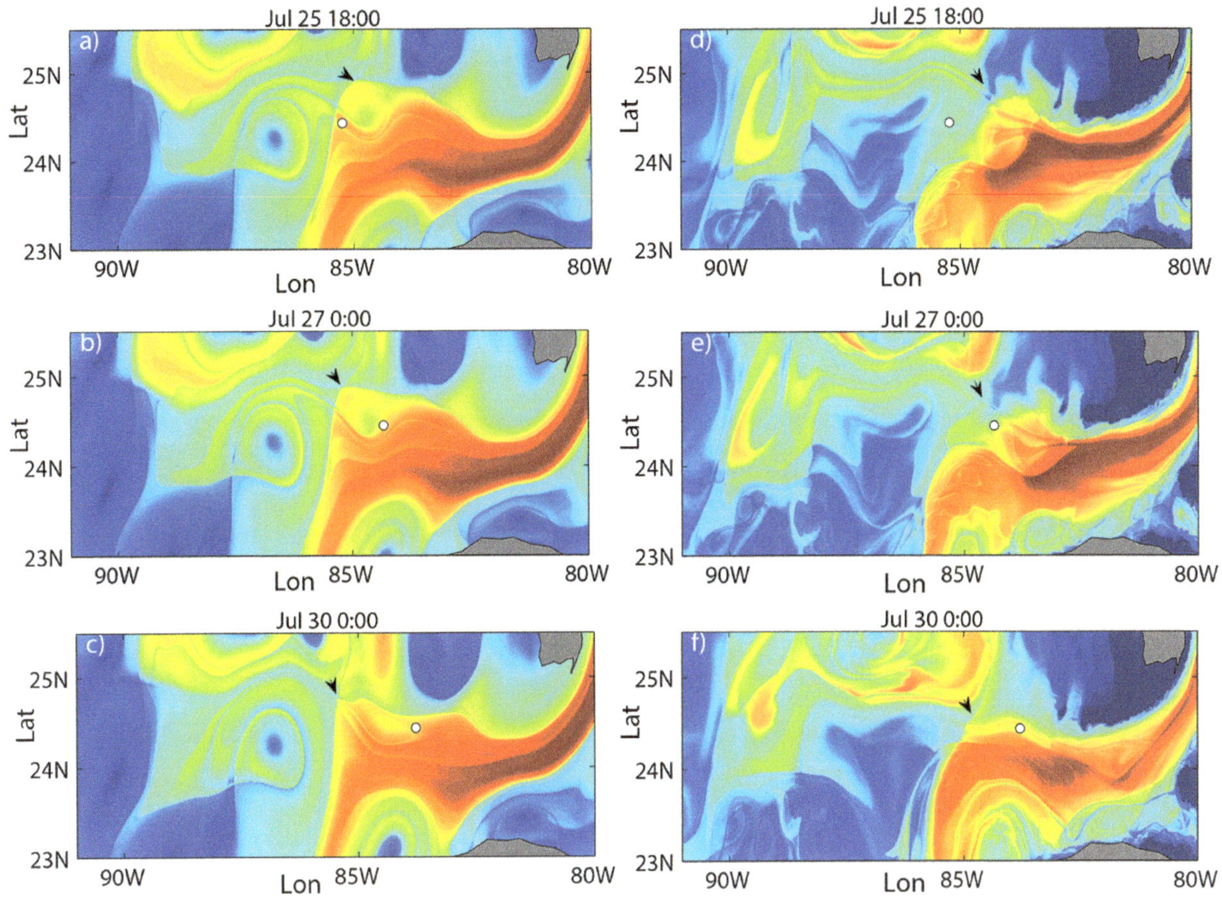

Figure 5. Sequence showing the evolution of the Lagrangian structures with AVISO data from **(a)** to **(c)** and with HYCOM data and from **(d)** to **(f)** for 25, 27 and 30 July respectively. Time is given as Universal Coordinated Time (UTC). The drifter "98945" is overlapped with these structures. It is observed how it is injected within the Gulf Stream through a lobe. The position of the hyperbolic trajectory is highlighted in each frame with the arrow.

Mancho et al. (2013) have shown, in the context of simple and controlled analytical examples, that Lagrangian descriptors present numerous advantages versus other Lagrangian tools common in the literature such as FTLEs or finite time averages. Rempel et al. (2013) have extended the comparison with FTLE to the context of 3-D turbulent flows. Some of the advantages found are listed next; LDs are computationally cheaper than FTLEs; on data sets the function M is less sensitive to the quality of the interpolation of the vector field. LDs have not been reported to produce ghost structures (i.e., false Lagrangian patterns), while there exists substantial literature reporting these issues for FTLE (Branicki and Wiggins, 2010; Farazmand and Haller, 2012; Rempel et al., 2013; Mancho et al., 2013). LDs are easier to program because they do not require the adjustment of any parameter except τ, which is directly related to the detail required in the manifolds. The maps of LDs systematically display sharper plots that facilitate the identification of Lagrangian coherent structures and lobe dynamics. For instance de la Cámara et al. (2012) identify routes of transport that cross the Antarctic polar jet. This has not been possible to obtain from FTLEs since they provide a blurred pattern, full of spurious structures. De la Cámara et al. (2013) provide a picture of the Antarctic polar jet that finds Rossby wave breaking evidence in the interior of the vortex, while these features have been unnoticed with other Lagrangian approaches. All these reasons justify our choice.

6 Results

In this section we describe the ocean Lagrangian skeleton and correlate it with the paths of several drifters in the Gulf of Mexico during the period under study. We compare side by side the outputs obtained from the HYCOM model and the AVISO data sets. At this point, it is not always clear which data set is more representative of the state of the ocean. Our analysis methodology is based on assuming that consistency between different data sources is a good metric for data quality. In particular, a sign of consistency is the agreement between the Lagrangian features and the paths followed by the

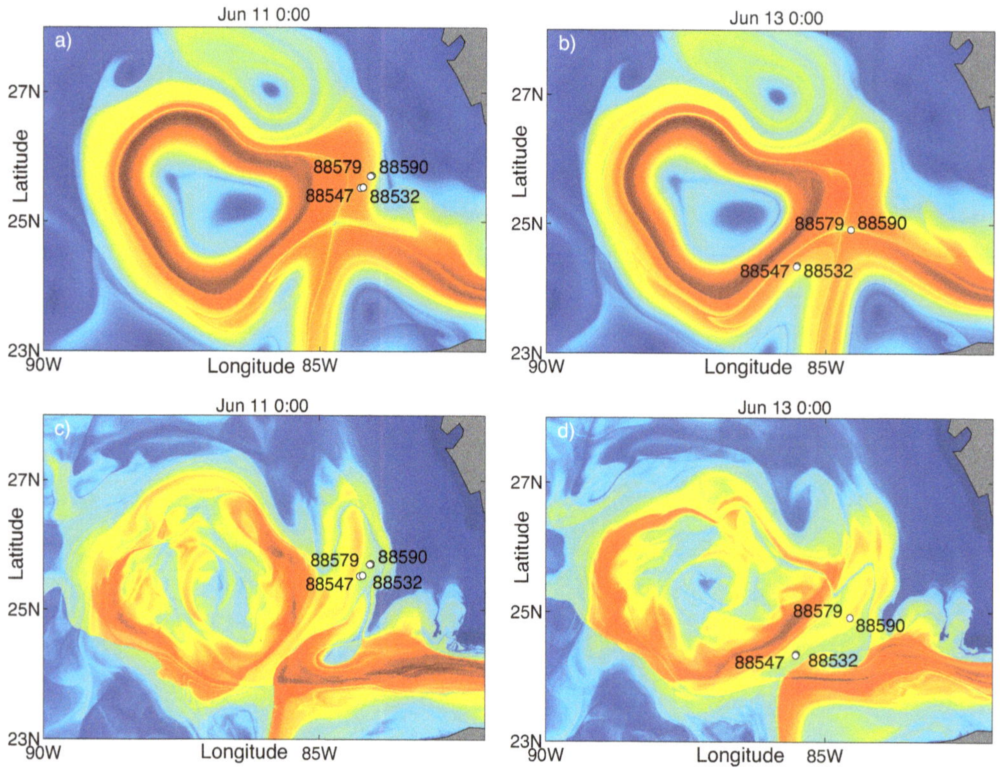

Figure 6. Sequence that shows the evolution of the Lagrangian structures with AVISO – **(a)** and **(b)** – and HYCOM – **(c)** and **(d)** – data. Drifters "88532", "88547", "88590", and "88579" are overlapped with these structures. The left column displays results for 11 June and the right one for 13 June.

drifters. Our focus is on this type of consistency because a comparison of individual trajectories of in situ drifters with the corresponding simulated trajectories of the respective velocities is not going to be close. This something expected in a chaotic system and has been reported for instance by Mancho et al. (2006a) and de la Cámara et al. (2010). Despite the fact that individual trajectories do not evolve close to each other, the underlying Lagrangian skeleton can be close or similar. Chaotic saddles are structurally stable objects and thus Lagrangian structures should be robust versus slight perturbations to the velocity field. For these reasons we base our analysis on Lagrangian features and not on individual trajectories.

During the period of interest, 36 drifters were tracked in the region of study. Our focus is on analyzing the agreement on typical dynamical features in the velocity field and the drifter motion. In the period under study, some drifters leave the domain under consideration rapidly, which is consistent with the observed behavior of the Gulf Stream. This is so in both HYCOM and AVISO data. In particular drifters 98944, 98934, 98942, 98904, 98936, 98941, 98910, and 98945 follow this fate. Figure 5 shows how the drifter 98945 is injected into the main stream through a dynamical object referred to as a "lobe". Lobes are water masses surrounded by a piece of a stable manifold and a piece of an unstable manifold. Close to hyperbolic points, lobes evolve forming filaments as described for instance in Mancho et al. (2008). This elongating structure is observed in Fig. 5 both for HYCOM and AVISO with the drifter inside it while it is swept along the current. The presence of the hyperbolic point is highlighted in the sequence.

On the first day of our study, 11 June, the Loop Current is pinched off forming a large anticyclonic Loop as reported by Sulman et al. (2013). Consistently, this scenario is observed both from AVISO and HYCOM data as observed from Fig. 6. It is noted though that the position of the main hyperbolic point is different at each data set. Figure 6 shows that the evolution of drifters 88532, 88547, 88590, and 88579 is consistent with the location of the hyperbolic trajectory in HYCOM but not in AVISO. In this latter case if drifters had followed the Lagrangian features highlighted in the figure, they should have evolved bending towards the east without crossing the invariant manifolds that intersect at that position. This inconsistency is not observed in the HYCOM data as the hyperbolic trajectory is shifted towards the southeast. Other examples exist during the period under study that confirm the efficiency of HYCOM data to correctly position the hyperbolic trajectory in a manner that justifies drifter's dispersion.

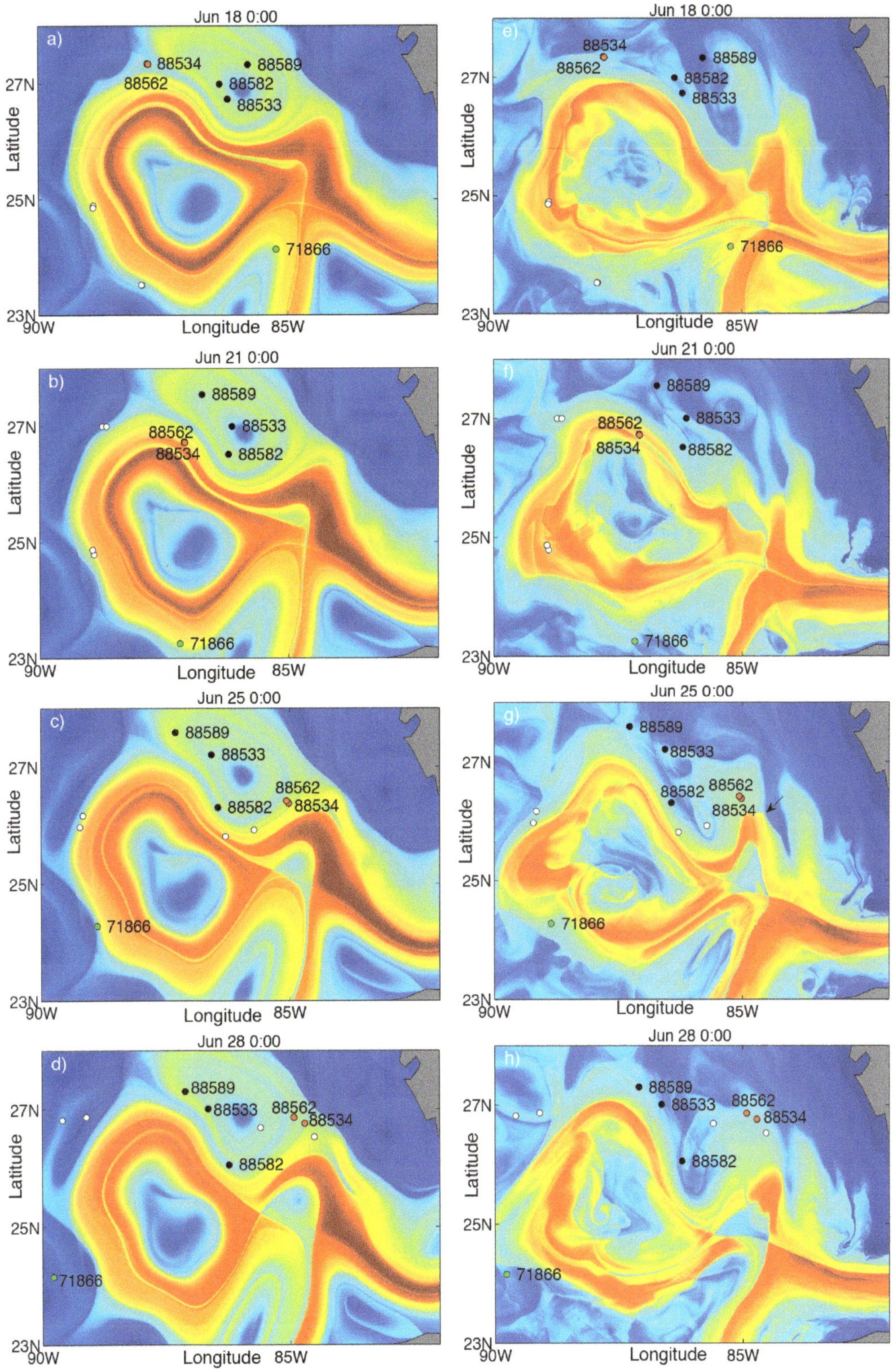

Figure 7. Sequence that shows the evolution of the Lagrangian structures with AVISO – **(a)** and **(d)** – and HYCOM – **(e)** and **(h)** – data. Drifters "88532", "88547", "88590", "88579", already depicted in Fig. 6, are represented without labels and in white as before. Drifters "88562" and "88534" are overlapped in red and drifter "71866" in green. Drifters "88589", "88533" and "88582" are in black. The first row displays results for 18 June, the second for 21 June, the third row for 25 June and the fourth row for 28 June. An arrow highlights the position of a hyperbolic trajectory in **(g)**.

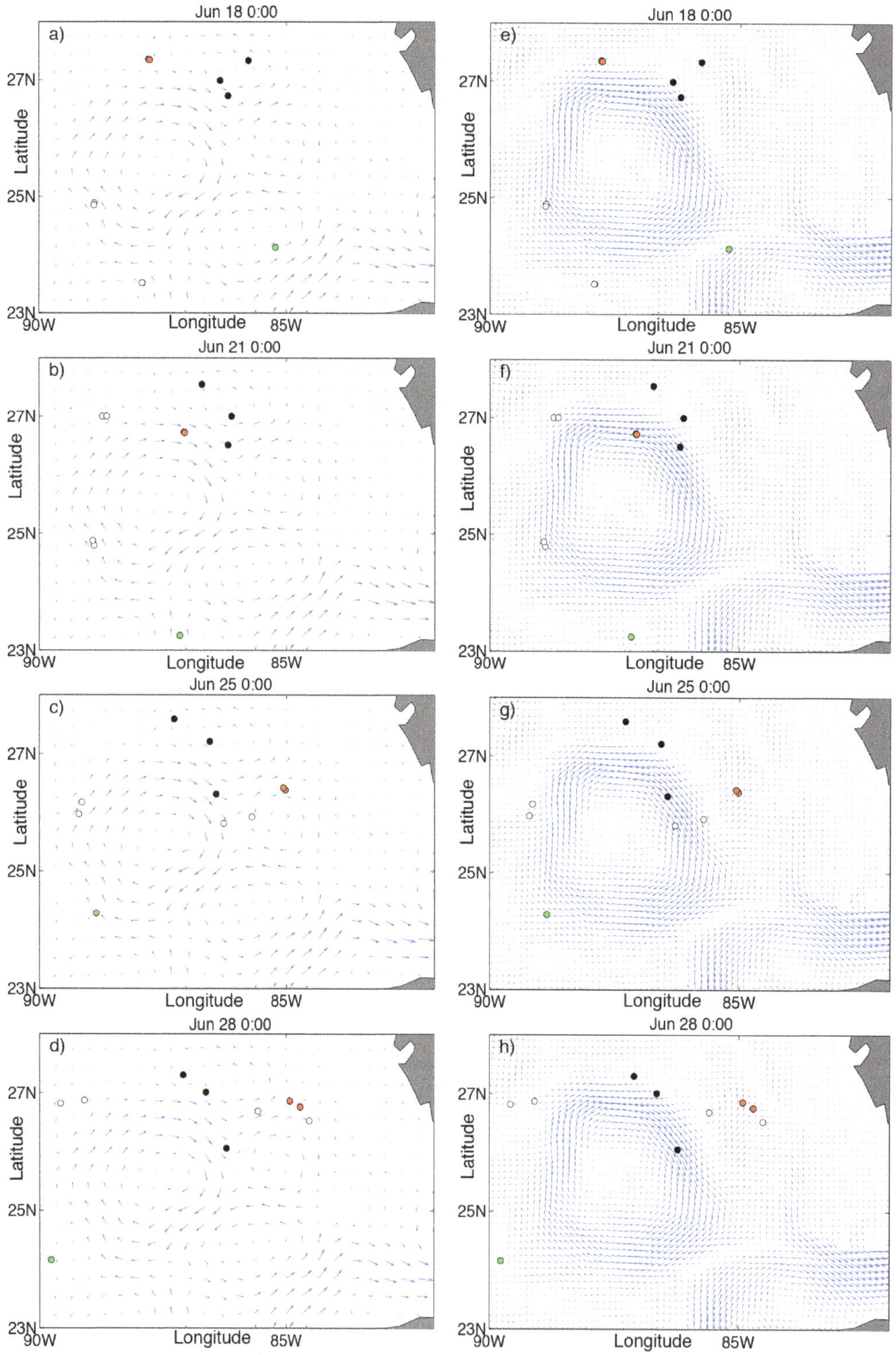

Figure 8. Sequence that shows the evolution of the Eulerian velocities from AVISO data from **(a)** to **(d)** and from HYCOM data from **(e)** to **(h)**. Drifters of Fig. 7 are depicted with the same color code but without identifiers. It is hard to understand drifter evolution directly from these Eulerian fields.

The above drifters, together with drifter 71866, that joins the Loop Current from the south, circulate clockwise in the stream consistently with the pictures provided both by HYCOM and AVISO. This is confirmed in Fig. 7. Drifters leave the loop, although without showing a clear connection with hyperbolic trajectories or lobes. This is so both for HYCOM and AVISO data.

The Loop Current, as explained in Sect. 3 and visible in Fig. 7, breaks into smaller eddies after 24 June in the HYCOM simulations and around 25 June in AVISO. The number of eddies though is not the same for both data. In particular at the northern part of the loop, AVISO displays clear eddy-like structures, while HYCOM does not. Figure 7 confirms that the evolution of drifters 88562 and 88534 from 18 June through 28 June is consistent with Lagrangian structures from both data, although Lagrangian structures are different in each case. Figure 7 shows four snapshots from AVISO and HYCOM for these drifters. They are trapped into the loop, and their position is displayed on 18 June, They evolve in both cases close to a manifold highlighted in the middle of the jet. On 25 June the drifter in AVISO seems to be trapped in an upper eddy, while HYCOM suggests that the drifter evolves within a lobe that elongates due to the presence of a hyperbolic point marked in the figure, which could cause the separation between drifters. Eventually, the position on 28 June is marked. Drifters are clearly within an eddy in the AVISO data, and this structure is unobserved in HYCOM.

The Lagrangian descriptors truly help to make an interpretation of the dynamics. Figure 8 shows the Eulerian fields that correspond to the same sequence of snapshots as seen in Fig. 7. The AVISO data are not provided in hourly intervals as is the case of the HYCOM data; thus what we show in this figure is the interpolation in time on the grid nodes of the velocity field supplied on a daily basis. First is observed that from the Eulerian point of view both the HYCOM model and AVISO present a similar structure, but the HYCOM model has a much higher resolution. From these pictures the perception of the eddy structures is much weaker, and the hyperbolic trajectories are not distinguished.

Many drifters, such as 88589 88582, 88533, 98905, 98938, 98909, 98914, 98945, 98916, 98915, 98911, and 98939 (see some of them in black in Fig. 7), follow approximate circular motions that seem consistently linked to eddies observed in AVISO data. However, there exist events for which the Lagrangian skeleton that connects the motions of drifters passing form one eddy to another is obscure. Sometimes it happens that the data do not provide clear Lagrangian patterns that justify these transactions. Mesoscale eddies seem to be more faded on HYCOM, as a lot of Lagrangian submesoscale structures distort them. However, this does not imply that the the motion of the drifters is inconsistent in this case.

From the point of view taken in this article, inconsistencies would indicate that the velocity field deduced from the altimetry or from mathematical models, despite being accurate for describing some events, may be not be precise enough for describing all of them. On the other hand, inconsistencies could be caused by drifters that are not truly Lagrangian tracers (Tallapragada and Ross, 2008), and thus their dynamics would not be forced to follow that of authentic Lagrangian tracers. We note however that this is not the perspective adopted in most of the Lagrangian studies in oceanic contexts (Griffa et al., 2013), in which drifter trajectories are even assimilated into the model considering them as the "truth" (Nencioli et al., 2011). A thorough revision of these questions is of great interest but is beyond the scope of this article.

7 Summary, conclusions, and outlook

In this article we have explored the ability of Lagrangian tools to extract information from oceanic data and their role in helping to interpret drifter paths on the ocean surface. We have found that both AVISO and HYCOM data do a reasonable job in revealing the Lagrangian mesoscale structures that govern the drifter motion.

The Lagrangian analysis is shown to be very useful in order to make an interpretation of drifter routes and highlighting differences between HYCOM and AVISO. The velocity fields are all rather "similar" based on an Eulerian comparison (i.e., just comparing the velocity fields). But they are different, based on Lagrangian diagnostics and the hidden structures, not displayed in the Eulerian fields, truly organize the trajectories and help to characterize the quality of the data, and their inaccuracies.

At a first look AVISO and HYCOM data are different in that HYCOM has a higher resolution and presents a noisier structure, which is also reflected in the Lagrangian patterns. AVISO tends to provide clean eddy-like structures, which in HYCOM are faded by overlapped features, though consistency with respect to drifter motions seems to be present in both cases. Both AVISO and HYCOM do well in providing Lagrangian structures that guide drifter motions, as for instance within the Gulf Stream or circular motions along the Loop Current. However, evidence is provided in which AVISO fails to correctly locate hyperbolic trajectories, while in HYCOM hyperbolic trajectories are consistent with drifters trajectories.

Acknowledgements. The research of C. Mendoza and A. M. Mancho is supported by MINECO grant MTM2011-26696. The research of S. Wiggins was supported by ONR grant no. N00014-01-1-0769. We acknowledge support from MINECO: ICMAT Severo Ochoa project SEV-2011-0087. Thanks are owed to CESGA, UPM and ICMAT for computing facilities.

Edited by: C. López

References

Andrade-Canto, F., Sheinbaum, J., and Zavala Sansón, L.: A Lagrangian approach to the Loop Current eddy separation, Nonlin. Processes Geophys., 20, 85–96, doi:10.5194/npg-20-85-2013, 2013.

Beron-Vera, F. J., Olascoaga, M. J., and Goni, G. J.: Surface Ocean Mixing Inferred from Different Multisatellite Altimetry Measurements, J. Phys. Oceanogr., 40, 2466–2480, 2010.

Bettencort, J. H., López, C., and Hernández-García, E.: Oceanic three-dimensional Lagrangian coherent structures: A study of a mesoscale eddy in the Benguela upwelling region, Ocean Modell., 51, 73–83, doi:10.1016/j.ocemod.2012.04.004, 2012.

Bleck, R.: An oceanic general circulation model framed in hybrid isopycnic-Cartesian coordinates, Ocean Modell., 4, 55–88, 2002.

Branicki, M. and Kirwan, A. D., Jr.: Stirring: The Eckart paradigm revisited, Int. J. Eng. Sci., 48, 1027–1042, 2010.

Branicki, M. and Wiggins, S.: An adaptive method for computing invariant manifolds in non-autonomous, three-dimensional dynamical systems, Physica D, 238, 1625–1657, 2009.

Branicki, M. and Wiggins, S.: Finite-time Lagrangian transport analysis: stable and unstable manifolds of hyperbolic trajectories and finite-time Lyapunov exponents, Nonlin. Processes Geophys., 17, 1–36, doi:10.5194/npg-17-1-2010, 2010.

Branicki, M., Mancho, A. M., and Wiggins, S.: A Lagrangian description of transport associated with a Front-Eddy interaction: application to data from the North-Western Mediterranean Sea, Physica D, 240, 282–304, doi:10.1016/j.physd.2010.09.003, 2011.

Chassignet, E. P., Hurlburt, H. E., Smedstad, O. M., Halliwell, G. R., Hogan, P. J., Wallcraft, A. J., Baraille, R., and Bleck, R.: The HYCOM (HYbrid Coordinate Ocean Model) data assimilative system, J. Mar. Syst.,65, 60–83, doi:10.1016/j.jmarsys.2005.09.016, 2007.

de la Cámara, A., Mechoso, C. R., Ide, K., Walterscheid, R., and Schubert, G.: Polar night vortex breakdown and large-scale stirring in the southern stratosphere, Clim. Dynam., 35, 965–975, 2010.

de la Cámara, A., Mancho, A. M., Ide, K., Serrano, E., and Mechoso, C.: Routes of transport across the Antarctic polar vortex in the southern spring., J. Atmos. Sci., 69, 753–767, doi:10.1175/JAS-D-11-0142.1, 2012.

de la Cámara, A., Mechoso, R., Mancho, A. M., Serrano, E., and Ide., K.: Quasi-horizontal transport within the Antarctic polar night vortex: Rossby wave breaking evidence and Lagrangian structures., J. Atmos. Sci., 70, 2982–3001, doi:10.1175/JAS-D-12-0274.1, 2013.

d'Ovidio, F., Fernández, V., Hernández-García, E., and López, C.: Mixing structures in the Mediterranean sea from finite-size Lyapunov exponents, Geophys. Res. Lett., 31, L12203, doi:10.1029/2004GL020328, 2004.

Farazmand, M. and Haller, G.: Computing Lagrangian Coherent Structures from variational theory, Chaos, 22, 013128, doi:10.1063/1.3690153, 2012.

Forristall, G. Z., Schaudt, K. J., and Cooper, C. K.: Evolution and Kinematics of a Loop Current Eddy in the Gulf of Mexico During 1985, J. Geophys. Res., 97, 2173–2184, 1992.

Froyland, G., Horenkamp, C., Rossi, V., Santitissadeekorn, N., and Gupta, A. S.: Three-dimensional characterization and tracking of an Agulhas Ring, Ocean Modell., 52–53, 69–75, 2012.

Griffa, A., Haza, A., Özgökmen, T. M., Molcard, A., Taillandier, V., Schroeder, K., Chang, Y., and Poulain, P. M.: Investigating transportpathwaysintheocean, Deep Sea Res. II, 85, 81–95, 2013.

Haller, G.: Finding finite-time invariant manifolds in two-dimensional velocity fields, Chaos, 10, 99–108, 2000.

Haller, G.: Distinguished material surfaces and coherent structure in three-dimensional fluid flows, Physica D, 149, 248–277, 2001a.

Haller, G.: Lagrangian structures and the rate of strain in a partition of two-dimensional turbulence, Phys. Fluids, 13, 3365–3385, 2001b.

Haller, G.: Lagrangian coherent structures from approximate velocity data, Phys. Fluids, 14, 1851–1861, 2002.

Haller, G. and Yuan, G.: Lagrangian coherent structures and mixing in two-dimensional turbulence, Physica D, 147, 352–370, 2000.

Haza, A. C., Griffa, A., Martin, P., Molcard, A., Özgökmen, T. M., Poje, A., Barbanti, R., Book, J., Poulain, P., Rixen, M., and Zanasca, P.: Model-based directed drifter launches in the Adriatic Sea: Results from the DART experiment, Geophys. Res. Lett., 34, L10605, doi:10.1029/2007GL029634, 2007.

Haza, A. C., Özgökmen, T. M., Griffa, A., Molcard, A., Poulain, P. M., and Peggion, G.: Transport properties in small-scale coastal flows: relative dispersion from VHF radar measurements in the Gulf of La Spezia, Ocean Dynam., 60, 861–882, 2010.

Hernández-Carrasco, I., López, C., Hernández-García, E., and Turiel, A.: How reliable are finite-size Lyapunov exponents for the assessment of ocean dynamics?, Ocean Modell., 36, 208–218, 2011.

Ide, K., Small, D., and Wiggins, S.: Distinguished hyperbolic trajectories in time-dependent fluid flows: analytical and computational approach for velocity fields defined as data sets, Nonlin. Processes Geophys., 9, 237–263, doi:10.5194/npg-9-237-2002, 2002.

Jones, C. K. R. T. and Winkler, S.: Invariant manifolds and Lagrangian dynamics in the ocean and atmosphere, in: Handbook of dynamical systems, 55–92, North-Holland, Amsterdam, 2002.

Kuznetsov, L., Toner, M., Kirwan Jr., A. D., Jones, C. K. R. T., Kantha, L. H., and Choi, J.: The Loop Current and adjacent rings delineated by Lagrangian analysis of the near-surface flow, J. Mar. Res., 60, 405–429 doi:10.1357/002224002762231151, 2002.

Lekien, F., Shadden, S., and Marsden, J. E.: Lagrangian coherent structures in n-dimensional systems, J. Math. Phys., 48, 065404, doi:10.1063/1.2740025, 2007.

Madrid, J. A. J. and Mancho, A. M.: Distinguished trajectories in time dependent vector fields, Chaos, 19, 013111, doi:10.1063/1.3056050, 2009.

Malhotra, N. and Wiggins, S.: Geometric Structures, Lobe Dynamics, and Lagrangian Transport in Flows with Aperiodic Time-Dependence, with Applications to Rossby Wave Flow, J. Nonlin. Sci., 8, 401–456, 1998.

Mancho, A. M., Small, D., Wiggins, S., and Ide, K.: Computation of Stable and Unstable Manifolds of Hyperbolic Trajectories in Two-Dimensional, Aperiodically Time-Dependent Vectors Fields, Physica D, 182, 188–222, doi:10.1016/S0167-2789(03)00152-0, 2003.

Mancho, A. M., Small, D., and Wiggins, S.: Computation of hyperbolic trajectories and their stable and unstable manifolds for oceanographic flows represented as data sets, Nonlin. Processes Geophys., 11, 17–33, doi:10.5194/npg-11-17-2004, 2004.

Mancho, A. M., Small, D., and Wiggins, S.: A comparison of methods for interpolating chaotic flows from discrete velocity data, Comput. Fluids, 35, 416–428, doi:10.1016/j.compfluid.2005.02.003, 2006a.

Mancho, A. M., Small, D., and Wiggins, S.: A tutorial on dynamical systems concepts applied to Lagrangian transport in oceanic flows defined as finite time data sets: Theoretical and computational issues, Phys. Rep., 437, 55–124, doi:10.1016/j.physrep.2006.09.005, 2006b.

Mancho, A. M., Hernández-García, E., Small, D., Wiggins, S., and Fernández, V.: Lagrangian transport through an ocean front in the North-Western Mediterranean Sea, J. Phys. Oceanogr., 38, 1222–1237, doi:10.1175/2007JPO3677.1, 2008.

Mancho, A. M., Wiggins, S., Curbelo, J., and Mendoza, C.: Lagrangian descriptors: A Method for Revealing Phase Space Structures of General Time Dependent Dynamical Systems, Commun. Nonlinear Sci., 18, 3530–3557, doi:10.1016/j.cnsns.2013.05.002, 2013.

Mendoza, C. and Mancho, A. M.: The hidden geometry of ocean flows, Phys. Rev. Lett., 105, 038501, doi:10.1103/PhysRevLett.105.038501, 2010.

Mendoza, C. and Mancho, A. M.: Review Article: "The Lagrangian description of aperiodic flows: a case study of the Kuroshio Current", Nonlin. Processes Geophys., 19, 449–472, doi:10.5194/npg-19-449-2012, 2012.

Mendoza, C., Mancho, A. M., and Rio, M.-H.: The turnstile mechanism across the Kuroshio current: analysis of dynamics in altimeter velocity fields, Nonlin. Processes Geophys., 17, 103–111, doi:10.5194/npg-17-103-2010, 2010.

Mezic, I., Loire, S., Fonoberov, V. A., and Hogan, P. A.: A New Mixing Diagnostic and Gulf Oil Spill Movement, Science, 330, 486–489, 2010.

Miller, P. D., Jones, C. K. R. T., Rogerson, A. M., and Pratt, L. J.: Quantifying transport in numerically generated velocity fields, Physica D, 110, 105–122, 1997.

Molcard, A., Poje, A. C., and Özgökmen, T. M.: Directed drifter launch strategies for Lagrangian data assimilation using hyperbolic trajectories, Ocean Modell., 12, 268–289, 2006.

Nencioli, F., d'Ovidio, F., Dogliogi, A. M., and Petrenko, A. A.: Surface coastal circulation patterns by in-situ detection of Lagrangian coherent structures, Geophys. Res. Lett., 38, L17604, doi:10.1029/2011GL048815, 2011.

Nencioli, F., d'Ovidio, F., Dogliogi, A. M., and Petrenko, A. A.: In situ estimates of submesoscale horizontal eddy diffusivity across an ocean front, J. Geophys. Res.-Oceans, 118, 7066–7080, doi:10.1002/2013JC009252, 2013.

Olascoaga, M. J.: Isolation on the West Florida Shelf with implications for red tides and pollutant dispersal in the Gulf of Mexico, Nonlin. Processes Geophys., 17, 685–696, doi:10.5194/npg-17-685-2010, 2010.

Pascual, A., Faugère, Y., Larnicol, G., and Traon, P. Y. L.: Improved description of the ocean mesoscale variability by combining four satellites altimeters, Geophys. Res. Lett., 33, L02611, doi:10.1029/2005GL024633, 2006.

Pierrehumbert, R. T.: Large-scale horizontal mixing in planetary atmospheres, Phys. Fluids A, 3, 1250–1260, 1991.

Prants, S. V., Budyansky, M. V., and Uleysky, M. Yu.: Lagrangian study of surface transport in the Kuroshio Extension area based on simulation of propagation of Fukushima-derived radionuclides, Nonlin. Processes Geophys., 21, 279–289, doi:10.5194/npg-21-279-2014, 2014.

Press, W., Teukolsky, S., Vetterling, W., and Flannery, B.: Numerical Recipes in Fortran 77, The Art of Scientific Computing, Cambridge University Press, Cambdridge, 2 Edn., 1999.

Rempel, E. L., Chian, A. C.-L., Brandenburg, A., Munuz, P. R., and Shadden, S. C.: Coherent structures and the saturation of a nonlinear dynamo, J. Fluid Mechan., 729, 309–329, 2013.

Samelson, R. and Wiggins, S.: Lagrangian Transport in Geophysical Jets and Waves: The Dynamical Systems Approach, Springer-Verlag, New York, 2006.

Shadden, S. C., Lekien, F., and Marsden, J. E.: Definition and properties of Lagrangian Coherent structures from finite-time Lyapunov exponents in two-dimensional aperiodic flows, Physica D, 212, 271–304, 2005.

Shadden, S. C., Lekien, F., Paduan, J. D., Chavez, F. P., and Marsden, J. E.: TheThe correlation between surface drifters and coherent structures based on high-frequency radar data in Monterey Bay, Deep Sea Res. II, 56, 161–172, 2009.

Sulman, M. H. M., Huntley, H. S., Lipphardt Jr., B. L., Jacobs, G., Hogan, P., and Kirwan Jr., A. D.: Hyperbolicity in temperature and flow fields during the formation of a Loop Current ring, Nonlin. Processes Geophys., 20, 883–892, doi:10.5194/npg-20-883-2013, 2013.

Tallapragada, P. and Ross, S. D.: Particle segregation by Stokes number for small neutrally buoyant spheres in a fluid, Phys. Rev. E, 78, 036 308, doi:10.1103/PhysRevE.78.036308, 2008.

Titaud, O., Brankart, J. M., and Verron, J.: On the use of Finite-Time Lyapunov Exponents and Vectors for direct assimilation of tracer images into ocean models, Tellus A, 63A, 1038–1051, 2011.

Wiggins, S.: The Dynamical Systems Approach to Lagrangian Transport in Oceanic Flows, Annu. Rev. Fluid Mech., 37, 295–328, 2005.

14

Assimilation of HF radar surface currents to optimize forcing in the northwestern Mediterranean Sea

J. Marmain[1,2], A. Molcard[1,2], P. Forget[1,2], A. Barth[3], and Y. Ourmières[1,2]

[1]Université de Toulon, UMR7294, CNRS/INSU, IRD – Mediterranean Institute of Oceanography (MIO), UM 110, 83957 La Garde, France
[2]Aix Marseille Université, CNRS/INSU, IRD, Mediterranean Institute of Oceanography (MIO), UM 110, 13288 Marseille, France
[3]GeoHydrodynamics and Environment Research (GHER), AGO/MARE, Univeristy of Liège, Liége, Belgium

Correspondence to: J. Marmain (julien.marmain@univ-tln.fr)

Abstract. HF radar measurements are used to optimize surface wind forcing and baroclinic open boundary condition forcing in order to constrain model coastal surface currents. This method is applied to a northwestern Mediterranean (NWM) regional primitive equation model configuration. A new radar data set, provided by two radars deployed in the Toulon area (France), is used. To our knowledge, this is the first time that radar measurements of the NWM Sea are assimilated into a circulation model. Special attention has been paid to the improvement of the model coastal current in terms of speed and position. The data assimilation method uses an ensemble Kalman smoother to optimize forcing in order to improve the model trajectory. Twin experiments are initially performed to evaluate the method skills. Real measurements are then fed into the circulation model and significant improvements to the modeled surface currents, when compared to observations, are obtained.

1 Introduction

The ocean coastal dynamics of the northwestern Mediterranean (NWM) Sea has been studied over the last decades using observations (e.g., Millot, 1999; Petrenko, 2003; Sammari et al., 1995; Allou et al., 2010), modeling (Echevin et al., 2003; Casella et al., 2011; Ourmières et al., 2011; Pairaud et al., 2011), or a combination of both (Schaeffer et al., 2011b; Guihou et al., 2013). The NWM is a micro-tidal sea (e.g., Alberola et al., 1995) and its dynamic is mainly governed by wind forcing and by the Northern Current (NC), which is the northern branch of the general circulation of the NWM Sea (Millot, 1999). The NC is a quasi-permanent density current that originates in the eastern part of the NWM and forms a large-scale loop. Bathymetric constraints operate on the northern, western and southwestern sides of its path. This geostrophic current can experience a strong variability in this particular region, due to the steep topography and the irregular coastline, as well as the atmospheric forcing. The NC varies in strength and width, its meandering can generate eddies, and its separation in the vicinity of Toulon (Fig. 1) controls water fluxes at the entrance of the Gulf of Lion (GoL). Several of the above studies were carried out in the GoL or off Nice. However, the dynamics of the NC in the intermediate area, typically the region off Toulon, is poorly documented. Since May 2010, an HF radar system has been deployed in this region in an attempt to fill this void, allowing the gathering of continuous and synoptic observations of surface currents up to 100 km off the coastline with typical temporal and spatial resolutions of 1 h and 3 km, respectively.

Scientific interest in the assimilation of HF radar measurements in hydrodynamical models has been growing over the last few years. Various assimilation methods have been developed based on nudging (Lewis et al., 1998), Kalman filtering using parametrized (Paduan and Shulman, 2004; Shulman and Paduan, 2009) or ensemble (Breivik and Sætra, 2001; Oke et al., 2002b; Barth et al., 2008a) error covariances, and the application of adjoint-based methods (Kurapov et al., 2003; Hoteit et al., 2009).

The method applied herein consists in optimizing the wind-forcing and lateral boundary-forcing components of a

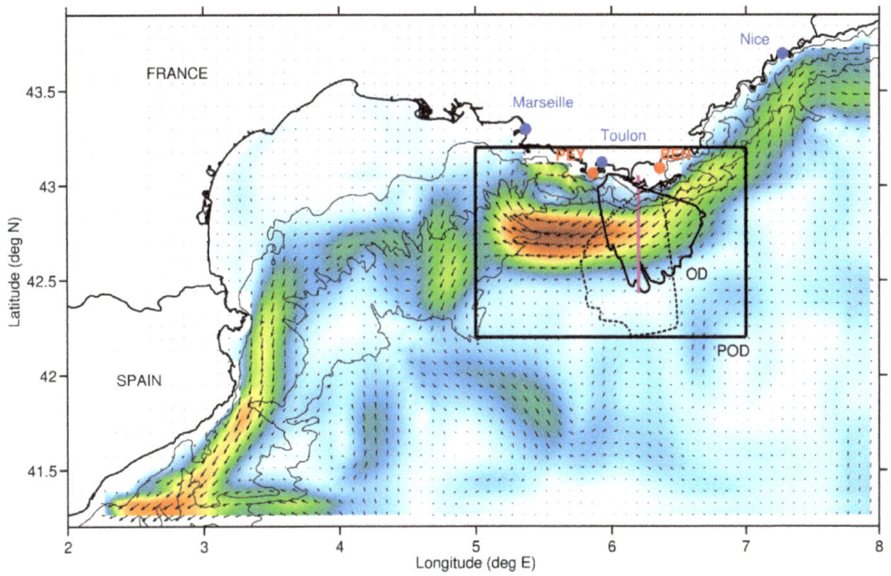

Figure 1. Depiction of the studied area. The larger area is the model domain (MD). Red points mark radar stations PEY and BEN. The black contour lines show areas where HF radar data are available more than 20 % over the assimilation period for PEY (full) and BEN (dashed) radars and is called the observation domain (OD). The black box sketches the pseudo-observation domain (POD). The meridional magenta line crossing the coastal current will be used for Hovmöller diagrams. Surface currents of free simulation averaged over the assimilation period are shown. Color bars range from 0 (white) to 0.5 (red) m s^{-1}. The 100, 1000 and 2000 m isobaths are drawn (black thin lines).

numerical regional primitive equation model in order to minimize differences between the model and the radar-derived surface currents.

Wind optimization is an important issue as, in coastal zones, surface wind errors can explain a large part of the model error regarding surface currents (He et al., 2004; Barth et al., 2008b). Such optimization can be achieved by correcting the surface wind field (Barth et al., 2011) or the parameters governing the wind–sea momentum transfer, e.g., the drag coefficient (Ten Brummelhuis et al., 1993; Peng et al., 2012).

Similarly, open boundary condition (OBC) optimization can be achieved by correcting the OBC values themselves or by correcting the OBC's formulation and discretization. This last approach has been addressed several times (e.g., Orlanski, 1976; Flather, 1976; Marchesiello et al., 2001; Blayo and Debreu, 2005). It consists in a numerical treatment of the oceanic boundaries allowing the removal of the perturbations generated inside the model domain without deteriorating the inner model solution (Røed and Cooper, 1986). Research in this field is still ongoing. Instead of pursuing this approach further, we shall address the first approach, i.e., the optimization of time-dependent OBC values. This has been achieved previously in barotropic models (e.g., Seiler, 1993; Ten Brummelhuis et al., 1993; Gunson and Malanotte-Rizzoli, 1996; Bogden et al., 1996; Taillandier et al., 2004) and more rarely in baroclinic models (e.g., Shulman et al., 1999; Taillandier et al., 2009).

Variational methods (e.g., Le Dimet and Talagrand, 1986) are often preferred to tackle forcing optimization problems. The adjoint model technique is the one most often used, but depending on the model it can sometimes be difficult to formulate (Shulman et al., 1999). Another way of solving optimization problems is by the application of a Kalman filter (e.g., Gelb, 1974). However, in its original form, this technique has a prohibitively high computational burden to be used in geophysical applications. The development of reduced rank methods such as ensemble methods (Evensen, 1994) allows for the reduction of this computational cost because only a limited number of model states ($\mathcal{O}(100)$) needs to be stored in order to obtain an accurate estimate of the error covariance matrices. Moreover, ensemble methods are efficient for dealing with error propagation in non-linear models.

Recently, the Ensemble Perturbation Smoother (EnPS) has been developed to optimize stationary tidal boundary conditions of a hydrodynamical model of the German Bight (Barth et al., 2010), a very shallow area (maximum depth of about 50 m) where winds and tides are the main sources of variability. The EnPS has also been used successfully to optimize surface wind estimates (Barth et al., 2011) in the same area, but the resulting model surface current was not directly evaluated. This beckons the following questions: firstly, whether optimizing wind can improve model surface currents, and secondly, whether it is possible to optimize non-stationary baroclinic OBC by applying the EnPS with the aim of improving model surface currents in a highly dynamic area.

These issues are addressed by using the EnPS in a regional baroclinic model of the NWM Sea and by using the HF radar-derived surface currents measured off Toulon.

Section 2 describes the hydrodynamical model configuration and the experimental data set. The data assimilation method is explained in Sect. 3. Results of wind optimization are shown in Sects. 4 and 5, showing the results of OBC optimization. Finally, results are discussed in the conclusion.

2 Model and data

2.1 Model

The simulations described below were obtained using the Regional Ocean Modeling System (ROMS, Shchepetkin and McWilliams, 2003, 2005), which solves the ocean primitive equations, featuring a free surface formulation and hydrostatic approximation. It uses terrain-following coordinates in the vertical dimension and curvilinear coordinates in the horizontal dimension. ROMS incorporates advanced features and high-order numerics, allowing efficient and robust resolution of mesoscale dynamics in the oceanic and coastal domains. ROMS has been widely applied to different oceanic regions for a variety of ocean dynamics studies (Penven et al., 2001; Capet et al., 2008; Casella et al., 2011). The model domain (MD) as depicted in Fig. 1 extends from 2.10 to 7.93° E and from 41.26 to 43.96° N. Horizontal resolution is 1/12°, and 32 vertical σ levels are employed with vertical refinement near the surface to obtain a satisfactory representation of the surface layer.

Two active open boundaries connecting the MD with the NWM basin are incorporated. The OBC treatment is a well-known ill-posed problem (e.g., Chapman, 1985; Marchesiello et al., 2001; Blayo and Debreu, 2005), and the goal here is to implement the optimal conditions in order to allow inflow from the open sea, to conserve mass and energy and to reduce any wave reflection phenomena. A radiation–relaxation condition for 3-D momentum and tracers is combined with a Flather condition for the 2-D momentum (Marchesiello et al., 2001), and a sponge layer is added to avoid spurious reflections. The typical relaxation coefficients have been adjusted to 1 and 10 days (for tracers and baroclinic velocities, respectively) for inflow and to 360 days for outflow.

The data for the relaxation boundary conditions were derived from the PSY2V4R2 operational configuration developed by MERCATOR-OCEAN (http://www.mercator-ocean.fr), based on the NEMO (Nucleus for European Modelling of the Ocean) model. The PSY2V4R2 configuration features a horizontal resolution of 1/12° with 50 vertical z levels, and includes a sequential data assimilation system (SAM2V1) based on the Kalman filter. Satellite sea surface temperature (SST) and sea level anomaly (SLA) are assimilated, in addition to available in situ temperature and salinity profiles. Daily averages of the velocity, mass fields and surface height at the MD open boundaries are used. The forcing is made off-line.

At the air–sea interface, the oceanic model is forced by the ALADIN operational regional model from Météo-France. ALADIN runs at a resolution of 1/10° and features data assimilation and state-of-the-art atmospheric physics (Fischer et al., 2005). The importance of such high spatial and temporal resolution in terms of oceanic response has been widely studied and demonstrated over the last few years, particularly to achieve a more realistic representation of the oceanic mesoscale circulation (Langlais et al., 2009; Schaeffer et al., 2011a).

The initial condition of the simulation is based on the daily averaged baroclinic and barotropic velocities, temperature, salinity and surface elevation derived from PSY2V4R2 for 1 July 2011. A 4 month spin-up period is considered to be adequate for reaching a state of equilibrium. The model output corresponds to 3-hour averaged fields.

The average surface current for the period 15 November–15 December 2012 is shown in Fig. 1. The model simulations give a maximum mean current of $0.5\,\mathrm{m\,s^{-1}}$, in accordance with the observations.

2.2 Observations

Since May 2010, HF radars have been monitoring surface currents off Toulon (Fig. 1). The radars employed are WERAs (Wellen Radar, Gurgel et al., 1999) and operate at 16.15 MHz with 50 kHz bandwidth. A radar system is installed at Fort Peyras in Cape Sicié (hereafter PEY) and operates in a monostatic configuration, i.e., transmitter and receiver are very close together ($\sim$ 100 m). The second system operates in a bistatic configuration. The transmitting antennas are those of the PEY radar, whereas the receiving antennas consist of a linear array of eight antennas installed near Cape Bénat (hereafter BEN), 40 km to the east of PEY. Integration time, i.e., the time over which the temporal signals are processed, is one hour. Velocity accuracy is $4\,\mathrm{cm\,s^{-1}}$, and the spatial sampling is 3 km in range. An azimuthal discretization of 2° was achieved through direction finding using the MUSIC processing technique (Schmidt, 1986). This technique is now in common use in the HF radar community (Lipa et al., 2006; Sentchev et al., 2013). The radial component of the current is measured along the look direction of the radar. With the given radar frequency, the radial velocity is theoretically measured at a depth of 74 cm (Stewart and Joy, 1974).

The MUSIC technique involves spatial holes, temporal gaps and false alarms (outliers) on radial velocity maps. For the present study, a series over a 30 day time period was used at every radar cell in order to remove outliers. Outlier removal was performed using the histogram of the temporal gradients of the current. Spatial holes and temporal gaps were not filled, in order to avoid the addition of spu-

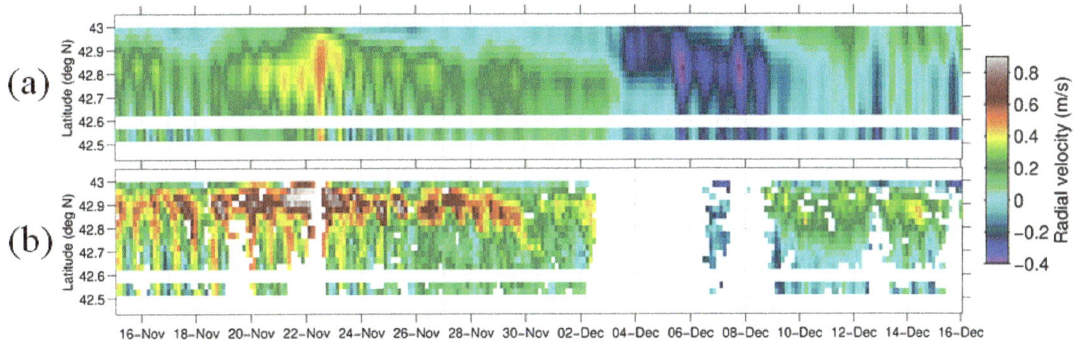

Figure 2. Hovmöller diagram of radial velocities from the PEY station along a south–north transect at 6.2° N (see Fig. 1), **(a)** from free simulation and **(b)** from observation.

rious and correlated information in radial velocities, which can result from interpolations. Only the radial component maps were taken into consideration, and not the current vectors that can be deduced by the geometric combination of the two sets of radar measurements. The first reason for this choice is to prevent the introduction of the geometric dilution of precision, which is inherent in the vector mapping process (Chapman et al., 1997). The second reason is to dispose of the most extended observation area and to avoid wasting valid observations.

HF radar measurements from PEY and BEN (y_{pey}, y_{ben}) gathered during the period 15 November 2011 00:00 UTC to 15 December 2011 23:00 UTC were taken into consideration for this study. A spatial mask rejecting the input from radar cells having a temporal coverage below 20 % over the period was applied to the data. The 20 % temporal coverage level is depicted in Fig. 1 and is hereafter referred to as the observed domain (OD). When measurements from both stations are available, variability in the total coverage is mainly due to the degradation of PEY performance owing to high wind speed.

Twenty-eight temperature and salinity profiles from conductivity–temperature–depth (CTD) measurements were also taken into consideration for the analysis of assimilation results. CTD measurements were carried out in the area within the geographical grid defined by coordinates 5.8–7.35° E and 42.6–43.3° N, over the period 11–14 December 2011, during a cruise undertaken in the framework of the European TOSCA (Tracking Oil Spill and Coastal Awareness network, http://www.tosca.med.eu) program. Note that this data set is independent data that can be used for validation of results.

2.3 Model validation

The Hovmöller diagram (time versus latitude) of both model-derived and observed radial velocities is depicted for the PEY station (Fig. 2). The radial velocity is the component of the total surface velocity of the current $\boldsymbol{U}$ in the radial direction. By convention its sign is positive (negative) for velocity going toward (away from) the station. High positive values in the PEY radial component are a distinct signature of the coastal current (Marmain et al., 2011). The model clearly underestimates the intensity of the current radial components, and the position of the maximum seems slightly shifted off-shore. However, the temporal variability is well simulated: the change in sign at the beginning of December, as well as the high-frequency modulations (near-inertial oscillations), and the maximum intensity observed from PEY around 23 November are reproduced well by the model.

The validity of the vertical structure of the model was evaluated using CTD measurements. Model temperature and salinity exhibit low departures from observations. According to the definition of the root mean square error (RMSE), Er and the bias B given in Sect. 3, Er = 0.780 °C and B = 0.002 °C for temperature and Er = 0.084 and B = −0.054 for salinity. Significant correlations appear (0.79 and 0.86 respectively). The main errors come from the inaccurate representation of the thermocline by the model. However, densities are in good agreement.

3 Data assimilation methodology

HF radar-derived velocities have already been used to correct M2 tidal boundary conditions (Barth et al., 2010) and surface wind fields (Barth et al., 2011) using the so-called EnPS, a fixed-interval sequential smoother (Cosme et al., 2012). In this study, we investigate, firstly, whether optimizing wind can improve model surface currents, and secondly, whether it is possible to optimize non-stationary baroclinic OBC by applying the EnPS with the aim of improving model surface currents in a highly dynamic area. Uncertainties in these atmospheric and oceanic fields are represented by an ensemble of wind forcing or an ensemble of OBC forcing based on perturbations of the original (or background) fields ($\mathbf{x}^{\text{b}}$).

3.1 Ensemble setup

The perturbation method is detailed in Barth et al. (2011). Briefly, the spatial structure of the forcing perturbations is obtained by using a Fourier decomposition of the forcing field. The Fourier coefficients correspond to the spatial field. For each temporal frequency subjected to perturbation, a complex random time series is constructed with a temporal correlation scale according to the perturbed frequency, with nil mean and unit variance (Evensen, 1994). Such a series is then multiplied by the corresponding Fourier coefficient and the resulting products are summed. The real part of this quantity corresponds to the forcing perturbation $\mathbf{x}_\mathrm{p}^l$, with l as the member index. $\mathbf{x}_\mathrm{p}^l$ is added to $\mathbf{x}^\mathrm{b}$, resulting in

$$\mathbf{x}^l = \mathbf{x}^\mathrm{b} + \alpha \mathbf{x}_\mathrm{p}^l. \tag{1}$$

With this method, forcing errors are assumed to have similar spatial and temporal scales to the original, non-perturbed forcing. By using this approach, it is assumed that the uncertainty in the forcing field is mainly caused by phase and amplitude errors of the underlying processes rather than errors in the variability. Perturbations are estimated using the time variability of the forcing field. $\mathbf{x}_\mathrm{p}^l$ is weighted by the α coefficient in order to achieve realistic perturbations. This approach ensures that a spatial structure is multiplied by a compatible temporal scale. The forcing ensemble must verify $\mathbf{x}^\mathrm{b} = \langle \mathbf{x}^l \rangle$, with $\langle . \rangle$ as the ensemble average operator.

An ensemble of 100 perturbations for zonal and meridional components of the wind vector $\boldsymbol{W}$ was generated over the MD and over a period extending from 1 November to 31 December 2011, taking into consideration temporal scales ranging from 3 h to 30 days. The minimum period is the temporal resolution of the original forcing and the maximum is set to 30 days, assuming that higher periods do not contain uncertainties.

An ensemble of 100 OBC perturbations was generated, taking into consideration temporal scales ranging from 1 to 30 days, over the same time period, for temperature, salinity, surface elevation, zonal velocities and meridional velocities (respectively, T^obc, S^obc, η^obc, U^obc, V^obc) at the eastern boundary of the MD. Only this boundary was taken into consideration because it is located upstream of the observation area along the NC flow. Perturbations were introduced in the northern part of the OBC (42.5–44° N) and over the sub-surface ocean layer (−200–0 m) as the prognostic variables at higher depths do not exhibit any significant correlation with the surface current in the OD. In fact, such a restriction on the perturbation domain can be seen as a spatial localization (Hamill et al., 2001; Houtekamer and Mitchell, 2001). Therefore, we consider that errors in the model solution originate only from the perturbed area.

For each forcing perturbation, the model was run using the same initial condition as the free simulation. A 15 day ensemble spin-up was selected to attain a satisfactory variance between surface current members. The ensemble simulations for the time period 15 November–15 December 2011 were taken into consideration.

3.2 Assimilation scheme

A general description of the EnPS (Ensemble Perturbation Smoother) is given in Appendix A. For more information, the reader is referred to Barth et al. (2010, 2011).

Since the EnPS is a non-sequential method, all available observations over the period of analysis are used to compute the optimal state. The time dimension is consequently embedded in the observation and optimization vectors. The model trajectory is not estimated directly. Rather, the optimal forcing trajectory is derived and then run in the model. Furthermore, the advantage of estimating the forcing first is that the final model trajectory satisfies (per construction) the model equations (and thus all non-linear balances between model variables), which is not the case if the model trajectory is analyzed directly, and correcting the forcing instead of the model state also avoids unrealistic transients (Malanotte-Rizzoli et al., 1989).

The optimized forcing $\mathbf{x}^\mathrm{a}$ is computed using the Kalman filter analysis step with a non-linear observation operator:

$$\mathbf{x}^\mathrm{a} = \mathbf{x}^\mathrm{b} + \mathbf{K}(\boldsymbol{y}^\mathrm{o} - h(\mathbf{x}^\mathrm{b})), \tag{2}$$

where $\mathbf{K}$ is the Kalman gain that varies according to ensemble and observation error covariance matrices (Appendix A). The non-linear observation operator $h(.)$ is applied to the forcing $\mathbf{x}$ and $h(\mathbf{x})$ represents the surface currents within the integration period.

The analyzed model trajectory is obtained in a final step, by performing a new run of the model using the optimized forcing field.

The analysis $\mathbf{x}^\mathrm{a}$ is unbiased if both background estimate $\mathbf{x}^\mathrm{b}$ and observations $\boldsymbol{y}^\mathrm{o}$ are unbiased. If the observation operator is non-linear, it can be verified that the analysis is unbiased using Eq. (3) (Barth et al., 2011):

$$< h(\mathbf{x}^l) >= h(< \mathbf{x}^l >). \tag{3}$$

The degree of linearity of the observation operator was verified using Eq. (3). The overall root mean square difference between both sides of Eq. (3) was computed for wind and OBC ensembles. A value of $0.03\,\mathrm{m\,s^{-1}}$ was found, which is small in comparison with other errors and bias. However, the standard analysis methods provide often useful results even if the background is biased.

The observation vector $\boldsymbol{y}^\mathrm{o}$ is associated with an observation error covariance matrix $\mathbf{R}$. For the sake of simplicity, $\mathbf{R}$ is assumed diagonal and it is also assumed that $\mathbf{R}$ is the sum of the measurement error covariance $R_\mathrm{obs}\mathbf{I}$ and the representativity error covariance $S_\mathrm{obs}^2\mathbf{I}$ where $\mathbf{I}$ is the identity matrix. R_obs represents the measurement errors that arise from noise effects imparted by data processing as well as by the radar itself. S_obs is introduced to characterize processes and scales present in the observation but not resolved by the model. The representativity error could be extended to include processes that cannot be corrected using these forcings only (Barth et al., 2011).

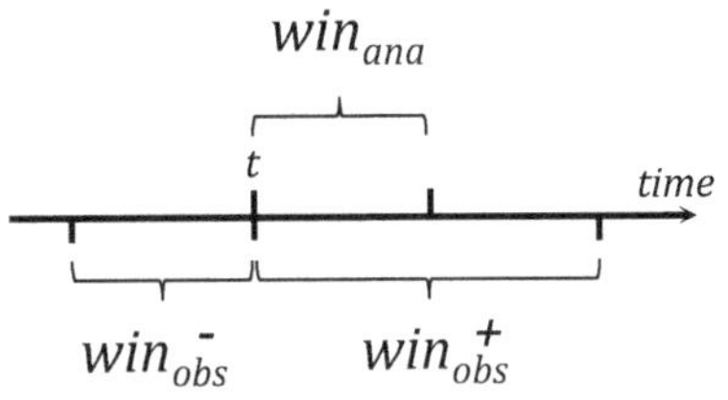

Figure 3. Schematic representation of the optimization window (win_{ana}), the backward ($\text{win}^{-}_{\text{obs}}$) observation window and the forward ($\text{win}^{+}_{\text{obs}}$) observation window. t is the beginning of win_{ana}.

To increase the effective dimension of the error subspace, to discard distant observations involving unrealistic correlations and to improve computational capabilities, the Kalman filter analysis step can be performed in reduced spatial domains, zone by zone, which is known as spatial localization (Hamill et al., 2001; Houtekamer and Mitchell, 2001). A similar approach can be taken in the temporal dimension.

Typically, the model integration period is divided into a series of shorter, non-overlapped, intervals called optimization windows with a duration of win_{ana}. Considering an observation window of duration win_{obs}, such as $\text{win}_{\text{ana}} = \text{win}_{\text{obs}}$, forcing fields within the time range $[t; +\text{win}_{\text{ana}}]$ are updated using all available observations in such observation windows, with t as the beginning of the optimization window (Barth et al., 2010, 2011). Here, the definition of the observation window was modified to take into account backward ($\text{win}^{-}_{\text{obs}}$) or forward ($\text{win}^{+}_{\text{obs}}$) observations within the time range $\left[t - \text{win}^{-}_{\text{obs}}; t + \text{win}^{+}_{\text{obs}}\right]$ (Fig. 3), considering the delay of the oceanic response to the lateral and surface forcing that may result in non-instantaneous correlations between forcing and surface currents (e.g., Seiler, 1993). This way, the standard EnPS is retrieved with $\text{win}^{+}_{\text{obs}} = \text{win}_{\text{ana}}$ and $\text{win}^{-}_{\text{obs}} = 0$.

3.3 Experimental setup

The assimilation methodology was first applied to a set of radar pseudo-observations in identical twin experiments (TE) to assess the method, and then applied to experimental radar observations.

In TE, forcing and model control states ($\mathbf{x}^{\text{true}}, h(\mathbf{x}^{\text{true}})$) are extracted from the ensemble members. These control states are used to generate a set of pseudo-radial velocity fields by adding a Gaussian noise. The free run corresponds to running the model using the original forcing. The assimilation of these pseudo-observations results in an analyzed forcing that can be used in the model to simulate the resulting analyzed model surface current. The true state is known and the skill of the assimilation scheme can be quantified.

Pseudo-observation values are given for a geographical grid defined by coordinates 5–7° E and 42.2–43.2° N, the pseudo observation domain (POD), and are defined as the projection of model surface velocities in the radial directions at each model node. The measurement noise is assumed to be nil.

The experimental observations consist in the radial velocity measurements derived from the two PEY and BEN radars (Sect. 2.2) with R_{obs} is $0.04\,\text{m}\,\text{s}^{-1}$, corresponding to the velocity accuracy.

3.4 Diagnostics

The skill of the assimilation method is evaluated by comparing the control state (control) to the free (b) or analyzed (a) states in terms of RMSE (e.g., Oke et al., 2002a). In the case of the twin experiments, the control state is the true one, true. In the real case experiments, a cross-validation-like technique is used (e.g., Brankart and Brasseur, 1996): the control state is a random set of observations, o, discarded at each time step from the initial set of observations.

Let n be the spatial index of dimension N, and k the temporal index of dimension K. The RMSE of a variable Φ is defined by

$$\text{Er}^{\text{a,b}} = \sqrt{\frac{1}{NK}\sum_{n=1}^{N}\sum_{k=1}^{K}\left(\Phi^{\text{a,b}}_{n,k} - \Phi^{\text{control}}_{n,k}\right)^2} \quad (4)$$

When the summation is applied to both spatial and temporal dimensions of Φ, Er corresponds to an overall scalar measurement of the assimilation performance. If the sum is computed in the spatial dimension only, the error is evaluated during the experiment time evolution ($\text{Er}_N(k)$), and when the average is calculated in the temporal dimension, the performance is evaluated in terms of spatial structure ($\text{Er}_K(n)$).

The bias B can also be computed to determine the mean difference between model solution and control:

$$B^{\text{a,b}} = \frac{1}{NK}\sum_{n=1}^{N}\sum_{k=1}^{K}\left(\Phi^{\text{a,b}}_{n,k} - \Phi^{\text{control}}_{n,k}\right) \quad (5)$$

The skill score "Sc" is an additional metric that can measure the performance of the data assimilation (e.g., Oke et al., 2002a), and is defined by

$$\text{Sc} = 1 - \left(\frac{\text{Er}^{\text{a}}}{\text{Er}^{\text{b}}}\right)^2. \quad (6)$$

A positive score "Sc" indicates that the assimilation is efficient, while a negative score indicates a degradation. In particular, Sc = 1 is obtained for a perfect analyzed field, and Sc = 0 reflects a nil impact on average.

Eventually, $\Delta = \text{Er}^{\text{b}}_K(n) - \text{Er}^{\text{a}}_K(n)$ will also be computed. $\Delta > 0$ ($\Delta < 0$) indicates an area of improvement (degradation).

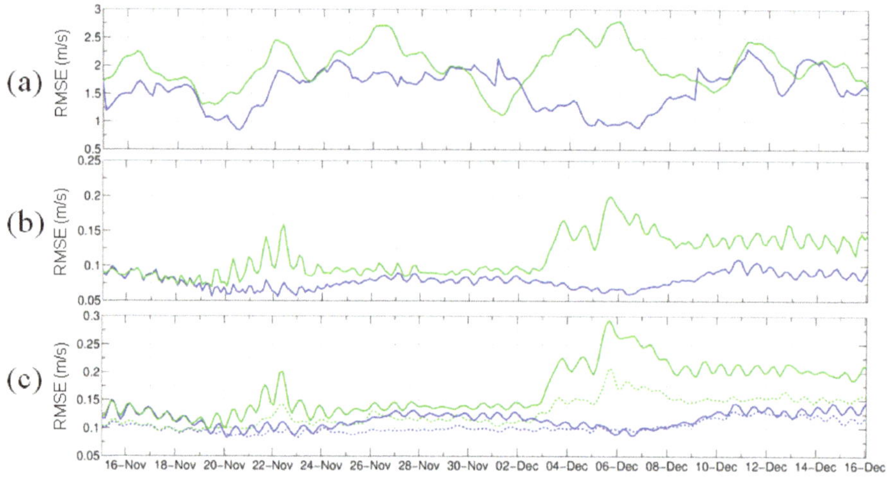

Figure 4. Root mean square error Er time evolution of **(a)** wind (***W***), **(b)** radial velocity (y) and **(c)** surface current (***U***) for *WindTE*, computed over the POD (full lines) or the MD (dashed lines). Er^b (Er^a) is in green (blue).

4 Impact of the wind optimization

4.1 Results of a twin experiment

The first sensitivity study concerns the wind forcing. Wind perturbations are introduced in such a way that the resulting wind ensemble represents the wind uncertainties. Insufficient wind perturbations may induce only a small variance of the model surface current ensemble and the model control space will be misrepresented by this ensemble of model states. On the contrary, excessive wind perturbations can generate unrealistic patterns in the wind ensemble or in the corresponding model ensemble. We found that, for the present model setting, an error in wind forcing corresponding to 50 % of its temporal variability ($\alpha = 0.5$) is the most efficient in terms of the overall RMSE between control and analyzed states for wind and model solutions during the assimilation period.

The representativity error, S_{obs}, is also investigated. We look for the value of S_{obs} that provides the most accurate optimized wind and model solutions for a given optimization and observation window. A high value of S_{obs} can give insufficient weighting to the observations relative to the background state, while a low value of S_{obs} can produce unrealistic optimized forcings. Therefore, the analysis of the forcing and the model solution must be investigated concurrently to find the optimal compromise. For the present experiment, S_{obs} was set at $1\,\mathrm{m\,s^{-1}}$.

The assimilation process also requires the selection of the duration of the temporal optimization window, $\mathrm{win}_{\mathrm{ana}}$, used to localize the analysis. This duration should minimize the discontinuities between two consecutive windows. Typically, the amplitude of these discontinuities must be on the same order as the temporal variability of the model, and again a compromise must be found to reduce the error of optimized forcing and the resulting model solution. 1 day was found to be adequate for $\mathrm{win}_{\mathrm{ana}}$.

Traditionally, observations are considered within the optimization window. However, depending on the application, the covariances between observations and ensemble simulations at a given time may be low due to a weak causality between forcing and model solution at the observation points. At such a stage we suggest taking into consideration a temporal shift between the observation and optimization windows. Given a time reference that corresponds to the beginning of $\mathrm{win}_{\mathrm{ana}}$, the observations before (after) this time are included in $\mathrm{win}_{\mathrm{obs}}^{-}$ ($\mathrm{win}_{\mathrm{obs}}^{+}$). The values $\mathrm{win}_{\mathrm{obs}}^{-} = 0$ day and $\mathrm{win}_{\mathrm{obs}}^{+} = 3$ days provide the best results in terms of the overall Er for wind and surface currents. The experiment performed with this data set is hereafter referred to as *WindTE*.

Figure 4a shows the improvement in wind field estimation using idealized radar measurements in terms of the time evolution of $\mathrm{Er}_N(k)$. The assimilation is effective over the total period except at some dates (e.g., 30 November–2 December 2011), representing only 8.6 % of the overall duration. These dates mainly correspond to weak covariances between surface currents on the POD and wind over the MD, corresponding to the presence of mesoscale circulation patterns that are not wind induced.

The resulting velocity field in the radial component shows an improvement during the assimilation period after a 3-day adjustment period (Fig. 4b). The overall Sc of radial velocities is 0.54. Oscillations can be observed to have a period of 17 h, which is close to the inertial period at the considered latitudes, but cannot be directly correlated with the wind solutions (Fig. 4a). The presence of such oscillations indicates that free and control surface current near-inertial oscillations are not in phase and that the wind optimization has only a limited impact on the near-inertial oscillations.

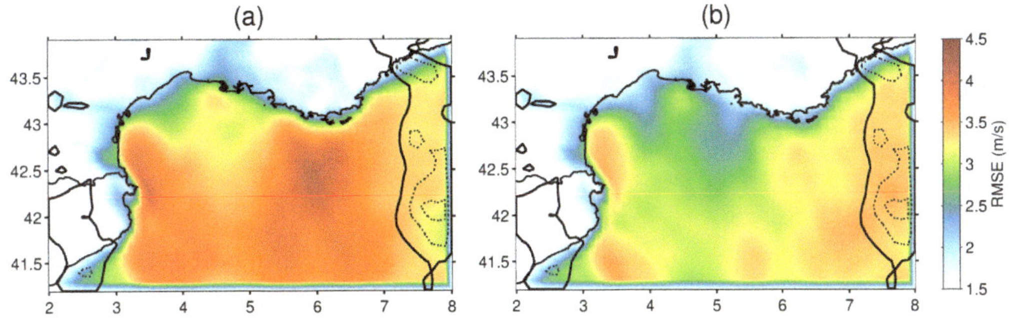

Figure 5. Spatial pattern of **(a)** Er^b and **(b)** Er^a of the wind ($\boldsymbol{W}$) computed over the assimilation period for *WindTE*. Δ is indicated by contour lines and only $0\,\mathrm{m\,s^{-1}}$ (bold black lines) and negative (-0.2 and $-0.1\,\mathrm{m\,s^{-1}}$, dashed black lines) values are shown.

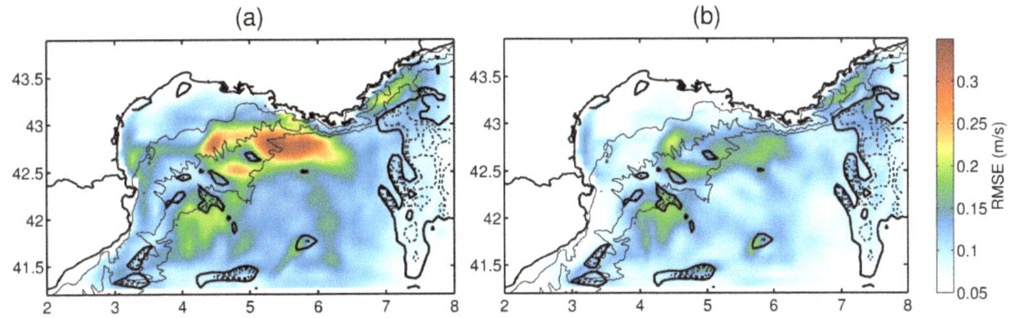

Figure 6. Spatial pattern of **(a)** Er^b and **(b)** Er^a of the surface current $\boldsymbol{U}$ for *WindTE*. Δ is indicated by contour lines and only $0\,\mathrm{m\,s^{-1}}$ (bold black lines) and negative (-0.03, -0.02 and $-0.01\,\mathrm{m\,s^{-1}}$, dashed black lines) values are shown. The 100, 1000 and 2000 isobaths are drawn (black thin lines).

The spatial patterns of $Er_K(n)$ for the forcing wind are depicted in Fig. 5, superimposed on the negative values of Δ to highlight the regions where the optimized wind is degraded compared to the original wind. Over most of the MD, assimilation of surface currents improves the wind estimation. The area in the eastern part of the MD, where wind estimation is degraded, corresponds to a region where original and control winds are in good agreement before assimilation.

The model response to the analyzed wind forcing in terms of Er of surface velocity field is shown in Fig. 6. The surface current vector field resulting from analysis is generally more accurate than the free run. Maximum improvement is observed at the entrance of the GoL and along the shelf break. Along the eastern open boundary the assimilated solution is less accurate, corresponding to the degradation of the analyzed wind forcing (Fig. 5).

Hovmöller diagrams of radial velocities along a south–north transect at 6.2° N across the POD are shown in Fig. 7 for the PEY radar station. The efficiency of the optimized wind forcing on the radial velocity model response is visible in the high intensity patterns. For the positive radial velocities (red), the assimilation reduces the overestimation of the free run in the period 20–24 November. Later in December, the free run clearly underestimates the radial velocities, and again the optimized wind intensifies these features, for outward components (4–10 December) and inward components (10–16 December). Moreover, the use of the optimized wind preserves the near-inertial oscillations. Results for the BEN radar station (not shown) are similar but weaker.

4.2 Results of a real case experiment

The *WindReal* experiment is performed to correct the same free model run with real radar observations, from the two PEY and BEN radar stations (Fig. 1). Observations are averaged over consecutive 3 h windows to improve computational capabilities.

The best compromise in terms of lowest overall Er and Sc is obtained for $\alpha = 0.3$, $S_{obs} = 1\,\mathrm{m\,s^{-1}}$, $win_{ana} = 1$ day, $win_{obs}^{-} = 0$ day and $win_{obs}^{+} = 3$ days, according to the findings from sensitivity tests performed on these parameters.

For this experiment, contrary to the TE where the true state is known in terms of wind and model solution, no ground truth data are available to validate the optimized wind forcing. Meteorological stations in the region did supply wind observations effectively, but all were assimilated into the ALADIN meteorological forcing, and no additional independent wind observations were available. To ensure realistically optimized winds, even if this study does not aim at improving wind fields, the original and optimized winds have been compared to in situ winds from four land stations located near Toulon and from two moorings located in the GoL and off

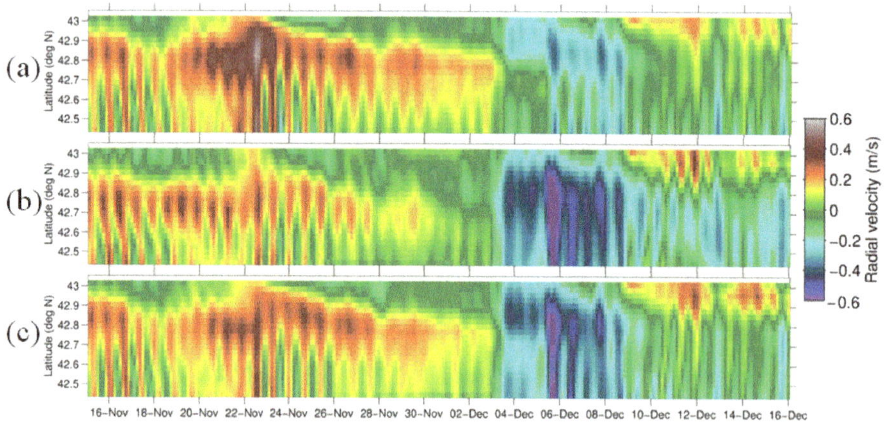

Figure 7. Hovmöller diagram of radial velocities from PEY along the south–north transect at 6.2° N (see Fig. 1), simulated by **(a)** the free simulation, **(b)** the control and **(c)** the analysis for *WindTE* simulation.

Table 1. Radial velocity statistics for the PEY, BEN and both (PEY + BEN) radar stations computed for the free, *WindReal* and *ObcReal* runs in relation to the observation over the entire assimilation period. B and Er are in m s^{-1}. Sc is without unit.

		Free	*WindReal*	*ObcReal*
PEY + BEN	B	−0.101	−0.051	−0.078
	Er	0.191	0.158	0.169
	Sc	–	0.31	0.23
PEY	B	−0.133	−0.074	−0.099
	Er	0.225	0.179	0.192
	Sc	–	0.36	0.27
BEN	B	−0.056	−0.022	−0.053
	Er	0.138	0.127	0.134
	Sc	–	0.13	0.06

Nice. No significant changes appear in correlations between observed and original or *WindReal* analyzed winds. The optimized meridional wind component approximates the original one. The optimized zonal wind component Er increases by 0.6 m s^{-1} compared to the original wind. The original and optimized wind maps show some differences, but the optimization does not introduce any unrealistic features.

The time evolution of $\mathrm{Er}_N(k)$ computed from the radial surface current components measured by the PEY and BEN radars is plotted in Fig. 8.

Current estimation by using wind optimization is better for the PEY station than for the BEN station, with an Sc of 0.36 and 0.13, respectively (Fig. 8b, c, and Table 1). The two radar stations have different viewing geometries, in particular regarding the NC vein, which is the most dynamically active region where errors are the highest (Fig. 9a, d). While the PEY radar station is well placed to identify the westward coastal current velocity corresponding to the largest component of the NC, the BEN station mainly measures its meandering and oscillations, as the NC flows perpendicularly to the look direction of the BEN radar.

Table 1 shows radial velocities Er, B and Sc for PEY, BEN and both radar stations. When both stations are considered together, Er reaches values of up to 0.15 m s^{-1} (Table 1), and the overall bias is halved.

The spatial patterns of $\mathrm{Er}_K(n)$ are limited to the radar regions (Fig. 9) to be compared with the observations. Free radial velocities from the PEY station show higher discrepancies with the observations (Fig. 9a) than free radial velocities from the BEN station (Fig. 9d). The assimilation performs well in most of the domain, especially north of 42.7° N. The best improvement (red areas of the figure) corresponds to the NC vein, the most energetic feature of the region where errors are the highest.

Hovmöller diagrams from free (Fig. 2a) and analyzed (Fig. 11a) radial velocities from the PEY station show that optimized wind forcing acts on the NC vein intensity. During the period 15–26 November 2011, the intensity of the NC vein is strengthened, better approximating the observations (Fig. 2b), and is weakened during the period 4–9 December 2011. A southward shift of the NC vein was also found during the period 10–16 November 2011.

Data assimilation performance reduces when moving southward and degradation regions may appear. This may be due to the scarcity of observations when moving away from the radar stations.

The norm of vector speed difference $\boldsymbol{U}^{\mathrm{b}} - \boldsymbol{U}^{\mathrm{a}}$ between free currents $\boldsymbol{U}^{\mathrm{b}}$ and analyzed currents $\boldsymbol{U}^{\mathrm{a}}$ averaged over the assimilation period and at the surface (Fig. 10a) and integrated over the first 200 m below the surface (Fig. 10c) shows that the corrections are primarily located on the surface in the NC vein all along the French coast and over the GoL shelf break and can occur over the whole MD. However, they vanish with depth.

Once we have verified that the wind optimization obtained through the assimilation of surface velocities does

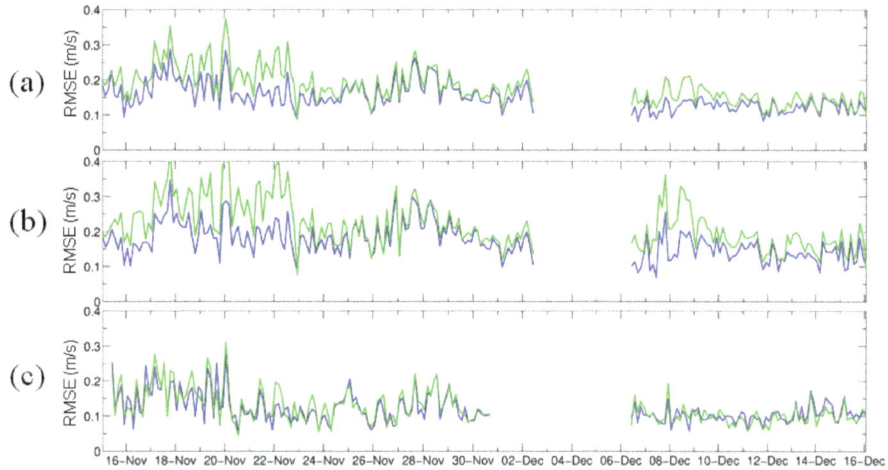

Figure 8. (a) Er time evolution of radial velocity from **(a)** both radar sites (y), **(b)** PEY (y_{pey}) and **(c)** BEN (y_{ben}) for *WindReal*. Er^{b} (Er^{a}) is in green (blue).

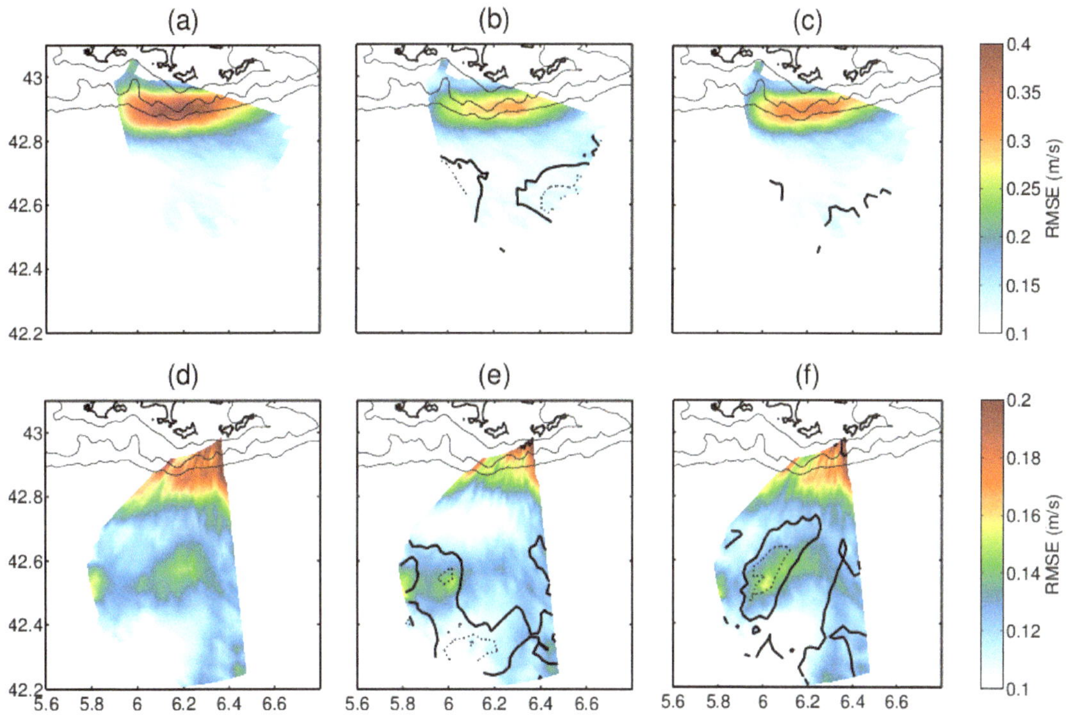

Figure 9. Spatial distributions of **(a)** $\text{Er}^{\text{b}}(y_{\text{pey}})$, **(b)** $\text{Er}^{\text{a}}(y_{\text{pey}})$ for *WindReal* and **(c)** $\text{Er}^{\text{a}}(y_{\text{pey}})$ for *ObcReal* (respectively **d**, **e**, **f** for y_{ben}). Δ is indicated by contour lines and only $0\,\text{m}\,\text{s}^{-1}$ (bold black lines) and negatives (-0.02 and $-0.01\,\text{m}\,\text{s}^{-1}$, dashed black lines) values are shown. Colorbars differ for PEY and BEN. The 100, 1000 and 2000 m isobaths are drawn (black thin lines).

not degrade the model solution in the observation area, the question of its impact off the OD, and in the whole water column, in terms of hydrological variables, arises. Temperature and salinity from CTD measurements and free/analyzed simulations are compared. The wind optimization method employed only has a weak effect on the temperature ($B = 0.017\,°$, $\text{Er} = 0.818\,°$) and salinity ($B = -0.049$, $\text{Er} = 0.081$) profiles. No significant changes in correlation between observed and free or *WindReal* analyzed temperature and salinity are noted. This result is not surprising, as the primary wind effect on the ocean dynamics is surface current response, and only for longer timescales should the turbulent processes transfer the surface instant impact to deeper layers. However, assimilation of radial velocities may not enable the corresponding wind forcing to impact the whole water column or any hydrological variables.

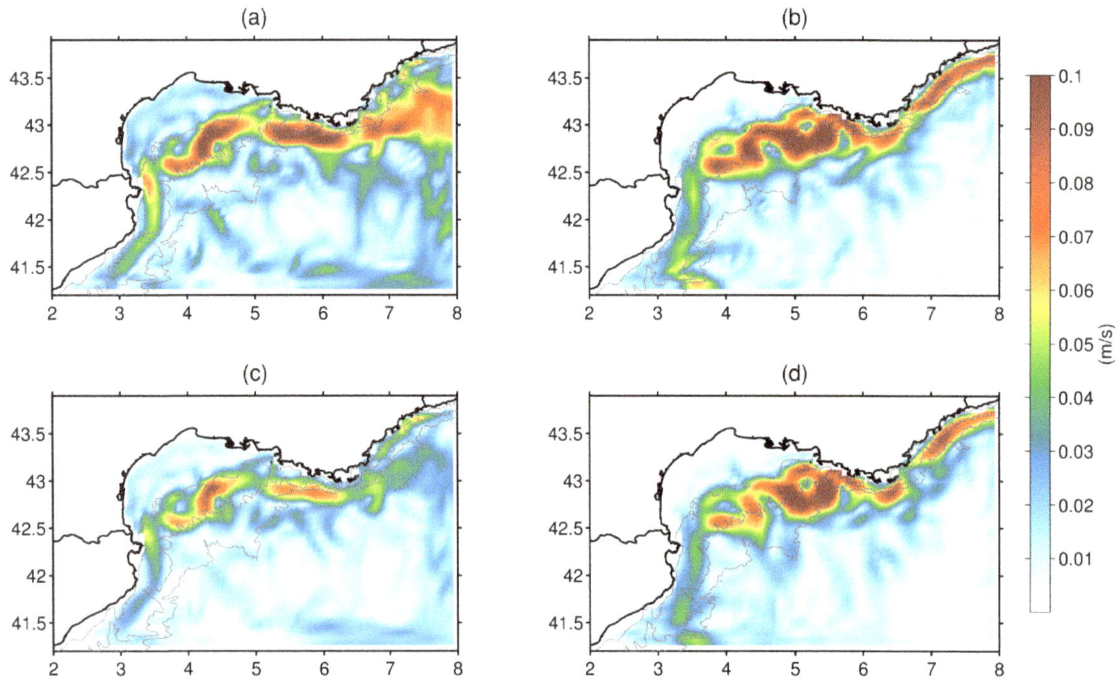

Figure 10. Vector speed difference computed over the assimilation period between free surface current and analyzed surface current from **(a)** *WindReal* or **(b)** *ObcReal*, and free current and analyzed current integrated over the first 200 m below the surface from **(c)** *WindReal* or **(d)** *ObcReal*. The 100, 1000 and 2000 m isobaths are drawn (gray lines).

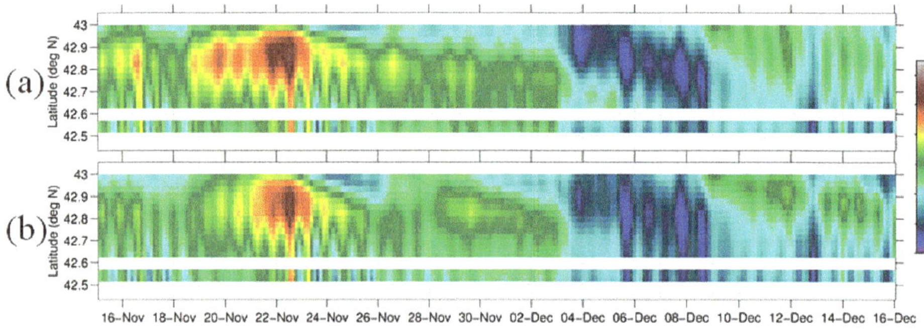

Figure 11. Hovmöller diagram of radial velocities from the PEY station along a south–north transect at 6.2° N (see Fig. 1) **(a)** from *WindReal* and **(b)** *ObcReal* analysis.

5 Impact of the OBC optimization

5.1 Results of a twin experiment

An attempt to correct the OBC values using the same method as for the wind correction is evaluated first on the TE. These experiments are performed with 1 day-smoothed atmospheric forcing to weaken the influence of the high-frequency wind variability on surface currents and focus on the OBC influence only.

The method requires the tuning of the same parameters as for wind optimization: α, S_{obs}, win_{ana}, $\text{win}^{+}_{\text{obs}}$, $\text{win}^{-}_{\text{obs}}$. The reference control state is a member of the ensemble with $\alpha = 1$. After several numerical tests, the best compromise in terms of the overall Er and Sc was found with $\alpha = 1.5$, $S_{\text{obs}} = 2\,\text{m}\,\text{s}^{-1}$, $\text{win}_{\text{ana}} = 1$ day, $\text{win}^{-}_{\text{obs}} = 0$ day and $\text{win}^{+}_{\text{obs}} =$ 15 days. The experiment performed with this data set is hereafter referred to as *ObcTE*.

It should be noted that the choice $\text{win}^{+,-}_{\text{obs}}$ is crucial. As the coastal circulation flows westward, $\text{win}^{+}_{\text{obs}}$ allows the temporal localization to take into account the delay necessary for propagating OBC information up to the observation area. This delay is about 15 days for an advection propagation of $25\,\text{cm}\,\text{s}^{-1}$ (e.g., Fig. 1). The nil value obtained for $\text{win}^{-}_{\text{obs}}$ denotes that no valuable information propagates from the observation area to the eastern OBC.

The assimilation method leads to a significant improvement in optimized OBC variables U^{obc} and η^{obc}. The method

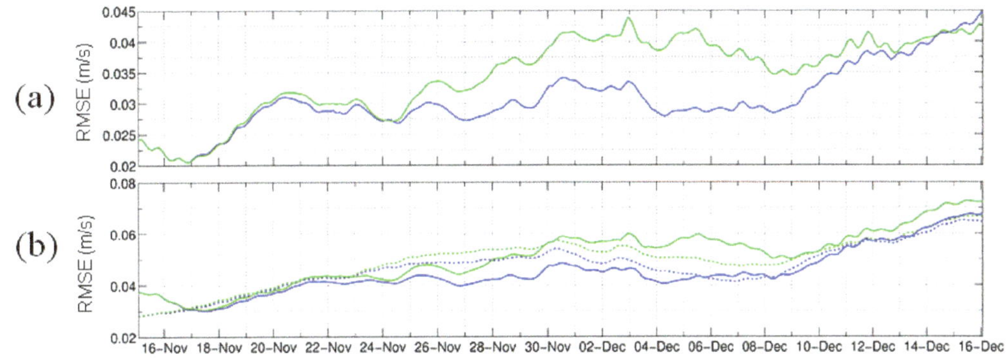

Figure 12. Time evolution of **(a)** $\mathrm{Er}^{\mathrm{a,b}}(y)$ and **(b)** $\mathrm{Er}^{\mathrm{a,b}}$ $(\boldsymbol{U})$ computed over the POD (full lines) or the MD (dashed lines) for *ObcTE*. Er^{b} (Er^{a}) is in green (blue).

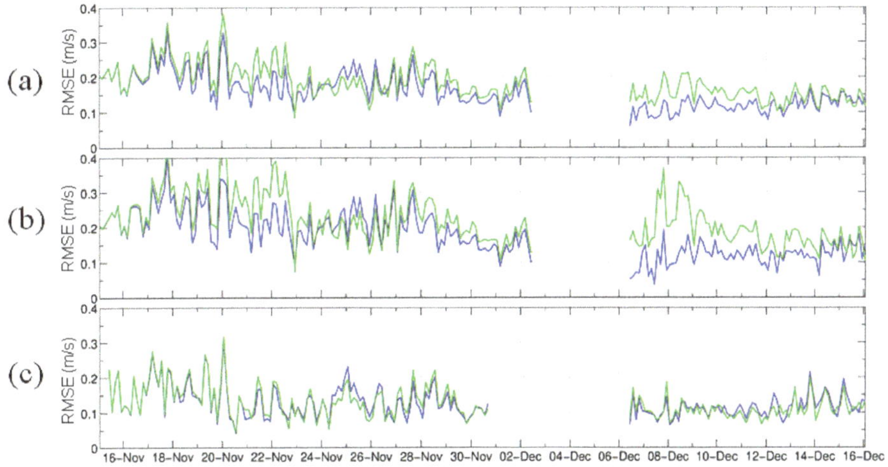

Figure 13. Temporal evolution of **(a)** $\mathrm{Er}^{\mathrm{a,b}}(y)$, **(b)** $\mathrm{Er}^{\mathrm{a,b}}(y_{\mathrm{pey}})$, and **(c)** $\mathrm{Er}^{\mathrm{a,b}}(y_{\mathrm{ben}})$ for *ObcReal*. Er^{b} (Er^{a}) is in green (blue).

Table 2. T^{obc}, S^{obc}, U^{obc}, V^{obc} and η^{obc} statistics computed for *ObcTE* between control and free or analyzed OBC values over the assimilation period. Er unit depends on physical parameters. Sc is without unit.

	Er^{b}	Er^{a}	Sc
T^{obc} (°C)	0.147	0.142	0.06
S^{obc}	0.014	0.014	0.06
U^{obc} ($\mathrm{m\,s^{-1}}$)	0.022	0.019	0.27
V^{obc} ($\mathrm{m\,s^{-1}}$)	0.017	0.019	−0.24
η^{obc} (m)	0.009	0.008	0.20

has only a limited effect on T^{obc} and S^{obc}, while V^{obc} is degraded (Table 2). This suggests that zonal velocity and surface elevation at the eastern boundary are the key parameters controlling model surface currents in the POD. The first parameter imposes the fluxes at the entrance of the MD, whereas the second parameter governs the geostrophic component of the NC, which is the main large-scale feature of the region (e.g., Birol et al., 2010). The degradation of the meridional velocity reflects the weak correlation that exists between surface currents in the POD and V^{obc}. The spatial pattern of $\mathrm{Er}_K^{\mathrm{a,b}}(n)$ at the eastern boundary of the model shows that the maximum improvement for U^{obc} and η^{obc} occurs north of 43.5° N (not shown).

In the POD, the optimized OBC values improve the model radial velocity (Fig. 12a). From 15 November to 18 November 2011, no effect on the surface currents is apparent. This corresponds to the time necessary for the OBC information to reach the POD. From 25 November 2011, $\mathrm{Er}_N^{\mathrm{a}}(k)$ is smaller than $\mathrm{Er}_N^{\mathrm{b}}(k)$, with differences of up to $1\ \mathrm{cm\,s^{-1}}$, except at the end of the time period. The lower efficiency of assimilation on surface currents from 14 December 2011 is a direct consequence of the degradation of U^{obc}. The overall Sc of radial velocities is 0.22. Similar comments can be made for the vectorial surface currents (Fig. 12b), except that no degradation occurs at the end of the time period.

Despite several regions of degradation, model radial surface currents are globally improved, as well as surface temperature and salinity (not shown).

5.2 Results of a real case experiment

The same free run and the same real radar observations as in *WindReal* are used to correct OBC values. In order to set the optimal value for α, the nearest observations were used to get an estimate of the forcing error relative to its temporal variability at the eastern boundary. α ranged from 0.5 to 1. In this study, the lowest overall Er were obtained for $\alpha = 1$. The value $S_{\text{obs}} = 2\,\text{m}\,\text{s}^{-1}$ was the best compromise for the representativity error. The other parameters are identical to those of the *ObcTE* experiment. The experiment performed with this data set is hereafter referred to as *ObcReal*.

Figure 13 shows the temporal evolution of $\text{Er}_N(k)$ for radial surface currents, for each individual radar site, and for both. When both stations are considered, the analyzed currents better approximate observations, with differences between $\text{Er}_N^{\text{a}}(k)$ and $\text{Er}_N^{\text{b}}(k)$ of up to $7\,\text{cm}\,\text{s}^{-1}$. The adjustment delay at the beginning of the period is needed for the optimized information to be propagated from the eastern boundary to the OD. Statistical analysis reveals that the absolute value of the overall B is reduced by $2.3\,\text{cm}\,\text{s}^{-1}$ for the analyzed surface currents compared to the free surface currents (Table 1). As in *WindReal*, greater improvements are obtained for the PEY station (Fig. 13b) and the impact of data assimilation is small on BEN radial velocities (Fig. 13c).

$\text{Er}_K(n)$ maps for PEY and BEN are shown in Fig. 9c, f. Modeled radar currents are mainly improved north of 42.8° N in the NC vein. For PEY, no significant degradation is observed. On the contrary, for BEN, areas of degradation up to $1.6\,\text{cm}\,\text{s}^{-1}$ are observed around 42.6° N, which corresponds to the southern edge of the NC. The analysis of vectorial surface currents (not shown here) shows that, using the optimized OBC, the NC vein is more marked due to higher surface velocities along the whole NC pathway.

Hovmöller diagrams of free (Fig. 2a) and analyzed (Fig. 11b) radial velocities from the PEY station show that the use of optimized OBC values has a similar effect on radial velocities to the use of optimized wind fields but to a lesser extent.

The structure of corrections is investigated by computing the norm of $\boldsymbol{U}^{\text{b}} - \boldsymbol{U}^{\text{a}}$ averaged over the assimilation period, at the surface (Fig. 10b) and integrated over the first 200 m below the surface (Fig. 10d). It appears that corrections are primarily located in the NC vein all along the French coast and over the GoL shelf break and affect a large part of the water column. This result proves clearly that the NC strongly depends on its upstream dynamic.

The effect of OBC optimization on the water column is also studied using CTD measurements. Correlations between observed and free or *OBCreal* analyzed temperature and salinity remain unchanged. A bias difference between *ObcReal* analyzed and free runs of 0.12 °C is observed in temperature values, which indicates that the effect of assimilation on the water column is more important than in *WindReal*. A slight degradation is obtained for salinity values, with a bias difference of −0.006.

6 Conclusions

For the first time, HF radar measurements in the NWM Sea are assimilated into a circulation model. The region is characterized by a steep topography and complex coastline, and its ocean dynamics includes a geostrophic coastal current associated with strong mesoscale activity. The EnPS is used to correct two different sources of error acting on the model surface current: high-resolution wind forcing and baroclinic open boundary conditions.

The performance of the assimilation method can be assessed by evaluating either the optimized forcing or the resulting model response. Since no independent wind measurements were available, the assessment of the wind optimization could be done only on the TEs, which clearly demonstrate the strong non-local impact of this method. A limited area of HF radar observations can be effective in large-scale wind correction, and it can reasonably be inferred that the resulting model response will be improved over the whole area. This method also improves surface current simulations through OBC optimization. This result is non-trivial because the OBC optimized area is located far from the observation area (150 km). The perturbation of OBC is done in the NC vein area and takes into account uncertainties regarding width, depth and speed of the NC vein since all variables are perturbed.

The results show that the correction of the wind forcing has a stronger impact on the resulting simulated surface current than the OBC forcing correction. *WindReal* Er is $1\,\text{cm}\,\text{s}^{-1}$ less than the *ObcReal* one, and in both cases, bias is reduced. Wind optimization effects are felt in surface currents over short time intervals contrarily to OBC optimization, whose improved values are propagated by model equations over time and space. In terms of temperature and salinity along the water column, mixed results are achieved. No significant improvements are obtained and areas of degradations can occur in temperature values for *WindReal* or salinity values for *ObcReal*.

The assimilation of HF radar-derived radial velocities acts on model surface currents over the whole of the MD. The main corrections are depicted by red values in Fig. 10, which represent the norm of $\boldsymbol{U}^{\text{b}} - \boldsymbol{U}^{\text{a}}$ averaged over the assimilation period. These corrections are primarily located in the NC vein all along the French coast and over the GoL shelf break. *WindReal* corrections are less significant off Nice but occur over the whole of the MD. When the same vector speed difference is applied to the 200 m-depth averaged currents, the correction introduced by *ObcReal* can affect model currents of the deeper water masses, whereas the correction introduced by *WindReal* mainly concerns model surface currents.

These results inspire confidence in the implementation of the method on OBC forcing optimization. When dealing with a domain with large open boundaries that strongly influences internal ocean dynamics, a small inconsistency from the outside will certainly trigger unrealistic features in the domain of interest. The availability of local monitoring systems in the area of interest could be used with this method, to correct the OBC input information, in terms of intensity or positioning of incoming/outcoming currents. In addition, the monitoring of the region off Toulon can be of value for the comprehension of the NWM sea circulation, in particular when considering the NC variability, which is still poorly documented. The recent development of HF radars on the Mediterranean coast through national and international projects should be extended to enhance the coastal monitoring and the assimilation of HF radar data.

The methodology was successfully applied and differences between observed and modeled surface currents were significantly reduced. However, further improvements can be envisaged by taking more aspects into consideration. Wind and OBC optimization have been tested independently and the impact on the model by simultaneous forcing optimizations should be considered in future work, which would require the use of a more realistic eddy-resolving model. Only the eastern boundary was addressed here, and the complementary treatment of the southern boundary may also improve the overall result. Furthermore, the regional model used in this study is quite coarse and may contain errors not directly linked to the forcing: for instance, the bathymetry and the coastline contour may not be well resolved, or the parametrization of the OBC treatment or the turbulent schemes may not be accurate enough when considering the assimilation of high-resolution surface currents derived from HF radar. The use of a high-resolution regional model configuration with an advanced physical parametrization (Ourmières et al., 2011; Guihou et al., 2013) could also lead to significant improvements.

Acknowledgements. The experiment at sea and the HF radar installation were supported by ECCOP-LEFE/IDAO (CNRS/INSU) and TOSCA (2G-MED09-425). The TOSCA project is cofinanced by the European Regional Development Fund, in the framework of the MED program. This study is also supported by LEFE-INSU GMMC. The calculations were performed using NEC-SX8 from GENCI-IDRIS resources (grant 2012011707). The authors also wish to thank Mercator Ocean and Météo-France for providing numerical data and observations. The MIO RADAR team (Y. Barbin, C. Quentin, J. Gaggelli, P. Guterman) is also acknowledged for its major contribution to radar field campaigns of this study. The authors thank the two anonymous referees for their fruitful comments.

Edited by: G. Desroziers

Appendix A: Ensemble Perturbation Smoother

EnPS was proposed by Barth et al. (2010) to optimize tidal boundary conditions and was successfully applied by Barth et al. (2011) to optimize surface wind forcing. Contrary to the classical sequential method, which intends to derive the best estimation of the model state, this method aims at optimizing forcing values. These forcing fields are then run once again in the model to get the final model solution. Thus, the model trajectory is corrected instead of the model state as in the ensemble smoother (Van Leeuwen, 2001), the 4D-LETKF (Hunt et al., 2004, 2007) or the AEnKF (Sakov et al., 2010). This method also ensures that the model solution satisfies the model equations exactly.

This non-sequential method is derived from the sequential EnKF scheme by including the time dimension in the observation vector (Sakov et al., 2010). Consequently, all observations in the optimization period are gathered in a single observation vector $\boldsymbol{y}^{\mathrm{o}}$, with an observation error covariance matrix $\mathbf{R}$.

The optimization vector $\boldsymbol{x}$, called the state vector in the classical method, contains all forcing and model parameters to be improved.

An ensemble of forcing $\mathbf{x}^{l}$, where l is the index of a member in the ensemble, is generated by adding a perturbation to the background (or original) forcing $\mathbf{x}^{\mathrm{b}}$. If perturbed forcings are time dependent, their temporal evolution is gathered in a single vector, in the same way as observations.

For each perturbed forcing, the model is run. Note that at this step ensemble members are not influenced by observations. For each ensemble member, the observed part of the model state $h(\mathbf{x}^{l})$ is extracted using a non-linear observation operator $h(.)$. Consequently, model dynamic is included in the observation operator. Each element of the ensemble $h(\mathbf{x}^{l})$ can be directly compared to the corresponding one in the observation vector $\boldsymbol{y}^{\mathrm{o}}$.

Error covariance matrices can be estimated from the use of an ensemble of forcing state and model state. Using the same notation as in Barth et al. (2010, 2011), we define the deviation of the ensemble members around the ensemble mean of forcing $\mathbf{S}$ and observed part of surface currents $\mathbf{E}$ as

$$(\mathbf{S})_l = (L-1)^{-\frac{1}{2}}(\mathbf{x}^{(l)} - \langle \mathbf{x} \rangle) \tag{A1}$$

$$(\mathbf{E})_l = (L-1)^{-\frac{1}{2}}(h(\mathbf{x}^{(l)}) - \langle h(\mathbf{x}) \rangle) \tag{A2}$$

where L is the number of ensemble members, the index l refers to the ensemble member ($l = 1,\ldots,N$) and $\langle . \rangle$ is the ensemble average operator. These matrices are scaled in such a way that $\mathbf{SE}^T$ and $\mathbf{EE}^T$ represent, respectively, ensemble forcing covariance and covariance between ensemble forcing and extracted observations:

$$\mathbf{SE}^T = \mathrm{cov}(\mathbf{x}^{\mathrm{b}}, h(\mathbf{x}^{\mathrm{b}})) \tag{A3}$$

$$\mathbf{EE}^T = \mathrm{cov}(h(\mathbf{x}^{\mathrm{b}}), h(\mathbf{x}^{\mathrm{b}})) \tag{A4}$$

Note that there is no need to form explicitly and to store the covariance matrices, because analysis is computed in the ensemble space (e.g., Nerger et al., 2005).

The optimal forcing $\mathbf{x}^{\mathrm{a}}$ that minimizes the analysis error is computed using Eq. (2), which is the Kalman filter analysis step with a non-linear observation operator (Chen and Snyder, 2007). It can be developed as

$$\mathbf{x}^{\mathrm{a}} = \mathbf{x}^{\mathrm{b}} + \mathbf{SE}^T(\mathbf{EE}^T + \mathbf{R})^{-1}(\mathbf{y}^{\mathrm{o}} - h(\mathbf{x}^{\mathrm{b}})) \tag{A5}$$

or using the Sherman–Morrison–Woodbury formula:

$$\mathbf{x}^{\mathrm{a}} = \mathbf{x}^{\mathrm{b}} + \mathbf{S}(\mathbf{E}^T\mathbf{R}^{-1}\mathbf{E} + 1)^{-1}\mathbf{E}^T\mathbf{R}^{-1}(\mathbf{y}^{\mathrm{o}} - h(\mathbf{x}^{\mathrm{b}})) \tag{A6}$$

The superscripts "a" and "b" refer to the analysis and the background estimates. As in Kalman filtering, background and observation errors are assumed to be unbiased, but in the case of a strongly non-linear observation operator, such assumptions are not sufficient to ensure that the analysis $\mathbf{x}^{\mathrm{a}}$ is unbiased. However, the analysis is unbiased if Eq. (3) is verified (Barth et al., 2011).

The model trajectory is corrected in a final step that consists in running the model with the optimized forcing $\mathbf{x}^{\mathrm{a}}$. Consequently, observations influence the analyzed model solution using an optimal combination of ensemble forcing.

References

Alberola, C., Rousseau, S., Millot, C., Astraldi, M., Font, J., Garcialafuente, J., Gasparini, G., Send, U., and Vangriesheim, A.: Tidal currents in the western mediterranean-sea, Oceanol. Acta, 18, 273–284, 1995.

Allou, A., Forget, P., and Devenon, J.-L.: Submesoscale vortex structures at the entrance of the Gulf of Lions in the Northwestern Mediterranean Sea, Cont. Shelf Res., 30, 724–732, 2010.

Barth, A., Alvera-Azcárate, A., and Weisberg, R. H.: Assimilation of high-frequency radar currents in a nested model of the West Florida Shelf, J. Geophys. Res.-Oceans (1978–2012), 113, C08033, doi:10.1029/2007JC004585, 2008a.

Barth, A., Alvera-Azcarate, A., and Weisberg, R. H.: A nested model study of the Loop Current generated variability and its impact on the West Florida Shelf, J. Geophys. Res., 113, C05009, doi:10.1029/2007JC004492, 2008b.

Barth, A., Alvera-Azcárate, A., Gurgel, K.-W., Staneva, J., Port, A., Beckers, J.-M., and Stanev, E. V.: Ensemble perturbation smoother for optimizing tidal boundary conditions by assimilation of High-Frequency radar surface currents-application to the German Bight, Ocean Sci., 6, 161–178, doi:10.5194/os-6-161-2010, 2010.

Barth, A., Alvera-Azcárate, A., Beckers, J.-M., Staneva, J., Stanev, E. V., and Schulz-Stellenfleth, J.: Correcting surface winds by assimilating High-Frequency Radar surface currents in the German Bight, Ocean Dynam., 61, 599–610, 2011.

Birol, F., Cancet, M., and Estournel, C.: Aspects of the seasonal variability of the Northern Current (NW Mediterranean Sea) observed by altimetry, J. Mar. Syst., 81, 297–311, 2010.

Blayo, E. and Debreu, L.: Revisiting open boundary conditions from the point of view of characteristic variables, Ocean Modell., 9, 231–252, 2005.

Bogden, P. S., Malanotte-Rizzoli, P., and Signell, R.: Open-ocean boundary conditions from interior data: Local and remote forcing of Massachusetts Bay, J. Geophys. Res.-Oceans (1978–2012), 101, 6487–6500, 1996.

Brankart, J.-M. and Brasseur, P.: Optimal analysis of in situ data in the Western Mediterranean using statistics and cross-validation, J. Atmos. Ocean. Technol., 13, 477–491, 1996.

Breivik, Ø. and Sætra, Ø.: Real time assimilation of HF radar currents into a coastal ocean model, J. Mar. Syst., 28, 161–182, 2001.

Capet, X., McWilliams, J., Molemaker, M., and Shchepetkin, A.: Mesoscale to submesoscale transition in the California Current System. Part I: Flow structure, eddy flux, and observational tests, J. Phys. Oceanogr., 38, 29–43, 2008.

Casella, E., Molcard, A., and Provenzale, A.: Mesoscale vortices in the Ligurian Sea and their effect on coastal upwelling processes, J. Mar. Syst., 88, 12–19, 2011.

Chapman, D. C.: Numerical treatment of cross-shelf open boundaries in a barotropic coastal ocean model, J. Phys. Oceanogr., 15, 1060–1075, 1985.

Chapman, R., Shay, L., Graber, H., Edson, J., Karachintsev, A., Trump, C., and Ross, D.: On the accuracy of HF radar surface current measurements: Intercomparisons with ship-based sensors, J. Geophys. Res.-Oceans (1978–2012), 102, 18737–18748, 1997.

Chen, Y. and Snyder, C.: Assimilating vortex position with an ensemble Kalman filter, Mon. Weather Rev., 135, 1828–1845, 2007.

Cosme, E., Verron, J., Brasseur, P., Blum, J., and Auroux, D.: Smoothing Problems in a Bayesian Framework and Their Linear Gaussian Solutions., Mon. Weather Rev., 140, 683–695, doi:10.1175/MWR-D-10-05025.1, 2012.

Echevin, V., Crepon, M., and Mortier, L.: Interaction of a coastal current with a gulf: application to the shelf circulation of the Gulf of Lions in the Mediterranean Sea, J. Phys. Oceanogr., 33, 188–206, 2003.

Evensen, G.: Sequential data assimilation with a nonlinear quasi-geostrophic model using Monte Carlo methods to forecast error statistics, J. Geophys. Res.-Oceans (1978–2012), 99, 10143–10162, 1994.

Fischer, C., Montmerle, T., Berre, L., Auger, L., and Ştefănescu, S. E.: An overview of the variational assimilation in the ALADIN/France numerical weather-prediction system, Q. J. Roy. Meteorol. Soc., 131, 3477–3492, 2005.

Flather, R.: A tidal model of the northwest European continental shelf, Mem. Soc. R. Sci. Liege, 10, 141–164, 1976.

Gelb, A.: Applied optimal estimation, The MIT press, 1974.

Guihou, K., Marmain, J., Ourmières, Y., Molcard, A., Zakardjian, B., and Forget, P.: A case study of the mesoscale dynamics in the North-Western Mediterranean Sea: a combined data–model approach, Ocean Dynam., 63, 1–16, , doi:10.1007/s10236-013-0619-z, 2013.

Gunson, J. R. and Malanotte-Rizzoli, P.: Assimilation studies of open-ocean flows: 1. Estimation of initial and boundary conditions, J. Geophys. Res.-Oceans (1978–2012), 101, 28457–28472, 1996.

Gurgel, K.-W., Antonischki, G., Essen, H.-H., and Schlick, T.: Wellen Radar (WERA): a new ground-wave HF radar for ocean remote sensing, Coast. Eng., 37, 219–234, 1999.

Hamill, T. M., Whitaker, J. S., and Snyder, C.: Distance-dependent filtering of background error covariance estimates in an ensemble Kalman filter, Mon. Weather Rev., 129, 2776–2790, 2001.

He, R., Liu, Y., and Weisberg, R. H.: Coastal ocean wind fields gauged against the performance of an ocean circulation model, Geophys. Res. Lett., 31, L14303, doi:10.1029/2003GL019261, 2004.

Hoteit, I., Cornuelle, B., Kim, S., Forget, G., Köhl, A., and Terrill, E.: Assessing 4D-VAR for dynamical mapping of coastal high-frequency radar in San Diego, Dynam. Atmos. Ocean., 48, 175–197, 2009.

Houtekamer, P. L. and Mitchell, H. L.: A sequential ensemble Kalman filter for atmospheric data assimilation, Mon. Weather Rev., 129, 123–137, 2001.

Hunt, B., Kalnay, E., Kostelich, E., Ott, E., Patil, D., Sauer, T., Szunyogh, I., Yorke, J., and Zimin, A.: Four-dimensional ensemble Kalman filtering, Tellus A, 56, 273–277, 2004.

Hunt, B. R., Kostelich, E. J., and Szunyogh, I.: Efficient data assimilation for spatiotemporal chaos: A local ensemble transform Kalman filter, Physica D: Nonlinear Phenomena, 230, 112–126, 2007.

Kurapov, A. L., Egbert, G. D., Allen, J., Miller, R. N., Erofeeva, S. Y., and Kosro, P.: The M 2 internal tide off Oregon: Inferences from data assimilation, J. Phys. Oceanogr., 33, 1733–1757, 2003.

Langlais, C., Barnier, B., and Fraunié, P.: Towards high resolution coastal and shelf circulation modeling: impact of the atmospheric forcing resolution in the Gulf of Lions, Mercator. Ocean Q. Newsl., 32, 5–13, 2009.

Le Dimet, F.-X. and Talagrand, O.: Variational algorithms for analysis and assimilation of meteorological observations: theoretical aspects, Tellus A, 38, 97–110, 1986.

Lewis, J. K., Shulman, I., and Blumberg, A. F.: Assimilation of Doppler radar current data into numerical ocean models, Cont. Shelf Res., 18, 541–559, 1998.

Lipa, B., Nyden, B., Ullman, D. S., and Terrill, E.: SeaSonde radial velocities: derivation and internal consistency, Ocean. Eng., IEEE J., 31, 850–861, 2006.

Malanotte-Rizzoli, P., Young, R. E., and Haidvogel, D. B.: Initialization and data assimilation experiments with a primitive equation model, Dynam. Atmos. Ocean., 13, 349–378, 1989.

Marchesiello, P., McWilliams, J. C., and Shchepetkin, A.: Open boundary conditions for long-term integration of regional oceanic models, Ocean Modell., 3, 1–20, 2001.

Marmain, J., Forget, P., and Molcard, A.: Characterization of ocean surface current properties from single site HF/VHF radar, Ocean Dynam., 61, 1967–1979, 2011.

Millot, C.: Circulation in the western Mediterranean Sea, J. Mar. Syst., 20, 423–442, 1999.

Nerger, L., Hiller, W., and Schröter, J.: A comparison of error subspace Kalman filters, Tellus A, 57, 715–735, 2005.

Oke, P. R., Allen, J., Miller, R., Egbert, G., Austin, J., Barth, J., Boyd, T., Kosro, P., and Levine, M.: A modeling study of the three-dimensional continental shelf circulation off Oregon. Part I: Model-data comparisons, J. Phys. Oceanogr., 32, 1360–1382, 2002a.

Oke, P. R., Allen, J. S., Miller, R. N., Egbert, G. D., and Kosro, P. M.: Assimilation of surface velocity data into a primitive equation coastal ocean model, J. Geophys. Res., 107, 3122, doi:10.1029/2000JC000511, 2002b.

Orlanski, I.: A simple boundary condition for unbounded hyperbolic flows, J. Comput. Phys., 21, 251–269, 1976.

Ourmières, Y., Zakardjian, B., Béranger, K., and Langlais, C.: Assessment of a NEMO-based downscaling experiment for the North-Western Mediterranean region: Impacts on the Northern Current and comparison with ADCP data and altimetry products, Ocean Modell., 39, 386–404, 2011.

Paduan, J. D. and Shulman, I.: HF radar data assimilation in the Monterey Bay area, J. Geophys. Res.-Oceans (1978–2012), 109, C07S09, doi:10.1029/2003JC001949, 2004.

Pairaud, I., Gatti, J., Bensoussan, N., Verney, R., and Garreau, P.: Hydrology and circulation in a coastal area off Marseille: Validation of a nested 3D model with observations, J. Mar. Syst., 88, 20–33, 2011.

Peng, S., Li, Y., and Xie, L.: Adjusting the wind stress drag coefficient in storm surge forecasting using an adjoint technique, J. Atmos. Ocean. Technol., 30, 590–608, doi:10.1175/JTECH-D-12-00034.1, 2012.

Penven, P., Roy, C., Brundrit, G., De Verdière, A. C., Fréon, P., Johnson, A., Lutjeharms, J., and Shillington, F.: A regional hydrodynamic model of upwelling in the Southern Benguela, South Afr. J. Sci., 97, 472–475, 2001.

Petrenko, A. A.: Variability of circulation features in the Gulf of Lion NW Mediterranean Sea. Importance of inertial currents, Oceanol. Acta, 26, 323–338, 2003.

Røed, L. and Cooper, C.: Open boundary conditions in numerical ocean models, in: Advanced physical oceanographic numerical modelling, 411–436, Springer, 1986.

Sakov, P., Evensen, G., and Bertino, L.: Asynchronous data assimilation with the EnKF, Tellus A, 62, 24–29, 2010.

Sammari, C., Millot, C., and Prieur, L.: Aspects of the seasonal and mesoscale variabilities of the Northern Current in the western Mediterranean Sea inferred from the PROLIG-2 and PROS-6 experiments, Deep. Sea Res. Part I, 42, 893–917, 1995.

Schaeffer, A., Garreau, P., Molcard, A., Fraunié, P., and Seity, Y.: Influence of high-resolution wind forcing on hydrodynamic modeling of the Gulf of Lions, Ocean Dynam., 61, 1823–1844, 2011a.

Schaeffer, A., Molcard, A., Forget, P., Fraunié, P., and Garreau, P.: Generation mechanisms for mesoscale eddies in the Gulf of Lions: radar observation and modeling, Ocean Dynam., 61, 1587–1609, 2011b.

Schmidt, R.: Multiple emitter location and signal parameter estimation, Antennas and Propagation, IEEE Trans., 34, 276–280, 1986.

Seiler, U.: Estimation of open boundary conditions with the adjoint method, J. Geophys. Res.-Oceans (1978–2012), 98, 22855–22870, 1993.

Sentchev, A., Forget, P., Barbin, Y., and Yaremchuk, M.: Surface circulation in the Iroise Sea (W. Brittany) from high resolution HF radar mapping, J. Mar. Syst., 109–110, S153–S168, doi:10.1016/j.jmarsys.2011.11.024, 2013.

Shchepetkin, A. F. and McWilliams, J. C.: A method for computing horizontal pressure-gradient force in an oceanic model with a nonaligned vertical coordinate, J. Geophys. Res.-Oceans (1978–2012), 108, 3090, doi:10.1029/2001JC001047, 2003.

Shchepetkin, A. F. and McWilliams, J. C.: The regional oceanic modeling system (ROMS): a split-explicit, free-surface, topography-following-coordinate oceanic model, Ocean Modell., 9, 347–404, 2005.

Shulman, I. and Paduan, J. D.: Assimilation of HF radar-derived radials and total currents in the Monterey Bay area, Deep. Sea Res. Part, 56, 149–160, 2009.

Shulman, I., Lewis, J. K., and Mayer, J. G.: Local data assimilation in the estimation of barotropic and baroclinic open boundary conditions, J. Geophys. Res.-Oceans (1978–2012), 104, 13667–13680, 1999.

Stewart, R. H. and Joy, J. W.: HF radio measurements of surface currents, Deep Sea Res., 21, 1039–1049, 1974.

Taillandier, V., Echevin, V., Mortier, L., and Devenon, J.-L.: Controlling boundary conditions with a four-dimensional variational data-assimilation method in a non-stratified open coastal model, Ocean Dynam., 54, 284–298, 2004.

Taillandier, V., Echevin, V., Mortier, L., and Devenon, J.-L.: Controlling the open ocean forcing of baroclinic coastal jets with a 4D-var method, Dynam. Atmos. Ocean., 48, 155–174, 2009.

Ten Brummelhuis, P., Heemink, A., and Van Den Boogaard, H.: Identification of shallow sea models, Int. J. Num. Method. Fluids, 17, 637–665, 1993.

Van Leeuwen, P. J.: An ensemble smoother with error estimates, Mon. Weather Rev., 129, 709–728, 2001.

15

Testing the detectability of spatio–temporal climate transitions from paleoclimate networks with the START model

K. Rehfeld[1,2,3], **N. Molkenthin**[1,2], **and J. Kurths**[1,2]

[1]Potsdam Institute for Climate Impact Research, P.O. Box 601203, 14412 Potsdam, Germany
[2]Department of Physics, Humboldt-Universität zu Berlin, Newtonstr. 15, 12489 Berlin, Germany
[3]Alfred-Wegener-Institut Helmholtz-Zentrum für Polar- und Meeresforschung, Telegrafenberg A43, 14473 Potsdam, Germany

Correspondence to: K. Rehfeld (krehfeld@awi.de)

Abstract. A critical challenge in paleoclimate data analysis is the fact that the proxy data are heterogeneously distributed in space, which affects statistical methods that rely on spatial embedding of data. In the paleoclimate network approach nodes represent paleoclimate proxy time series, and links in the network are given by statistically significant similarities between them. Their location in space, proxy and archive type is coded in the node attributes.

We develop a semi-empirical model for **S**patio-**T**emporally **A**utoco**R**related **T**ime series, inspired by the interplay of different Asian Summer Monsoon (ASM) systems. We use an ensemble of transition runs of this START model to test whether and how spatio–temporal climate transitions could be detectable from (paleo)climate networks. We sample model time series both on a grid and at locations at which paleoclimate data are available to investigate the effect of the spatially heterogeneous availability of data. Node betweenness centrality, averaged over the transition region, does not respond to the transition displayed by the START model, neither in the grid-based nor in the scattered sampling arrangement. The regionally defined measures of regional node degree and cross link ratio, however, are indicative of the changes in both scenarios, although the magnitude of the changes differs according to the sampling.

We find that the START model is particularly suitable for pseudo-proxy experiments to test the technical reconstruction limits of paleoclimate data based on their location, and we conclude that (paleo)climate networks are suitable for investigating spatio–temporal transitions in the dependence structure of underlying climatic fields.

1 Introduction

A growing number of paleoclimate records from environmental archives are available for past climate reconstruction. Fundamentally, this should increase the potential for successful reconstructions of the spatial and temporal features of past climatic changes, and thus enhance the general understanding of the climate system.

The paleoclimate network approach (Rehfeld et al., 2013), illustrated in Fig. 1, is a tool adapted to the challenges in environmental data analysis. As in the climate network approach (Tsonis et al., 2006; Donges et al., 2009b), *nodes* are identified with positions for which climate time series are available, and *links* are drawn between the nodes if statistically significant associations are found. The *adjacency matrix* **A** is then a sparse binary matrix with the (i, j)th entry being non-zero if (and only if) the time series representing nodes i and j are *significantly* associated. Network statistics can then reflect global and local characteristics of the underlying data: the importance of a node, for example, can be measured by its degree, i.e., how many links the individual node has, or more abstract measures such as betweenness centrality (Opsahl et al., 2010; Barthélemy, 2011). Complex networks have been used to investigate the behavior of the climate system from reanalysis data (Donges et al., 2009a, 2011; Gozolchiani et al., 2011; Steinhaeuser et al., 2010; Tsonis et al., 2010; Yamasaki et al., 2008) or recent observations (Ge-Li and Tsonis, 2009; Malik et al., 2010, 2011). The network methodology can, however, not be applied directly to paleoclimate data.

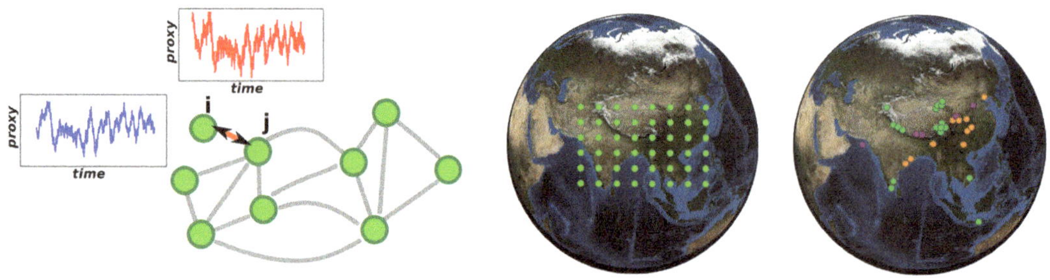

(a) Network construction (b) Climate network (c) Paleoclimate network

Figure 1. **(a)** Schematic illustration of paleoclimate network construction: nodes in the network represent paleoclimate archives with proxy time series. If the time series similarity estimate between a node pair is significant, these two nodes are considered to be linked. The connectedness of the obtained graph reflects the underlying dependencies created by climate processes. While the **(b)** climate network approach is based on gridded and dense observations of climatic parameters such as temperature or precipitation, the **(c)** paleoclimate network combines paleoclimate proxy records that are heterogeneously and sparsely distributed on the Earth's surface.

The first reason is that the availability of paleoclimate data is heterogeneous in time, which implies that standard time series similarity measures are not applicable, since bias effects distort the results (Schulz and Stattegger, 1997; Rehfeld et al., 2011, 2013). For regular data the standard definitions for mutual information (MI, Cover and Thomas, 2006) or Pearson cross correlation (CC, Chatfield, 2004) can be used to obtain correlation matrices for the network. For paleoclimatic data sets, adapted techniques that are more robust against time series irregularity, such as Gaussian kernel-based cross correlation (Rehfeld et al., 2011), mutual information (Rehfeld et al., 2013) or the event synchronization function (ESF, Rehfeld and Kurths, 2014) need to be employed.

The effect of the spatially heterogeneous node distribution on network measures has also not received much attention so far, beyond the studies of Heitzig et al. (2010) and Rheinwalt et al. (2012). In particular for node-based network measures that depend on non-local network topology, changes in network structure could cause non-trivial bias effects. A systematic investigation of the effects of spatial sampling on spatially embedded (climate) networks can be found in the paper of Molkenthin et al. (2014b).

In this paper we first review the paleoclimate network approach and identify potentially suitable network measures. Then we investigate how different network measures reflect distinct changes (transitions) in the underlying climate fields. To this end we develop and use the semi-empirical simplified model **START** (name derived from **S**tream **T**ransported **A**utocor**R**related **T**ime series, or **S**patio-**T**emporally **A**utoco**R**related **T**ime series) to simulate characteristic changes in a spatially extended domain. The model is based on an approximated solution of the advection–diffusion equation, which describes how temperature fluctuations are dissipated through stationary flow fields (Molkenthin et al., 2014a). A single forcing parameter can be varied to change the underlying flow, resulting in distinctly different fields, and in distinctly different climate network evolution patterns (Tupikina et al., 2014). Transitions, in this context, are large-scale dependence changes due to changes in the atmospheric flow patterns. Using this computationally efficient and reduced model we test which climate network measures are suitable for characterizing the transition in the underlying climate field. In particular we evaluate if it is possible to infer the degree of interaction between different regions in the network and how spatially heterogeneously distributed nodes affect the analysis of spatio–temporal dynamics.

2 The paleoclimate network approach (PAN)

A schematic illustration of a paleoclimate network is given in Fig. 1a. Its nodes are given by paleoclimate proxy archives, its links by significant statistical association between the archives' time series. As a method, climate networks ask the question "How dependent, linearly or nonlinearly, are the climate changes in place A on climate changes in another place B – and vice versa?", rather than "Was the temperature in A strongly linearly correlated with the temperature in B at the same point in time?", which is the case for standard empirical orthogonal function analysis (EOFs were used, for example, in Mayewski et al., 2004; Sinha et al., 2011; Yi et al., 2011). If, say, a proxy for local temperature in China co-varied significantly with reconstructed rainfall in India, this is caused by either (a) a common driving phenomenon (for example through the North Atlantic Oscillation, Wu et al., 2009 or solar forcing, Agnihotri et al., 2002), (b) local convective phenomena (for example internal ASM dynamics (Wang et al., 2010)), (c) an artifact in the reconstruction, for example non-climate-related common trends in the time series, or (d) a false positive (Rehfeld and Kurths, 2014).

Consider a paleoclimate network as a graph $G = (V, E)$ on a set of N vertices or *nodes* V, which are connected by a set of edges or *links* E. The *nodes* in graph G are embedded in space. Nodes have certain differing *properties* that include their *position* on the Earth's surface, the corresponding *type* of paleoclimate archive (e.g., tree, stalagmite, marine sediment) and *proxy* type (e.g., isotope ratios, lithogenic grain size or annual ring width). There exist time series whose archive sources are (a) distributed over a large area (Yi et al., 2011), or are (b) not Earth-bound, as for insolation (Steinhilber et al., 2009). Pragmatically, in the former case, the nodes should be placed in the center of the considered region. The latter should be considered as a node in a different *subnetwork* (see the illustration in Fig. 2 and the paragraph on subnetworks below).

Technically, each node is associated with at least one time series of an environmental proxy. If age uncertainty is considered, an ensemble of observation time vectors $\boldsymbol{T}_i$, $i = 1, \ldots, N_{\mathrm{ens}}$ is generated and all ensemble members are considered equally likely realizations of the proxy time series of this node. For modeled network results, an ensemble of model realizations for the observation times t can be considered. The spatial distribution of nodes is also time dependent: while high-resolution archives (such as trees, Anchukaitis et al., 2006; Sano et al., 2011; Singh et al., 2009, but increasingly also stalagmites, Kennett et al., 2012; Hu et al., 2008; Tan et al., 2009; Fleitmann, 2004) predominantly cover more recent periods at annual timescales, others (e.g., marine or lacustrine sediment sequences, stalagmites or ice cores) might grow at very slow rates, and over the course of millennia (Dykoski et al., 2005; Rodbell et al., 1999; Fleitmann et al., 2007; Wang et al., 2001). A node will only be incorporated into the network evaluation for a time window W if it fulfills the minimal sampling requirements of > 50 observations per window.

In the considered paleoclimate networks, *links* are undirected and weighted. A link between node i and node j exists, if the link *weight*, given by the link strength, is greater than zero (Rehfeld and Kurths, 2014). Generally, links in the paleoclimate network approach can not be assigned by simple thresholding of a similarity value $S(X, Y)$ when the time series are irregular, of different length and/or age-uncertain (Rehfeld and Kurths, 2014). Therefore, significant similarity is established using surrogate data. The surrogate time series have the same temporal resolution as the original time series, but the observed proxy values are replaced using autocorrelated noise (c.f. Rehfeld et al., 2011; Rehfeld and Kurths, 2014). Using N_{sim} different similarity measures S with different characteristics and algorithms increases the robustness of the link detection, as the proxy-climate relationship might be nonlinear, weak or even erratic (Rehfeld and Kurths, 2014).

Fundamentally, for each pair of nodes i and j, time series ensemble member k and similarity measure l, a similarity $\mathbf{S}_{i,j}^{l,k}$ is calculated. N_{sur} autocorrelated but mutually uncorrelated time series surrogates are employed to obtain a distribution of surrogate similarity values $\mathbf{S}_{i,j}^{*l,k}$. The *fundamental adjacency matrix* entry $\mathbf{A}_{i,j}^{l,k}$ consequently results from

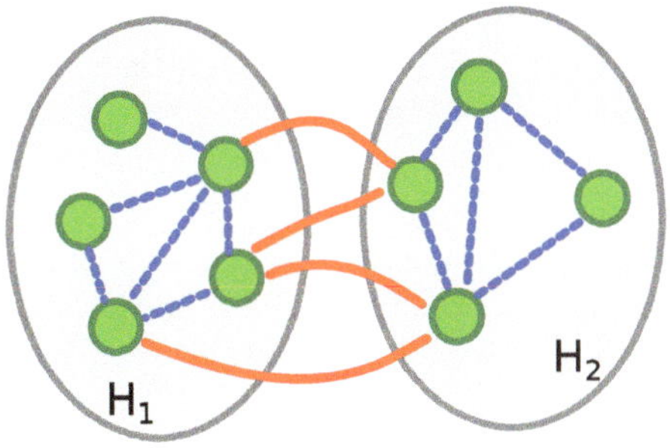

Figure 2. To test (paleo)climatic hypotheses, the considered domain can be split and the nodes in the different subdomains associated with different sub-networks. This offers the possibility of differentiating statistically between associations within sub-networks (autolinks, in blue) and those connecting different sub-domains (cross links, in red).

thresholding the similarity value $\mathbf{S}_{i,j}^{l,k}$ using the chosen critical values $\mathbf{S}^*(q_{\mathrm{low}})$ and $\mathbf{S}^*(q_{\mathrm{hi}})$, corresponding to the quantiles $q_{\mathrm{low}} = 0.05$ and $q_{\mathrm{hi}} = 0.95$ of the distribution $\mathbf{S}_{i,j}^{*l,k}$:

$$\mathbf{A}_{i,j}^{l,k} = 1 \text{ if } \mathbf{S}_{i,j}^{l,k} < \mathbf{S}^*(q_{\mathrm{low}}) \text{ or } \mathbf{S}_{i,j}^{l,k} > \mathbf{S}^*(q_{\mathrm{hi}}) \qquad (1)$$

for asymmetric measures that distinguish between positive and negative similarity (e.g., CC), and

$$\mathbf{A}_{i,j}^{l,k} = 1 \text{ if } \mathbf{S}_{i,j}^{l,k} > \mathbf{S}^*\left(q'_{\mathrm{hi}}\right) \qquad (2)$$

for symmetric measures that consider only an association strength (e.g., MI and the ESF), where $q'_{\mathrm{hi}} = 0.90$.

The weight of a link between nodes i and j is given by the ratio of the number of significant associations between them for all N_{ens} ensemble realizations and N_{sim} similarity measures. This is summarized in the link weight matrix **LW**:

$$\mathbf{LW}(i, j) = \frac{\sum_{k=1}^{N_{\mathrm{sim}}} \sum_{l=1}^{N_{\mathrm{ens}}} \mathbf{A}_{i,j}^{l,k}}{N_{\mathrm{ens}} \cdot N_{\mathrm{sim}}}. \qquad (3)$$

2.1 Subnetworks

The nodes in the paleoclimate network G have different properties (e.g., archive type or geographic origin from a specific region) that may influence its role within the network. To investigate regional dynamics, these nodes are considered here as lying in different *subnetworks*.

A subnetwork H_1 is formed by a subset of nodes $V(H_1)$ and links $E(H_1)$ from network G, where all nodes in $V(H_1)$ fulfill a certain property (e.g., geographic location in the region of 60–100° eastern longitude and 0–40° northern latitude) and all links between these nodes. If, and how, the nodes are sorted into subnetworks depends on the research question that is being asked. Splitting the domain as above is motivated by the different ASM subsystems that are thought to influence the regions west and east of 100° E. To investigate the dependence of the ASM on solar irradiance, a proxy record of insolation (e.g., Steinhilber et al., 2009) could then be considered as a node in a separate subnetwork.

Links within the subnetwork, $E(H_1)$, are *internal links*, and links from nodes $V(H_1)$ to nodes in another subnetwork H_2, $V(H_2)$, are *cross links* $E(H_1, H_2)$. Thus, the overall link set is the unity of internal and cross links between the subnetworks, $E(G) = \cup_{i,j} E(H_i, H_j)$, similar to Donges et al. (2011).

2.2 PAN measures

An abundance of graph-theoretical measures, i.e., statistics that are supposed to reflect characteristic node, link or network properties (Tsonis et al., 2006; Gozolchiani et al., 2011), is available. In the paleoclimate context, data-based challenges require an adaptation and careful evaluation of commonly employed (climate) network measures, because of age uncertainty and temporal and spatial heterogeneity. Here we test how well spatio–temporal changes are reflected in network measures, and to what extent they are influenced by spatially heterogeneous node distribution. Network measures considered for the test in the following include (a) the average *link density* (or connectivity), (b) the *regional degree* and *cross link ratio*, and (c) shortest path *betweenness* centrality.

In the following paragraphs we derive and review these basic network measures.

2.2.1 Average link density

For general complex networks, the link density, or connectivity, of a graph G with N_{no} nodes is simply the ratio of realized links between the nodes vs. the number of possible links

$$\mathrm{LD}'(G) = \frac{\sum_{i,j} \mathbf{A}_{i,j}}{(N_{no} - 1)\, N_{no}}, \tag{4}$$

which is between zero and one.

However, the number of nodes of a paleoclimate network can vary if the minimal overlap between the time series is not always fulfilled. Therefore the node number may differ between the ensemble realizations $N_{no} = N^l_{no}$, and this results in a link density that depends on the realization number l and the link weight LW:

$$\mathrm{LD}(l) = \frac{\sum_{i,j} \mathbf{LW}^l_{i,j}}{\left(N^l_{no} - 1\right) N^l_{no}}. \tag{5}$$

This expression is averaged to obtain the *average link density* for the considered ensemble of time series,

$$\mathrm{LD} = \frac{\sum_{l=1}^{N_{ens}} \mathrm{LD}(l)}{N_{ens}}. \tag{6}$$

As all time series in this study are consistently sampled the network is thresholded such that the 20 % strongest links are considered significant.

2.2.2 Cross link probability

Assume network G consists of N^G_{no} nodes and N^G_{ed} edges. Let us partition this network into nodes in two subnetworks, say, H_1 and H_2 with $N^{H_1}_{no}$ or $N^{H_2}_{no}$ nodes each such that $N^G_{no} = N^{H_1}_{no} + N^{H_2}_{no}$. Accordingly, the sum of edges is partitioned into the sum of edges within H_1 and H_2, $N^{H_1}_{ed}$ and $N^{H_2}_{ed}$, and edges from H_1 to H_2, N^{1-2}_{ed}.

The relative frequency of *realized* cross edges,

$$P'_{1-2} = \frac{N^{1-2}_{ed}}{N^{H_1}_{no} \cdot N^{H_2}_{no}}, \tag{7}$$

gives the *cross link probability* P_{1-2}

$$P_{1-2} = \frac{\sum_{i \in H_1, j \in H_2} \mathbf{LW}_{i,j}}{N^{H_1}_{no} \cdot N^{H_2}_{no}} \tag{8}$$

with a link weight matrix **LW** and $\mathbf{LW}_{i,j} \in [0, 1]$.

2.2.3 Cross link ratio

The cross link ratio CLR(H_1, H_2, G) is given by the cross link probability divided by the overall link probability

$$\mathrm{CLR}(H_1, H_2, G) = \frac{P_{1-2}}{\mathrm{LD}}. \tag{9}$$

2.2.4 Average and regional node strength

In classical complex network theory, the *degree* D of a node i is a measure of the presumed importance of a node, given by the number of its links to all other nodes $j = 1, \ldots, N_{no}$, $j \neq i$:

$$D'(i) = \sum_{i, j \neq i} \mathbf{A}(i, j). \tag{10}$$

We consider the links to be weighted (cf. Opsahl et al., 2010, and references therein), and therefore the notion of the *degree* of a node was replaced by that of a *node strength*, also called the *vertex strength* (Gozolchiani et al., 2011; Opsahl et al., 2010). Using link weights this gives a *node strength* D^l

$$D^l(i) = \sum_{i, j \neq i} \mathbf{LW}^l_{i,j} \tag{11}$$

for each ensemble realization l.

2.2.5 Shortest path betweenness centrality

Betweenness centrality has been regarded as a measure of local dynamical information flow (Opsahl et al., 2010;

Barthélemy, 2011). The shortest path betweenness of a node k is calculated from the number of *shortest paths* σ_{ij} between all nodes i and j, and the number of these paths that pass through node k, $\sigma_{ij}(k)$:

$$\mathrm{BC}'(k) = \sum_{j \neq k} \frac{\sigma_{ij}(k)}{\sigma_{ij}}. \tag{12}$$

If the node number changes over time, we can standardize the betweenness to obtain a measure that is independent under such changes:

$$\mathrm{BC}(k) = \frac{\mathrm{BC}'(k)}{(N_{\mathrm{no}} - 1)\,(N_{\mathrm{no}} - 2)}. \tag{13}$$

3 START: spatio–temporally autocorrelated time series

3.1 Model philosophy

START is a simple semi-empirical model for the propagation of climate variability through flows in a spatially extended domain. In the current implementation it is a statistical toy model with three independent spatial components that react differently to applied external forcing. Asian Summer Monsoon (ASM) dynamics are determined by the interplay between the Indian Summer Monsoon (ISM) and the East Asian Summer Monsoon (EASM). The predominant regions of influence, and main wind directions, of ISM, EASM and the continental westerlies are given in Fig. 3. They can be roughly divided at 100° eastern longitude (Wang et al., 2010). The dependence – and interplay – of the subsystems on each other is a topic of ongoing research (Wang et al., 2010; Yihui and Chan, 2005; Cao et al., 2012).

Assuming that we can view the model domain in summer as a region with three main wind systems (cf. Fig. 3) that each extend only zonally, i.e., laterally *or* longitudinally. While the true wind pattern might be significantly more complex, we argue that this could be viewed as a statistical decomposition of the mean summer surface wind field into simplified lateral/longitudinal components $V_X(\boldsymbol{p})$, $V_Y(\boldsymbol{p})$ and $V_Z(\boldsymbol{p})$. Each of these fields are assumed to be a Gaussian-modulated unidirectional front with a velocity at position $\boldsymbol{p}$ and time point t

$$v_X(t, \boldsymbol{p}, m_X, W) = m_X\, e^{-(p_x - p_{0,x})^2/2W}, \tag{14}$$

with a full width at a half maximum of $2\sqrt{W \log 2}$. The maximal amplitude of the velocity, m_X,

$$m_X(t, F) = m_X = B_X + \alpha F, \tag{15}$$

is found in the center of the Gaussian front, as in Fig. 3. Here, B_X is the baseline strength of the component's flow, and α is its amplitude, or susceptibility to the external forcing, represented by the parameter F, $F = [-1, \ldots, 1]$. The velocities

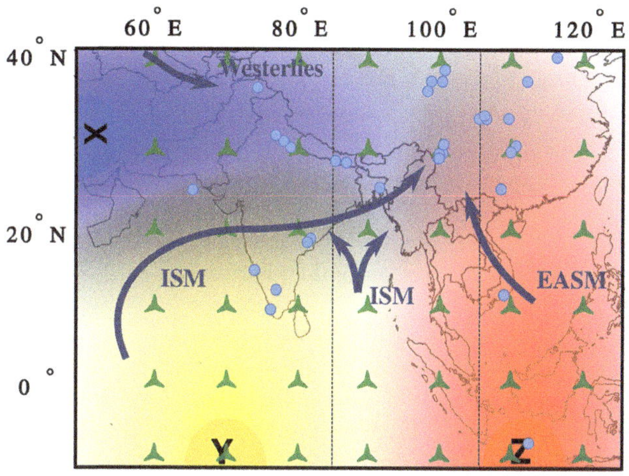

Figure 3. Map showing the main wind directions of the Indian and East Asian summer monsoon systems. Inflow corridors are modeled as sources of variability: the Indian Summer Monsoon (ISM) with a longitudinal (X, in blue) and a latitudinal component (Y) and the East Asian Summer Monsoon (Z, EASM) with a latitudinal component. The dynamics in the model, governed by the respective strengths of the source flows, are sampled at the grid locations (triangles) and where paleoclimate data is available (circles). Dashed lines bracket the intermediate domain.

for sources Y and Z are defined analogously, and the chosen values are given in Table 2.

Each of the fields originates from a source at a position $\boldsymbol{p}_{\mathrm{src}}$, and each of the sources X, Y and Z is associated with a climate process X_t, Y_t and Z_t that represents the annual mean of a hypothetical climate variable, for example surface temperature or precipitation anomalies, in the year t. By definition the model is restricted to modeling inter-annual variability, and the construction rationale is that local variability can be modeled as a superposition of variability mediated by atmospheric flows and local variability. The amount of dynamical information about the climatic process at the source that flows along one of these fields, say, from source X to a point at a position $\boldsymbol{p}$ in its region of influence, is approximated by a *variance factor* $f_X(\boldsymbol{p}, F)$. By construction, the square of the variance factor is proportional to the amount of variance shared between the source X and the time series at point $\boldsymbol{p}$: $f_X^2(\boldsymbol{p}, F) \propto \sigma_X(\boldsymbol{p})$.

We assume three sources for the underlying flow system, where random climate variability originates and which is then transported via advection and diffusion along the paths. The position and transmission direction of the sources and the observation points are illustrated schematically in Fig. 3. At each point in the ASM region a local time series of climate variability is computed as the sum of the noise contributions from each of the three sources. These components are scaled with a factor that quantifies the amount of information that is preserved from the source to the point of observation:

$$R_i = f_X(F,i)R_X + f_Y(F,i)R_Y + f_Z(F,i)R_Z + R_{\mathrm{noise}} \tag{16}$$

where R_i is the signal at point i obtained from the superposition of R_X, the signal of the longitudinal ISM component, R_Y, the latitudinal ISM component and R_Z, the signal of the EASM source. R_{noise} is local observation noise. The factors $f_X(F, i)$, $f_Y(F, i)$ and $f_Z(F, i)$ scale the contributions to the overall variance of the time series obtained at point i at time t and a potential forcing F.

3.2 Derivation of the scaling factors

The system we are looking at is a two-dimensional boundaryless fluid of constant diffusivity χ with a stationary flow described by the velocity field $\boldsymbol{v}(\boldsymbol{x})$. Temperature transport in the system is governed by the advection–diffusion equation, which states how the change in temperature over time is determined by the spatial temperature change and the velocity:

$$\frac{\partial T}{\partial t} = \chi \, \Delta T - \nabla \cdot (\boldsymbol{v}(\boldsymbol{x}) T), \tag{17}$$

which is obtained by inserting the advective and diffusive flux

$$\boldsymbol{j} = \boldsymbol{j}_{\text{diff}} + \boldsymbol{j}_{\text{adv}} = -\chi \, \nabla T + \boldsymbol{v} T \tag{18}$$

into the sourceless continuity equation for temperature

$$\frac{\partial T}{\partial t} = -\nabla \boldsymbol{j}. \tag{19}$$

$T(\boldsymbol{p}, t)$ is the temperature value at position $\boldsymbol{p}$ at time t.

To compute the scaling factors we approximate and solve Eq. (17) with temperature spikes along the source origin line, perpendicular to the flow direction. These spikes are propagated along a Gaussian-modulated velocity field, and the scaling factors are determined from the remaining spike height at position $\boldsymbol{p}$. We use this temperature δ peak as a tracer of the flow. The initial condition is a Gaussian-shaped temperature front of unit height (in the x or y direction)

$$T(\boldsymbol{p}, 0) = e^{-\frac{(p-p_0)^2}{s}}, \tag{20}$$

where s is the Gaussian front width and p_0 the position of the source. Local temperature is computed as a function

$$T(p, t) = \sqrt{\frac{s}{s + 4\chi t}} e^{-\frac{(p-p_0-vt)^2}{s+4\chi t}} \tag{21}$$

of time and space. For the constant v this can be solved analytically. The diffusion constant χ is set to unity. We neglect the derivative of the velocity field but replace $\boldsymbol{v}$ by $\boldsymbol{v}(\boldsymbol{p})$ and thereby get an approximate solution for velocity fields with a slow spatial variation. We also assume that this velocity field depends on a given forcing F. As this can be seen as a statistical description of how one original disturbance would dissipate over space, we use it to define the local variance factors: $f_X^*(T, i) = T(i, t_{\max})$ with the front in the y direction and $p_0 = y_X$, $f_Y^*(T, i) = T(i, t_{\max})$ with the front in the x direction and $p_0 = x_Y$, $f_Z^*(T, i) = T(i, t_{\max})$ with the front in the x direction, and $p_0 = x_Z$.

The factors depend on the observation position $\boldsymbol{p}$, source positions $\boldsymbol{p}_X$, $\boldsymbol{p}_Y$ and $\boldsymbol{p}_Z$ and the velocity component in flow direction, $v_X(t, F)$:

$$f_X^*(\boldsymbol{p}, F) = f(\boldsymbol{p}_X, \boldsymbol{p}, v_X, \chi, s). \tag{22}$$

3.3 Obtaining the START time series

At each point in space and for each time point t the factors are standardized by the factor sum $g(t, F)$,

$$g(t) = f_X^*(F) + f_Y^*(F) + f_Z^*(F) + f_{D,t}^*, \tag{23}$$

i.e., $f_X(F) = \frac{f_X^*(F)}{g(t)}$. Here, the processes X_t, Y_t and Z_t are uncoupled AR(1) processes of unit variance, and with a persistence time of $\tau = 6.4$ years. This value for τ was chosen as compatible with the order of magnitude estimated for ASM paleoclimate proxy records spanning the last millennium (Rehfeld et al., 2011). At each point p in the model domain, the local climate "history" $S_i(\boldsymbol{p})$ is computed as a superposition of source and noise terms (Eq. 16).

3.4 Model setup

START generates synthetic time series at locations in its integration domain. The interdependences between the time series depend on their relative position in the considered flows and the given forcing. Although each time series is distinct, as the influence of the different components is location dependent, time series located in close proximity are similar, and the amount of variance shared with the components' sources is both location and forcing dependent.

The transient forcing model runs are sampled for two spatial sampling types, a grid and a data set of locations of paleoclimate records (c.f. Table 1 and Fig. 3) throughout the ASM domain. The regional span of the spatial sampling and the node numbers is comparable, although the archive locations are spaced closer at the center of the ASM domain, while the grid also samples the areas over the Indian Ocean.

Networks are computed for twenty 200-year-long time slices of a 4000 year transient simulation, during which the model forcing parameter, F, was increased consistently from its minimum value, -1, to 1. To ensure the robustness of the intended spatial inference against estimation errors for these relatively short time series, 100 simulation realizations were analyzed separately.

4 Validation of PAN using START

To validate the PAN methodology we run START with different forcing parameters and investigate the resulting network topology. In transient runs we vary the parameter continuously. Three stages are distinguished in particular: "ISM

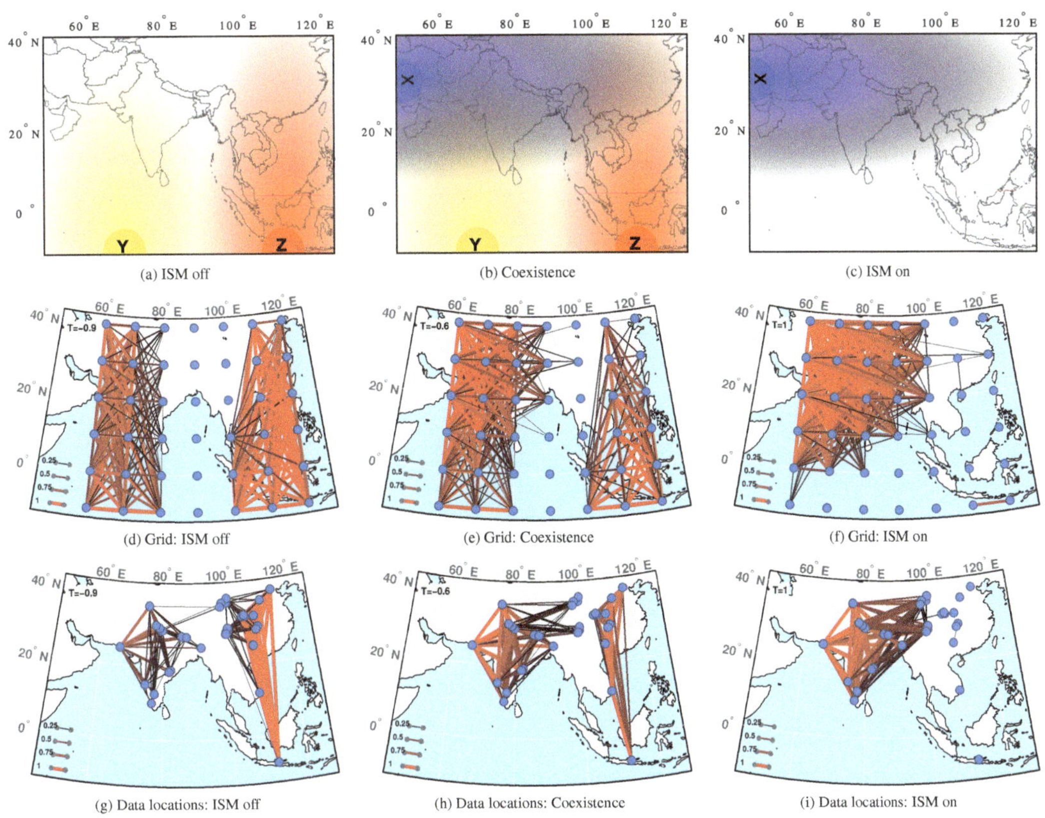

Figure 4. Extreme points of the modeled Asian monsoon dynamics: schematic illustration of input variance factors **(a)–(c)** and networks reconstructed on grid **(d)–(f)** as well as the actual data positions **(g)–(i)**. Only the 20 % strongest links are shown.

Table 1. Spatial setup for the START model experiments.

	Grid	Heterogeneous
Number of nodes	42	36
Regional span [latitude]	[−10; 40]	[−7; 39.5]
Regional span [longitude]	[60; 120]	[66; 115.5]
Node distribution	Regular grid	Paleoclimate record origins, c.f. Rehfeld et al. (2013)

off", equivalent to forcing $F = -1$, "Coexistence" with $F = 0$ and "ISM on" with $F = 1$ (c.f. Fig. 4).

The velocities in the Gaussian-shaped fronts in Eq. (22) are modulated through the forcing parameter F in the START model. Consequently the amount of variance conserved along a flow and the synchronizing reach of the three "wind components" change with changing forcing, as illustrated schematically in Fig. 4a–c. The components are tuned to react proportionally to the effected forcing, but compete at each point in space due to the standardization in Eq. (23). Therefore, the fraction of variance explained by the components in each location changes in a nonlinear way. With parameter F, the reach of the X component increases while the other components lose relevance.

4.1 Topology of the observed networks

The networks obtained from the 20 % strongest time series correlations are given in Fig. 4d–f for the grid-based time series, and in Fig. 4g–i for the proxy locations. In the first, "ISM off" case in Fig. 4d and g the latitudinal components Y and Z are strong, resulting in two clearly separated network components. The Y component covers the longitudes 60–80° E, while the Z component covers the longitudes 100–120° E, and no strong links appear between the two.

As the forcing increases, a "coexistence" stage is reached: the longitudinal X component strengthens, and the latitudinal components Y and Z lose in relation, as they have opposite sensitivities to forcing (c.f. Table 2). As Fig. 4e and h show, the reconstructed network is still split, as the strongest links are in the core region of the Y and Z flow parts. However, strong links, originating far in the west, connect to the 90° E grid points, and weaker links extend even beyond. The Z-component half of the network has retracted southward, in agreement with the decreased relative forcing strength.

Table 2. START source and flow attributes.

	Source	Position	Strength B	Ampl.	Width W
ISM (long. component)	X	(30, 55)	85	$\alpha = 70$	1200
ISM (lat. component)	Y	(−15, 70)	10	$\beta = -5$	100
EASM	Z	(−15, 112.5)	30	$\gamma = -12$	100
White noise	N	all	–	Const.	–

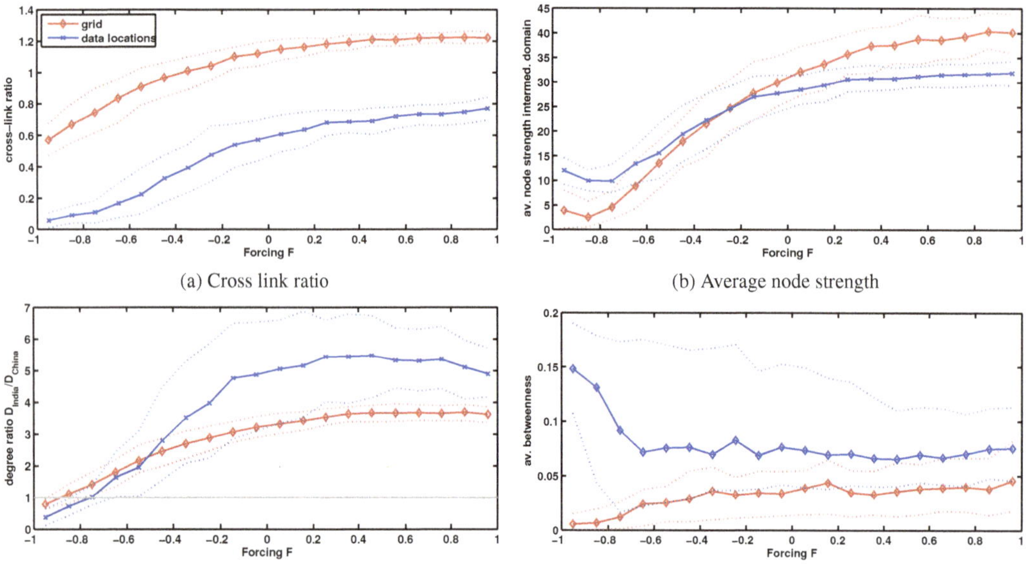

(a) Cross link ratio

(b) Average node strength

(c) Regional degree ratio

(d) Shortest path betweenness

Figure 5. Most network measures reflect the increasing change in network structure occurring with changing ISM dominance in START **(a)–(d)**, but they show different sensitivities. The spatial distribution of node positions influences the absolute value of the network measure, but the general trends are consistent for the underlying grid and the heterogeneous data positions. Broken lines indicate the 10 and 90 % quantiles of the ensemble results.

With maximal forcing the full "ISM only" stage is reached in Fig. 4f and i. The X component dominates, and the reconstructed network has only a single core region extending to and across 100° longitude. The changing strength of the synchronizing components due to the varied forcing is reflected by the extent of the synchronized regions in which the grid-point time series show the strongest similarities. Although differences exist, the networks reconstructed from grid and heterogeneous locations bear a large resemblance and relate to the underlying forcing structure.

4.2 Validation of the network measures

Network measures are statistical estimators that reflect properties of individual nodes as well as regional or global characteristics of the network. The spatio–temporal changes, visible as the START model is driven with different forcing, should effect consistent transitions in measures suitable for the investigation of spatio–temporal changes. Furthermore, these transitions should also be detectable under varying node distributions, i.e., for a grid structure as well as for heterogeneous node distribution. The gradually increased forcing results in a transition from a bimodal, laterally synchronized network with separated components to a widely connected state. Figure 5 shows the network measure expectation values from 100 transient START forcing runs.

4.2.1 Cross link ratio

In the grid-based network reconstruction the *cross link ratio* (Fig. 5a), calculated as the ratio of cross link probability over overall link probability, increases monotonously for the grid-based networks and saturates well before maximal forcing is reached. It crosses the threshold of equal probabilities, transitioning from values of 0.6 to 1.2. This means that while in the early, bimodal, stage the cross link density is about 60 % of the average link density, it reaches 120 % when the synchronized region spans the whole network. Thus, at low forcing the cross link density is significantly *lower* than the overall link density, with the network being effectively partitioned. At high forcing the co-varying region spans across the former separate parts, resulting in a *higher* cross link density than average link density. For the data-based network, the expected path of transition for the heterogeneous network differs from that observed for the grid topology by a negative offset, but parallels it otherwise.

4.2.2 Average node strength in the intermediate domain

The *intermediate domain* (85–105° E) between the defined *model* ISM/EASM core regions is not synchronized by an external source of variability for low forcing values. Thus, nodes within this domain are expected to have no, or few, connections, resulting in weak node strength, as illustrated in Fig. 5b. This changes, however, as the longitudinal X component strengthens with increasing forcing, when these nodes fall into its region of influence. Please note that this does not tie in exactly with the actual ISM/EASM transition region, which is situated farther to the east. This adaptation was necessary because the implemented flow paths in START are modeled for simplicity either laterally *or* longitudinally, and the core interaction region around 100° E should be reachable for both flows. Indeed, as shown in Fig. 5b, the average node strength in the intermediate domain rises from a low level to full connectivity (node strength equal to $N_{\mathrm{no}} - 1$) for the grid-based network measures after an initial short decline. The amplitude of the change is, however, lower for the heterogeneous locations.

4.2.3 Regional node strength ratio

The regional node strength ratio (Fig. 5c) highlights the different regional degrees in subregions of the network. Here it is computed using Eq. (10) as the ratio of the average node strengths in India vs. that in China. If both subnetworks are equally well connected within the overall network, their average node strength should be similar, and the node strength ratio should equal unity, as indicated by the gray line in Fig. 5c.

Starting off at equally well-connected subdomains, the western, "Indian" part of the network gains importance with increased forcing and the node strength ratio settles after a short growth slightly below 4 for the grid-based network. Thus, nodes in "India" are associated with four times the link weight when compared to those in "China". For the heterogeneous sampling scheme the line of equal node strength is crossed *later* than for the gridded data. The following incline, contrastingly, is sharp and the plateau reached gives fivefold node strength for "India" vs. "China", accompanied by a large uncertainty in this estimate. This, c.f. Fig. 3, is consistent with the stronger representation of "China" in the used grid.

4.2.4 Shortest path betweenness

Shortest path betweenness centrality is a measure developed to infer the presumed relative importance of nodes and regions (Gozolchiani et al., 2011; Opsahl et al., 2010; Barthélemy, 2011; Donges et al., 2009a). Furthermore, Donges et al. (2009b) found that "betweenness centrality allows to measure the importance of localized regions on the earth's surface for the transport of dynamical information within a climatological field in the long term mean", and stated that "information is transported by advective processes, where the assumption of information traveling on shortest paths can be substantiated by extremalization principles". As such it should, in principle, also be an interesting measure for paleoclimate network applications.

As a node-based measure it is not possible to compare it directly when using different spatial sampling schemes. Still, the characteristic dynamical changes should be reflected in regional properties. Based on this presumption, betweenness centrality estimates for nodes in the intermediate and central region (85–105° E) are averaged to obtain a domain estimate. The results, shown in Fig. 5d, are inconsistent for the first segment of the transition experiment, in which the largest dynamical changes occur. On the one hand, the betweenness estimates for the grid *in*crease slowly, albeit with a comparatively large uncertainty. On the other hand, the estimates for the heterogeneous locations *de*crease initially from large initial values to then stay on a flat plateau.

4.3 Comparison to flow network results

For comparison we also computed the networks analytically, directly from the flow using a continuously defined cross-correlation analog and a solution of the ADE from Molkenthin et al. (2014a). Figure 6 shows the results obtained for the cross link ratio and the regional degree ratio, where the average node strengths West/East of the 100° longitude boundary are compared. Both grid-based and data location-based results agree well with those obtained from the START time series for the cross link ratio, showing the gradual increase in the cross link density as the forcing parameter increases. The regional degree ratio shows substantially more variability than for START, though an overall positive trend is apparent. This may be due to the fact that the flow networks, by construction, have an overall higher connectivity due to a higher correlation level than the START networks. Please note that by construction the absolute forcing values are not equivalent for START and the flow networks and therefore the results can not be compared quantitatively. The main construction difference is that flow networks do not have designated fluctuation sources, but average over all possible source locations. By contrast, sources in START (e.g., X, Y and Z in Fig. 3) are the origin of variability. Links closer to the sources are more likely than those far away from them in START.

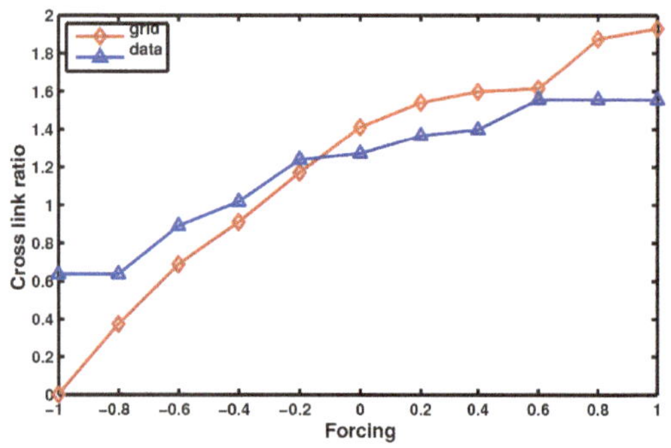

(a) Cross link ratio

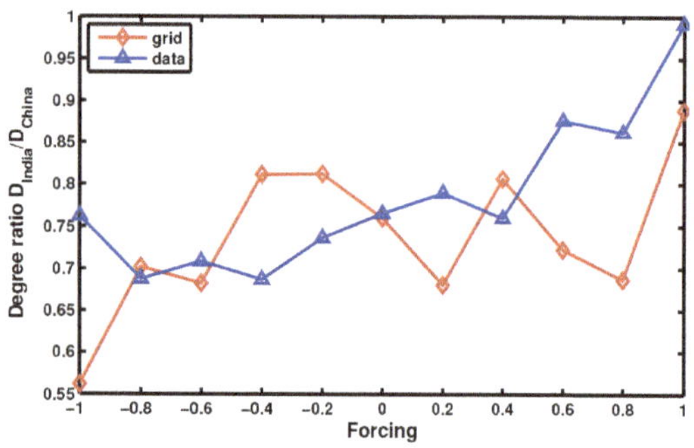

(b) Regional degree ratio

Figure 6. Cross link ratio **(a)** and average regional degree ratio **(b)** for networks computed directly from the flow data (Molkenthin et al., 2014a, b). Note that due to the necessarily different implementations the results can not be quantitatively compared to those obtained with START.

5 Discussion

5.1 Network measures

The cross link ratio reflects the transition in the model simulations well. While the paths for different node distributions differ substantially in their amplitude, baseline and speed of increase, both show the overall increase in subdomain-connecting cross links. The initial decline observed for the data topology results from spurious cross links, as the overall similarity level (not shown) is much lower than at higher forcing levels. These cross links appear randomly at low forcing, but disappear if the link density is chosen more conservatively. Heterogeneous sampling apparently also does not strongly affect the node strength in the intermediate domain. Both sampling schemes reflect the transition of these nodes from being irrelevant to heavily tied to the rest of the network.

The regional node strength ratio is more difficult to interpret. The expected transition path for the relative average strength of nodes in India vs. that of those in China is significantly different for differing sampling schemes. The general feature, however, that repeats itself is that proximity to the source clearly results in higher node strength. There is nevertheless large uncertainty associated with the transition path for the data-based network.

The betweenness centrality estimates are inconsistent for the different sampling schemes, and the error margins spread widely. This inconclusiveness could be due to hypersensitivity of the measure and the comparatively low number of nodes in our network.

5.2 Regional changes and inter-regional information flow

Three major regions are relevant for START dynamics: the ISM region, the intermediate region and the EASM region. While the first and the latter start off as independent subdomains of the network, they are connected by the longitudinal ISM component at increased forcing. The distinct dynamical features include (i) bimodality vs. later unimodality, (ii) increasing size of the spatially synchronized region and (iii) increased flow of dynamical information through the increasing strength of the longitudinal ISM component.

The initial *bimodality* of the network is directly visible in the network, and in the low cross link ratio and average node strength in the intermediate domain. Although the absolute values of the between cross link ratio are not on the same scale as those for the grid, a significant increase occurs. Thus, the change in model dynamics can be inferred if the node topology does not change and if the sampling bias (defined here as a systematic offset due to an other-than-regular node topology) can be quantified. Node strength in the transitional region shows smaller sampling-dependent deviations and could thus be a more robust measure of the importance of intermediate regions. The regional degree ratio shows, however, that spatially biased sampling can have large effects: as only one node of the paleoclimate network samples close to the EASM source, and most of the rest of the clusters far north of it, the synchronizing influence of this source quickly disappears as the ISM region grows. Located at the fringe of the ISM component, these nodes also get less ISM input than all its westwards neighbors. This results in few links and an under-representation of the Chinese part of the network, caused by a combination of model shortcomings and data sparsity. Such effects have to be addressed before a comparison of networks with changing node architecture is possible. The increasing strength of the ISM component is directly visible by its growth in the reconstructed network, and additionally reflected in the increasing cross link ratio. The source region of the modeled pathway is robustly characterized by a higher node degree. Increasing information flow from the ISM to the EASM core region is indecipherable using shortest path betweenness, which is sensitive to sampling changes and spurious links. Nevertheless, provided the potentially changing node topology is addressed, the cross link ratio could be a sufficient measure to quantify changes in inter-regional dependences.

What remains to be investigated is the influence that varying node numbers – in space and time – have on estimated network measures. This could be done by simulating constant START "climate dynamics" and removing nodes iteratively. The relevance of the spatial position of a node can then be assessed by the discrepancy between the network measures estimated with/without it in comparison to the expected value for high-resolution sampling.

6 Conclusions

We conclude that it is also possible to reconstruct the spatio–temporal changes in the semi-empirical ASM model for heterogeneous node distribution in space, though the results are limited to qualitative statements if the node topology changes. Model dynamics are reflected in the reconstructed network, specifically its link strength distribution, and in network measures such as regional average node strength and node strength in the transitional zone, the intermediate domain outside the latitudinal source influence.

Inter-regional information flow, or, in the context of the START model, spatial distribution of variance, cannot be inferred using shortest path betweenness, as it is found to be too sensitive to irregularities, and no clear dynamical signature can be found in the transition experiments. The cross link ratio is a better alternative, though sampling biases have to be taken into account for its analysis.

Spatial heterogeneity, in general, has strong effects, both on the reconstructed network and on the network measures. It manifests itself in (i) biases in network measures that can be negative (cross link ratio) as well as positive (regional degree ratio), (ii) increased variance in the estimates (betweenness centrality, regional degree ratio), and (iii) the amplification of effects due to node clustering (regional degree, reconstructed network). Therefore, if networks with varying spatial sampling are investigated, care has to be taken to perform adequate significance tests to ensure that spurious sampling effects can be distinguished from real climate processes.

The developed ASM model START is a toy model that can not be expected to reflect actual monsoon dynamics. In reality, local, global and external forcing influences local climate processes. In the START model world, information transfer and local climate processes are governed solely by physical flows. Processes external to the ASM domain are not considered but can, in nature, lead to increased correlation in the whole or parts of the ASM region. One of the desired features to improve realism, for example, would be the inclusion of regional sources of variance, i.e., by region-dependent noise terms. Then the propagation of information could be considered serially, and causality-sensitive directed measures (Granger causality, ESF) could be tested. In the model's simplicity, however, also lies its strength, because it is possible to interpret the results with respect to its dynamics, a task that is much more complicated if such pseudo-proxy experiments are conducted with actual global climate models (GCMs) (Smerdon, 2012; von Storch et al., 2004; Mann and Rutherford, 2002). Unlike GCMs, START is computationally inexpensive; therefore large ensembles of time series for pseudo-proxy experiments can be generated. Using START, hypotheses concerning local vs. global drivers of climate dynamics can be tested directly based on the paleoclimate data, because START models the propagation of local climate variability through advection and diffusion. For large-scale dynamical and coupled GCMs with their multitude of output variables and parameters, cause and effect are more difficult to discern. Thus, it provides a good opportunity to assess whether and how spatio–temporal dynamics of a given paleoclimate data set are affected by age uncertainty, spatio–temporal heterogeneity and sparsity.

Acknowledgements. The authors thank Liubov Tupikina, Veronika Stolbova and Norbert Marwan for stimulating discussions. Funding by the German Science Foundation (DFG graduate school 1539), the German Federal Ministry for Education and Research (BMBF project PROGRESS) and via Helmholtz grant VG-900NH is gratefully acknowledged.

Edited by: J. Davidsen

References

Agnihotri, R., Dutta, K., Bhushan, R., and Somayajulu, B.: Evidence for solar forcing on the Indian monsoon during the last millennium, Earth Planet. Sc. Lett., 198, 521–527, doi:10.1016/S0012-821X(02)00530-7, 2002.

Anchukaitis, K. J., Evans, M. N., Kaplan, A., Vaganov, E. A., Hughes, M. K., Grissino-Mayer, H. D., and Cane, M. A.: Forward modeling of regional scale tree-ring patterns in the southeastern United States and the recent influence of summer drought, Geophys. Res. Lett., 33, 2–5, doi:10.1029/2005GL025050, 2006.

Barthélemy, M.: Spatial networks, Phys. Rep., 499, 1–101, doi:10.1016/j.physrep.2010.11.002, 2011.

Cao, J., Hu, J., and Tao, Y.: An index for the interface between the Indian summer monsoon and the East Asian summer monsoon, J. Geophys. Res., 117, 1–9, doi:10.1029/2012JD017841, 2012.

Chatfield, C.: The analysis of time series: an introduction, 6th Edn., CRC Press, Florida, USA, 2004.

Cover, T. and Thomas, J.: Elements of information theory, 2nd Edn., John Wiley & Sons, Inc., Hoboken, New Jersey, 2006.

Donges, J. F., Zou, Y., Marwan, N., and Kurths, J.: The backbone of the climate network, Europhys. Lett., 87, 48007, doi:10.1209/0295-5075/87/48007, 2009a.

Donges, J. F., Zou, Y., Marwan, N., and Kurths, J.: Complex networks in climate dynamics, Eur. Phys. J. Spec. Top., 174, 157–179, doi:10.1140/epjst/e2009-01098-2, 2009b.

Donges, J. F., Schultz, H. C. H., Marwan, N., Zou, Y., and Kurths, J.: Investigating the topology of interacting networks, Eur. Phys. J. B, 84, 635–651, doi:10.1140/epjb/e2011-10795-8, 2011.

Dykoski, C., Edwards, R., Cheng, H., Yuan, D., Cai, Y., Zhang, M., Lin, Y., Qing, J., An, Z., and Revenaugh, J.: A high-resolution, absolute-dated Holocene and deglacial Asian monsoon record from Dongge Cave, China, Earth Planet. Sc. Lett., 233, 71–86, doi:10.1016/j.epsl.2005.01.036, 2005.

Fleitmann, D.: Palaeoclimatic interpretation of high-resolution oxygen isotope profiles derived from annually laminated speleothems from Southern Oman, Quaternary Sci. Rev., 23, 935–945, doi:10.1016/j.quascirev.2003.06.019, 2004.

Fleitmann, D., Burns, S. J., Mangini, A., Mudelsee, M., Kramers, J., Villa, I., Neff, U., Al-Subbary, A. A., Buettner, A., Hippler, D., Matter, A., and Alsubbary, A.: Holocene ITCZ and Indian monsoon dynamics recorded in stalagmites from Oman and Yemen (Socotra), Quaternary Sci. Rev., 26, 170–188, doi:10.1016/j.quascirev.2006.04.012, 2007.

Ge-Li, W. and Tsonis, A. A.: A preliminary investigation on the topology of Chinese climate networks, Chinese Phys. B, 18, 5091–5106, 2009.

Gozolchiani, A., Havlin, S., and Yamasaki, K.: Emergence of El Niño as an Autonomous Component in the Climate Network, Phys. Rev. Lett., 107, 1–5, doi:10.1103/PhysRevLett.107.148501, 2011.

Heitzig, J., Marwan, N., Zou, Y., and Donges, J. F.: Consistently weighted measures for complex network topologies, Europ. Phys. J. B, 85, 1–16, doi:10.1140/epjb/e2011-20678-7, 2010.

Hu, C., Henderson, G., Huang, J., Xie, S., Sun, Y., and Johnson, K.: Quantification of Holocene Asian monsoon rainfall from spatially separated cave records, Earth Planet. Sc. Lett., 266, 221–232, doi:10.1016/j.epsl.2007.10.015, 2008.

Kennett, D. J., Breitenbach, S. F. M., Aquino, V. V., Asmerom, Y., Awe, J., Baldini, J. U. L., Bartlein, P., Culleton, B. J., Ebert, C., Jazwa, C., Macri, M. J., Marwan, N., Polyak, V., Prufer, K. M., Ridley, H. E., Sodemann, H., Winterhalder, B., and Haug, G. H.: Development and Disintegration of Maya Political Systems in Response to Climate Change, Science, 338, 788–791, doi:10.1126/science.1226299, 2012.

Malik, N., Marwan, N., and Kurths, J.: Spatial structures and directionalities in Monsoonal precipitation over South Asia, Nonlin. Processes Geophys., 17, 371–381, doi:10.5194/npg-17-371-2010, 2010.

Malik, N., Bookhagen, B., Marwan, N., and Kurths, J.: Analysis of spatial and temporal extreme monsoonal rainfall over South Asia using complex networks, Clim. Dynam., 39, 971–987, doi:10.1007/s00382-011-1156-4, 2011.

Mann, M. E. and Rutherford, S.: Climate reconstruction using Pseudoproxies, Geophys. Res. Lett., 29, 1501, doi:10.1029/2001GL014554, 2002.

Mayewski, P. A., Rohling, E. E., Curt Stager, J., Karlén, W., Maasch, K. A., David Meeker, L., Meyerson, E. A., Gasse, F., van Kreveld, S., Holmgren, K., Lee-Thorp, J., Rosqvist, G., Rack, F., Staubwasser, M., Schneider, R. R., and Steig, E. J.: Holocene climate variability, Quaternary Res., 62, 243–255, doi:10.1016/j.yqres.2004.07.001, 2004.

Molkenthin, N., Rehfeld, K., Marwan, N., and Kurths, J.: Networks from Flows – from Dynamics to Topology, Sci. Reports, 4, 4119, doi:10.1038/srep04119, 2014a.

Molkenthin, N., Rehfeld, K., Stolbova, V., Tupikina, L., and Kurths, J.: On the influence of spatial sampling on climate networks, Nonlin. Processes Geophys., 21, 651–657, doi:10.5194/npg-21-651-2014, 2014b.

Opsahl, T., Agneessens, F., and Skvoretz, J.: Node centrality in weighted networks: Generalizing degree and shortest paths, Soc. Networks, 2010.

Rehfeld, K. and Kurths, J.: Similarity estimators for irregular and age-uncertain time series, Clim. Past, 10, 107–122, doi:10.5194/cp-10-107-2014, 2014.

Rehfeld, K., Marwan, N., Heitzig, J., and Kurths, J.: Comparison of correlation analysis techniques for irregularly sampled time series, Nonlin. Processes Geophys., 18, 389–404, doi:10.5194/npg-18-389-2011, 2011.

Rehfeld, K., Marwan, N., Breitenbach, S. F. M., and Kurths, J.: Late Holocene Asian Summer Monsoon dynamics from small but complex networks of palaeoclimate data, Clim. Dynam., 41, 3–19, doi:10.1007/s00382-012-1448-3, 2013.

Rheinwalt, A., Marwan, N., Kurths, J., Werner, P., and Gerstengarbe, F.-W.: Boundary effects in network measures of spatially embedded networks, Europhys. Lett., 100, 28002, doi:10.1209/0295-5075/100/28002, 2012.

Rodbell, D., Seltzer, G., and Anderson, D.: An ∼15,000-year record of El Niño-driven Alluviation in Southwestern Ecuador, Science, 283, 516–520, doi:10.1126/science.283.5401.516, 1999.

Sano, M., Ramesh, R., Sheshshayee, M. S., and Sukumar, R.: Increasing aridity over the past 223 years in the Nepal Himalaya inferred from a tree-ring $\delta^{18}O$ chronology, Holocene, 32, 245–251, doi:10.1177/0959683611430338, 2011.

Schulz, M. and Stattegger, K.: SPECTRUM: spectral analysis of unevenly spaced paleoclimatic time series, Comput. Geosci., 23, 929–945, doi:10.1016/S0098-3004(97)00087-3, 1997.

Singh, J., Yadav, R. R., and Wilmking, M.: A 694-year tree-ring based rainfall reconstruction from Himachal Pradesh, India, Clim. Dynam., 33, 1149–1158, doi:10.1007/s00382-009-0528-5, 2009.

Sinha, A., Berkelhammer, M., Stott, L., Mudelsee, M., Cheng, H., and Biswas, J.: The leading mode of Indian Summer Monsoon precipitation variability during the last millennium, Geophys. Res. Lett., 38, 2–6, doi:10.1029/2011GL047713, 2011.

Smerdon, J. E.: Climate models as a test bed for climate reconstruction methods: pseudoproxy experiments, Wiley Interdiscip. Rev. Clim. Chang., 3, 63–77, doi:10.1002/wcc.149, 2012.

Steinhaeuser, K., Chawla, N. V., and Ganguly, A. R.: An exploration of climate data using complex networks, ACM SIGKDD Explor. Newsl., 12, 25, doi:10.1145/1882471.1882476, 2010.

Steinhilber, F., Beer, J., and Fröhlich, C.: Total solar irradiance during the Holocene, Geophys. Res. Lett., 36, 1–5, doi:10.1029/2009GL040142, 2009.

Tan, L., Cai, Y., Cheng, H., An, Z., and Edwards, R. L.: Summer monsoon precipitation variations in central China over the past 750 years derived from a high-resolution absolute-dated stalagmite, Palaeogeogr. Palaeocl., 280, 432–439, doi:10.1016/j.palaeo.2009.06.030, 2009.

Tsonis, A. A., Swanson, K. L., and Roebber, P. J.: What Do Networks Have to Do with Climate?, B. Am. Meteorol. Soc., 87, 585–595, doi:10.1175/BAMS-87-5-585, 2006.

Tsonis, A. A., Wang, G., Swanson, K. L., Rodrigues, F. A., and Costa, L. D. F.: Community structure and dynamics in climate networks, Clim. Dynam., 37, 933–940, doi:10.1007/s00382-010-0874-3, 2010.

Tupikina, L., Rehfeld, K., Molkenthin, N., Stolbova, V., Marwan, N., and Kurths, J.: Characterizing the evolution of climate networks, Nonlin. Processes Geophys., 21, 705–711, doi:10.5194/npg-21-705-2014, 2014.

von Storch, H., Zorita, E., Jones, J. M., Dimitriev, Y., González-Rouco, F., and Tett, S. F. B.: Reconstructing past climate from noisy data, Science, 306, 679–82, doi:10.1126/science.1096109, 2004.

Wang, Y., Liu, X., and Herzschuh, U.: Asynchronous evolution of the Indian and East Asian Summer Monsoon indicated by Holocene moisture patterns in monsoonal central Asia, Earth-Sci. Rev., 103, 135–153, doi:10.1016/j.earscirev.2010.09.004, 2010.

Wang, Y. J., Cheng, H., Edwards, R. L., An, Z. S., Wu, J. Y., Shen, C. C., and Dorale, J. A.: A high-resolution absolute-dated late Pleistocene Monsoon record from Hulu Cave, China, Science, 294, 2345–2348, doi:10.1126/science.1064618, 2001.

Wu, Z., Wang, B., Li, J., and Jin, F.: An empirical seasonal prediction model of the East Asian summer monsoon using ENSO and NAO, J. Geophys. Res., 114, 1–13, doi:10.1029/2009JD011733, 2009.

Yamasaki, K., Gozolchiani, A., and Havlin, S.: Climate Networks around the Globe are Significantly Affected by El Niño, Phys. Rev. Lett., 100, 228501, doi:10.1103/PhysRevLett.100.228501, 2008.

Yi, L., Yu, H., Ge, J., Lai, Z., Xu, X., Qin, L., and Peng, S.: Reconstructions of annual summer precipitation and temperature in north-central China since 1470 AD based on drought/flood index and tree-ring records, Climatic Change, 110, 469–498, doi:10.1007/s10584-011-0052-6, 2011.

Yihui, D. and Chan, J. C. L.: The East Asian summer monsoon: an overview, Meteorol. Atmos. Phys., 89, 117–142, doi:10.1007/s00703-005-0125-z, 2005.

16

The Rossby wave extra invariant in the dynamics of 3-D fluid layers and the generation of zonal jets

A. M. Balk

Department of Mathematics, University of Utah, Salt Lake City, UT 84112, USA

Correspondence to: A. M. Balk (balk@math.utah.edu)

Abstract. We consider an adiabatic-type (approximate) invariant that was earlier obtained for the quasi-geostrophic equation and the shallow water system; it is an extra invariant, in addition to the standard ones (energy, enstrophy, momentum), and it is based on the Rossby waves. The presence of this invariant implies the energy transfer from small-scale eddies to large-scale zonal jets.

We show that this extra invariant can be extended to the dynamics of a three-dimensional (3-D) fluid layer on the beta plane. Combined with the investigation of other researchers, this 3-D extension implies enhanced generation of zonal jets.

For a general physical system, the presence of an extra invariant (in addition to the energy–momentum and wave action) is extremely rare. We summarize the unique conservation properties of geophysical fluid dynamics (with the beta effect) that allow for the existence of the extra invariant, and argue that its presence in various geophysical systems is a strong indication that the formation of zonal jets is indeed related to the extra invariant.

Also, we develop a new, more direct, way to establish extra invariants (without using cubic corrections). For this, we introduce the small denominator lemma.

1 Introduction

Alternating zonal jets are found in the large-scale dynamics of oceans and atmospheres under broad conditions. Clearly seen on Jupiter, zonal jets are also unambiguously observed in Earth's oceans (Maximenko et al., 2005). The significance of zonal jets also stems from the well-known mathematical analogy between rotation and magnetic field. Similar equations and similar jets also appear in the dynamics of magnetized plasmas; these jets serve as transport barriers in tokamaks, improving conditions for the controlled nuclear fusion with magnetic confinement (e.g., Manz et al., 2012).

The emergence of zonal jets has been and continues to be intensively studied in both geophysical and plasma contexts. The main purpose of the present paper is to draw attention to the fact that the equations used to describe zonal jets have an unusual adiabatic-like (approximate) invariant (in addition to the usual energy–momentum); this makes the inverse cascade anisotropic and leads to the energy transfer towards zonal jets (see Sect. 4.2). As will be clear below (Sect. 2), this extra invariant is very rare: almost all other physical systems do not have anything analogous to such an invariant. The presence of this conservation in various geophysical systems (within the beta-plane approximation) is a strong indication that the extra invariant indeed plays a role in the emergence of zonal jets.

The extra invariant is adiabatic-like, meaning that the invariant is conserved approximately over a long time. Usually, in the theory of dynamical systems, the adiabatic invariance is due to the slowness of some parameter variation (e.g., Landau and Lifshitz, 1976). However, here the adiabatic-like invariance is due to the smallness of wave amplitudes: it is assumed that the dynamics is weakly nonlinear and is dominated by the triad interactions. It should be noted that the equations describing zonal jets have anisotropy only in linear terms, so strongly nonlinear regimes cannot lead to the zonal–meridional asymmetry. Of course, there are intermediate situations when linear and nonlinear terms have comparable magnitudes. Then real or numerical experiments should be used to see the role of the extra invariant in the formation of zonal jets; we hope that the present paper gives qualitatively the right picture. It is also unknown whether the extra invariant is related to the conservation of potential vorticity, which is an essential feature of geophysical systems. Often, besides weak nonlinearity, we also require the slowness of some parameter variation in *space* (which usually holds with high accuracy).

At this point, it is beneficial to recall two seemingly unrelated problems (Sects. 1.1 and 1.2). Then in Sect. 2, we describe the extra invariant and summarize the corresponding unique conservation properties of geophysical systems that allow for its existence. This summary includes the extension of the extra invariant to 3-D fluid layers, which is derived in the present paper (Sect. 3). We see how this invariant implies the energy cascade towards zonal jets in Sect. 4.

1.1 Boltzmann's collision invariants

Introducing his kinetic equation, L. Boltzmann (1875) investigated the general form of the thermodynamic equilibrium (Maxwell distribution). Boltzmann studied rarefied gas when the main interaction between the molecules are binary collisions. In each collision, momenta of two molecules before ($\boldsymbol{p}_1$, $\boldsymbol{p}_2$) and after ($\boldsymbol{p}_3$, $\boldsymbol{p}_4$) the collision satisfy the momentum–energy conservation:

$$\boldsymbol{p}_1+\boldsymbol{p}_2=\boldsymbol{p}_3+\boldsymbol{p}_4, \qquad \boldsymbol{p}_1^2+\boldsymbol{p}_2^2=\boldsymbol{p}_3^2+\boldsymbol{p}_4^2. \tag{1}$$

These relations imply the conservation of the total momentum and energy

$$\boldsymbol{P}=\int \boldsymbol{p}\,N_{\boldsymbol{p}}\,\mathrm{d}\boldsymbol{p}, \qquad E=\int p^2\,N_{\boldsymbol{p}}\mathrm{d}\boldsymbol{p}, \tag{2}$$

where $N_{\boldsymbol{p}}$ is the particle distribution function (the number of molecules with momentum $\boldsymbol{p}$). The total number of molecules

$$N=\int N_{\boldsymbol{p}}\,\mathrm{d}\boldsymbol{p} \tag{3}$$

is also conserved, which corresponds to the obvious relation $1+1=1+1$ for each binary collision.

Boltzmann posed the following question: does there exist another function $\varphi(\boldsymbol{p})$ that is also conserved in binary collisions, so that the relations in Eq. (1) imply the relation

$$\varphi(\boldsymbol{p}_1)+\varphi(\boldsymbol{p}_2)=\varphi(\boldsymbol{p}_3)+\varphi(\boldsymbol{p}_4)\ \ ? \tag{4}$$

In other words, does there exist $\varphi(\boldsymbol{p})$, such that Eq. (4) holds for any four vectors $\boldsymbol{p}_1$, $\boldsymbol{p}_2$, $\boldsymbol{p}_3$, $\boldsymbol{p}_4$, bound by Eq. (1)? Then Boltzmann's kinetic equation would additionally conserve the integral

$$I=\int \varphi(\boldsymbol{p})\,N_{\boldsymbol{p}}\,\mathrm{d}\boldsymbol{p}. \tag{5}$$

It is supposed that the function $\varphi(\boldsymbol{p})$ is linearly independent of the five functions: 1, three components of $\boldsymbol{p}$, and $\boldsymbol{p}^2$, so that the quantity (5) is independent of quantities (2) and (3).

If there were such a function $\varphi(\boldsymbol{p})$, then the thermodynamic equilibrium would be different. However, Boltzmann (1875) found that there is no such a function $\varphi(\boldsymbol{p})$ (provided it is smooth enough; see also Sercignani, 1990).

1.2 Zakharov–Schulman test of integrability: degenerative dispersion laws

In a different development, Zakharov and Schulman (1980, 1988) posed the question "When a given nonlinear wave system is integrable?" They argued that, to be integrable, the system should possess an additional conservation law, besides the energy and the momentum (and, sometimes, the wave action).

Suppose our nonlinear wave system is given in the Fourier representation as dynamics of wave amplitudes $a_{\boldsymbol{k}}(t)$

$$\begin{aligned} i\frac{\partial}{\partial t}a_{\boldsymbol{k}_1} =& \omega(\boldsymbol{k}_1)a_{\boldsymbol{k}_1}+\frac{1}{2}\int V(\boldsymbol{k}_1,\boldsymbol{k}_2,\boldsymbol{k}_3) \\ & a_{\boldsymbol{k}_2}a_{\boldsymbol{k}_3}\delta(\boldsymbol{k}_1-\boldsymbol{k}_2-\boldsymbol{k}_3)\,\mathrm{d}\boldsymbol{k}_2\,\mathrm{d}\boldsymbol{k}_3+\ldots, \end{aligned} \tag{6}$$

so that the time derivative of a wave amplitude (with some wave vector $\boldsymbol{k}_1$) is determined by the contributions of all other waves. The dispersion law $\omega(\boldsymbol{k})$ determines the linear part of Eq. (6), and the coupling coefficient $V(\boldsymbol{k}_1,\boldsymbol{k}_2,\boldsymbol{k}_3)$ determines quadratic nonlinearity. The dots stand for other nonlinear terms; these are other quadratic terms (namely, the ones containing aa^* and a^*a^*; * denotes complex conjugation), as well as higher order nonlinearity terms. The system is obviously translationally symmetric with respect to shifts in time (since all coefficients are time-independent), so it conserves the energy. The system is also translationally symmetric with respect to shifts in space, due to the presence of the delta function of the wave vectors (it just provides a way to write the convolution), so the system conserves momentum.

Zakharov and Schulman assumed that the wave amplitudes are chosen to be canonical Hamiltonian variables, so that the dynamical Eq. (6) has the form

$$i\frac{\partial}{\partial t}a_{\boldsymbol{k}}=\frac{\delta\mathcal{H}}{\delta a_{\boldsymbol{k}}^*} \tag{7}$$

with some Hamiltonian $\mathcal{H}$ expanded in a series over powers of wave amplitudes

$$\begin{aligned} \mathcal{H}=&\int \omega(\boldsymbol{k})\,a_{\boldsymbol{k}}a_{\boldsymbol{k}}^*\,\mathrm{d}\boldsymbol{k} \\ &+\frac{1}{2}\int V(\boldsymbol{k}_1,\boldsymbol{k}_2,\boldsymbol{k}_3)\,a_{\boldsymbol{k}_1}^*a_{\boldsymbol{k}_2}a_{\boldsymbol{k}_3}\delta(\boldsymbol{k}_1-\boldsymbol{k}_2-\boldsymbol{k}_3)\,\mathrm{d}\boldsymbol{k}_2\,\mathrm{d}\boldsymbol{k}_3 \\ &+\{c.c.\}+\{\ldots\text{other nonlinear terms}\ldots\} \end{aligned} \tag{8}$$

($\{c.c.\}$ means the complex conjugated term). Then the momentum is

$$\boldsymbol{P}=\int \boldsymbol{k}a_{\boldsymbol{k}}a_{\boldsymbol{k}}^*\,\mathrm{d}\boldsymbol{k}, \tag{9}$$

and the energy is the Hamiltonian. Zakharov and Schulman investigated when the wave system has another conservation law:

$$I = \int \varphi(\boldsymbol{k}) a_{\boldsymbol{k}} a^*_{\boldsymbol{k}} \mathrm{d}\boldsymbol{k} + \frac{1}{2} \int F(\boldsymbol{k}_1, \boldsymbol{k}_2, \boldsymbol{k}_3) a^*_{\boldsymbol{k}_1} a_{\boldsymbol{k}_2} a_{\boldsymbol{k}_3} \delta(\boldsymbol{k}_1 - \boldsymbol{k}_2 + \boldsymbol{k}_3) \mathrm{d}\boldsymbol{k}_2 \mathrm{d}\boldsymbol{k}_3 + \{c.c.\} + \{\ldots\text{other nonlinear terms}\ldots\} \quad (10)$$

(with undetermined φ, F, and so on); φ is supposed to be linearly independent of $\boldsymbol{k}$ and ω, so that I is independent of the momentum and energy. They found that in order for Eq. (10) to be conserved by the dynamics in Eq. (7)–(8), the following *alternative* should hold. At least one of the two conditions are to be satisfied:

1. The function $\varphi(\boldsymbol{k})$ is conserved in the resonant triad interactions: the equation

$$\varphi(\boldsymbol{k}_1) = \varphi(\boldsymbol{k}_2) + \varphi(\boldsymbol{k}_3) \quad (11)$$

   holds for all vectors $\boldsymbol{k}_1$, $\boldsymbol{k}_2$, and $\boldsymbol{k}_3$ bound by the resonance relations

$$\boldsymbol{k}_1 = \boldsymbol{k}_2 + \boldsymbol{k}_3, \qquad \omega(\boldsymbol{k}_1) = \omega(\boldsymbol{k}_2) + \omega(\boldsymbol{k}_3). \quad (12)$$

2. Or the resonant triad interaction is impossible. This can happen in two ways: Eq. (12) can have no solutions (like it happens for the gravity waves); or the nonlinear coupling coefficient $V(\boldsymbol{k}_1, \boldsymbol{k}_2, \boldsymbol{k}_3)$ can identically vanish on the resonance manifold defined by Eqs. (12).

Introducing the classical scattering matrix, Zakharov and Schulman (1988) obtained other necessary conditions for the conservation of Eq. (10). These are equations on $F(\boldsymbol{k}_1, \boldsymbol{k}_2, \boldsymbol{k}_3)$ and other kernels of higher order nonlinear terms in Eq. (10). They demonstrated (1980; 1988) their arguments on several integrable systems, in particular, the Kadomtsev–Petviashvili equation.

They also called the dispersion law $\omega(\boldsymbol{k})$ – which admits the additional conservation in Eq. (11) – *degenerative*.

The Zakharov–Schulman Eq. (11) and (12) for nonlinear waves are similar to Boltzmann's Eqs. (4) and (1) for rarefied gas, but waves provide a bigger variety:

a. The main interactions can involve any number of waves, in particular, only three waves (like in Eqs. 11–12).

b. There are many forms of the dispersion law, not just the quadratic function $\omega(\boldsymbol{k}) = \boldsymbol{k}^2$ (like in Boltzmann's case).

c. The physically interesting media (in which the waves propagate) can have different dimensions of $d = 1, 2, 3$.

2 Unique conservation properties of large-scale geophysical fluid dynamics

2.1 Quasi-geostrophic equation

The Rossby wave dispersion law turned out to be degenerative (Balk et al., 1991; Balk, 1991).

When dispersion law has the form

$$\omega(\boldsymbol{k}) = -\frac{\beta p}{\alpha^2 + k^2} \qquad \left[\boldsymbol{k} = (p, q), \quad k^2 = p^2 + q^2\right] \quad (13)$$

(α and β are parameters), the resonance relations in Eq. (12) imply Eq. (11) with $\varphi(\boldsymbol{k})$ equal to the function

$$\eta(\boldsymbol{k}) \equiv \arctan \frac{\alpha\left(q + p\sqrt{3}\right)}{k^2} - \arctan \frac{\alpha\left(q - p\sqrt{3}\right)}{k^2} \quad (14)$$

(Balk, 1991). The conservation of the function in Eq. (14) in triad resonances implies that the quasi-geostrophic equation

$$\left(\Delta\psi - \alpha^2\psi\right)_t + \beta\psi_x + \psi_y\Delta\psi_x - \psi_x\Delta\psi_y = 0 \quad (15)$$

possesses an approximate invariant

$$I = \frac{1}{2} \int \frac{1}{p} \eta(\boldsymbol{k}) \, |\mathcal{Q}_{\boldsymbol{k}}(t)|^2 \, \mathrm{d}p\mathrm{d}q, \quad (16)$$

with $\mathcal{Q}_{\boldsymbol{k}}(t)$ being the Fourier transform of

$$\mathcal{Q}(x, y, t) = \Delta\psi - \alpha^2\psi. \quad (17)$$

The function in Eq. (14) is conserved exactly in *resonant triad* interactions, but the integral in Eq. (16) is conserved approximately in *all* interactions (including triads that are far from being resonant). (If typical length scale L and parameter β are normalized to 1, and ε is a characteristic nondimensional wave amplitude, then the invariant in Eq. (16) has a magnitude of $O(\varepsilon^2)$, while its variation $\Delta I \equiv I(t) - I(0)$ has a smaller magnitude of $\Delta I = O(\varepsilon^3)$ on long time intervals $O(1/\varepsilon)$; see Balk and van Heerden, 2006, for details.)

Ferapontov (1992) realized that the problem of collision invariants (for rarefied gas, Sect. 1.1) and the problem of degenerative dispersion laws (for nonlinear waves, Sect. 1.2) can be posed as a problem of *web geometry*, which has been developing since the 1920s (Blaschke, 1932).

In particular, using the web geometry results, it is possible to show (Balk and Ferapontov, 1998) that the function in Eq. (14) is *unique*: any function $\varphi(\boldsymbol{k})$, conserved in triad resonances in Eq. (12) – with the Rossby dispersion law $\omega(\boldsymbol{k})$ – is a linear combination of $\boldsymbol{k}$, $\omega(\boldsymbol{k})$, and $\eta(\boldsymbol{k})$.

The connection to the web geometry also shows how exceptional the systems with extra invariants are. As one can guess, it is rare that four Eqs. (11) and (12) (the first equation in (12) consists of two scalar equations) in the six-dimensional space of wave vectors $\boldsymbol{k}_1$, $\boldsymbol{k}_2$, $\boldsymbol{k}_3$ determine a manifold of dimension 3. And it is indeed extremely rare (see Balk and Ferapontov, 1998, for details).

Even though Eq. (11) follows from the resonance relations in Eq. (12) (Eq. 11 is only linearly independent of Eq. 12), the integral (Eq. 16) is completely independent of the standard invariants (energy–momentum), because the integrand contains $\mathcal{Q}_k(t)$.

The presence of the extra invariant in Eq. (16) implies the energy transfer from small-scale eddies to large-scale zonal jets (Balk, 2005).

This is often but *not always* the case: the extra invariant does not lead to zonal jets, when the energy source is on large scales (i.e., the forcing scale k_f is long-wave, $k_f \ll \alpha$). In the latter situation, the energy accumulates in the sectors of polar angles $\theta = \arctan(q/p)$ such that $60° < |\theta| < 90°$. However, in the present paper we do not need to deal with this, since we are concerned with the opposite situation $k \gg \alpha$ for all k. As $\boldsymbol{k} \to \infty$, the extra invariant function in Eq. (14) and the dispersion law in Eq. (13) turn out to be asymptotically proportional (the same proportionality in all directions $\boldsymbol{k}/k$!); a certain linear combination of Eqs. (14) and (13), which cancels the common asymptotics, gives an extra invariant

$$\frac{5}{4\alpha^4}\left[\frac{\eta(\boldsymbol{k})}{2\sqrt{3}\,\alpha}+\frac{\omega(\boldsymbol{k})}{\beta}\right] \sim \frac{p^3\left(p^2+5q^2\right)}{k^{10}}, \quad (k \gg \alpha). \quad (18)$$

This linear combination decreases much faster with $|\boldsymbol{k}|$ than both functions in Eq. (14) and (13); the energy transfer towards zonal jets turns out to be more pronounced in the shortwave case (see Balk, 2005, for details of long- and shortwave limits).

Nazarenko and Quinn (2009) demonstrated the conservation of the extra invariant in Eq. (18) in the numerical simulations of the quasi-geostrophic equation.

2.2 Shallow water system

The extra invariant in Eq. (16) was extended (Balk et al., 2011) to the shallow water system on the beta plane

$$\begin{aligned} u_t + u u_x + v u_y - f(y) v &= -g H_x, \\ v_t + u v_x + v v_y + f(y) u &= -g H_y, \\ H_t + (H u)_x + (H v)_y &= 0, \end{aligned} \quad (19)$$

where (u, v) is the horizontal fluid velocity, H the fluid height (flat bottom is assumed), g the gravity acceleration, and $f(y)$ the Coriolis parameter. The shallow water system in Eq. (19) adiabatically conserves the integral in Eq. (16), with $\mathcal{Q}_k(t)$ being the Fourier transform of the perturbational potential vorticity (e.g., Gill, 1982)

$$\mathcal{Q}(x, y, t) = v_x - u_y - f(y)h, \quad (20)$$

where $h = (H - H_0)/H_0$ is the fractional relative height, which measures the deviation of the fluid surface height $H(x, y, t)$ from its unperturbed value H_0.

The extension was possible due to the following two unique conservation properties of Eq. (19).

First, the shallow water dynamics in Eq. (19) – besides Rossby waves – contains the inertia–gravity (IG) waves. The former could (in general) leak energy to the latter, thereby destroying the conservation of the extra invariant in Eq. (16) – which is based only on Rossby waves. However, the coupling coefficient of triad interaction between the IG and the Rossby waves vanishes, provided the Coriolis parameter is constant. If $f(y)$ varies sufficiently slowly (i.e., $\beta(y) \equiv f'(y)$ is sufficiently small), then the extra invariant slowly changes due to $\beta \neq 0$, but within its non-conservation due to the quartet interactions (recall that the invariant is only adiabatically conserved and can slowly change).

Second, the beta plane in Eq. (19) can be not only midlatitudinal but also close to the Equator. As we approach the Equator, the Coriolis parameter decreases, and the quasi-geostrophic approximation becomes invalid. Correspondingly, in the perturbation expansions, the Coriolis parameter f enters the denominators. However, the specific form of Eq. (19) makes it possible to refine (Balk et al., 2011) the Rossby mode, so that the numerators (in the perturbation expansion) are equally small in the limit $f \to 0$.

2.3 3-D fluid layers

2.3.1 Motivation

The application of the extra invariant in Eq. (16) to the real oceans and atmospheres of rotating planets encounters the following natural question. In a fluid layer of finite depth H, the shallow water motions (which are two-dimensional, 2-D) can generate – due to nonlinearity – the 3-D inertia waves. One can argue that the length of a typical Rossby wave significantly exceeds the depth H, so the energy exchange between Rossby waves and inertia waves can be disregarded due to the disparity of scales. But the question remains for *short* Rossby waves, whose wavelength can be comparable to the depth H; they could effectively interact with 3-D modes. Even if the energy of short Rossby waves were small, they could gradually leak significant amount of energy during long time evolution.

At the same time, the extra conservation seems to be essentially tied to the two-dimensionality or, more generally, to low dimension. So far, extra invariants have been found in two situations: (1) triads in 2-D media (see Eq. 12) and (2) quartets in 1-D media (Balk and Ferapontov, 1993), e.g.,

$$p_1 + p_2 = p_3 + p_4, \quad \omega(p_1) + \omega(p_2) = \omega(p_3) + \omega(p_4) \quad (21)$$

(p is the wave number in the 1-D media). In both situations, a wave with some fixed wave vector (say, $\boldsymbol{k}_1$ or p_1) participates in only one-dimensional family of resonant interactions. Let us elaborate. The triad resonance relations in Eq. (12) for 2-D wave vectors define manifold of dimension 3. A wave with a fixed wave vector (say, $\boldsymbol{k}_1$) can be involved with only one-dimensional family of resonant triads. The quartet resonance manifold in Eq. (21) has a dimension of two; a fixed wave vector (say p_1) participates again in one-dimensional family of resonant interactions.

For 3-D media, the triad resonance relations in Eq. (12) define manifold of dimension 5; a fixed wave vector (say, $\boldsymbol{k}_1$) is involved in two-dimensional family of resonance triads. There is an example of extra invariants for triad interaction in 3-D media (Balk, 1997), but it seems somewhat artificial.

Nevertheless, the extra invariant in Eq. (16) can be extended to the 3-D dynamics. This is due to the special property of the fluid equations, which was found a long time ago (Greenspan, 1969).

2.3.2 Problem

Since the short Rossby waves are the prime candidates for the energy exchange with 3-D modes, we consider the 3-D dynamics in the well-known classical situation (e.g., Landau and Lifshitz, 1987) of ideal incompressible (with constant density ρ_0) rotating fluid

$$\boldsymbol{v}_t + (\boldsymbol{v}\cdot\nabla)\boldsymbol{v} + \boldsymbol{f}\times\boldsymbol{v} = -\nabla\Pi, \qquad \nabla\cdot\boldsymbol{v} = 0, \tag{22}$$

where $\boldsymbol{v} = (u, v, w)$ is the fluid velocity; Π is the fluid pressure (centrifugal force included) divided by ρ_0. Euler's equation (Eq. 22) involves Coriolis force, with $\boldsymbol{f}$ being the double angular velocity. The fluid is assumed to occupy a layer between two parallel horizontal planes, rotating about vertical axis z. The boundary condition is vanishing of the vertical fluid velocity at the solid plane boundaries

$$w(x, y, 0, t) = w(x, y, H_0, t) = 0. \tag{23}$$

The Coriolis parameter slowly changes with the meridional coordinate y:

$$\boldsymbol{f} = (0, 0, f), \quad f = f_0 + \beta y. \tag{24}$$

The system in Eq. (22) describes the dynamics of inertia waves and vortical mode (e.g., Landau and Lifshitz, 1987, §14); the latter – due to the dependence of f on y – represents *short* Rossby waves with dispersion law

$$\Omega(\boldsymbol{k}) = -\beta\frac{p}{k^2} \quad \left[\boldsymbol{k} = (p, q)\right].$$

Greenspan (1969) discovered that on the f plane (when $\beta = 0$), the inertia modes do not produce triad-resonant response on the geostrophic mode. In other words, in the equation for the geostrophic mode (which has zero frequency), the coupling coefficient between this mode and two inertia modes (with frequencies ω_m and ω_n) vanishes whenever $\omega_m + \omega_n = 0$. It is reasonable to think that for nonzero – but sufficiently small – β, the triad resonant response is sufficiently small, so the second condition of Zakharov–Schulman's alternative (Sect. 1.2) is realized approximately: $V(\boldsymbol{k}_1, \boldsymbol{k}_2, \boldsymbol{k}_3)$ on the triad resonance manifold is not zero but "small". Then, the rapidly rotating 3-D fluid dynamics appears to have the invariants of the 2-D shallow water dynamics. In Sect. 3 this is derived explicitly.

2.3.3 Result

Let L be the length scale, U_0 the velocity scale, and $f_0 = f(0)$ the reference value of the Coriolis parameter. We use two small parameters:

- First, the Coriolis parameter $f(y)$ is a slow function: the variation $\Delta f \equiv f - f_0 \sim \beta L$ is small compared to f_0

  $$\frac{\beta L}{f_0} \ll 1\,.$$

  This condition means that L is much less than the Earth radius R_0, since at midlatitudes, $\beta \sim f_0/R_0$.

- Second, we consider weakly nonlinear dynamics, when the nonlinear terms are small compared to the linear terms. This means nonlinearity being small compared to the beta effect, namely,

  $$\epsilon = \frac{U_0}{\beta L^2} \ll 1\,.$$

  We need this, even though the Euler equation (Eq. 22) contains a linear term with f. Indeed, later (Sect. 3.2), we consider Eq. (33) for the vertical component of vorticity. We integrate Eq. (33) over the fluid depth, and the f term disappears (due to the incompressibility $u_x + v_y + w_z = 0$ and boundary condition in Eq. (23)). Then the only remaining linear term contains β, and the typical ratio of nonlinear terms to linear term is ϵ.

  Let us estimate ϵ for large-scale dynamics in Earth's oceans: flow at latitude $\approx 30°$ with $U_0 \approx 5\,\mathrm{cm\,s^{-1}}$ and $L \approx 100\,\mathrm{km}$ has $\epsilon \approx 1/4$. Is this sufficiently small to have the extra invariant? Perhaps, the weakly nonlinear dynamics is just a useful model that cannot be fully justified but nevertheless leads to plausible results (see also discussion of weak nonlinearity in the Introduction).

At this point, we assume that the fluid equations are written in dimensionless form with timescale $1/f_0$ and length scale L (in other words, the units are chosen such that $f_0 = L = 1$). Then the dimensionless Coriolis parameter is $f = 1 + \beta y$ with dimensionless $\beta \ll 1$; the dimensionless typical fluid velocity is $U_0 = \epsilon\beta$.

We will derive in Sect. 3 that the dynamics in Eq. (22)–(24) possesses an invariant

$$I = \frac{1}{2}\int \frac{p^2\left(p^2 + 5q^2\right)}{\left(p^2 + q^2\right)^5}\,|\zeta_{\boldsymbol{k}}(t)|^2 \mathrm{d}\boldsymbol{k}\,, \tag{25}$$

where $\boldsymbol{k} = (p, q)$ is the horizontal wave vector, and $\zeta_{\boldsymbol{k}}(t)$ is the Fourier transform of z-averaged vertical vorticity

$$\zeta(x, y, t) = \frac{1}{H_0}\int_0^{H_0} (v_x - u_y)\,\mathrm{d}z\,. \tag{26}$$

We will derive the following estimate for the variation of the integral in Eq. (25):

$$\Delta I \equiv I(t) - I(0) = O\left(\epsilon^{2.5}\beta^2\right) + O\left(\epsilon^2\beta^{2.5}\right), \qquad (27)$$

and over time $t = O(\epsilon^{-1}\beta^{-1})$. At the same time, the invariant I has the order $\epsilon^2\beta^2$, so $\Delta I \ll I$. The period of Rossby waves is $O(\beta^{-1})$, so I is conserved during time intervals of many – namely $O(\epsilon^{-1})$ – Rossby wave periods. (An estimate stronger than Eq. (27) can be obtained with further assumptions.)

3 Derivation

3.1 Normal mode expansion

To obtain the extra invariant in Eq. (25), let us expand the solution of Eq. (22) in normal modes of linearized system with $\beta = 0$:

$$\begin{aligned} u_t - f_0 v &= -\Pi_x, \\ v_t + f_0 u &= -\Pi_y, \\ w_t &= -\Pi_z, \\ u_x + v_y + w_z &= 0, \\ w(x,y,0,t) = w(x,y,H_0,t) &= 0. \end{aligned} \qquad (28)$$

(In our dimensionless units $f_0 = 1$, but we keep f_0 just to see physical meaning.) There are normal modes of two types: (1) the vortical mode, representing the Taylor–Proudman column

$$u = -\frac{1}{f_0}\Pi_y, \quad v = \frac{1}{f_0}\Pi_y, \quad w = 0, \quad \Pi_z = 0, \qquad (29)$$

(which is t and z independent), and (2) the inertial waves (standing in the z direction)

$$\begin{bmatrix} u \\ v \\ w \\ \Pi \end{bmatrix} = \begin{bmatrix} \hat{u}\cos mz \\ \hat{v}\cos mz \\ \hat{w}\sin mz \\ \hat{\Pi}\cos mz \end{bmatrix} e^{i(px+qy-\omega t)}, \qquad (30)$$

where $m = n\pi/H_0$ with *non-zero* integer n. Substitution of Eq. (30) into Eq. (28) gives the dispersion relation

$$\omega^2 K^2 = f_0^2 m^2 \quad \left(K^2 = p^2 + q^2 + m^2\right), \qquad (31)$$

and the polarization vector

$$\begin{bmatrix} \hat{u} \\ \hat{v} \\ \hat{w} \\ \hat{\Pi} \end{bmatrix} = \begin{bmatrix} -iqf_0 - \omega p \\ ipf_0 - \omega q \\ i\omega k^2/m \\ f_0^2 - \omega^2 \end{bmatrix}.$$

The capital $\boldsymbol{K}$ denotes the full three-dimensional wave vector, $\boldsymbol{K} = (p,q,m)$, while the small case $\boldsymbol{k}$ stands for the horizontal 2-D vector, $\boldsymbol{k} = (p,q)$. The general solution of Eq. (28) is a linear combination of all normal modes with some coefficients $\Psi_{\boldsymbol{k}}$ and $A_{\boldsymbol{K}}$:

$$\begin{bmatrix} u(x,y,z,t) \\ v(x,y,z,t) \\ w(x,y,z,t) \\ \Pi(x,y,z,t) \end{bmatrix} = \int \mathrm{d}p\,\mathrm{d}q\, e^{i(px+qy)} \qquad (32)$$

$$\left\{ \Psi_{\boldsymbol{k}} \begin{bmatrix} -iq \\ ip \\ 0 \\ f_0 \end{bmatrix} + \sum_m A_{\boldsymbol{K}} \begin{bmatrix} (-iqf_0 - \omega p)\cos mz \\ (ipf_0 - \omega q)\cos mz \\ (i\omega k^2/m)\sin mz \\ (f_0^2 - \omega^2)\cos mz \end{bmatrix} e^{-i\omega t} \right\}.$$

Here $\omega \equiv f_0 m/K$; the summation in Eq. (32) is over $m = \frac{\pi}{H_0}n$, $n = \pm1, \pm2, \ldots$; $\Psi_{-\boldsymbol{k}} = \Psi^*_{\boldsymbol{k}}$, $A_{-\boldsymbol{K}} = A^*_{\boldsymbol{K}}$.

The solution of the full (nonlinear and with non-zero β) system in Eq. (22) can also be expanded in the form (32), but the coefficients $\Psi_{\boldsymbol{k}}$ and $A_{\boldsymbol{K}}$ would depend on time t.

3.2 Equation for the vertical vorticity in terms of normal modes

From Eq. (22) we have

$$\begin{aligned} &(v_x - u_y)_t + (uv_x + vv_y + wv_z)_x \\ &\quad - (uu_x + vu_y + wu_z)_y + f(u_x + v_y) + \beta v = 0. \end{aligned} \qquad (33)$$

Now we substitute Eq. (32) into Eq. (33) and integrate over z (from 0 to H_0), taking into account that cosines $\cos mz$ integrate to zero. To shorten writing, any subscript wave vector $\boldsymbol{k}_j$ or $\boldsymbol{K}_j$ will be replaced by its label subindex j (e.g., $\Psi_j \equiv \Psi_{\boldsymbol{k}_j}$, $A_j \equiv A_{\boldsymbol{K}_j}$). This is unambiguous: index j stands for $\boldsymbol{K}_j$ if the corresponding quantity – like A – depends on the three-dimensional wave vector $\boldsymbol{K}$; j stands for $\boldsymbol{k}_j$ if the corresponding quantity – like Ψ – depends on the two-dimensional wave vector $\boldsymbol{k}$. In addition, $-j$ stands for $-\boldsymbol{K}_j$ or $-\boldsymbol{k}_j$; $\mathrm{d}_{23} \equiv \mathrm{d}\boldsymbol{k}_2\,\mathrm{d}\boldsymbol{k}_3 \equiv \mathrm{d}p_2\mathrm{d}q_2\,\mathrm{d}p_3\mathrm{d}q_3$.

Thus, Eq. (33) becomes

$$\begin{aligned} k_1^2\dot{\Psi}_1 &= \beta i p_1 \Psi_1 + \frac{1}{2}\int W_{-1,2,3}\,\Psi_2\Psi_3\,\mathrm{d}_{23} \\ &\quad + \frac{1}{2}\int \sum_{m_2,m_3} V_{-1,2,3}\, e^{-i(\omega_2+\omega_3)t}\, A_2 A_3\,\mathrm{d}_{23}. \end{aligned} \qquad (34)$$

The integrals in Eq. (34) have been symmetrized with respect to transposition of indexes 2 and 3: this gives the factor 1/2 in front of the integrals and the symmetric coupling coefficients ($W_{-1,2,3} = W_{-1,3,2}$ and $V_{-1,2,3} = V_{-1,3,2}$)

$$\begin{aligned} W_{1,2,3} &\equiv W_{\boldsymbol{k}_1,\boldsymbol{k}_2,\boldsymbol{k}_3} \\ &= \Delta_{1,2,3}\left(k_3^2 - k_2^2\right)\delta(\boldsymbol{k}_1 + \boldsymbol{k}_2 + \boldsymbol{k}_3), \end{aligned} \qquad (35)$$

$$\begin{aligned} V_{1,2,3} &\equiv V_{\boldsymbol{K}_1,\boldsymbol{K}_2,\boldsymbol{K}_3} \\ &= \left[\left(f_0\Delta_{1,2,3} - i\omega_3\boldsymbol{k}_1\cdot\boldsymbol{k_3}\right)\left(f_0\boldsymbol{k}_1\cdot\boldsymbol{k_2} - i\omega_2\Delta_{123}\right)\right. \\ &\quad \left. - \left(f_0\Delta_{1,2,3} + i\omega_2\boldsymbol{k}_1\cdot\boldsymbol{k_2}\right)\left(f_0\boldsymbol{k}_1\cdot\boldsymbol{k_3} + i\omega_3\Delta_{123}\right)\right] \\ &\quad \left(\delta_{m_2-m_3} + \delta_{m_2+m_3}\right)\delta(\boldsymbol{k}_1 + \boldsymbol{k}_2 + \boldsymbol{k}_3). \end{aligned} \qquad (36)$$

Here $\delta(\cdot)$ is the Dirac delta function, and δ_m is the Kronecker symbol (which is 1 if $m=0$ and 0 if $m \neq 0$):

$$\Delta_{1,2,3} = p_2 q_3 - p_3 q_2,$$

which – because of the delta function – has the following obvious symmetries:

$$\Delta_{1,2,3} = \Delta_{2,3,1} = \Delta_{3,1,2} = -\Delta_{1,3,2} = -\Delta_{3,2,1} = -\Delta_{2,1,3}.$$

That is, $\Delta_{1,2,3}$ is symmetric with respect to the three cyclic permutations of its indexes and anti-symmetric with respect to the other three permutations. Software MATHEMATICA has been used to obtain V in the form of Eq. (36).

3.3 Adiabatic conservation

We intend to find invariant

$$I = \frac{1}{2} \int X(\boldsymbol{k})\, \Psi_{\boldsymbol{k}}\, \Psi_{-\boldsymbol{k}}\, \mathrm{d}\boldsymbol{k}, \quad [X(\boldsymbol{k}) = X(-\boldsymbol{k})] \tag{37}$$

(with undetermined kernel X), which are approximately conserved over a long time $O(\epsilon^{-1}\beta^{-1})$. This is the nonlinear time of wave interactions (since the wave amplitudes are of the order of fluid velocity, which is $O(\epsilon\,\beta)$). To show that $I(t)$ stays almost constant, it would be enough to show that its time derivative $\dot{I}$ is sufficiently small. However, the latter is not the case: $I(t)$ oscillates in time (similar to adiabatic invariants in the theory of dynamical systems). The change $\Delta I \equiv I(t) - I(0)$ is "small", but $\dot{I}$ is "big". The usual way to deal with such a situation has been the following. The quadratic integral in Eq. (37) is supplemented by cubic corrections, e.g., for the dynamics in Eq. (34):

$$I^{\text{suppl.}} = \frac{1}{2} \int X(\boldsymbol{k})\, \Psi_{\boldsymbol{k}}\, \Psi_{-\boldsymbol{k}}\, \mathrm{d}\boldsymbol{k} + \int Y_{123} \Psi_1 \Psi_2 \Psi_3 \mathrm{d}_{123} + \int Z_{123} \Psi_1 A_2 A_3 \mathrm{d}_{123} \tag{38}$$

(X, Y, Z are undetermined kernels); $\dot{I}^{\text{suppl.}}$ is required to be sufficiently small. At the end, the cubic corrections can be dropped, since they turn out to be within the variation of $\Delta I^{\text{suppl.}} \equiv I^{\text{suppl.}}(t) - I^{\text{suppl.}}(0)$.

Here we develop a new, more direct, approach (without cubic corrections).

Let us, first, apply the transformation

$$\Psi_{\boldsymbol{k}} = \psi_{\boldsymbol{k}}\, e^{-i\Omega(\boldsymbol{k})t}; \tag{39}$$

it removes the linear term in Eq. (34),

$$\dot{\psi}_1 = \frac{1}{2k_1^2} \int W_{-1,2,3}\, e^{i(\Omega_1 - \Omega_2 - \Omega_3)t}\, \psi_2\, \psi_3\, \mathrm{d}_{23} + \frac{1}{2k_1^2} \int \sum_{m_2, m_3} V_{-1,2,3}\, e^{i(\Omega_1 - \omega_2 - \omega_3)t}\, A_2\, A_3\, \mathrm{d}_{23}, \tag{40}$$

and leaves the invariant (37) in the old form

$$I = \frac{1}{2} \int X(\boldsymbol{k})\, \psi_{\boldsymbol{k}}\, \psi_{-\boldsymbol{k}}\, \mathrm{d}\boldsymbol{k}.$$

Its time derivative is

$$\dot{I} = \dot{I}^{\mathrm{R}} + \dot{I}^{\mathrm{i}}, \tag{41}$$

where

$$\dot{I}^{\mathrm{R}} = \frac{1}{2} \int \mathrm{d}_{123}\, \frac{1}{k_1^2} W_{123}\, e^{-i(\Omega_1 + \Omega_2 + \Omega_3)t}\, \psi_1\, \psi_2\, \psi_3,$$

$$\dot{I}^{\mathrm{i}} = \frac{1}{2} \int \mathrm{d}_{123} \sum_{m_2, m_3} X_1 \frac{1}{k_1^2} V_{123}\, \psi_1 e^{-i(\Omega_1 + \omega_2 + \omega_3)t}\, A_2\, A_3;$$

$\dot{I}^{\mathrm{R}}$ and $\dot{I}^{\mathrm{i}}$ are respectively contributions to $\dot{I}$ resulting from the interaction between three Rossby waves and from the interaction between two inertia waves and one Rossby wave. In both integrals in Eq. (41), we have replaced $\boldsymbol{k}_1$ by $-\boldsymbol{k}_1$ and used that the dispersion law Ω is an odd function ($\Omega_{-1} = -\Omega_1$). Our goal is to estimate the contributions to ΔI resulting from $\dot{I}^{\mathrm{R}}$ and $\dot{I}^{\mathrm{i}}$, and thereby show that $\Delta I \ll I$.

3.4 Small denominator lemma

The essence of the extra invariant conservation can be expressed by the lemma below. Basically, we have

$$\frac{\mathrm{d}I}{\mathrm{d}\tau} = \int_{-\infty}^{\infty} F(\chi, \varepsilon\tau)\, e^{i\chi\tau}\, \mathrm{d}\chi, \tag{42}$$

and need to estimate the total variation $\Delta I \equiv I(\tau) - I(0)$ over a long time interval $0 \leq \tau \leq \varepsilon^{-1}$.

For instance, in the case of Eq. (41), the variable χ results from the exponent, and the function F is the result of all integrations while χ is held fixed (later, when applying the lemma, we will define χ and F exactly). The function F slowly depends on time τ, which is signified by a small parameter ε (this corresponds to the slow variation of wave amplitudes). The integral in Eq. (42) is well convergent at $\chi \to \pm\infty$ (corresponding to the vanishing of wave amplitudes as $\boldsymbol{k}_1$, $\boldsymbol{k}_2$, and $\boldsymbol{k}_3$ approach infinity). It is important that the integration interval in Eq. (42) includes point $\chi = 0$, so that the resonance (at $\chi \to 0$) is present.

In general, I exhibits secular growth associated with the resonance $\chi \to 0$, and the adiabatic conservation of I is out of the question. Indeed, assume for simplicity that F is time independent ($\varepsilon = 0$); then

$$\Delta I = \int F(\chi) \frac{e^{i\chi\tau} - 1}{i\chi}\, \mathrm{d}\chi\,; \tag{43}$$

here $i\chi$ is the "small denominator" ("small divisor"), corresponding to the resonance $\chi \to 0$. According to Eq. (43), for large τ, the increment ΔI grows proportional to τ, unless the small denominator is somehow cancelled.

For LEMMA,
Suppose, there are constants M and L such that

$$\int_{-\infty}^{\infty} \left| \frac{F(\chi, \mathcal{T})}{\chi} \right| d\chi \leq M, \tag{44}$$

$$\int_{-\infty}^{\infty} \left| F(\chi, \mathcal{T}'') - F(\chi, \mathcal{T}') \right| d\chi \leq L|\mathcal{T}'' - \mathcal{T}'| \tag{45}$$

for all positive $\mathcal{T}, \mathcal{T}', \mathcal{T}''$. Then

$$|\Delta I| \leq (2M + L)\varepsilon^{-1/2} \quad \text{for} \quad 0 \leq \tau \leq \varepsilon^{-1}. \tag{46}$$

Comment 1. The convergence of integral in Eq. (44) implies cancellation (at least partial) of the small denominator χ.
Comment 2. The constants M and L are $\mathcal{T}$-independent, but can depend on ε, which will be used in the lemma applications below.
Proof. We split the time interval $0 \leq \tau \leq \varepsilon^{-1}$ into many subintervals

$$[\tau_{j-1}, \tau_j], \quad j = 1, 2, \ldots J \qquad \left(\tau_0 = 0, \quad \tau_J = \varepsilon^{-1}\right),$$

of length $\varepsilon^{-1/2}$; there are $J = \varepsilon^{-1/2}$ of them. (To make this number integer, we can consider an appropriate subsequence of ε, or we can use the integer part $J = [\varepsilon^{-1/2}] + 1$ with slightly shorter subintervals.) On each subinterval

$$\frac{dI}{d\tau} = \int \{F(\chi, \varepsilon\tau_j) + [F(\chi, \varepsilon\tau) - F(\chi, \varepsilon\tau_j)]\} e^{i\chi\tau} d\chi;$$

Integrating this from τ_{j-1} to τ_j, we have

$$\Delta I_j \equiv I(\tau_j) - I(\tau_{j-1}) = \int d\chi \frac{F(\chi, \varepsilon\tau_j)}{i\chi} \left(e^{i\chi\tau_j} - e^{i\chi\tau_{j-1}}\right) + \int_{\tau_{j-1}}^{\tau_j} d\tau \int d\chi \left[F(\chi, \varepsilon\tau) - F(\chi, \varepsilon\tau_j)\right] e^{i\chi\tau}.$$

In the latter integral, $|\varepsilon\tau - \varepsilon\tau_j| \leq \varepsilon^{1/2}$. According to estimates in Eq. (44)–(45),

$$|\Delta I_j| \leq 2M + \int_{\tau_{j-1}}^{\tau_j} L\varepsilon^{1/2} d\tau = 2M + L.$$

Now, $|\Delta I| \leq \sum_{j=1}^{J} |\Delta I_j| \leq (2M + L)J$, and we find Eq. (46).

3.5 Interaction between three Rossby waves

Now we estimate ΔI resulting from $\dot{I}^{\mathrm{R}}$ in Eq. (41).

The periods of Rossby waves are $O(\beta^{-1})$, and we introduce function

$$\sigma(\boldsymbol{k}) = -\frac{p}{k^2}, \quad \text{so that} \quad \Omega(\boldsymbol{k}) = \beta\sigma(\boldsymbol{k}). \tag{47}$$

Rossby wave amplitude $\psi_{\boldsymbol{k}}$ evolves on the nonlinear timescale $O(\epsilon^{-1}\beta^{-1})$ (recall the transformation in Eq. (39)). Therefore, to apply the lemma of Sect. 3.4, we assume

$$\chi = -(\sigma_1 + \sigma_2 + \sigma_3), \quad \tau = \beta t, \quad \varepsilon = \epsilon.$$

Now,

$$\frac{dI^{\mathrm{R}}}{d\tau} = \frac{1}{\beta}\dot{I}^{\mathrm{R}} = \int F(\chi, \epsilon\tau) e^{i\chi\tau} d\chi, \quad \text{with} \tag{48}$$
$$F = \frac{1}{2\beta} \int d_{123} \delta(\sigma_1 + \sigma_2 + \sigma_3 + \chi) \frac{X_1}{k_1^2} W_{123} \psi_1 \psi_2 \psi_3.$$

We symmetrize the integral for F with respect to permutations of indexes 1, 2, and 3 (this, in particular, gives an additional factor 1/3 in front of the integral) and use Eq. (35) for the kernel W, taking into account that $\Delta_{1,2,3}$ is symmetric with respect to the cyclic permutations of its indexes

$$F = \frac{1}{6\beta} \int d_{123}\, \delta(\boldsymbol{k}_1 + \boldsymbol{k}_2 + \boldsymbol{k}_3)\, \Delta_{1,2,3}\, \psi_1 \psi_2 \psi_3 \, \delta(\sigma_1 + \sigma_2 + \sigma_3 + \chi) \left[\frac{X_1}{k_1^2}\left(k_3^2 - k_2^2\right) + \frac{X_2}{k_2^2}\left(k_1^2 - k_3^2\right) + \frac{X_3}{k_3^2}(k_2^2 - k_1^2)\right]. \tag{49}$$

Now we will determine special functions $X(\boldsymbol{k})$ so that the integral (44) converges as $\chi \to 0$. Due to the dispersion relation, $k_j^2 = -p_j/\sigma_j$ $(j = 1, 2, 3)$, so the square bracket in Eq. (49) equals

$$[\ldots] = \frac{1}{\sigma_1\sigma_2\sigma_3}\left[\frac{p_1 X_1}{k_1^4}(p_3\sigma_2 - p_2\sigma_3) + \frac{p_2 X_2}{k_2^4}(p_1\sigma_3 - p_3\sigma_1) + \frac{p_3 X_3}{k_3^4}(p_2\sigma_1 - p_1\sigma_2)\right]. \tag{50}$$

We have $p_1 + p_2 + p_3 = 0$ due to the delta function in Eq. (49); when $\chi = 0$, then $\sigma_1 + \sigma_2 + \sigma_3 = 0$, and the three pairs of parentheses in Eq. (50) turn out equal. Now the square bracket in Eq. (49) is

$$[\ldots] = \frac{p_2\sigma_1 - p_1\sigma_2}{\sigma_1\sigma_2\sigma_3}\left[\frac{p_1 X_1}{k_1^4} + \frac{p_2 X_2}{k_2^4} + \frac{p_3 X_3}{k_3^4}\right]; \tag{51}$$

it vanishes if the function

$$\varphi(\boldsymbol{k}) = \frac{p\, X(\boldsymbol{k})}{k^4}$$

is conserved in resonant triad interactions

$$\left.\begin{array}{l} \boldsymbol{k}_1 + \boldsymbol{k}_2 + \boldsymbol{k}_3 = 0 \\ \Omega_1 + \Omega_2 + \Omega_3 = 0 \end{array}\right\} \Rightarrow \quad \varphi_1 + \varphi_2 + \varphi_3 = 0. \tag{52}$$

Then the integral in Eq. (44) is convergent. Since $\psi_{\boldsymbol{k}} = O(\epsilon\beta)$, the constants M and L are $\frac{1}{\beta} O(\epsilon^3\beta^3) = O(\epsilon^3\beta^2)$, and, according to the lemma, $\Delta I = O(\epsilon^{2.5}\beta^2)$.

Remark. $\Omega(\boldsymbol{k})$ and $\varphi(\boldsymbol{k})$ are odd functions of the wave vector $\boldsymbol{k}$, so (implication (52)) is equivalent to the implication of Eqs. (12) $\Rightarrow$ (11).

There are *five* possible functions $\varphi(\boldsymbol{k})$ that satisfy Eq. (52). Three of them are obvious and common:

$$\varphi(\boldsymbol{k}) = \Omega(\boldsymbol{k}), \quad \varphi(\boldsymbol{k}) = p, \quad \varphi(\boldsymbol{k}) = q. \tag{53}$$

The other two (Balk, 1991) are specific for short Rossby waves:

$$\varphi(\boldsymbol{k}) = \xi^{\text{short}}(\boldsymbol{k}) \equiv \frac{pq}{k^4}; \tag{54}$$

$$\varphi(\boldsymbol{k}) = \eta^{\text{short}}(\boldsymbol{k}) \equiv \frac{p^3(p^2+5q^2)}{k^{10}}. \tag{55}$$

Similar to Balk et al. (2011), we discard the third and the fourth of them as unphysical: the third gives singular $X(\boldsymbol{k}) = qk^4/p$ (on the line $p=0$), and the fourth is even, giving odd $X(\boldsymbol{k})$, which contradicts to the bracketed condition in Eq. (37). The first function in Eq. (53) corresponds to the energy of the quasi-geostrophic mode; the second corresponds to the zonal momentum (which is the enstrophy).

The fifth function (Eq. 55) gives the extra invariant (cf. Eq. 18).

3.6 Interaction between one Rossby wave and two inertia waves

Now, we estimate the variation ΔI resulting from $\dot{I}^{\mathrm{i}}$ in Eq. (41).

The inertia wave amplitude $A_{\boldsymbol{k}}(t)$ evolves faster than the Rossby wave amplitude $\psi_{\boldsymbol{k}}(t)$. The latter evolves on the nonlinear timescale $O(\epsilon^{-1}\beta^{-1})$. The former have linear response to the Rossby waves. Recall that we used normal modes for $\beta=0$, so $A_{\boldsymbol{k}}$ and $\psi_{\boldsymbol{k}}$ are independent in linearized dynamics if $\beta=0$ only. If $\beta\neq 0$, the amplitude $A_{\boldsymbol{k}}(t)$ evolves on timescale $O(\beta^{-1})$. Therefore, we use the lemma of Sect. 3.4 with

$$\chi = -(\omega_2+\omega_3), \quad \tau = t, \quad \varepsilon = \beta.$$

Now,

$$\frac{\mathrm{d}I^{\mathrm{i}}}{\mathrm{d}t} = \int F(\chi, \beta t)\, e^{i\chi t}\,\mathrm{d}\chi\,, \tag{56}$$

where

$$F = \frac{1}{2}\int \mathrm{d}_{123} \sum_{m_2,m_3} \delta(\omega_2+\omega_3+\chi)\,\frac{X_1}{k_1^2} V_{123}\Psi_1 A_2 A_3.$$

To apply the lemma, we show that the integral in Eq. (44) converges as $\chi\to 0$. Indeed, when $\chi=0$, then $\omega_2+\omega_3=0$, which implies $|\omega_2|=|\omega_3|$. Because of Kronecker deltas in Eq. (36), we have $|m_2|=|m_3|$. Therefore, – by the dispersion relation in Eq. (31) – $K_2=K_3$, and $k_2=k_3$. Now,

$$\boldsymbol{k}_2 = -\boldsymbol{k}_1-\boldsymbol{k}_3 \;\Rightarrow\; k_2^2 = k_1^2+k_3^2+2\boldsymbol{k}_1\cdot\boldsymbol{k}_3,$$
$$\boldsymbol{k}_3 = -\boldsymbol{k}_1-\boldsymbol{k}_2 \;\Rightarrow\; k_3^2 = k_1^2+k_2^2+2\boldsymbol{k}_1\cdot\boldsymbol{k}_2.$$

Thus $\boldsymbol{k}_1\cdot\boldsymbol{k}_3 = \boldsymbol{k}_1\cdot\boldsymbol{k}_2$, and in the square bracket of Eq. (36), the four pairs of parentheses are pairwise identical. Therefore, $V_{1,2,3}=0$.

The wave amplitudes Ψ and A are $O(\epsilon\beta)$, and, therefore, the constants M and L are $O(\epsilon^3\beta^3)$. Now, according to the lemma, $\Delta I = O(\epsilon^3\beta^{2.5})$ on any time interval of length β^{-1}. We can split the time interval $0\le t\le\epsilon^{-1}\beta^{-1}$ into subintervals of length $O(\beta^{-1})$; there are $O(\epsilon^{-1})$ of them. Thus, $\Delta I = \frac{1}{\epsilon}O(\epsilon^3\beta^{2.5})$ on time interval $0\le t\le\epsilon^{-1}\beta^{-1}$.

Combining this estimate and the estimate found in the previous Sect. 3.5, we find Eq. (27).

Remark. The estimate in Eq. (27) holds for any solution of the 3-D layer dynamics in Eq. (22)–(24). If the wave turbulence were considered, the estimate would hold for any realization, even the least probable; on average (or for a typical realization) the estimate is stronger.

4 Emergence of zonal jets

4.1 Invariants

According to the expansion in Eq. (32), the vertical vorticity in Eq. (26) is

$$\zeta(x,y,t) = -\int k^2\Psi_{\boldsymbol{k}}(t)\, e^{i(px+qy)}\mathrm{d}\boldsymbol{k} \tag{57}$$

(since the cosines integrate to zero); in the Fourier representation, Eq. (57) means $\zeta_{\boldsymbol{k}} = -k^2\Psi_{\boldsymbol{k}}$. So, we can express integral in Eq. (37) in terms of $\zeta_{\boldsymbol{k}}$:

$$I = \frac{1}{2}\int \frac{1}{p}\varphi(\boldsymbol{k})\,|\zeta_{\boldsymbol{k}}|^2\,d\boldsymbol{k}. \tag{58}$$

Thus, we have three invariants:

1. The energy of the quasi-geostrophic mode (the energy of inertia waves is separate)

$$E = \frac{1}{2}\int \frac{1}{k^2}\,|\zeta_{\boldsymbol{k}}|^2\,\mathrm{d}\boldsymbol{k}\,. \tag{59}$$

2. The enstrophy

$$\text{Enstrophy} = \frac{1}{2}\int |\zeta_{\boldsymbol{k}}|^2\,\mathrm{d}\boldsymbol{k}\,. \tag{60}$$

3. The extra invariant

$$I = \frac{1}{2}\int \frac{1}{p}\eta^{\text{short}}(\boldsymbol{k})\,|\zeta_{\boldsymbol{k}}|^2\,\mathrm{d}\boldsymbol{k} \tag{61}$$

with $\eta^{\text{short}}(\boldsymbol{k})$ given in Eq. (55). Though the function in Eq. (55) has non-integrable singularity at $k=0$, the energy spectrum usually vanishes at $k\to 0$, and the integral in Eq. (61) converges. The expression in Eq. (61) is the same as Eq. (25).

4.2 Anisotropic inverse cascade

In rapidly rotating 3-D fluids, the energy from inertia waves is transferred towards the quasi-geostrophic mode (Smith and Waleffe, 1999; Cambon et al., 2004; Staplehurst et al., 2008; Duran-Matute et al., 2013).

The conservation of the energy in Eq. (59) and the enstrophy in Eq. (60) implies the inverse cascade of the two-dimensional quasi-geostrophic energy (Rhines, 1975; Pedlosky, 1987; Vallis, 2006). This is because the energy spectrum spreads in the Fourier space (unless the initial conditions are very specially arranged); the spreading happens due to (Rhines, 1975) the decay instability (when a wave decays into the other two waves of a resonant triad), as well as due to the resonant generation of higher harmonics.

We assume that the inverse energy cascade takes place. Then the presence of the third, extra, invariant in Eq. (61) implies the energy transfer towards large-scale *zonal* flow, not just to large scales in general, but specifically into the region of the $\boldsymbol{k}$ plane around the q axis. This is the only way to keep the extra invariant approximately constant.

Indeed, let $E_{\boldsymbol{k}} = |\zeta_{\boldsymbol{k}}|^2/k^2$ be the energy spectrum of the quasi-geostrophic mode. According to Eqs. (59) and (61), the dynamics preserves integrals

$$E = \frac{1}{2}\int E_{\boldsymbol{k}}\,\mathrm{d}\boldsymbol{k} \quad \text{and} \quad I = \frac{1}{2}\int \phi(\boldsymbol{k}) E_{\boldsymbol{k}}\,\mathrm{d}\boldsymbol{k}, \tag{62}$$

where

$$\phi(\boldsymbol{k}) = \frac{\eta^{\mathrm{short}}(\boldsymbol{k})}{-\Omega(\boldsymbol{k})/\beta} = \frac{p^2(p^2+5q^2)}{k^8} \tag{63}$$

is shown in Fig. 1. The ratio ϕ is small for large k, and large near the origin, except for the vicinity of the q axis. Therefore, the energy transferring from large k towards the origin – via the inverse cascade – must accumulate near the q axis (which corresponds to zonal jets). Such an anisotropic inverse cascade is similar to the one in geostrophic dynamics (see Balk, 2005, for precise estimates). The reasoning for the generation of zonal flow tacitly assumes that there is a source of small-scale energy. We see that the formation of zonal jets in 3-D dynamics is enhanced: because of the *two-dimensionalization*, the energy of 3-D inertia waves also transfers into zonal jets.

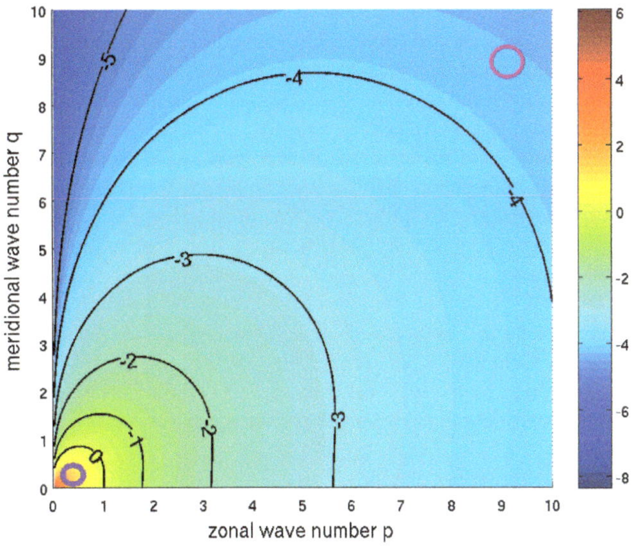

Fig. 1. The values of the ratio $\phi = \frac{\eta^{\mathrm{short}}}{-\Omega/\beta}$ (in logarithmic scale) as a function of the wave vector $\boldsymbol{k} = (p, q)$. The numbers of the color bar are the values of $\log_{10}(\phi)$. The black curves are the level lines $\phi(\boldsymbol{k}) = 10^n$ for six integers $n = 0, -1, -2, -3, -4, -5$. All the contour lines pass through the origin tangent to the q axis. The figure shows that the inverse cascade transfers energy from small-scale eddies to large-scale zonal jets. Indeed, let us pose a question: is it possible that the energy from the area marked by the pink circle (in the upper right corner) – via the inverse cascade – ends up in the purple circle near the origin? This is clearly impossible when both integrals $\int E_{\boldsymbol{k}}\mathrm{d}\boldsymbol{k}$ and $\int \phi(\boldsymbol{k})E_{\boldsymbol{k}}\mathrm{d}\boldsymbol{k}$ are preserved. If such transfer were to occur, it would lead to a significant increase of the extra invariant: the value of the ratio ϕ in the purple circle is more than six orders of magnitude bigger than its value in the pink circle. The only possibility for the inverse cascade is that the energy should end up near the q axis where the ratio ϕ is also small, similar to its values at large wave numbers (away from the q axis), so that the energy and extra invariant in Eq. (62) can be both conserved. The energy concentration near the q axis means the fluid should have velocity mostly parallel to the x axis, which is zonal flow.

5 Conclusion

The system in Eq. (22) is known to conserve the energy and helicity. We have shown that, in the presence of the beta effect ($\beta \equiv f'(y) \neq 0$), there are three adiabatic-type invariants: the energy (Eq. 59), the enstrophy (Eq. 60), and the extra invariant (Eq. 61), all being based on the Rossby waves. The presence of these invariants – together with the *two-dimensionalization* – implies the transfer of energy towards zonal jets (Sect. 4.2).

We have also summarized the unique conservation properties of the geophysical fluid dynamics (with the beta effect) that allow for the presence of the Rossby wave extra invariant in the various systems (Sect. 2). It seems possible to extend the extra invariant to the stratified fluid dynamics that involves several Rossby modes with different Rossby radii of deformation (e.g., to the systems with several shallow layers). This is plausible because the coupling coefficients of resonant triad interactions between the barotropic mode and the other Rossby modes vanish, provided the densities of the layers are sufficiently close (Soomere, 2003).

Acknowledgements. I wish to thank L. Smith for her suggestion to consider 3-D dynamics. As well, I am grateful to E. Ferapontov for introducing to me Boltzmann's problem of collision invariants (Sect. 1.1) many years ago and to P. Weichman for the valuable possibility to discuss the extra invariant. I am also thankful for discussions of this research at the workshop "Nonlinear processes in atmospheric and oceanic flows", the Instituto de Ciencias Matemáticas (ICMAT), Madrid, Spain, July 2012. The provided support is gratefully acknowledged. The publication of the present paper is funded by the ICMAT Severo Ochoa project SEV-2011-0087. Likewise, I am grateful for discussions during the program Stochastic Flows and Climate Modeling, Aspen Center for Physics, Colorado, USA, June 2012. The present work was supported in part by the National Science Foundation under grant no. PHYS-1066293 and the hospitality of the Aspen Center.

Edited by: E. Hernández-García

References

Balk, A. M.: A new invariant for Rossby wave systems, Phys. Lett. A, 155, 20–24, 1991.

Balk, A. M.: New conservation laws for the interaction of nonlinear waves, SIAM Rev., 39, 68–94, 1997.

Balk, A. M.: Angular distribution of Rossby wave energy, Phys. Lett. A, 345, 154–160, 2005.

Balk, A. M. and Ferapontov, E. V.: Invariants of 4-wave interactions, Physica D, 65, 274–288, 1993.

Balk, A. M. and Ferapontov, E. V.: Invariants of wave systems and web geometry, in: Nonlinear waves and weak turbulence, edited by: Zakharov, V. E., T. Am. Math. Soc., Ser. 2, vol. 182, 1–30, 1998.

Balk, A. M. and van Heerden, F.: Conservation style of the extra invariant for Rossby waves, Physica D, 223, 109–120, 2006.

Balk, A. M., Nazarenko, S. V., and Zakharov, V. E.: New invariant for drift turbulence, Phys. Lett. A, 152, 276–280, 1991.

Balk, A. M., van Heerden, F., and Weichman, P. B.: Rotating shallow water dynamics: Extra invariant and the formation of zonal jets, Phys. Rev. E, 83, 046320, doi:10.1103/PhysRevE.83.046320, 2011.

Blaschke, W.: Topological differential geometry, Chicago University Press, USA, 1932.

Boltzmann, L.: Über das warmegleichgewicht von gasen, auf welche äußere kräfte wirken, Wiener Berichte, 72, 427–457, 1875 (in German).

Cambon, C., Rubinstein, R., and Godeferd, F. S.: Advances in wave turbulence: rapidly rotating flows, New J. Phys., 6, 73, 29 pp., 2004.

Duran-Matute, M., Flór, J.-B., Godeferd, F. S., and Jause-Labert, C.: Turbulence and columnar vortex formation through inertial-wave focusing, Phys. Rev. E, 87, 041001(R), doi:10.1103/PhysRevE.87.041001, 2013.

Ferapontov, E. V.: Web geometry and mathematical physics, in: Geometry and Algebra of Multidimensional Three-Webs, authored by: Akivis, M. A. and Shelekhov, A. M., Kluwer Academic Publishers, Norwell, MA, 310–323, 1992.

Gill, A. E.: Atmosphere–Ocean Dynamics, Academic Press, New York, 1982.

Greenspan, H. P.: On the nonlinear interaction of inertial modes, J. Fluid Mech., 36, 257–264, 1969.

Landau, L. D. and Lifshitz, E. M.: Mechanics, Course of theoretical physics, v. 1, 3rd Edn., Butterworth-Heinemann, New York, 1976.

Landau, L. D. and Lifshitz, E. M.: Fluid Mechanics, Course of theoretical physics, v. 6, 2nd Edn., Butterworth-Heinemann, New York, 1987.

Manz, P., Xu, G. S., Wan, B. N., Wang, H. Q., Guo, H. Y., Cziegler, I., Fedorczak, N., Holland, C., Muller, S. H., Thakur, S. C., Xu, M., Miki, K., Diamond, P. H., and Tynan, G. R.: Zonal flow triggers the L-H transition in the Experimental Advanced Superconducting Tokamak, Phys. Plasmas, 19, 072311, doi:10.1063/1.4737612, 2012.

Maximenko, N., Bang, B., and Sasaki, H.: Observational evidence of alternating zonal jets in the world ocean, Geophys. Res. Lett., 32, L12607, doi:10.1029/2005GL022728, 2005.

Nazarenko, S. and Quinn, B.: Triple Cascade Behavior in Quasigeostrophic and Drift Turbulence and Generation of Zonal Jets, Phys. Rev. Lett., 103, 118501, doi:10.1103/PhysRevLett.103.118501, 2009.

Pedlosky, J.: Geophysical Fluid Dynamics, Springer, New York, 1987.

Rhines, P. B.: Waves and turbulence on a beta plane, J. Fluid Mech., 69, 417–443, 1975.

Sercignani, C.: Are there more than 5 linearly-independent collision invariants for the Boltzmann equation?, J. Stat. Phys., 58, 817–823, 1990.

Smith, L. M. and Waleffe, F.: Transfer of energy to two-dimensional large scales in forced, rotating three-dimensional turbulence, Phys. Fluids, 11, 1608–1622, 1999.

Soomere, T.: Coupling coefficients and kinetic equation for Rossby waves in multi-layer ocean, Nonlin. Processes Geophys., 10, 385–396, doi:10.5194/npg-10-385-2003, 2003.

Staplehurst, P. J., Davidson, P. A., and Dalziel, S. B.: Structure formation in homogeneous freely decaying rotating turbulence, J. Fluid Mech., 598, 81–105, 2008.

Vallis, G. K.: Atmospheric and Oceanic Fluid Dynamics, Cambridge University Press, Cambridge, UK, 2006.

Zakharov, V. E. and Schulman, E. I.: Degenerative dispersion laws, motion invariants, and kinetic equations, Physica D, 10, 192–202, 1980.

Zakharov, V. E. and Schulman, E. I.: On additional motion invariants of classical Hamiltonian wave systems, Physica D, 29, 283–320, 1988.

On the influence of spatial sampling on climate networks

N. Molkenthin[1,2], K. Rehfeld[1,3], V. Stolbova[1,2], L. Tupikina[1,2], and J. Kurths[1,2]

[1]PIK Potsdam Institute of Climate Impact Research, P.O. Box 601203, 14412 Potsdam, Germany
[2]Department of Physics, Humboldt-Universität zu Berlin, Newtonstr. 15, 12489 Berlin, Germany
[3]Alfred-Wegner Institute for Polar and Marine Research, Telegrafenberg A43, 14473 Potsdam, Germany

Correspondence to: N. Molkenthin (molkenthin@pik-potsdam.de)

Abstract. Climate networks are constructed from climate time series data using correlation measures. It is widely accepted that the geographical proximity, as well as other geographical features such as ocean and atmospheric currents, have a large impact on the observable time-series similarity. Therefore it is to be expected that the spatial sampling will influence the reconstructed network. Here we investigate this by comparing analytical flow networks, networks generated with the START model and networks from temperature data from the Asian monsoon domain. We evaluate them on a regular grid, a grid with added random jittering and two variations of clustered sampling. We find that the impact of the spatial sampling on most network measures only distorts the plots if the node distribution is significantly inhomogeneous. As a simple diagnostic measure for the detection of inhomogeneous sampling we suggest the Voronoi cell size distribution.

1 Introduction

Complex networks are used in many areas of physics to describe systems as different as the brain, the internet and social interactions (Boccaletti et al., 2006; Barthélemy, 2011). While the first applications were discrete by nature, complex networks have recently also found use in continuous systems like climate (Tsonis et al., 2006; Donges et al., 2009; Hlinka et al., 2013) . Additionally in climate networks, spatial embedding plays a big role. While the properties of some spatially embedded networks, like those discussed in Barthélemy (2011) and Kosmidis et al. (2008), have been analyzed, their construction mechanisms are very different from that of climate networks. In those applications, the system's spatial description is reduced to a number of discrete points, between which correlations are evaluated to construct a network (Malik et al., 2011; Rehfeld et al., 2012; Stolbova et al., 2014). This reveals a variety of interesting structures, many of which can be explained with geographical features. Spatial proximity often leads to large correlations, but mountains for example, can act as a barrier and prohibit exchange. Regions with a dominant wind or ocean flow can exhibit longer links. With all this, an effect of the sampling locations has to be expected.

Aliasing effects can occur if the sampling steps are larger than the half-width of investigated phenomena and can distort the actual signal substantially (Unser, 2000; Dippé and Wold, 1985). Spatial sampling effects have been discussed (Heitzig et al., 2012), where they proposed weighted measures to deal with differently sized nodes. In climate networks the related issue of temporal sampling and irregularity are treated in Rehfeld et al. (2011) and the role of boundary effects was investigated in Rheinwalt et al. (2012). While previous studies have mostly used grids for the locations of the measurements (Yamasaki et al., 2008; Tsonis et al., 2010), in this paper we study the effects of the spatial sampling itself on two typical model example systems and one set of observational data. In contrast to previous studies, we look at the sampling effects compared to the underlying system, rather than just differently sized regions. The first example is a model that generates networks directly from flows, introduced in Molkenthin et al. (2013). The second example is the START model, introduced in Rehfeld et al. (2013), in which artificial autocorrelated time series are transported by a flow field. In these models the node locations can be chosen freely, making them ideal for studying the effects of spatial sampling. The third example is the analysis of temperature data from the Asian monsoon area.

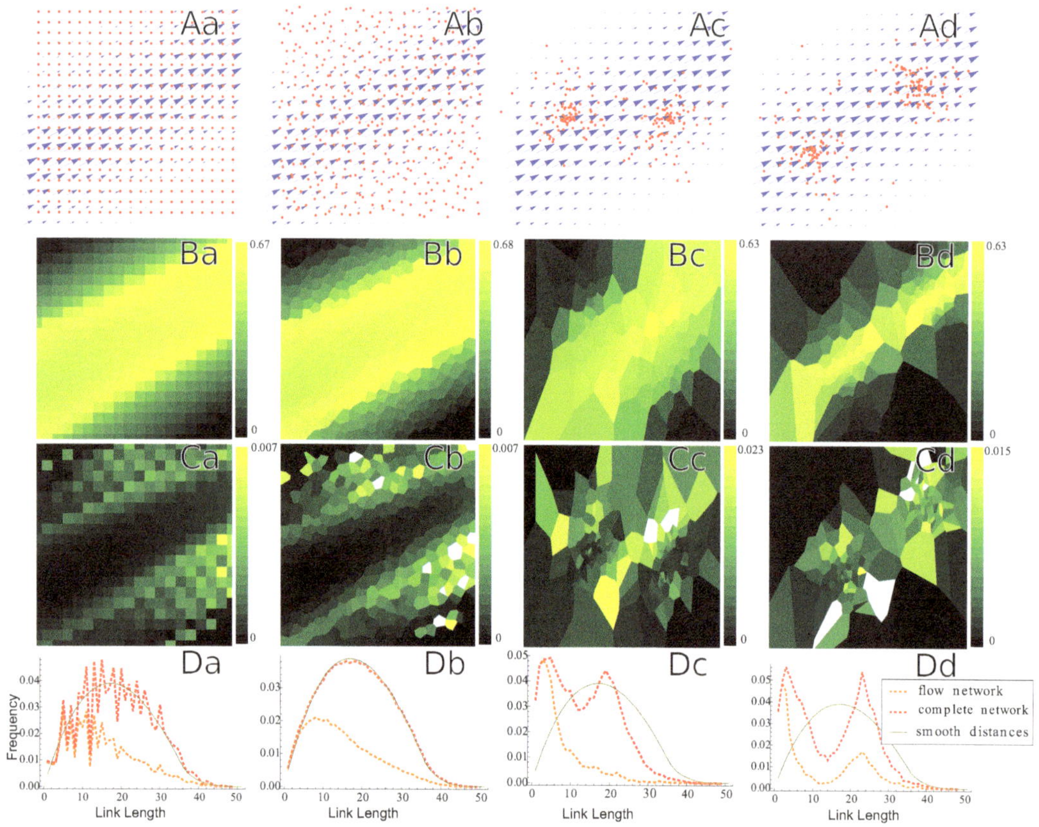

Figure 1. A diagonal flow is sampled with **(Aa)** a grid, **(Ab)** a jittered grid, **(Ac)** two clusters on opposite sides of the flow, and **(Ad)** two clusters inside the flow. The networks are constructed with a link density of 40 %. Row **(B)** shows the degree, row **(C)** the betweenness and row **(D)** the link length distribution.

We compute networks of a regular grid, a grid with a uniformly random jitter and clustered node distributions and compare their degree, betweenness and link length distribution to see how the spatial sampling affects common network measures.

2 Spatial effects in flow networks

Flow networks as introduced in Molkenthin et al. (2013) are constructed using analytical solutions of the advection-diffusion-equation (ADE), instead of time series in the definition of the correlation. The correlation measure is based on the scalar product between the temperature profile of a single δ-peak's evolution due to advection and diffusion, evaluated at each pair of nodes. We construct these networks and analyze the network measures degree, betweenness and link length distribution of the resulting networks and compare the influence of the spatial sampling on the result. Flow networks for a diagonal flow pattern are constructed for the node distributions shown in row A of Fig. 1, on a regular 20×20 grid, a jittered 20×20 grid and for two versions of Gaussian clustering. The jittered sampling pattern was generated by uniformly drawing a point from each grid cell. For the clustering, we used a Gaussian distribution around two central points on opposite sides of the flow (Ac) and along the flow (Ad) with 200 nodes. We plotted the node distribution in the flow, node degrees, node betweenness and the link length distribution for each of the sampling patterns.

Degree: we observe that the degree (shown in row B in Fig. 1) is higher in the middle stripe, where the flow velocity is highest. The degree is not very sensitive to the spatial sampling. In all four plots the degree is clearly highest where the absolute velocity is highest. Clustered sampling can lead to a skewed shape of the stripe, or underestimate its width if the outer region is poorly sampled.

Betweenness: the betweenness is shown in row C in Fig. 1. Grid and grid plus jitter show similar patterns: the betweenness is highest in the transition zone between the fast and slow flowing areas. This structure is barely visible in the clustered sampling plots. In the clusters on opposite sides, two stripes are visible, but the lower one is located in the center of the flow, rather than at its side, in the middle between the two clusters.

This indicates that rather than showing a transition zone of the flow, it emphasizes a region that is poorly sampled but central in the flow. In the plot of the two clusters in the flow,

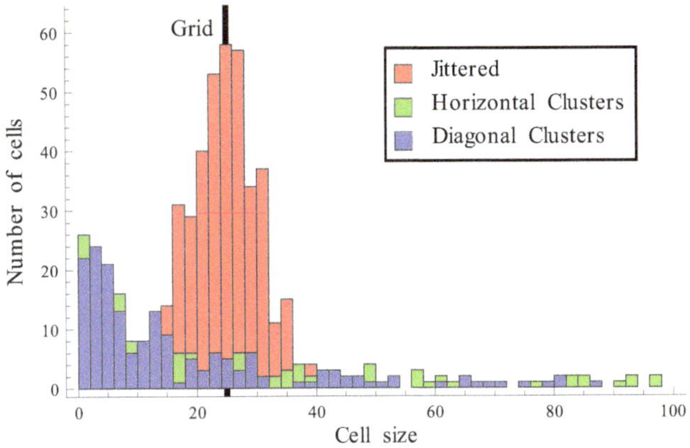

Figure 2. Tile size distribution for jittered and clustered sampling. The jittered sampling is peaked around the grid cell size, with clustered sampling results in many small and few very large tiles. Note that grid and jittered grid have twice as many nodes as the clustered sampling.

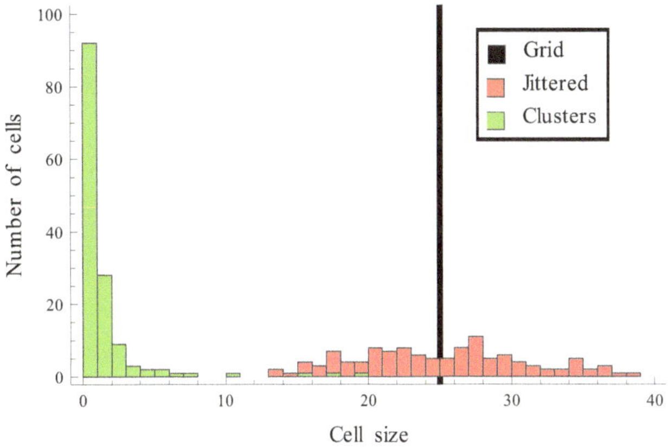

Figure 3. Tile size distribution for the three node distributions in START.

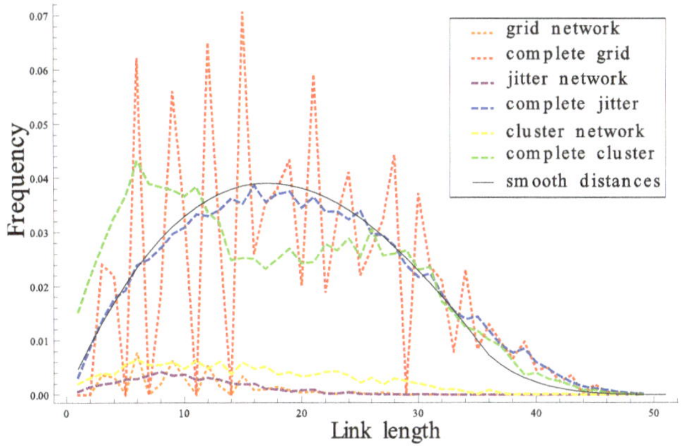

Figure 4. Link length distribution for the three node distributions in START.

the outer regions are poorly sampled, therefore the transition is not visible.

Voronoi tessellation analysis: a Voronoi tessellation assigns a cell to each node, such that every point in the cell is closer to that node than to any other. If the nodes are uniformly distributed in space, the resulting cell sizes will have a clear peak around $A_{\text{sam}}/N_{\text{nod}}$, where A_{sam} is the size of the total sampled area and N_{nod} is the number of nodes. However, coarse homogeneous sampling may not be able to make meaningful statements about processes happening on a finer spatial scale. The differences in the sampling can be quantified using the area size distributions of their Voronoi tessellations (Barthélemy, 2011) as shown in Fig. 2.

In the grid all tiles are exactly the same size, so the histogram would be one very sharp peak. In the case of the grid with jitter the peak is broadened. The two clustered distributions peak around much smaller values.

Link length distribution: the link length distribution, shown in row D in Fig. 1, is spiky for the grid and considerably smoother for the jittered node distribution. In clustered sampling, the link length distributions show two peaks. The beige line shows the distance distribution of pairs of points in a continuous square for comparison.

The regular arrangement of points in the grid leads to an overrepresentation of some specific link lengths (i.e. multiples of the grid constant), while other node-node distances are excluded by the node distribution. The distance distribution of pairs of points in a continuous square is the continuous analogue to the link length distribution of the fully connected graph of nodes in the discretized network description. Since climate is a continuous phenomenon, an appropriate discretization should well approximate that distance distribution. The distance distribution is obtained by integrating the distances between all pairs of points x and y, which is equivalent to integrating over the area of the square and all circle segments that lie in the square of circles of radius d. Normalizing the resulting function of d one gets the distance distribution. It coincides well with the link length distribution of the fully connected nodes of the jittered sampling. The fully connected grids' link length distribution is more spiky but varies around the same curve, while the clustered node sampling leads to a biased link length distribution with one maximum for links within each cluster and another maximum of links connecting the clusters.

Compared to the link length distributions of the fully connected networks, the actual flow networks only express a subset of links that is dominated by short ranged links. In the clustered case Fig. 1Dc this eliminates the inter-cluster peak, because the clusters are not connected by a flow. In the clustered case Fig. 1Dd however, this second peak persists, suggesting that links of length 30 are preferred, even though this effect is entirely a sampling effect and not a property of the flow.

Degree distribution: furthermore we find (results not shown) that the spatial sampling has little impact on the degree distribution. For all sampling types the degree

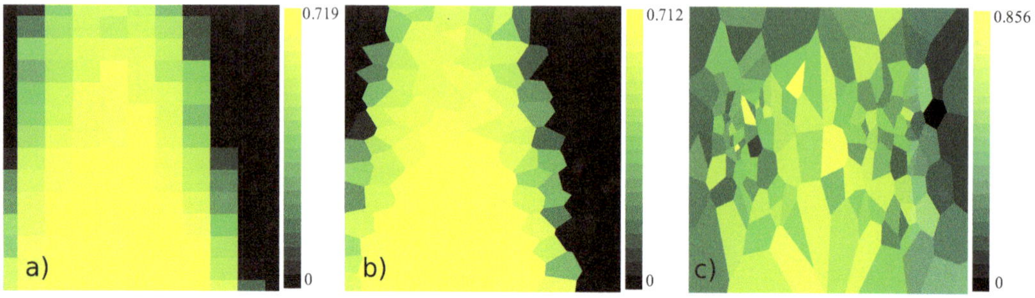

Figure 5. Degree of the network of the flow generated with START for grid **(a)**, grid with jitter **(b)** and two clusters **(c)**.

distribution was constant with one peak, the position of which depended on the value of the threshold.

Link number with threshold: the total number of links is a decreasing function of the threshold (results not shown). For both, grid and grid with jitter, this function decreases smoothly rather than in steps.

3 Spatial sampling in networks from the START model

Next we consider gridded, jittered and clustered networks using the START (Stream Transported Auto Regressive Temperature) model described in Rehfeld et al. (2013).

It is used to simulate time series at each node and construct the network by computing the correlations for each pair. The 20 % strongest correlations are used to set the links in the network. The START model assumes three sources of random information that are transported by three independent flows. The ADE is used to compute variance factors $f_X(\boldsymbol{p}, F)$ that approximate the influence of source X on position $\boldsymbol{p}$ at forcing F. At each point, the signal is computed as the sum of the contributions from the three sources, scaled with the corresponding variance factor. A local noise contribution is also added:

$$R_i = f_X(i, F)R_X + f_Y(i, F)R_Y + f_Z(i, F)R_Z + R_{\text{noise}}. \quad (1)$$

Here we use a forcing where only one of the flows is active for better comparison to the diagonal flow in the flow networks.

Despite being also constructed according to advection and diffusion, START is significantly different from the flow networks. Firstly, instead of averaging over all possible initial peak positions, there is a particular source location. This results in a decay of signal strength with increasing distance from the source.

Figure 5 shows the node degrees for the resulting networks, which drop off with distance from the source, unlike the flow networks, where the degree only depended on the velocity. We can see that, while the jittered node distribution shows the same structure as the grid, using clustered node locations leads to a distortion of the pattern.

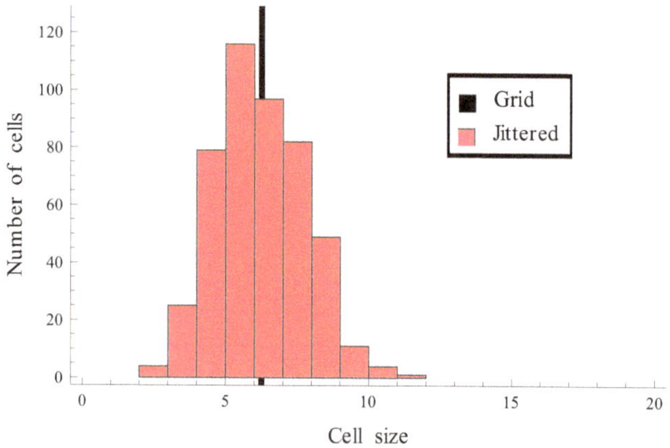

Figure 6. Tile size distribution for the jittered tile size distribution of the reanalysis data together with the grid line.

Additionally we can see that due to the addition of random fluctuations at each point, the degree in the stagnant and faraway regions goes to zero.

The link length distribution is similar to the link length distributions for the flow networks (see Fig. 4). The Tile size distribution (see Fig. 3) shows, that grid and jittered node distribution are homogeneous, while the clusters are not. The betweenness (results not shown) shows the same tongue-like structure as the degree. This can be explained with the random fluctuations reducing the dynamic areas to only one, which is the region inside the flow, while everything else is essentially unconnected.

4 Spatial sampling effects in data from the Asian monsoon domain

The data analysis was done on NCEP/NCAR (Kalnay et al., 1996; NCEP/NCAR, 2014) reanalysis daily surface temperature anomalies for the summer monsoon months (June to September) for the years 1971–2010 with a resolution of $2.5° \times 2.5°$, which results in a grid constant $a = 280$ km. The network was constructed from the anomaly time series using a Pearson correlation. We threshold the correlation matrix using a link density of the 5 % strongest correlations and obtain

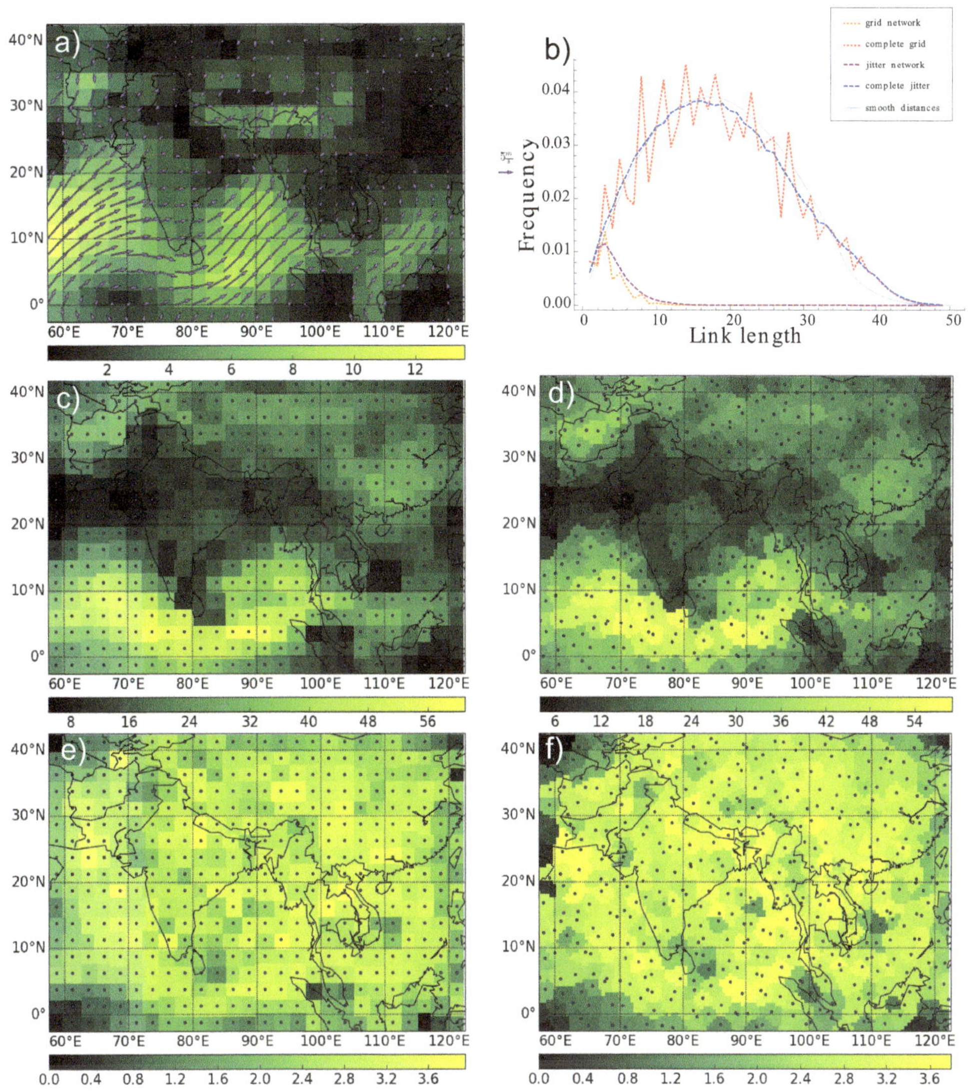

Figure 7. The correlation network of temperature data from NCEP/NCAR on a grid and a jittered grid. **(a)** Absolute mean wind velocity, **(b)** link length distributions, **(c)** degree of the grid network, **(d)** degree of the jittered network, **(e)** betweenness of the grid network, **(f)** betweenness of the jittered network.

degree and betweenness as shown in Fig. 7. The geographical coordinates of each node are marked by a point at the center of each cell. To investigate sampling effects, we added a jitter in longitude and latitude at each grid point with a uniformly random number between $\pm a/2$. The tile size distributions are shown in Fig. 6.

The new time series are constructed as Gaussian weighted averages of all of the original time series on the grid. The weights depend on the euclidean distance between the grid points and the new location.

Figure 7 shows (a) the annual average of the surface wind speeds, where the arrow displays the direction and the color shows the absolute value, which is computed by averaging over longitudinal and latitudinal components separately and taking the absolute value of the resulting vectors. Subfigure (b) shows the link length distribution. The degree is presented for grid and jittered sampling in the second row (c, d). The bottom row shows the betweenness for grid and jittered sampling (e, f).

The node degree (Fig. 7c and d) is highest over the ocean and lowest over the Indian subcontinent. This coincides with the absolute wind velocities, which are highest over the ocean, Bangladesh and the Himalayas. This supports the theory that connections in the climate network are often caused by an atmospheric or ocean flow between the node locations. On a daily scale, temperature fluctuations are transported along the flow, leading to a high correlation of nodes connected by a high velocity. Despite the high wind speeds over the Himalayas, however, the degrees there are not elevated. This may be because the plateau is relatively secluded from the rest of the network due to its high altitude. Consequently, even if the link density inside the secluded area were very high, the degree would be low, as connections to other regions are less likely.

We find that the effect of spatial sampling on the degree patterns is small and the overall patterns are not affected by the node distribution. This coincides well with our findings from the theoretical approaches in Sects. 2 and 3, as the jittered node distribution is still homogeneous as shown in the Voronoi tile size distribution in Fig. 6.

The shortest-path betweenness (e, f) is more sensitive to the spatial distribution of the nodes and the spatial resolution. Therefore, few clear patterns can be seen, and most regions have a betweenness between 2.5 and 3.5. The jittered network shows higher boundary effects in the west. There are three regions with a particularly low betweenness in both the gridded and the jittered version of the network at (80–85° E, 5–10° N), (100° E, 0–5° N) and (110° E, 10–15° N), these coincide with small regions with a low degree.

5 Discussion and conclusions

We have analyzed the influence of the spatial distribution of nodes on the network topology for three examples: flow networks, the START model and surface temperature correlation networks. We find that the degree is robust, while the betweenness reacts more sensitively to the spatial sampling. All these examples are related to climate, yet the results can be generalized for all networks coming from discretizations of any continuous system.

For a smooth link length distribution, the node-to-node separation has to cover all possible distances. A grid does not fulfill that, as all distances are of the form $d(x_1, x_2) = a\sqrt{n^2 + m^2}$, where a is the grid constant and n and m are integer numbers. Distances that can not be represented in this way are not present in a grid.

While the effects of the area sizes themselves are discussed in detail in Heitzig et al. (2012) and can be removed using their proposed consistently weighted network measures, this study shows that the problem goes further, when considering the underlying physical system. While the node size distribution in the two clustered sampling versions discussed in Sect. 2 is very similar, their relative position to the underlying flow is not, and they have very different distorting effects on the network structure.

In future work it might be interesting to compare different jitter realizations quantitatively, for example using the common component evolution function (CCEF) as introduced in Tupikina et al. (2013).

So in summary, we found that, as long as the node distribution in space is sufficiently homogeneous, the exact spatial sampling chosen has little impact on the topology of the network in all cases we analyzed. Only in cases of significantly inhomogeneous sampling, distortion and misleading structures arose. It is therefore important to discuss the sampling and its impact when analyzing spatially sampled data. As a simple test of the spatial sampling we suggest looking at the Voronoi size distribution. In a homogeneous sampling this will have a clear peak around A_{sam}/N_{nod}, where A_{sam} is the size of the total sampled area and N_{nod} is the number of nodes. If the peak is shifted or very spread out, that suggests poor spatial sampling.

Acknowledgements. The authors acknowledge funding by the German Science Foundation (DFG graduate school 1539), the German Federal Ministry for Education and Research (BMBF project PROGRESS, support code 03IS2191B) and the support of the LINC project (No. 289447) funded by EC's Marie-Curie ITN program (FP7-PEOPLE-2011-ITN). K. Rehfeld was supported by the Helmholtz grant VG-900NH. We would like to thank Norbert Marwan for discussion.

Edited by: M. Small

References

Barthélemy, M.: Spatial networks, Phys. Rep., 499, 1–101, doi:10.1016/j.physrep.2010.11.002, 2011.

Boccaletti, S., Latora, V., Moreno, Y., Chavez, M., and Hwang, D.: Complex networks: Structure and dynamics, Phys. Rep., 424, 175–308, doi:10.1016/j.physrep.2005.10.009, 2006.

Dippé, M. and Wold, E.: Antialiasing through stochastic sampling, ACM Siggraph Computer Graphics, 19, 69–78, 1985.

Donges, J. F., Zou, Y., Marwan, N., and Kurths, J.: Complex networks in climate dynamics, Eur. Phys. J. Spec. Top., 174, 157–179, doi:10.1140/epjst/e2009-01098-2, 2009.

Heitzig, J., Donges, J., Zou, Y., Marwan, N., and Kurths, J: Node-weighted measures for complex networks with spatially embedded, sampled, or differently sized nodes, Eur. Phys. J. B, 85, 1–21, doi:10.1140/epjb/e2011-20678-7, 2012.

Hlinka, J., Hartman, D., Vejmelka, M., Novotná, D., and Paluš, M.: Non-linear dependence and teleconnections in climate data: sources, relevance, nonstationarity, Clim. Dynam., doi:10.1007/s00382-013-1780-2, in press, 2013.

Kalnay, E., Kanamitsu, M., Kistler, R., Collins, W., Deaven, D., Gandin, L., Iredell, M., Saha, S., White, G., Woollen, J., Zhu, Y., Chelliah, M., Ebisuzaki, W., Higgins, W., Janowiak, J., Mo, K. C., Ropelewski, C., Wang, J., Leetmaa, A., Reynolds, R., Jenne, R., and Joseph, D.: The NCEP/NCAR 40-year reanalysis project, B. Am. Meteorol. Soc., 77, 437–440, 1996.

Kosmidis, K., Havlin, S., and Bunde, A.: Structural properties of spatially embedded networks, Europhys. Lett., 488, 129, doi:10.1038/488129b, 2008.

Malik, N., Bookhagen, B., Marwan, N., and Kurths, J.: Analysis of spatial and temporal extreme monsoonal rainfall over South Asia using complex networks, Clim. Dynam., 39, 971–987, doi:10.1007/s00382-011-1156-4, 2011.

Molkenthin, N., Rehfeld, K., Marwan, N., and Kurths, J.: Networks from flows – from dynamics to topology, Sci. Reports, doi:10.1038/srep04119, in press, 2013.

NCEP/NCAR: NCEP/NCAR, http://www.erls.noaa.gov/psd, last access: February 2014.

Rehfeld, K., Marwan, N., Heitzig, J., and Kurths, J.: Comparison of correlation analysis techniques for irregularly sampled time series, Nonlin. Processes Geophys., 18, 389–404, doi:10.5194/npg-18-389-2011, 2011.

Rehfeld, K., Marwan, N., Breitenbach, S. F. M., and Kurths, J.: Late Holocene Asian summer monsoon dynamics from small but complex networks of paleoclimate data, Clim. Dynam., 41, 3–19, doi:10.1007/s00382-012-1448-3, 2012.

Rehfeld, K., Molkenthin, N., and Kurths, J.: Spatio-temporal climate transitions from paleoclimate networks, Nonlin. Process Geophys., accepted, 2013.

Rheinwalt, A., Marwan, N., Kurths, J., Werner, P., and Gerstengarbe, F.-W.: Boundary effects in network measures of spatially embedded networks, Europhys. Lett., 100, 28002, doi:10.1209/0295-5075/100/28002, 2012.

Stolbova, V., Martin, P., Bookhagen, B., Marwan, N., and Kurths, J.: Topology and seasonal evolution of the network of extreme precipitation over the Indian subcontinent and Sri Lanka, Nonlin. Process Geophys., in review, 2014.

Tsonis, A. A., Swanson, K. L., and Roebber, P. J.: What Do Networks Have to Do with Climate?, B. Am. Meteorol. Soc., 87, 585–595, doi:10.1175/BAMS-87-5-585, 2006.

Tsonis, A. A., Wang, G., Swanson, K. L., Rodrigues, F. A., and Costa, L. D. F.: Community structure and dynamics in climate networks, Clim. Dynam., 37, 933–940, doi:10.1007/s00382-010-0874-3, 2010.

Tupikina, L., Rehfeld, K., Molkenthin, N., Stolbova, V., Marwan, N., and Kurths, J.: Detecting evolution of networks using spatial-temporal autocorrelation function, Nonlin. Process Geophys., accepted, 2013.

Unser, M.: Sampling-50 years after Shannon, Proc. IEEE, 88, 569–587, doi:10.1109/5.843002, 2000.

Yamasaki, K., Gozolchiani, A., and Havlin, S.: Climate Networks around the Globe are Significantly Affected by El Niño, Phys. Rev. Lett., 100, 228501, doi:10.1103/PhysRevLett.100.228501, 2008.

18

Full-field and anomaly initialization using a low-order climate model: a comparison and proposals for advanced formulations

A. Carrassi[1], R. J. T. Weber[1,2], V. Guemas[1,3], F. J. Doblas-Reyes[1,4], M. Asif[1], and D. Volpi[1,5]

[1]Institut Català de Ciències del Clima (IC3), Barcelona, Spain
[2]Department of Physics, University of Heidelberg, Heidelberg, Germany
[3]Centre National de Recherches Météorologiques/Groupe d'Etude de l'Atmosphère Météorologique, UMR3589, Toulouse, France
[4]Institució Catalana de Recerca i Estudis Avançats (ICREA), Barcelona, Spain
[5]Department of Mathematics, University of Reading, Reading, UK

Correspondence to: A. Carrassi (alberto.carrassi@ic3.cat)

Abstract. Initialization techniques for seasonal-to-decadal climate predictions fall into two main categories; namely full-field initialization (FFI) and anomaly initialization (AI). In the FFI case the initial model state is replaced by the best possible available estimate of the real state. By doing so the initial error is efficiently reduced but, due to the unavoidable presence of model deficiencies, once the model is let free to run a prediction, its trajectory drifts away from the observations no matter how small the initial error is. This problem is partly overcome with AI where the aim is to forecast future anomalies by assimilating observed anomalies on an estimate of the model climate.

The large variety of experimental setups, models and observational networks adopted worldwide make it difficult to draw firm conclusions on the respective advantages and drawbacks of FFI and AI, or to identify distinctive lines for improvement. The lack of a unified mathematical framework adds an additional difficulty toward the design of adequate initialization strategies that fit the desired forecast horizon, observational network and model at hand.

Here we compare FFI and AI using a low-order climate model of nine ordinary differential equations and use the notation and concepts of data assimilation theory to highlight their error scaling properties. This analysis suggests better performances using FFI when a good observational network is available and reveals the direct relation of its skill with the observational accuracy. The skill of AI appears, however, mostly related to the model quality and clear increases of skill can only be expected in coincidence with model upgrades.

We have compared FFI and AI in experiments in which either the full system or the atmosphere and ocean were independently initialized. In the former case FFI shows better and longer-lasting improvements, with skillful predictions until month 30. In the initialization of single compartments, the best performance is obtained when the stabler component of the model (the ocean) is initialized, but with FFI it is possible to have some predictive skill even when the most unstable compartment (the extratropical atmosphere) is observed.

Two advanced formulations, least-square initialization (LSI) and exploring parameter uncertainty (EPU), are introduced. Using LSI the initialization makes use of model statistics to propagate information from observation locations to the entire model domain. Numerical results show that LSI improves the performance of FFI in all the situations when only a portion of the system's state is observed. EPU is an online drift correction method in which the drift caused by the parametric error is estimated using a short-time evolution law and is then removed during the forecast run. Its implementation in conjunction with FFI allows us to improve the prediction skill within the first forecast year.

Finally, the application of these results in the context of realistic climate models is discussed.

1 Introduction

State estimation theory in geosciences is commonly referred to as data assimilation (DA), (Daley, 1991). This term encompasses the entire sequence of operations that, starting from the observations of a system, and possibly from additional statistical or dynamical information (i.e., the model), provides the best possible estimate of its state (Kalnay, 2003). This estimate, the analysis, is then used for diagnostic purposes or as initial condition for predictions with environmental numerical models. Another notable application of DA, having paramount importance in climate science, is in the production of reanalyses, multiyear global state-of-the-art gridded representations of the atmosphere/ocean generated by the same model and the same DA method (Dee et al., 2011). The ultimate goal of DA is to give a dynamically consistent reconstruction of all the elements of the climate system. By improving the initial condition, through a better use of the observations and model, DA has dramatically contributed to enhance the forecast skill in weather and ocean prediction in the last decades, and is nowadays regarded with attention from the seasonal-to-decadal (s2d) community interested in improving the initialization procedures.

At the basis of this growing interest there is the hope that optimizing the simultaneous use of model and observational information at the initialization step will help to improve the prediction skill in all those circumstances, and for those time horizons, when a nonnegligible portion of the forecast uncertainty is explained by the internal climate variability. It is known that this component accounts for a significant amount of the total prediction uncertainty at global scale for up to a decade in the future (and even longer at regional scales), and tends to monotonically decay afterward, dominated by the uncertainty associated with the model and the projected scenario (Smith et al., 2007; Hawkins and Sutton, 2009). Seasonal to decadal predictions, in contrast to continental-scale projections of climate change, are initialized using observations of the current climate state. Skillful (initialized) seasonal forecasts are nowadays run in several operational climate services worldwide; a recent review on the development and current status of the seasonal forecasting practice can be found in Doblas-Reyes et al. (2013a). Using global climate models and simulated observations, several studies have identified and assessed the benefits of initializing decadal predictions (see e.g., Latif et al., 2006; Dunstone and Smith, 2010, and references therein) and set the basis for the design of decadal prediction practice using real observations (see e.g., Smith et al., 2007; Keenlyside et al., 2008; van Oldenborgh et al., 2012; Hazeleger et al., 2013).

Current initialization techniques for seasonal-to-decadal climate predictions fall into two main categories; namely full-field initialization (FFI) and anomaly initialization (AI). In FFI the initial model state is replaced by the best possible available estimate of the real state. By doing so the initial error is efficiently reduced but, due to the unavoidable presence of model deficiencies, once the model is let free to run a prediction, its trajectory drifts away from the observations no matter how small the initial error is (e.g., Stockdale, 1997). This problem is partly overcome with the AI where the aim is to forecast future anomalies by assimilating observed climate anomalies on an estimate of the model mean climate. In this way, the initial model state is kept on (or closer to) its own attractor (e.g., Smith et al., 2007). The large variety of experimental setups, models and observational networks adopted in the studies to date makes it difficult to draw firm conclusions on the respective advantages and drawbacks of FFI and AI, let alone identifying distinctive lines for improvement. The lack of a unified mathematical framework adds an additional difficulty toward the design of adequate initialization strategies that fit the desired forecast horizon, observational network and model at hand.

Comprehensive comparisons between FFI and AI using state-of-the-art coupled climate models for seasonal-to-multiyear time horizons, have recently appeared (Magnusson et al., 2012; Smith et al., 2013; Hazeleger et al., 2013). These studies represent a first attempt to analyze their impact using exactly the same observational and model setup and are therefore of central importance in guiding future development of initialized climate prediction systems. Overall, these results have indicated that initialization systematically improves over a climatology at seasonal timescale, with a slight, but clear, better skill for FFI (Magnusson et al., 2012; Smith et al., 2013). The picture is far less clear on multiyear horizons. Some advantage of AI was reported in Smith et al. (2013) at global scales using the MetOffice decadal prediction system, while using the EC-Earth v2.3 atmosphere–land–ocean–sea ice model, Hazeleger et al. (2013) have found a somewhat better skill for FFI in areas such as the North Atlantic. This behavior is thought to be related to the difficulties of generating physically consistent initial sea-ice and ocean conditions.

The present study aims to contribute to the current debate on which initialization algorithm, and under which conditions, is the most suitable for seasonal-to-decadal prediction. We propose using the notation and concepts of data assimilation theory to outline a unified formalism from which FFI and AI can be derived. This furthermore allows for identifying specific features such as the initialization method sensitivity to model and observational accuracy. FFI and AI are studied in a range of different observational and model error scenarios, helping to clarify under which conditions one approach outperforms the other. Two advanced formulations are then introduced and discussed. The first, named least-square initialization (LSI), is aimed to improve the fit to the observations allowing for their informational content, usually restricted to the observational space in standard FFI or AI, to be propagated to the entire model domain. LSI uses a least-square initialization update to merge observation with the model, and the required (unknown) model error covariance is estimated using the covariance of the model anomalies. The estimation and reduction of the drift associated with parametric model error is the objective of the second novel formulation proposed in this study: the exploring parameter uncertainty (EPU) method provides an online correction of the drift on the basis of a linear and short-time approximation of its evolution. EPU works during the forecast run and as such can be used in combination with any initialization scheme, either FFI or AI.

The numerical analysis in this work is carried out using a prototype of climate dynamics simple enough to reduce computational cost and allow for robust statistical inferences. These conditions are not easily met when dealing with realistic models and observational scenarios. In this study we aim at gaining additional insight on FFI and AI, to advise and support the big ongoing research effort in the development of initialized decadal prediction with state-of-the-art climate models. Moreover, the capabilities of the proposed LSI and EPU methods are better understood using a more controllable experimental setup before thinking of a possible implementation in a more realistic context.

The paper is organized as follows. In Sect. 2 the general problem is posed and the basic assumptions are described. Section 3 describes FFI and AI using the data assimilation formalism, while LSI and EPU are presented in Sect. 4. The low-order climate model is described in Sect. 5 and the results are given in Sect. 6. Final Conclusions are drawn in Sect. 7.

2 Posing the problem

Let us write the prognostic climate model under the form of an autonomous dynamical system:

$$\frac{\mathrm{d}\boldsymbol{x}}{\mathrm{d}t} = \boldsymbol{F}(\boldsymbol{x}, \boldsymbol{\lambda}) \tag{1}$$

with $\boldsymbol{x}$ and $\boldsymbol{\lambda}$ being the state and parameter vectors of dimensions I and P, respectively. The parameters can also have a time dependence, such as is the case of the impact of solar activity or the anthropogenic greenhouse gases, which has been dropped here to simplify the notation.

The model (1) is used to describe the unknown Earth's climate system evolution; we will call our target, the real world state, "nature". We suppose that nature can be formally expressed under the form of a system similar to model (1), such as

$$\frac{\mathrm{d}\boldsymbol{x}^{\mathrm{nat}}}{\mathrm{d}t} = \mathbf{F}\left(\boldsymbol{x}^{\mathrm{nat}}, \boldsymbol{\lambda}^{\mathrm{nat}}\right) + \mathbf{G}\left(\boldsymbol{x}^{\mathrm{nat}}, \boldsymbol{\lambda}^{\mathrm{nat}}\right) \tag{2}$$

with $\boldsymbol{x}^{\mathrm{nat}}$ and $\boldsymbol{\lambda}^{\mathrm{nat}}$ being the unknown nature state and parameters. In this formulation, the model and the nature are assumed to span the same phase space, of dimension I. As a consequence, model error can only originate from parametric uncertainty, $\delta\boldsymbol{\lambda} = \boldsymbol{\lambda} - \boldsymbol{\lambda}^{\mathrm{nat}}$, and from the presence of the extra term, $\mathbf{G}$, accounting for all processes not properly described by the model, with $|\mathbf{G}| \ll |\mathbf{F}|$ in some proper norm (Nicolis, 2003). Note furthermore that the model is assumed to describe all scales of motion: model error coming from unresolved scales, a common drawback in geophysical modeling, is not considered in this formulation.

Observations, $\boldsymbol{y}_i^{\mathrm{o}} = \boldsymbol{y}^{\mathrm{o}}(t_i)$, are assumed to be available at equally spaced times $t_i = i\tau$, $i = 0, 1, \ldots$; in real applications, the vector of observations has a much smaller dimension than the model state vector, so that $O \ll I$ with O being the dimension of $\boldsymbol{y}_i^{\mathrm{o}}$, i.e., the number of observations available at each observation time. Moreover, unless specified otherwise, these observations are affected by some error represented as a white Gaussian noise, $\boldsymbol{\epsilon}^{\mathrm{o}}$, with zero mean and standard deviation $\boldsymbol{\sigma}^{\mathrm{o}}$, so that $\boldsymbol{\epsilon}^{\mathrm{o}} \in \mathcal{N}(0, \boldsymbol{\sigma}^{\mathrm{o}})$, with $\mathcal{N}$ standing for the Gaussian distribution and

$$\boldsymbol{y}^{\mathrm{o}} = \mathcal{H}\left(\boldsymbol{x}^{\mathrm{nat}}\right) + \boldsymbol{\epsilon}^{\mathrm{o}}, \tag{3}$$

$\mathcal{H}$ being the observation operator mapping from nature to the observation phase space. Note finally that in this formulation the observational error, $\boldsymbol{\epsilon}^{\mathrm{o}}$, accounts for both the instrumental and the representativity error connected with the specification of the operator $\mathcal{H}$ (Cohn, 1997; Janjić and Cohn, 2006).

3 Initialization methods

The initialization procedures start typically from a model state obtained after a long spin-up run. By using terms and concepts borrowed from a data assimilation context (see e.g., Kalnay, 2003), the model state at the end of the transient spin-up (the control) can be interpreted as a *background* field, embedding all information about the real system (nature) prior to the assimilation of the observation; it will be hereafter indicated as $\boldsymbol{x}^{\mathrm{b}}$. Similarly, the initial condition obtained at the end of the initialization procedure is interpreted as the *analysis* field, and indicated as $\boldsymbol{x}^{\mathrm{a}}$ in the following. Here, the additional assumption of a linear observation operator is made; this implies that the observation operator is given under the form of the $O \times I$ matrix, $\mathbf{H}$.

3.1 Full-field initialization (FFI)

This approach consists of substituting the model state with the available observations at the initialization times. Using the data assimilation notation FFI can be formally written as

$$\boldsymbol{x}^{\mathrm{a}} = \boldsymbol{x}^{\mathrm{b}} + \mathbf{H}^{\mathrm{T}}\left[\boldsymbol{y}^{\mathrm{o}} - \mathbf{H}\boldsymbol{x}^{\mathrm{b}}\right] \tag{4}$$

with $\mathbf{H}^{\mathrm{T}}$ being the transpose of the linearized observation operator; note that in this case the entries of the matrix $\mathbf{H}$ are equal to one in correspondence with the observed variables and grid points, and zero elsewhere. By left multiplying both sides of Eq. (4) by $\mathbf{H}$, we retrieve the usual FFI formula in which the control is replaced by the observations at observational locations, and left unchanged elsewhere.

Alternatively to the direct replacement of the observed values, FFI can also be implemented using a nudging approach.

Nudging is an empirical data assimilation technique, consisting of adding a term to the prognostic equations that, acting like an extra-coupling term, nudges the model trajectory toward a consistent representation of the unknown system's evolution intended to be estimated (Hoke and Anthes, 1976). The coupling strength is expressed as a relaxation timescale and is usually chosen on the basis of the properties of the variable to be nudged: it has to be small enough to avoid dynamical shocks but large enough for the correction to counteract actively the error growth. Advanced formulations of the nudging approach have been proposed recently (Auroux and Blum, 2008), and the use of the nudging procedure using unstable modes of the dynamics has been also explored using simple chaotic systems (Yang et al., 2006).

The FFN can be written as follows:

$$\frac{\mathrm{d}\boldsymbol{x}}{\mathrm{d}t} = \mathbf{F}(\boldsymbol{x}, \boldsymbol{\lambda}) + \mathbf{H}^{\mathrm{T}}\mathbf{N}\left[\boldsymbol{y}^{\mathrm{o}}_{\mathrm{nudg}} - \mathbf{H}\boldsymbol{x}\right] \qquad t \le t^{\mathrm{init}} \tag{5}$$

with $\mathbf{N}$ being the diagonal $O \times O$ nudging matrix, with the dimension equal to that of the observation vector and measured in units of time^{-1}. The matrix $\mathbf{N}$ contains, along its diagonal, the relaxation time rate of each nudged/observed variable. Any specific setups of a nudging scheme are equivalent to specifying the entries of the matrices $\mathbf{H}$ and $\mathbf{N}$: the former indicates the observed variables, the latter the corresponding relaxation times. Note that the nudging observation vector $\boldsymbol{y}^{\mathrm{o}}_{\mathrm{nudg}}$, for which the time dependence has been omitted for clarity in Eq. (5), coincides with the observation $\boldsymbol{y}^{\mathrm{o}}$ at observation times, t_i, and with their time interpolation between two successive observation times. The initial condition, $\boldsymbol{x}^{\mathrm{a}}$, is then obtained by integrating Eq. (5) up to the initialization time (i.e., the start date), t^{init}. FFN has recently been used in Magnusson et al. (2012) to nudge SST (sea surface temperature) with the ECMWF (European Centre for Medium-Range Weather Forecasts) climate model.

3.2 Anomaly initialization (AI)

In AI (Smith et al., 2007) the model state at initial time is replaced by the observed anomaly (i.e., the difference between the current observation and its long-term average[1]) plus the model average (i.e., the simulated climate). To write the AI equation using the data assimilation formalism, we introduce the pseudo-observation vector:

$$\boldsymbol{y}^{\mathrm{pso}} = \boldsymbol{y}^{\mathrm{o}} - \left(\overline{\boldsymbol{y}^{\mathrm{o}}} - \mathbf{H}\overline{\boldsymbol{x}}\right), \tag{6}$$

with the overbar indicating the long-term average. The pseudo-observations are equal to the observations minus the difference between the observed and modeled climate. The AI equation can then be written as

$$\begin{aligned}\boldsymbol{x}^{\mathrm{a}} &= \boldsymbol{x}^{\mathrm{b}} + \mathbf{H}^{\mathrm{T}}\left[\boldsymbol{y}^{\mathrm{pso}} - \mathbf{H}\boldsymbol{x}^{\mathrm{b}}\right] \\ &= \boldsymbol{x}^{\mathrm{b}} + \mathbf{H}^{\mathrm{T}}\left[\boldsymbol{y}^{\mathrm{o}} - \overline{\boldsymbol{y}^{\mathrm{o}}} + \mathbf{H}\overline{\boldsymbol{x}} - \mathbf{H}\boldsymbol{x}^{\mathrm{b}}\right].\end{aligned} \tag{7}$$

The observation operator $\mathbf{H}$ is defined as in Eq. (4) and, as above, by left multiplying both sides of Eq. (7) by $\mathbf{H}$, we see that at observational locations, the background is replaced by the pseudo-observations and left unchanged elsewhere.

To get the expression for anomaly nudging, AN, we can proceed as for Eq. (5) and after introducing the pseudo-observations we have

$$\begin{aligned}\frac{\mathrm{d}\boldsymbol{x}}{\mathrm{d}t} &= \mathbf{F}(\boldsymbol{x}, \boldsymbol{\lambda}) + \mathbf{H}^{\mathrm{T}}\mathbf{N}\left[\boldsymbol{y}^{\mathrm{pso}}_{\mathrm{nudg}} - \mathbf{H}\boldsymbol{x}\right] = \mathbf{F}(\boldsymbol{x}, \boldsymbol{\lambda}) \\ &\quad + \mathbf{H}^{\mathrm{T}}\mathbf{N}\left[\boldsymbol{y}^{\mathrm{o}}_{\mathrm{nudg}} - \overline{\boldsymbol{y}^{\mathrm{o}}}_{\mathrm{nudg}} + \mathbf{H}\overline{\boldsymbol{x}} - \mathbf{H}\boldsymbol{x}\right] \qquad t \le t^{\mathrm{init}}.\end{aligned} \tag{8}$$

Similar to FFN, the initial condition, $\boldsymbol{x}^{\mathrm{a}}$, is obtained by integrating Eq. (8) up to t^{init}. AN has been used in Smith et al. (2007) and Smith et al. (2013) to nudge oceanic variables of the MetOffice global climate model.

3.3 FFI and AI: some properties

FFI and AI are easy to implement and do not require any hints on the relative accuracy of the estimators, the background and observations, entering the initialization procedure. These features make them very attractive in the context of seasonal-to-decadal prediction with global numerical models whose typical size and complexity limits the use of more advanced initialization strategies. FFI and AI produce different initial states, and one approach can outperform the other depending on a number of competing factors such as the accuracy and type of the observational network, the amplitude of the model biases and the desired forecast horizon.

It is interesting to consider the error-scaling properties of FFI and AI, with respect to the level of accuracy of the information sources, model and observations. To this end, let us derive an expression for the initial condition (analysis) error, $\boldsymbol{\epsilon}^{\mathrm{a}} = \boldsymbol{x}^{\mathrm{a}} - \boldsymbol{x}^{\mathrm{nat}}$, by formally subtracting the unknown nature's state, $\boldsymbol{x}^{\mathrm{nat}}$, from Eqs. (4) and (7). Assuming for simplicity that the full system is observed (i.e., $\mathbf{H} = \mathbf{I}$, with $\mathbf{I}$ being the $I \times I$ identity matrix) and after rearranging we have

$$\boldsymbol{\epsilon}^{\mathrm{a}}_{\mathrm{FFI}} = \boldsymbol{\epsilon}^{\mathrm{o}}, \qquad \boldsymbol{\epsilon}^{\mathrm{a}}_{\mathrm{AI}} = \boldsymbol{\epsilon}^{\mathrm{o}} + \overline{\boldsymbol{\epsilon}^{\mathrm{b}}} \tag{9}$$

with $\boldsymbol{\epsilon}^{\mathrm{b}} = \boldsymbol{x}^{\mathrm{b}} - \boldsymbol{x}^{\mathrm{nat}}$ being the background error.

The expressions (Eq. 9) highlight a main feature of these methods; namely, their dependence on the accuracy of the observations and, for AI, the role of the model bias, $\overline{\boldsymbol{\epsilon}}^{\mathrm{b}}$, as the error level in the idealized limit $\boldsymbol{\epsilon}^{\mathrm{o}} \to 0$. By taking the average of Eq. (9), and using the property of unbiased observations, we get the expressions for the mean initial error:

$$\overline{\boldsymbol{\epsilon}}^{\mathrm{a}}_{\mathrm{FFI}} = 0, \qquad \overline{\boldsymbol{\epsilon}}^{\mathrm{a}}_{\mathrm{AI}} = \overline{\boldsymbol{\epsilon}}^{\mathrm{b}}, \tag{10}$$

[1] According to the WMO definition, at least 30 years are necessary to compute a climatology, called long-term average in the text.

whereas for the root-mean-square error at analysis time ($RMSE^a$) we have:

$$RMSE^a_{FFI} = \sigma^o, \qquad RMSE^a_{AI} = \sigma^o \sqrt{1 + \frac{\overline{\epsilon}^{b\,2}}{\sigma^{o\,2}}}. \tag{11}$$

$RMSE^a_{FFI}$ scales linearly with the observational error, while the behavior of $RMSE^a_{AI}$ depends on the ratio $\frac{\overline{\epsilon}^{b\,2}}{\sigma^{o\,2}}$ that measures the relative accuracy of model–observations. In realistic circumstances this ratio is (much) larger than 1 ($\overline{\epsilon}^{b\,2} \gg \sigma^{o\,2}$, implying the observational error being much lower than model error) and $RMSE^a_{AI}$ behaves almost independently on the accuracy of the observations. As a consequence and in contrast to $RMSE^a_{FFI}$, which can in principle be arbitrarily reduced by intervening on the observational accuracy, a significant reduction of $RMSE^a_{AI}$ can only be achieved by improving the model. Moreover, when $\frac{\overline{\epsilon}^{b\,2}}{\sigma^{o\,2}} \approx 1$, $RMSE^a_{AI}$ behaves almost quadratically with σ^o so that for the same reduction of σ^o, $RMSE^a_{AI}$ will decrease at a slower rate than $RMSE^a_{FFI}$.

It is worth mentioning that we have intentionally focused our discussion on the actual (unknown) error with the purpose of highlighting the response of FFI and AI to observation and model error. In real applications one usually computes errors using observations, possibly taken from a data set independent from the one used in the initialization. If we had proceeded similarly and had used pseudo-observations for the AI analysis, the two schemes would have shown similar scaling properties with respect to the observational error. Nevertheless this approach would have hidden the presence of the model bias, whose role we wanted to stress here.

Arguing on how the accuracy of the initial conditions will impact the prediction skill at seasonal-to-decadal timescales is far more difficult. In contrast to weather forecast practice (with horizons of 2 weeks) where model error is often neglected and much attention is placed toward an efficient control of chaotic error growth (see e.g., Palatella et al., 2013), in seasonal-to-decadal prediction the growth of initial error is relatively less important than the bias caused by model deficiencies. The unavoidable presence of model error causes the so called "model drift", in which a forecast initialized close to the observed state will eventually drift toward the model climate, i.e., the model attractor, following a nonlinear state-dependent evolution that can appear as quasi-erratic before stabilizing. With AI, although at the price of larger errors, the initial state is kept close to the model attractor and the drift is mitigated. This makes the mean forecast error less time dependent and, as argued by Magnusson et al. (2012), the use of standard a posteriori bias correction techniques is more robust. Nevertheless, although the bias correction post-processing for forecasts initialized with AI has the advantage of being independent from the forecast times, it still requires running long and computationally expensive transient simulations before reaching model equilibrium.

FFI is more prone to suffer initial dynamical shocks caused by the displacement of the model state to the observed values that can be out of the model attractor. These initial shocks are smaller, but still present, in the AI, since the model is forced to deviate from its mean state by only the amount of the observed anomalies. However shocks can also be generated by inconsistencies or geographical mismatches between the model climate and observed anomalies (Magnusson et al., 2012). By smoothly moving the model trajectory toward the observations, the nudging approach has the potential to reduce this problem (see e.g., Smith et al., 2013).

4 Advanced initialization procedures

In this section we introduce and discuss two advanced initialization methods based on data assimilation theory. The first approach, referred to as LSI, is thought to improve the fit to the observations. The second, named EPU, is designed to estimate and reduce the model drift. LSI and EPU are described here in the context of FFI, however they can equivalently be applied in the framework of AI with only minor modifications.

4.1 Least-square initialization – LSI

In FFI the observations are fit as if they were perfect and the model state is replaced by the observations in all variables and geographical locations where measurements are available, no matter how accurate these observations are with respect to the model solution. The information is not transferred from observed-to-unobserved areas and the observational impact is confined to the measurement locations, despite the possible presence of physically relevant spatial correlations. Note however that, in the practice of climate prediction with realistic models, this issue is partly mitigated by using either a full and homogeneous reanalysis field of all model variables over its entire domain, or by smoothly nudging toward the observations data set when this represents only a subset of the model variables/domain (Meehl et al., 2013; Doblas-Reyes et al., 2013b).

In data assimilation, statistical knowledge about the accuracy of each piece of information entering the analysis update is used to determine the relative weights of their contribution according to some criteria of optimality (Daley, 1991). Using a least-square framework, the background and the observations are linearly combined in order to minimize the expected analysis error variance (Jazwinski, 1970). All errors are assumed to be Gaussian and represented by the first two statistical moments only: the mean and the covariance. The background and observation error covariances are referred hereafter to as **B** and **R**, according to a standard notation in data assimilation literature (Ide el al., 1997).

The least-square analysis update, common to most data assimilation procedures, reads (Jazwinski, 1970)

$$x^{a} = x^{b} + \mathbf{B}\mathbf{H}^{T}\left[\mathbf{H}\mathbf{B}\mathbf{H}^{T} + \mathbf{R}\right]^{-1}\left[y^{o} - \mathbf{H}x^{b}\right]. \quad (12)$$

When $\mathbf{B}$ and $\mathbf{R}$ are the correct error statistics, x^a would coincide with the best linear unbiased estimate and the analysis solution minimizes the associated posterior error covariance. The matrix $\mathbf{B}$ plays the crucial role of spreading out the observational information content throughout the entire model domain according to the assumed forecast error structure; in practice the analysis correction is confined within the subspace spanned by the range of $\mathbf{B}$. Classical data assimilation methods, such as optimal interpolation or 3DVar (Daley, 1991), rely on a statistical estimation of the covariance matrices, $\mathbf{B}$ and $\mathbf{R}$. Nevertheless, while a relatively robust estimate of the latter can be easily obtained especially when it is related to the observational accuracy alone, more problems arise to estimate $\mathbf{B}$. The approach that has been almost universally adopted since the 1990s is known as the NMC (National Meteorological Center)-method (Parrish and Derber, 1992). With the NMC $\mathbf{B}$ is estimated using the difference between two forecasts verified at the same time, let us say the 24 and 12 h forecasts; this matrix is then used as a proxy for the true (and unknown) forecast error matrix. And it is not based on observations, that are usually sparsely distributed over the globe, the NMC method provides a proper representation of the global error structure (Kalnay, 2003), and it is also consistent with the hypothesis of uncorrelated forecast and observation error used in Eq. (12). Finally, it is worth mentioning the research efforts carried out in the last decades that led to the development of the so-called ensemble-based data assimilation algorithms in which the forecast error covariance matrix, estimated on the basis of an ensemble of model trajectories, is made time dependent, a very desirable feature when dealing with chaotic systems (see e.g., Evensen, 2009).

In the initialization of climate predictions, the NMC method cannot be directly applied because the background field here is a climatological (control) model solution output after a sufficiently long spin-up, (see Eqs. 4, 7). In the LSI method described here the actual $\mathbf{B}$ is approximated using the model climate covariance, which is then incorporated in the analysis update (12). The model covariance can be estimated rather accurately, and the robustness of this estimate is limited only by the availability of enough computational resources to make long runs feasible. Under this assumption, the proxy of the background error covariance matrix reads

$$\mathbf{B}^{m} = \alpha\overline{(x-\overline{x})(x-\overline{x})^{T}}, \quad (13)$$

where x is a solution of the model (1), α a scalar tuning parameter and, as above, the overbar refers to a long-term average. The full-field least-square initialization (FF-LSI) equation then becomes

$$x^{a} = x^{b} + \mathbf{B}^{m}\mathbf{H}^{T}\left[\mathbf{H}\mathbf{B}^{m}\mathbf{H}^{T} + \mathbf{R}\right]^{-1}\left[y^{o} - \mathbf{H}x^{b}\right]. \quad (14)$$

Equation (13) embeds the idea of interpreting the model anomalies as forecast errors (the actual $\mathbf{B}$ reads in fact $\mathbf{B} = \overline{(\epsilon^{b})(\epsilon^{b})^{T}}$) and we expect $\mathbf{B}^m$ to reproduce at least the spatial structure of the error covariance. The parameter α compensates for the deviation of the amplitude of the simulated covariance from the error covariance and helps to adapt the weighting of the background term in Eq. (14). The occurrence of nature's statistical modifications (i.e., a climate change) that have not been tracked by the model during the reference averaging period can be a further limiting factor for the accuracy of Eq. (13). However, FF-LSI is easy to implement and has the potential to be beneficial in all the situations when only a portion of the model state vector is observed.

A very similar approach was originally introduced by Smith and Murphy (2007) to generate a global ocean analysis based on sparsely distributed observations of temperature and salinity. Using a state-of-the-art coupled climate model they demonstrate that using model-based covariance allows for successfully propagating information from data-rich to data-poor areas and significantly improved the model representation of the observed variability. The LSI procedure follows a similar line, although our attention here is in the propagation of information between different model compartments, such as the atmosphere and the ocean, and the experiments described later are designed for that purpose. Another difference between the two approaches stands in the definition of the background field, x^b. In LSI this is a forecast-field solution of the model equation at analysis time, while Smith and Murphy (2007) uses a climatology field. Our choice has the advantage of incorporating fresh, time-dependent, information about the system, potentially propagating forward the signal of the most recent observations. However it would in principle require a corresponding time-dependent background error covariance that is very difficult to get unless sophisticated data assimilation, such as the ensemble Kalman filter (Evensen, 2009), is adopted. The use of a climatological background avoids this issue but at the price of a static a priori picture of the system's state. Finally note that, instead of using the tuning coefficient α, in Smith and Murphy (2007) the misrepresentation of the actual $\mathbf{B}$ is treated by limiting the use of Eq. (14) to the set of observations that lead to a decrease of the posterior analysis error variance; the latter is computed using observations when available and turns the global solution (14) into a series of local analyses centered around each grid point.

4.2 Exploring parameter uncertainty – EPU

The second novel formulation discussed in this study is named full-field initialization with exploring parameter uncertainty (FFI-EPU), and is designed to reduce the model drift caused by parametric error. EPU is an online bias correction method based on a linear, and short-time, approximation of the bias evolution; the procedure is said to be online because the estimation and correction of the bias is done during the prediction run. A similar approach, in the offline mode, has been introduced in the context of sequential data assimilation by Carrassi et al. (2008), on the basis of the theory of deterministic model error dynamics given in Nicolis (2003).

An equation for the evolution of the estimation error, $\delta \boldsymbol{x}$, can be obtained by linearizing the model equation, (1), along one of its solutions. By expanding Eq. (1) to the first order in $\delta \boldsymbol{x}$ and $\delta \boldsymbol{\lambda}$, we obtain

$$\frac{\mathrm{d}\delta \boldsymbol{x}}{\mathrm{d}t} \approx \frac{\partial \mathbf{F}}{\partial \boldsymbol{x}}|_{x,\lambda}\delta \boldsymbol{x} + \frac{\partial \mathbf{F}}{\partial \boldsymbol{\lambda}}|_{x,\lambda}\delta \boldsymbol{\lambda}. \tag{15}$$

The solution of Eq. (15), with initial condition $\delta \boldsymbol{x}_0$, reads

$$\delta \boldsymbol{x} \approx \mathbf{M}_{t,t_0}\delta \boldsymbol{x}_0 + \int_{t_0}^{t} \mathbf{M}_{t,\tau}\delta \boldsymbol{\mu}(\tau)\mathrm{d}\tau \tag{16}$$

with $\mathbf{M}_{t,t_0} = e^{\int_{t_0}^{t}\frac{\partial \mathbf{F}}{\partial x}|_x \mathrm{d}t}$ being the tangent linear model of Eq. (1) between t_0 and t, and $\delta \boldsymbol{\mu} = \frac{\partial \mathbf{F}}{\partial \boldsymbol{\lambda}}|_{x,\lambda}\delta \boldsymbol{\lambda}$. The vector $\delta \boldsymbol{\mu}$ embeds all information about the model error through the parametric error $\delta \boldsymbol{\lambda}$ and the functional dependence, $\frac{\partial \mathbf{F}}{\partial \boldsymbol{\lambda}}$, of the dynamics on the uncertain parameters. Assuming that the initial error is unbiased, after taking the average of (16), we get an estimate of the model bias evolution, the drift, as

$$\boldsymbol{b}(t) = < \delta \boldsymbol{x}(t) > \approx < \int_{t_0}^{t} \mathbf{M}_{t,\tau}\delta \boldsymbol{\mu}(\tau)\mathrm{d}\tau > . \tag{17}$$

Expression (17) provides the time evolution of the bias under the hypothesis of linearity aforementioned. Nevertheless, it cannot be solved in the case of realistic climate models, mainly because of the huge dimension of the systems involved, and some approximations are required.

From Eq. (17) we see that the bias evolution (the drift) is fully correlated in time. Therefore, an approximation suitable for realistic applications can be obtained by expanding it in a Taylor series up to the first order in time, so that:

$$\boldsymbol{b}(t) \approx < \delta \boldsymbol{\mu}_0 > [t - t_0] = < \frac{\partial \mathbf{F}}{\partial \boldsymbol{\lambda}}|_{x,\lambda}\delta \boldsymbol{\lambda} > [t - t_0]. \tag{18}$$

Equation (18) gives the short-time evolution of the bias; its accuracy is related to the linear hypothesis made above and to the duration of the short-time linear regime. The extent of duration of this regime is known to be proportional to the largest (in absolute value) Lyapunov exponent of the dynamics (Nicolis, 2003). For the large class of dissipative chaotic systems to which environmental models belong, the largest Lyapunov exponent is usually the most negative.

The basic idea behind EPU is to use Eq. (18) to estimate and remove the bias from the forecast at lead time $t - t_0$. Let us define the bias correction time interval, ΔT_{Bias}, as the time period over which the bias evolution law (18) is used and the bias correction is applied. Every ΔT_{Bias} the bias is estimated and removed from the forecast at time t_i according to:

$$\begin{aligned} \boldsymbol{x}^{\mathrm{un}}(t_i) &= \boldsymbol{x}_i^{\mathrm{un}} = \boldsymbol{x}_i - \boldsymbol{b}_i = \boldsymbol{x}_i - \mathbf{C}_i \Delta T_{\mathrm{Bias}}, \\ &= \boldsymbol{x}_i - \frac{\partial \mathbf{F}}{\partial \boldsymbol{\lambda}}|_{x_{i-1},\lambda}\delta \boldsymbol{\lambda}_i \Delta T_{\mathrm{Bias}}, \\ t_i &= i\Delta T_{\mathrm{Bias}} \quad i = 1, 2, \ldots, \end{aligned} \tag{19}$$

where the suffix "un" stands for unbiased, and the compact form $\boldsymbol{x}(t_i) = \boldsymbol{x}_i$ is used to simplify the notation. In Eq. (19), $\boldsymbol{\lambda}$ represents the value of the parameter used in the model along its entire run, while $\delta \boldsymbol{\lambda}_i$ the (estimate of) parametric error at time t_i.

The use of Eq. (19) as a bias correction approach requires the estimation of the operator $\mathbf{C}_i = \frac{\partial \mathbf{F}}{\partial \boldsymbol{\lambda}}|_{x_{i-1},\lambda}\delta \boldsymbol{\lambda}_i$. The first term of $\mathbf{C}$, $\frac{\partial \mathbf{F}}{\partial \boldsymbol{\lambda}}$, gives the model functional dependence on the uncertain parameters, i.e., the model sensitivity; it depends on time through the dependence on the system's state and allows us to project the error from the parameter to the state vector space. It is a rectangular matrix having as many rows and columns as the dimension of the state vector and of the parametric error respectively, and can be computed at any time along the model integration, the only practical concern being the computational constraints arising when using large numerical models.

The second term, the parametric error, $\delta \boldsymbol{\lambda}$, is unknown. To cope with this we introduce a guess strategy. Two hypotheses are formulated: (1) model users have identified a (possibly limited) set of relevant parameters suspected to be uncertain, and, (2) a range of possible parameter values, $\Delta\Lambda = [\boldsymbol{\lambda}_{\min}, \boldsymbol{\lambda}_{\max}]$. The parametric error at time t_i, $\delta \boldsymbol{\lambda}_i$, is then sampled according to

$$\delta \boldsymbol{\lambda}_i \in \begin{cases} \mathcal{U}\left(0, \boldsymbol{\lambda}^{\max} - \overline{\boldsymbol{\lambda}}\right), & \text{if } \boldsymbol{\lambda} > \overline{\boldsymbol{\lambda}} \\ \mathcal{U}\left(\boldsymbol{\lambda}^{\min} - \overline{\boldsymbol{\lambda}}, 0\right), & \text{if } \boldsymbol{\lambda} < \overline{\boldsymbol{\lambda}} \end{cases}$$

$$t_i = i\Delta T_{\mathrm{Bias}}, \quad i = 1, 2, \ldots, \tag{20}$$

with $\mathcal{U}(a,b)$ being the uniform distribution in the interval (a,b), $\overline{\boldsymbol{\lambda}}$ the mean value of the range $\Delta\Lambda$ and $\boldsymbol{\lambda}$ the parameter used in the model. In practice $\overline{\boldsymbol{\lambda}}$ plays the role of the most probable parameter value and is used in Eq. (20) to discriminate between over-/underestimation of the unknown $\boldsymbol{\lambda}^{\mathrm{nat}}$. Note that, even in the favorable situation $\boldsymbol{\lambda}^{\mathrm{nat}} \in \Delta\Lambda$, given that $\boldsymbol{\lambda}^{\mathrm{nat}} \neq \overline{\boldsymbol{\lambda}}$ in general, it is possible that the guess strategy (20) erroneously selects positive/negative parametric error ($\delta \boldsymbol{\lambda}_i > 0/\delta \boldsymbol{\lambda}_i < 0$, i.e., over-/underestimation) when the true parametric error is actually of the opposite sign ($\boldsymbol{\lambda} - \boldsymbol{\lambda}^{\mathrm{nat}} < 0/\boldsymbol{\lambda} - \boldsymbol{\lambda}^{\mathrm{nat}} > 0$).

In the experiments described in Sect. 6 we will put ourselves in this situation and study the capability of EPU to reduce the model drift of predictions initialized using FFI. FFI-EPU is thus given by the FFI update (4) in conjunction with the EPU bias correction technique (19)–(20) during the forecast run.

5 The low-order climate model

The low-order climate model is based on the Lorenz 3-variables model (Lorenz, 1963), and has been introduced by Peña and Kalnay (2004). To mimic the simultaneous presence of different dynamical compartments having different timescales, three similar copies of the original Lorenz model are coupled to simulate the extratropical and tropical atmosphere and the ocean. The equations read

$$\begin{aligned}
\frac{dx_e}{dt} &= \sigma(y_e - x_e) - c_e(Sx_t + k_1),\\
\frac{dy_e}{dt} &= rx_e - y_e - x_e z_e + c_e(Sy_t + k_1),\\
\frac{dz_e}{dt} &= x_e y_e - bz_e,\\
\\
\frac{dx_t}{dt} &= \sigma(y_t - x_t) - c(SX + k_2) - c_e(Sx_e + k_1),\\
\frac{dy_t}{dt} &= rx_t - y_t - x_t z_t + c(SY + k_2) + c_e(Sy_e + k_1),\\
\frac{dz_t}{dt} &= x_t y_t - bz_t + c_z Z,\\
\\
\frac{dX}{dt} &= \tau\sigma(Y - X) - c(x_t + k_2),\\
\frac{dY}{dt} &= \tau(rX - Y - SXZ) + c(y_t + k_2),\\
\frac{dZ}{dt} &= \tau(SXY - bZ) - c_z z_t.
\end{aligned} \tag{21}$$

The atmospheric variables are denoted with the lower-case variables, with the subscripts e/t referring to the extratropical/tropical atmosphere; the ocean variables are denoted with capital letters. The two atmospheres are coupled through the variables x and y at a strength given by the parameter c_e; the tropical atmosphere and ocean are coupled through all variables with a strength given by the parameters c, for the x and y, and c_z for the z component. The parameters, $\sigma = 10$, $r = 28$, and $b = 8/3$, are the same as in the classical 3-variable Lorenz model (Lorenz, 1963)); the "uncentering" parameters $k_1 = 10$ and $k_2 = -11$, taken from Peña and Kalnay (2004), introduce a sort of phase lag between model compartments. The parameters S and τ modulate the amplitude and timescale of the ocean: they are chosen as in Peña and Kalnay (2004) to be $S = 1$ and $\tau = 0.1$ respectively, implying that the ocean variables will have the same amplitude as the atmospheres' but a slower rate by one order of magnitude.

In all subsequent experiments the nature climate evolution is represented by a solution of Eq. (21) with the coupling strength parameters set to the standard values, $c_e = 0.08$ and $c = c_z = 1$. This configuration implies that the tropical atmosphere and the ocean are strongly coupled, while the two atmospheres are only weakly coupled. Numerical solutions of Eq. (21) are obtained using a second-order Runge–Kutta scheme with time steps of $\delta t = 0.01$. According to Peña and Kalnay (2004), the model (21) represents an ENSO (El Niño–Southern Oscillation)-like configuration with an almost slave, small amplitude, atmosphere whose regime changes are modulated by the slow ocean component. Following their convention, a simulated year is made to correspond to an ocean regime, and the system oscillates between the normal regime, lasting between 3 and 12 years, and an El Niño regime, lasting only 1 year (equivalent to 240 time steps in the present experimental setup), (Peña and Kalnay, 2004). With the standard values for the coupling parameters given above, the model (Eq. 21) is chaotic with two positive Lyapunov exponents, $\gamma_{1,2}^{\text{nat}} = 0.9063, 0.3150$, while the third corresponds to the null one.

To simulate parametric errors originating at the level of the coupling between the different model compartments, we have modified simultaneously the tropical atmosphere/ocean coupling parameters c^{m} and c_z^{m}, with the superscript m standing for "model", in the range 0.1–1.5 with steps of 0.1. We have assumed that the coupling between the two atmospheres is known, so that $c_e^{\text{m}} = c_e$. To place ourselves in the situation in which the model is able to reproduce the qualitative behavior of nature, we have restricted our analysis to the 109 parameter combinations for which the model stability properties, as measured by the first three Lyapunov exponents, do not differ too much from those of nature. We further assume that the amplitude of the natural variability and its spectrum is well reproduced within each model component. This is not true and it might lead to an overestimation of the performance of AI in our conceptual model.

Figure 1 shows, with shadow bars, the distributions of the model bias and of the first three Lyapunov exponents relative to these 109 configurations; the RMS bias is defined by taking the square root of the mean quadratic biases on all of the 9 model variables, each one normalized with respect to its own variance:

$$\text{RMS bias} = \frac{\sum_{i=1}^{9} \sqrt{\frac{(\bar{x}_i - \bar{x}_i^{\text{nat}})^2}{(\sigma_i^{\text{nat}})^2}}}{9}.$$

The distribution of RMS bias values appears relatively homogeneous with a marked peak between 0.2 and 0.25. As expected, the range of values of the first Lyapunov exponent, γ_1, does not depart too much from $\gamma_1^{\text{nat}} = 0.9063$, given that the two perturbed coupling parameters, c and c_z, do not directly affect the dynamics of the extratropical atmosphere to which γ_1 is associated. In contrast, the Lyapunov exponents γ_2 and γ_3 display a much larger variability; in particular, most of the coupling configurations give rise to a less/more unstable tropical atmosphere/ocean with respect to the standard "nature" configuration. This is due to the form of the coupling between these two model compartments and to the role of the coefficients c and c_z: note that in a few cases γ_3 becomes positive, and the null exponent corresponds to the fourth one.

To illustrate how the statistical properties of the model (21) respond to a change in the parameters, c and c_z, the dependence of the variance versus the mean for each model variable and for all of the 109 model parameter variations is shown in Fig. 2; note that the values of the mean and the variance are normalized with respect to those of nature. The figure clearly reveals the complex interplay between the first two moments of the model probability density function, typical of nonlinear dynamics, and points out to the difficulty in estimating the climate variance of a model in a regime of bias change (i.e., a climate change). This figure illustrates the dependence of the climate variance on its mean state whereas the hypothesis of independence of those is often made in climate prediction. In particular, the classical anomaly initialization method relies on such a hypothesis. According to this figure, a scaling of the variance should be applied also in AI.

6 Results

In this section we describe the results of the numerical analysis. We proceed by showing first the results of the comparison between FFI and AI, while LSI and EPU are discussed in the two subsequent subsections.

6.1 FFI and AI comparison

The experimental setup is as follows. We have worked under the standard observation system simulation experiment (OSSE, Bengtsson et al., 1981) configuration in which the simulated nature evolution is sampled at discrete times to generate the series of simulated observations. The setup has then been structured following the typical hindcast format of climate prediction studies. A 40-year-long hindcast period is considered and the observations/start dates are distributed homogeneously every month during the first 30 years, making a total of 360 start dates. The effective number of independent start dates (Van Storch and Zwiers, 2001), N_{eq}, is around 241, 154, and 127 for the extratropics, tropics and ocean respectively; N_{eq} has been calculated using the novel approach given in Guemas et al. (2014) (their Eq. 7). Note that in the CMIP5 (Coupled Model Intercomparison Project) initiative (Taylor et al., 2012), at most we have 52 decadal predictions, with some forecasting systems having only 10. The "nature" and the "control" runs are integrated, after a long spin-up of 60 000 time steps, during the hindcast period. The observations are generated by sampling the nature trajectory and then adding a Gaussian white noise, $\boldsymbol{\epsilon}^{\mathrm{o}}$, with zero mean and standard deviation, $\boldsymbol{\sigma}^{\mathrm{o}}$, so that $\boldsymbol{\epsilon}^{\mathrm{o}} \in \mathcal{N}(0, \boldsymbol{\sigma}^{\mathrm{o}})$. Ten-year-long predictions are initialized, at each start date, by using either FFI, Eq. (4), or AI, Eq. (7), with either observations of each model compartment independently (**H** with dimension 3×9) or of the whole system (**H** equal to the identity 9×9 matrix). The difference between the model and observed mean states, $\mathbf{H}\overline{\boldsymbol{x}} - \overline{\boldsymbol{y}^{\mathrm{o}}}$, entering the AI equation (7), is estimated using the sample of 360 observations/start dates, for the case of observing the whole system (i.e., $\mathbf{H} = \mathbf{I}$). The distribution of this "estimated bias" is shown superimposed onto the "real bias" distribution in the top-right panel in Fig. 1: their comparison helps to visualize the difference between the two and quantify the error in the bias estimate used in AI. We see that the estimated biases reproduce relatively well the actual distribution, although some discrepancies are evident in particular in relation with the position of the respective largest peaks, slightly shifted toward higher values in the estimated bias distribution.

In most of the discussion that follows, we make use of the RMS skill score (RMSSS) to measure the skill of the initialized predictions. RMSSS is defined as:

$$\mathrm{RMSSS} = \left(1 - \frac{\mathrm{RMSE}^{\mathrm{Init}}}{\mathrm{RMSE}^{\mathrm{UnInit}}}\right) \cdot 100\,[\%],$$

where $\mathrm{RMSE}^{\mathrm{Init/UnInit}}$ refers to the RMS error for the initialized and uninitialized predictions respectively; the latter corresponds to the control run. The RMSE is computed using the anomalies with respect to the mean error over all start dates. In Fig. 3 the RMSSS, for both FFI and AI, is shown as a function of time over a 10-year-long prediction. The RMSSS is calculated by averaging the RMSE over all start dates and is displayed after a monthly averaging of the variables; the observation error standard deviation of each variable is set to $\sigma^{\mathrm{o}} = 1.5\,\%$ of the corresponding system's natural variability (i.e., the square root of the system's climate variance). Note that in a realistic prediction framework, observation error is meant to incorporate also the representativity error arising from the mismatch between the model and the observation-resolved scales. These errors are not present in our experimental setup, and this legitimates the choice of a relatively small σ^{o}. Two cases of coupling error are considered, $c^{\mathrm{m}} = 0.8, 0.3$ and $c_z^{\mathrm{m}} = 0.9, 1.2$: these configurations give rise to biases falling in the peak and in the rightmost extreme of the real-bias distribution (Fig. 1), while the spectrum of the first three Lyapunov exponents is $\gamma_{1,2,3} = 0.9036, 0.1895, 0.0$, and $\gamma_{1,2,3} = 0.9032, 0.2162, 0.0153$, respectively. By comparing these values with those of nature given above, we see that both configurations have a smaller second exponent and the second configuration possesses three positive exponents.

The first clear message in Fig. 3 is that the largest, long-lasting RMSSSs are obtained when the whole system is simultaneously initialized (black curves). Note however that FFI performs slightly better with a clear predictive skill

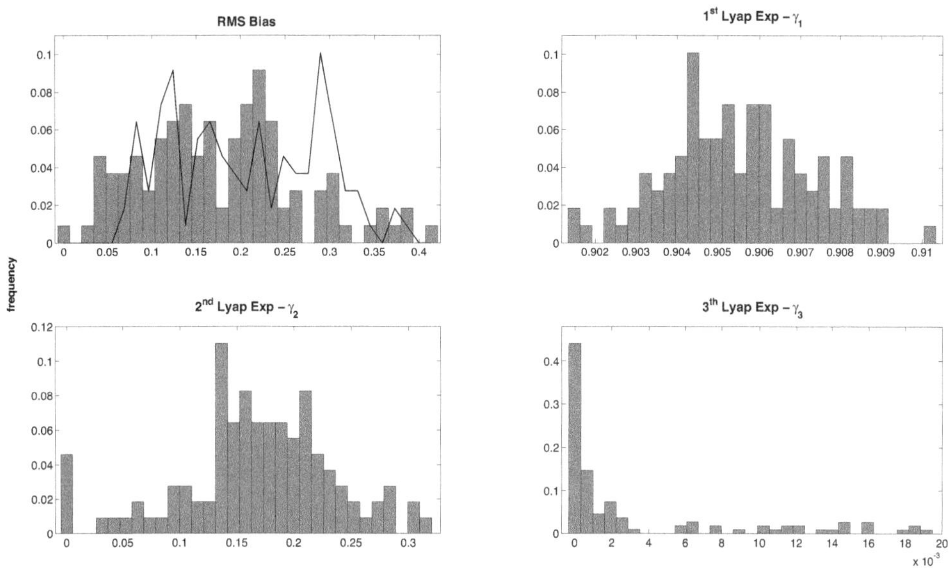

Fig. 1. Distribution of the RMS bias (top-left panel) and of the first, second and third Lyapunov exponents. The solid line in the top-left panel shows the distribution of the estimated RMS bias used in the AI.

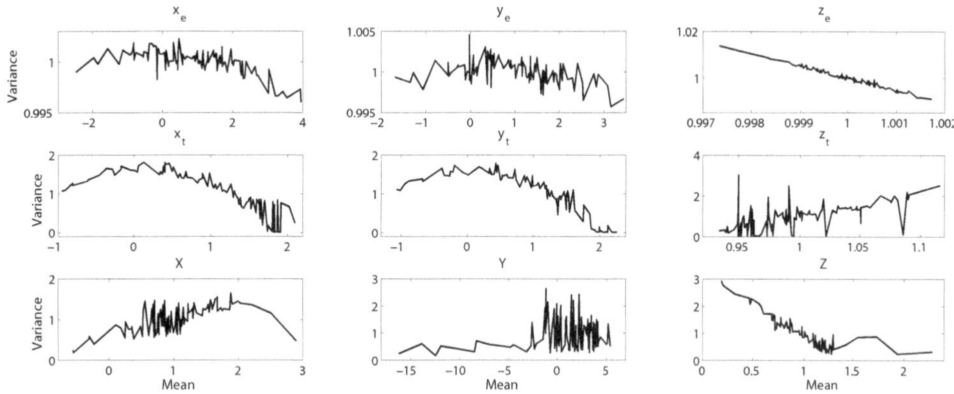

Fig. 2. Variance of model as a function of its mean for the sample of 109 model configurations considered and for the 9 model variables.

(RMSSS > 20 %) until around month 30, while in AI the drop of the RMSSS after the initialization is more abrupt and skillful forecasts last less than 20 months.

The second aspect concerns the role of the ocean as the first driver of the system's predictive skill. When only the ocean is initialized, significant skill is observed (RMSSS > 15 %) until the 20th month. The behavior of FFI and AI is similar, with only very marginal advantages for FFI. The largest initial error reduction obtained with FFI does not seem to be effective in this case to systematically outperform AI: a slow drift (not shown) is present that masks the benefit of this initial error reduction. Having the slowest error growth rate, the ocean behaves like the system's memory. Initializing this compartment is thus the most efficient way to improve prediction skill and move forward the predictive horizon.

The performance when only one of the "atmospheres" is initialized is considered now. When the extratropical atmosphere is initialized with FFI, the RMSSS is large (RMSSS > 10 %) until month 20 in both model configurations. The lack of a similar behavior in AI suggests that this gain in predictability is due to the larger reduction of the actual initial error obtained with FFI. The extratropical atmosphere is in fact the most unstable model compartment, the one with the faster error growth: reducing the initial error counteracts the effects of this growth, and helps to increase the time horizon of skillful predictions. This effect is less pronounced in AI, where the initial error is not much reduced and is sized as the model bias (see Eq. 11 and related comments). Finally, when only the tropics are initialized, the prediction skill for both FFI and AI maintains very low levels.

The behavior of FFI and AI as a function of the observational accuracy at different prediction ranges is analyzed

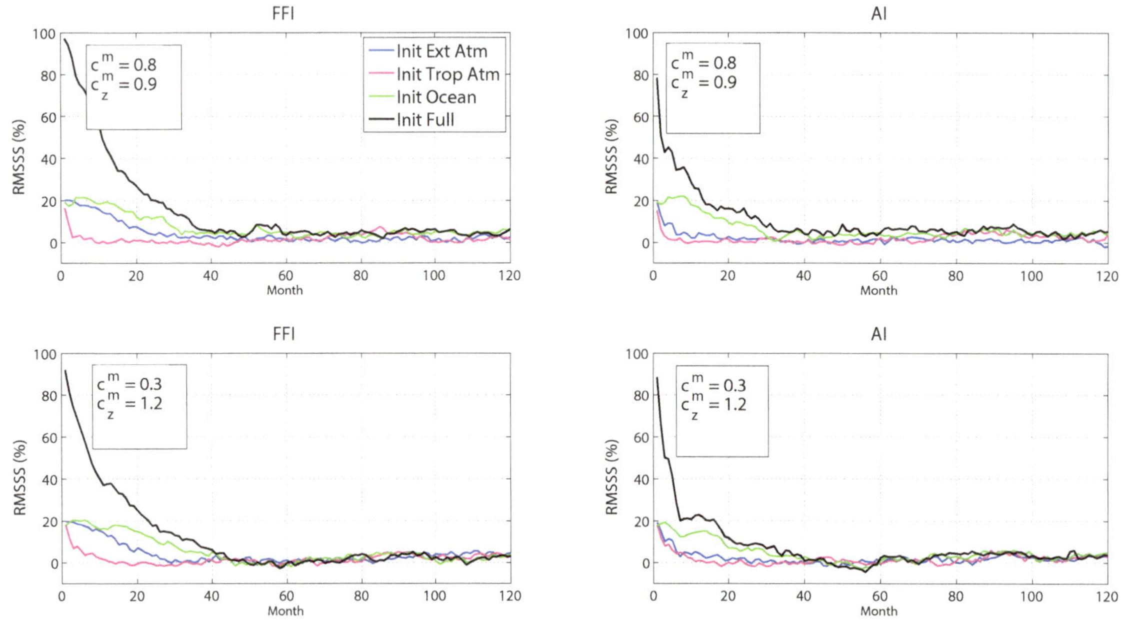

Fig. 3. RMSSS as a function of the forecast lead time for FFI (left panels) and AI (right panels). Top/bottom panels refer to the case $c^{\mathrm{m}} = 0.8$, $c_z^{\mathrm{m}} = 0.9/c^{\mathrm{m}} = 0.3$, $c_z^{\mathrm{m}} = 1.2$, respectively. Different colors represent the case of the initialization of the full system (black), ocean (green), tropical atmosphere (purple) and extratropical atmosphere (blue).

in Fig. 4. In these experiments the full system is observed, and the parametric error is set as in the first configuration ($c^{\mathrm{m}}/c_z^{\mathrm{m}} = 0.8/0.9$); results (not shown) for the case ($c^{\mathrm{m}}/c_z^{\mathrm{m}} = 0.3/1.2$) are qualitatively equivalent. Figure 4 has to be interpreted in the light of the discussion in Sect. 3.3 on the dependence of $\mathrm{RMSE}^{\mathrm{a}}_{\mathrm{FFI/AI}}$ on the observation accuracy σ^{o}. It illustrates that this dependence is propagated and is also found at longer lead times. Note for instance that, in contrast to AI, FFI prediction skill appears clearly related to the observational accuracy up to the third year. A closer inspection of the left-bottom panel reveals that such a behavior, although at a lower extent, is present even in the longer prediction horizon (years 4–5) as long as observations are sufficiently accurate ($\sigma^{\mathrm{o}} < 0.03$). However in the longest lead time, and for the horizon years 4–5 when $\sigma^{\mathrm{o}} > 2.5\,\%$, AI shows slightly better skills, a behavior in qualitative agreement with results of Smith et al. (2013) with the MetOffice climate model.

In a similar way, Fig. 5 analyzes the behavior of FFI and AI in relation with the model bias. As above the full system is observed and the same averaging periods are considered and the observational error standard deviation is set to $\sigma^{\mathrm{o}} = 2.5\,\%$. In contrast to Fig. 4, we see here that while FFI is almost insensitive to the model bias, AI displays a marked dependence with a performance that systematically improves when the bias is reduced, that is to say when the model is improved. Furthermore, the two approaches converge to similar RMSSS levels in the limit of long lead times. It is worth mentioning that the independence of the FFI skill on the model accuracy contradicts what is observed in realistic weather and climate prediction practices. We believe this is due to two reasons. First, limiting the source of model error to parameter uncertainty marks a key difference with respect to realistic s2d contexts where model error is often connected to multiple simultaneous causes, including unresolved scales and numerical discretization errors. Second, the model configurations considered in this study, with errors in the coupling parameters, do not give rise to strong initial shocks when initialized; results (not shown) with error in the model forcing confirm this conjecture.

Together, Figs. 4 and 5 give an illustration of the response of FFI and AI to the variation of the two ingredients on which they are based, and the behavior of the forecast error seems to be quite consistent with the conjecture done in Sect. 3.3 in relation with the analysis error. Even if obtained with a very simple model, but with a quite large ensemble of start dates, these results point clearly to key differences between FFI and AI. FFI seems to be favorable when a good observational network is at hand and its skill is expected to improve with a refinement of the observations. However, the performance of AI is related to the model accuracy and its overall skill is expected to increase in coincidence with model upgrades.

6.2 LSI – numerical results

In this section we study the performance of FF-LSI in comparison with standard FFI. The observational error standard deviation is equal to $\sigma^{\mathrm{o}} = 1.5\,\%$ and the error covariance $\mathbf{R}$ is diagonal and contains the correct observational error variance. The background error covariance matrix $\mathbf{B}^{\mathrm{m}}$ is

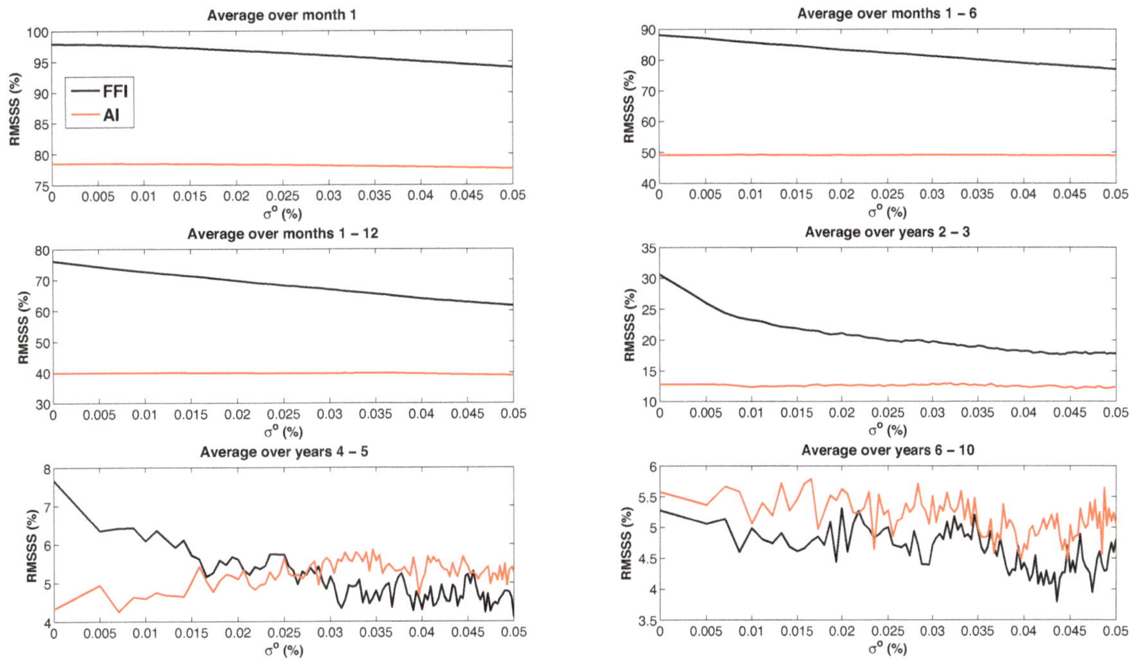

Fig. 4. RMSSS as a function of the standard deviation of the observational error, σ^{o}, expressed as a fraction of the system's natural variability. The six panels refer to six different averaging periods and are indicated in the corresponding labels. FFI (black line), AI (red line).

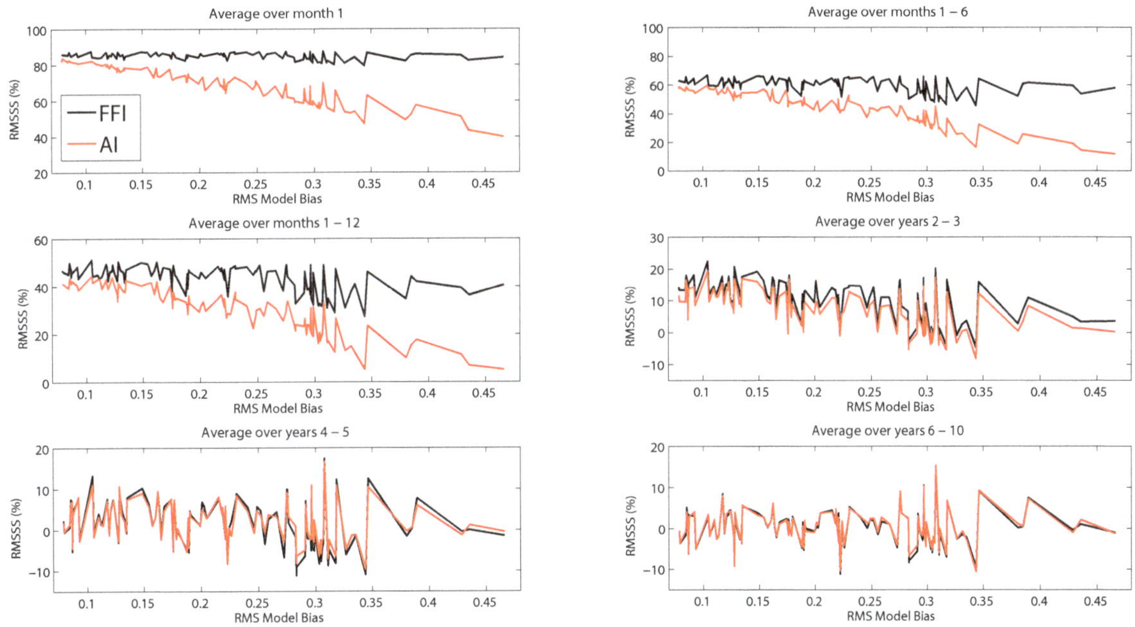

Fig. 5. RMSSS as a function of the RMS model bias. The six panels refer to six different averaging periods and are indicated in the corresponding labels. FFI (black line), AI (red line).

estimated using Eq. (13) over 50 years of an uninitialized model run.

The first result of LSI is given in Fig. 6, and shows the RMSSS as a function of the scaling parameter α in Eq. (13); similar to Figs. 4 and 5, the different panels refer to the different prediction horizons. The four observational scenarios are considered and model error is set as in the first configuration, c^{m}, $c_z^{\mathrm{m}} = 0.8, 0.9$; results do not qualitatively differ if the second configuration is considered. The range of values is $10^{-5} \leq \alpha \leq 15$ and the x axis is displayed in log scale. The RMSSS values of the FFI are also shown for comparison (dotted lines).

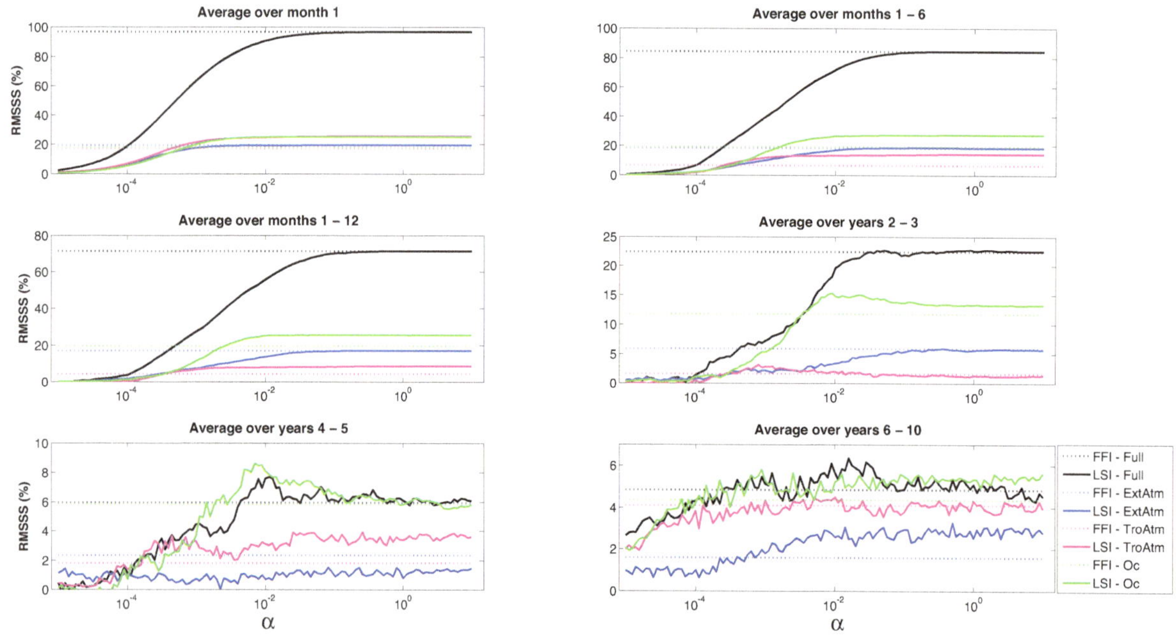

Fig. 6. RMSSS as a function of the tuning coefficient α for LSI (see text). The six panels refer to six different averaging periods and are indicated in the corresponding labels. Different colors represent the case of the initialization of the full system (black), ocean (green), tropical atmosphere (purple) and extratropical atmosphere (blue). The values of the standard FFI are displayed for reference (dots).

For very small α values the LSI analysis, Eq. (14), tends to over-trust the background field, the observations are practically ignored and the RMSSS converges to the small values associated with uninitialized predictions. By increasing α the RMSSS increases monotonically and eventually exceeds the FFI skill level and stabilizes on a plateau afterward.

The fact that the best performances are obtained for large enough α values and that no improvements are observed when it is further increased is connected to the use of observations that are significantly more accurate than the background. The relevant aspect in the discussion about the potentials of LSI is that, when only a portion of the full system's state is observed, its RMSSS converging level is higher than for FFI. This indicates that the benefit of LSI comes by the spatial correlations embedded in $\mathbf{B}^{\mathrm{m}}$ that make it possible to propagate the observation's informational content, otherwise restricted to the observation subspace alone, to the entire model's domain. In fact when α is large, the observations are fitted as if they were perfect, and the model state vector is replaced by the observations at their locations. Out of these points the information is propagated according to the off-diagonal elements of $\mathbf{B}^{\mathrm{m}}$, those that account for the correlations between different variables and model compartments. This interpretation is also supported by the fact that when a diagonal $\mathbf{B}^{\mathrm{m}}$ is used the FF-LSI's RMSSS (not shown) converges to the same values as FFI in the limit of large α, meaning that as in FFI the observations are perfectly fit but no correction is applied out of observational locations.

The adequacy of the approximation (Eq. 13) in describing the subspace spanned by the range of the actual $\mathbf{B}$ is diagnosed in Fig. 7, which shows the percentage of the explained variance of each of the eigenvectors of $\mathbf{B}$ and $\mathbf{B}^{\mathrm{m}}$ (Fig. 7a), as well as the scalar product of each pair of them (Fig. 7b), $\boldsymbol{v}_i(\mathbf{B}) \times \boldsymbol{v}_i(\mathbf{B}^{\mathrm{m}})$, with $\boldsymbol{v}_i(\mathbf{B})$ and $\boldsymbol{v}_i(\mathbf{B}^{\mathrm{m}})$ being the ith eigenvector of $\mathbf{B}$ and $\mathbf{B}^{\mathrm{m}}$ respectively, $\times$ indicates the scalar product, and $i = 1, \ldots, 9$. The percentage of variance explained by the eigenvector $\boldsymbol{v}_i$ is calculated as $\mathrm{Exp_{var}}(\boldsymbol{v}_i) = \frac{\lambda_i}{(\sum_i \lambda_i)} \cdot 100[\%]$, with λ_i being the ith eigenvalue. From Fig. 7 we see that the variance distribution over the eigenmodes is accurately reproduced and that, except for the third and fourth that are almost perpendicular to each other, all remaining eigenvectors are almost fully aligned.

6.3 EPU – numerical results

Similarly to the previous section we study here the performance of the drift correction method EPU. EPU is applied during the forecast run once the prediction is initialized using FFI; these experiments are hereafter referred to as FFI-EPU. We compare its performance with standard FFI in the absence of drift correction procedure.

We have first studied FFI and FFI-EPU in a set of 109 experiments using the model configurations described in Sect. 5.1; this gives the range of parameter $\Delta\Lambda$, while the bias correction time interval is set equal to one time step, $\Delta T_{\mathrm{Bias}} =$ time step. Figure 8 shows the mean and standard deviation of the RMSE relative to the 109 configurations, for

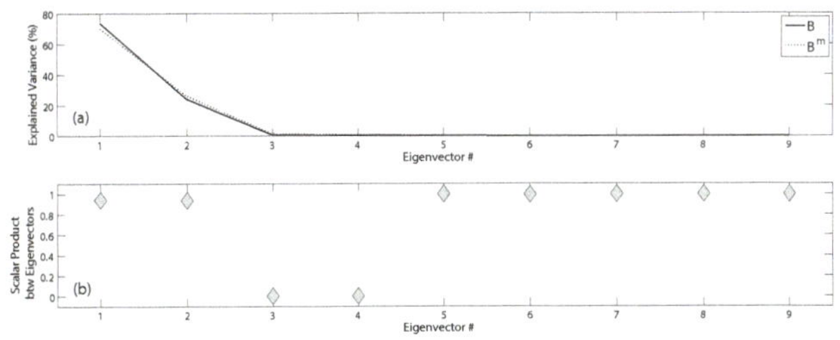

Fig. 7. **(a)** Explained variance over each eigenvector of $\mathbf{B}^{\mathrm{m}}$ used in LSI (dotted line; see text) and the actual background error covariance matrix **B** (solid line). **(b)** Scalar product between pairs of eigenvectors of $\mathbf{B}^{\mathrm{m}}$ and **B**.

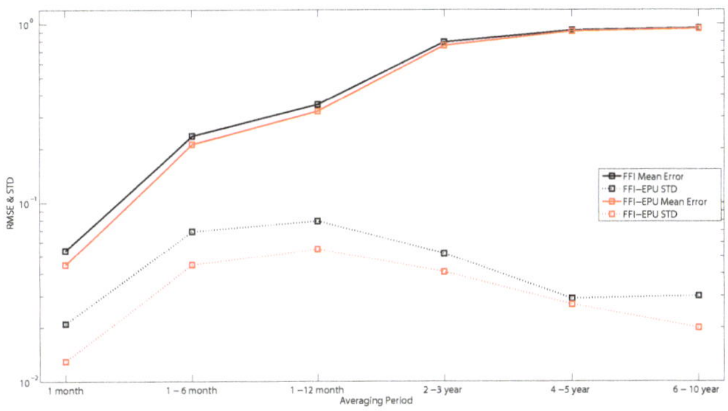

Fig. 8. Mean and standard deviation of the distribution of the RMSE relative to the 109 configurations, for FFI (black) and FFI-EPU (red), as a function of the averaging forecast period.

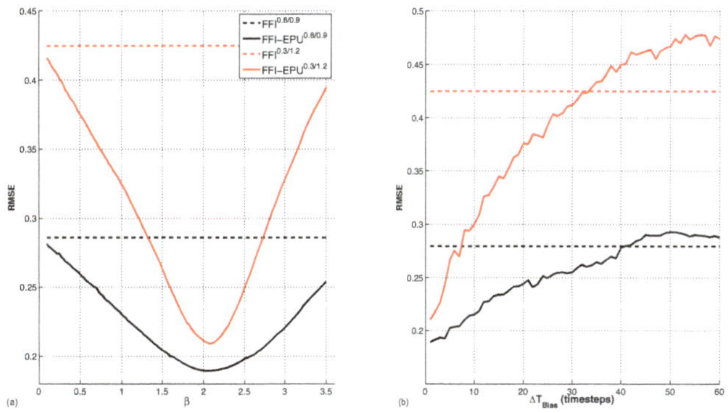

Fig. 9. **(a)** RMSE of FFI-EPU (solid line) over the first year of prediction as a function of the width of the sampling distribution (Eq. 20) scaled with the parameter β (see text) for the cases $c^{\mathrm{m}}=0.8$, $c_z^{\mathrm{m}}=0.9/c^{\mathrm{m}}=0.3$, $c_z^{\mathrm{m}}=1.2$. **(b)** RMSE of FFI-EPU over the first year of prediction as a function of the bias correction interval ΔT_{Bias}, for the case $c^{\mathrm{m}}=0.8$, $c_z^{\mathrm{m}}=0.9$ (black line) and $c^{\mathrm{m}}=0.3$, $c_z^{\mathrm{m}}=1.2$ (red line). FFI (dashed lines) is superimposed for reference in both panels.

FFI (black) and FFI-EPU (red), as a function of the averaging forecast period. Results reveal the clear benefit of using EPU within the forecast year. The relative improvements over FFI are equal to 17, 10, and 8 % in the first month and in the first 6 and 12 months respectively. A certain minor advantage is also found afterward, with relative improvements equal to 4, 0.6, and 0.2 % in the forecast years 2–3, 4–5, and 6–10. The distributions (not shown) of the RMSE for FFI-EPU in the six averaging periods are all shifted toward smaller values with respect to FFI. This behavior is in part observed by looking at the corresponding standard deviations (dotted lines in Fig. 8) that are, except for the period of 4–5 years, lower than for FFI. In Fig. 8 EPU appears to be more skillful within the time horizon of 1 year, suggesting its potential usage in the context of seasonal forecasting. Errors in the sampling procedure and, most importantly, the deviation of the drift dynamics from the linear assumption on which EPU is built seems to limit its benefit on longer timescales.

The impact of the uncertainty on the width of the uniform distributions in Eq. (20) and the length of the bias correction time interval, ΔT_{Bias}, the two factors controlling the EPU setup, is investigated in Fig. 9a. Both model configurations, $c^{\mathrm{m}}=0.8/c_z^{\mathrm{m}}=0.9$ and $c^{\mathrm{m}}=0.3/c_z^{\mathrm{m}}=1.2$, are considered. The RMSE is calculated over the first year of prediction. The error on the width of the sampling distribution reflects the uncertainty about the actual range of possible parameter values. To simulate this circumstance, we have used the actual parameter, $\boldsymbol{\lambda}^{\mathrm{nat}}$, and modulated the distribution width using a scalar coefficient β, such that $\delta\lambda_i \in \mathcal{U}\left(0, \beta\left(\boldsymbol{\lambda}-\boldsymbol{\lambda}^{\mathrm{nat}}\right)\right)$. As expected, the best performances are with $\beta \approx 2$, as in this situation the mean of the sampling distribution coincides with the actual parametric error. However the important remark is that FFI-EPU outperforms FFI systematically for all values of β in both model configurations.

The sensitivity of FFI-EPU to the length of bias correction interval, ΔT_{Bias}, is analyzed in the right panel of Fig. 9b. The same two model configurations are considered, and are displayed with the black and red lines respectively (see corresponding labels in the figure panel); similarly the RMSE of FFI is shown with dashed lines. To focus on the impact of ΔT_{Bias} alone, we have set the scaling parameter $\beta=2$. As expected, the accuracy of FFI-EPU decreases by increasing ΔT_{Bias}, as the limit of duration of the short-time regime of the error growth is approached. It is remarkable that, with these particular model configurations, FFI-EPU outperforms FFI in a large range of ΔT_{Bias}. Note moreover that the improvements over FFI are kept as long as $\Delta T_{\mathrm{Bias}} < 40$ and $\Delta T_{\mathrm{Bias}} < 30$ time steps for the first and second model configurations respectively. If these time intervals are taken as measures of the duration of the short-time regime, it is worth highlighting that the numerical result $\Delta T_{\mathrm{Bias}}^{c^{\mathrm{m}}=0.8/c_z^{\mathrm{m}}=0.9} > \Delta T_{\mathrm{Bias}}^{c^{\mathrm{m}}=0.3/c_z^{\mathrm{m}}=1.2}$ is consistent with the amplitudes of the largest (in absolute) value of the Lyapunov exponents, $|\gamma_9^{c^{\mathrm{m}}=0.8/c_z^{\mathrm{m}}=0.9}| < |\gamma_9^{c^{\mathrm{m}}=0.3/c_z^{\mathrm{m}}=1.2}|$, in agreement with the theory of deterministic model error dynamics mentioned in Sect. 4.2 (Nicolis, 2003).

The results in Fig. 9 are quite encouraging about the robustness of EPU. Although the extent of their validity with more realistic models cannot be assessed with the simple dynamics used here, they at least motivate studying its performance in the situations where large uncertainties on the optimal parameter range are present, or with large climate models where reducing the frequency of its application is desirable to decrease the associated computational cost.

7 Conclusion and Discussion

This study provides an outlook of the initialization techniques for seasonal-to-decadal predictions using concepts and notations borrowed from the data assimilation context. Our first purpose was to give a comprehensive comparison between full-field and anomaly initialization (FFI and AI), the two main classes of initialization methods used in major climate prediction institutions, with the aim of contributing to the current debate on their respective advantages and drawbacks. The second objective of this investigation was the introduction of two methods to improve the initialization procedures; namely, the least-square initialization (LSI) and the exploring parameter uncertainty (EPU).

Following a standard procedure for the analysis update in most data assimilation schemes, LSI operates a least-square fit between observations and model with weights related to the respective assumed level of accuracy. By optimally combining model and observations at the initialization step, the informational content of the observations is propagated to the entire model domain and no longer confined to the locations of the measurement, as in classical FFI or AI. The required model error covariance entering the LSI initialization is estimated using the covariance of the model anomalies taken over a long uninitialized run, in a similar manner to what was done in Smith and Murphy (2007). Using LSI, the model's initial state out of the observational locations is modified according to the structure of the model statistics. This allows to reduce the initial errors in areas otherwise totally unconstrained by the observations and has the potential to mitigate the dynamical shocks resulting from discontinuities caused by pushing the model state to observation values and leaving it unchanged elsewhere.

EPU is an online drift correction method in which the drift caused by the parametric error is estimated using a short-time evolution law and is then removed during the forecast run. It requires the computation of the model equation's first derivative with respect to the parameters assumed to be uncertain (i.e., the model parameter sensitivity), and a hypothesis about the assumed range of possible parameter values. Being based on a short-time approximation, EPU deteriorates when the length of the time interval over which it is applied is longer than the duration of the short-time regime. This duration is certainly connected to the model sensitivity to the specific set of parameters under consideration and is model dependent. The ideal situation would be to implement EPU at every time step. Nevertheless, although the computational demand of EPU is very low, reducing the frequency of its application (by increasing the drift correction interval) would reduce the computational cost.

The comparison between FFI and AI as well as the performances of LSI and EPU have been studied using an idealized coupled climate model based on the classical Lorenz 3 variables system (Lorenz, 1963)). The model, introduced by Peña and Kalnay (2004), possesses three compartments characterized by different timescales and amplitudes, and are taken as proxies for extratropical/tropical atmosphere and the ocean. The low model dimension and its low degree of complexity allowed us to run a large ensemble of trajectories (360 start dates) using a standard hindcast experimental setup over 30 years and monthly start dates.

The use of the data assimilation formalism helped us to rewrite the initialization updates in a general form and highlight specific features such as their sensitivity to observation and model accuracy. The analysis of their error-scaling properties suggests the use of FFI when a good observational network is available and reveals the direct relation of its skill with the observational accuracy. The skill of AI appears, however, mostly related to the model quality, and clear increases of skill can only be expected in coincidence with model upgrades. The numerical results confirm this behavior and reveal that, in contrast to AI, FFI prediction skill appears clearly related to the observational accuracy up to the third forecast year. As opposed to this, AI displays a marked dependence on the model bias with a performance that systematically improves when the model is improved.

We have compared FFI and AI in experiments in which either the full system or each of the model compartments was initialized independently. When the full system is observed the best performances are obtained in FFI and AI, although the former shows clearly better and longer-lasting improvements, with skillful predictions (RMSSS $> 20\,\%$) until month 30, whereas in AI the RMSSS falls below this level between months 15 and 20. In the initialization of single compartments, the best performance (with no remarkable differences between FFI and AI), is obtained when the stabler component of the model (the ocean) is initialized. Results show also that with FFI it is possible to have some predictive skill when the most unstable compartment (the extratropics) is observed. This behavior is not mirrored in AI, which supports the idea that the gain in skill is obtained by the efficient initial error reduction of FFI. However, in agreement with results from a state-of-the-art climate model (Smith et al., 2013), AI slightly outperforms FFI after the fourth year.

Finally, FFI was compared with FF-LSI and FFI-EPU. Results show that using LSI improves the performance of FFI in all the situations when only a portion of the system's state is observed. This proves that the performance of LSI is based on an efficient propagation of information from data-covered to data-uncovered areas and to some extent reduces the initial error also far from the observational locations. Estimating the background error covariance matrix using the statistics of the anomalies has therefore proved to be a viable choice. These results, in agreement with Smith and Murphy (2007), extend them to the case of the estimation of cross-covariances between different model compartments (atmospheres and ocean). This feature is extremely relevant for the development of coupled data assimilation systems, where the aim is to simultaneously update all coupled subsystems based on observations of one compartment alone.

The use of EPU has clearly improved the skill of FFI within the first forecast year; later, the limits of accuracy of the linear and short-time assumptions at the basis of EPU are approached, and only minor advantages over FFI are recorded. Results have also demonstrated the robustness of EPU with respect to the two factors determining its implementation: the length of the drift-correction interval and the accuracy of our knowledge of the actual range of possible parameters.

This study is intended as a proof of concept in which initialization algorithms are analyzed in a simplified context in which the respective advantages and drawbacks could be easily highlighted. Results of the two proposed advanced formulations, LSI and EPU, encourage their application with models, and observational setups, of increasing complexity. The extent to which the conclusions drawn in the present work can be applied to a realistic climate prediction exercise is a central topic of research the authors are currently undertaking. Two main lines of research have been undertaken as follow-up activities. In the context of a state-of-the-art climate model we are studying the effect of initializing different areas using FFI in a multiyear prediction horizon. Similarly, we are investigating the capability of EPU to estimate and deal with the drift caused by a very limited set of coupling parameters, on seasonal timescales. A crucial issue in this type of realistic applications will be the estimate of the drift caused by a parametric error having spatial dependence. We are also committed to extend the analysis presented in this study to a larger set of uncertain parameters and to study the use of LSI in coupled climate models of intermediate complexity.

Acknowledgements. The authors acknowledge the ICMAT Severo Ochoa project SEV-2011-0087 for covering the publication costs. A. Carrassi was financed through the IEF Marie Curie Project INCLIDA of the FP7. R. J. T.Weber is supported by the PROMOS grant of the DAAD (German Academic Exchange Service). This work was supported by the EU-funded SPECS (FP7-ENV-2012-308378) and the MINECO-funded PICA-ICE (CGL2012-31987) projects.

Edited by: J. Duan

References

Auroux, D. and Blum, J.: A nudging-based data assimilation method: the Back and Forth Nudging (BFN) algorithm, Nonlin. Processes Geophys., 15, 305–319, doi:10.5194/npg-15-305-2008, 2008.

Bengtsson, L., Ghil, M., and Källen, E.: (Eds.): Dynamic Meteorology: Data Assimilation Methods, Springer Verlag, New York/Heidelberg/Berlin, 330 pp., 1981.

Carrassi, A., Vannitsem, S., and Nicolis, C.: Model error and sequential data assimilation: A deterministic formulation, Q. J. Roy. Meteor. Soc., 134, 1297–1313, 2008.

Cohn, S. E.: An Introduction to Estimation Theory, J. Meteorol. Soc. Jpn., 75, 257–288, 1997.

Daley, R.: Atmospheric Data Analysis, Cambridge University Press, Cambridge, ISBN: 9780521382151, 1991.

Dee, D. P., Uppala, S. M., Simmons, A. J., Berrisford, P., Poli, P., Kobayashi, S., Andrae, U., Balmaseda, M. A., Balsamo, G., Bauer, P., Bechtold, P., Beljaars, A. C. M., van de Berg, L., Bidlot, J., Bormann, N., Delsol, C., Dragani, R., Fuentes, M., Geer, A. J., Haimberger, L., Healy, S. B., Hersbach, H., Holm, E. V., Isaksen, L., Kallberg, P., Kohler, M., Matricardi, M., McNally, A. P., Monge-Sanz, B. M., Morcrette, J. J., Park, B. K., Peubey, C., de Rosnay, P., Tavolato, C., Thpaut, J. N., and Vitart, F.: The era-interim reanalysis: configuration and performance of the data assimilation system, Q. J. Roy. Meteor. Soc., 137, 553–597, 2011.

Doblas-Reyes, F. J., García-Serrano, J., Lienert, F., Pintò Biescas, A., and Rodrigues, L. R. L.: Seasonal climate predictability and forecasting: status and prospects, WIREs Clim. Change, 4, 245–268, doi:10.1002/wcc.217, 2013a.

Doblas-Reyes, F. J., Andreu-Burillo, I., Chikamoto, Y., García-Serrano, J., Guemas, V., Kimoto, M., Mochizuki, T., Rodrigues, L. R. L., and van Oldenborgh, G. J.: Initialized near-term regional climate change prediction, Nature Communications, 4, 1715, doi:10.1038/ncomms2704.2900-2915, 2013b.

Dunstone, N. J. and Smith, D. M.: Impact of atmosphere and subsurface ocean data on decadal climate prediction, Geophys. Res. Lett., 37, L02709, doi:10.1029/2009GL041609, 2010.

Evensen, G.: Data Assimilation – The Ensemble Kalman Filter, 2nd Edn., Springer, Cambridge, ISBN: 978-3-642-03710-8, 2009.

Guemas, V., Auger, L., and Doblas-Reyes, F.: Hypothesis testing for auto-correlated short climate time series, J. Appl. Meteorol. Clim., 53, 637–651, doi:10.1175/JAMC-D-13-064.1, 2014.

Hawkins, E. and Sutton, R.: The potential to narrow uncertainty in regional climate prediction, B. Am. Meteorol. Soc., 90, 1095–1107, doi:10.1175/2009BAMS2607.1, 2009.

Hazeleger, W., Guemas, V., Wouters, B., Corti, S., Andreu-Burillo, I., Doblas-Reyes, F. J., Wyser, K., and Caian, M.: Multiyear climate predictions using two initialization strategies, Geophys. Res. Lett., 40, 1794–1798, doi:10.1002/grl.50355, 2013.

Ide, K., Courtier, P., Ghil, M., and Lorenc, A.: Unified notation for data assimilation: Operational, sequential and variational, J. Meteorol. Soc. Jpn., 75, 181–189, 1997.

Janjić, T. and Cohn, S. E.: Treatment of Observation Error due to Unresolved Scales in Atmospheric Data Assimilation, Mon. Weather Rev., 134, 2900–2915, 2006.

Jazwinski, A. H.: Stochastic Processes and Filtering Theory, Academic Press, New York, 1970.

Kalnay, E.: Atmospheric Modeling, Data Assimilation and Predictability, Cambridge University Press, Cambridge, UK, 2003.

Keenlyside, N., Latif, M., Jungclaus, J., Kornblueh, L., and Roeckner, E.: Advancing decadal-scale climate prediction in the North Atlantic sector, Nature, 453, 84–88, 2008.

Latif, M., Collins, M., Pohlmann, H., and Keenlyside, N.: A review of predictability studies of Atlantic sector climate on decadal time scales, J. Climate, 19, 5971–5987, 2006.

Lorenz, E. N.: Deterministic Nonperiodic Flow, J. Atmos. Sci., 20, 130–141, 1963.

Magnusson, L., Alonso-Balmaseda, M., Corti, S., Molteni, F., and Stockdale, T.: Evaluation of forecast strategies for seasonal and decadal forecasts in presence of systematic model errors, Clim. Dynam., 41, 2393–2409, doi:10.1007/s00382-012-1599-2, 2012.

Meehl, G. A., Goddard, L., Kirtman, B., Branstator, G., Danabasoglu, G., Hawkins, E., Kumar, A., Rosati, T., Smith, D., Sutton, R., Boer, G., Burgman, R., Carson, C., Corti, S., Karspeck, A., Keenlyside, N., Kimoto, M., Matei, D., Mignot, J., Msadek, R., Navarra, A., Pohlmann, H., Rienecker, M., Schneider, E., Tebaldi, C., Teng, H., van Oldenborgh, G., Vecchi, G., and Yeager, S.: Decadal Climate Prediction: An Update from the Trenches, B. Am. Meteorol. Soc., 95, 243–267, doi:10.1175/BAMS-D-12-00241.1, 2014.

Nicolis, C.: Dynamics of model error: Some generic features, J. Atmos. Sci., 60, 2208–2218, 2003.

Palatella, L., Carrassi, A., and Trevisan, A.: Lyapunov vectors and assimilation in the unstable subspace: theory and applications, J. Phys. A: Math. Theor., 46, 254020, doi:10.1088/1751-8113/46/25/254020, 2013.

Parrish, D. F. and Derber, J. C.: The National Meteorological Center's Spectral Statistical-Interpolation Analysis System, Mon. Weather Rev., 120, 1747–1763, 1992.

Peña, M. and Kalnay, E.: Separating fast and slow modes in coupled chaotic systems, Nonlin. Processes Geophys., 11, 319–327, doi:10.5194/npg-11-319-2004, 2004.

Smith, D. M. and Murphy, J. M.: An objective ocean temperature and salinity analysis using covariances from a global model, J. Geophys. Res., 112, C02022, doi:10.1029/2005JC003172, 2007.

Smith, D. M., Cusack, S., Coleman, A. W., Folland, C. K., Harris, G. R., and Murphy, J. M.: Improved surface temperature prediction for the coming decade from a global climate model, Science, 317, 796–799, 2007.

Smith, D. M., Eade, R., and Pohlmann, H.: A comparison of full-field and anomaly initialization for seasonal to decadal climate prediction, Clim. Dynam., 41, 3325–3338, doi:10.1007/s00382-013-1683-2, 2013.

Stockdale, T. N.: Coupled ocean-atmosphere forecasts in the presence of climate drift, Mon. Weather Rev., 125, 809–818, 1997.

Taylor, K. E., Stouffer, R. J., and Meehl, G. A.: An overview of cmip5 and the experimental design, B. Am. Meteorol. Soc., 93, 485–498, doi:10.1175/BAMS-D-11-00094.1, 2012.

van Oldenborgh, G. J., Doblas-Reyes, F. J., Wouters, B., and Hazeleger, W.: Skill in the trend and internal variability in a multi-model decadal prediction ensemble, Clim. Dynam., 38, 1263–1280, doi:10.1007/s00382-012-1313-4, 2012.

Von Storch, H. and Zwiers, F. W.: Statistical Analysis in Climate Research, Cambridge University Press, Cambridge, 484 pp., 2001.

Yang, S.-C., Baker, D., Li, H., Huff, M., Nagpal, G.. Okereke, E., Villafane, J., Kalnay, E., and Duane, G.: Data assimilation as synchronization of truth and model: experiments with the 3-variable Lorenz system, J. Atmos. Sci., 63, 2340–2354, 2006.

19

One-dimensional modelling of upper ocean mixing by turbulence due to wave orbital motion

M. Ghantous and A. V. Babanin

Swinburne University of Technology, Hawthorn, Victoria, Australia

Correspondence to: M. Ghantous (mghantous@swin.edu.au)

Abstract. Mixing of the upper ocean affects the sea surface temperature by bringing deeper, colder water to the surface. Because even small changes in the surface temperature can have a large impact on weather and climate, accurately determining the rate of mixing is of central importance for forecasting. Although there are several mixing mechanisms, one that has until recently been overlooked is the effect of turbulence generated by non-breaking, wind-generated surface waves.

Lately there has been a lot of interest in introducing this mechanism into ocean mixing models, and real gains have been made in terms of increased fidelity to observational data. However, our knowledge of the mechanism is still incomplete. We indicate areas where we believe the existing parameterisations need refinement and propose an alternative one.

We use two of the parameterisations to demonstrate the effect on the mixed layer of wave-induced turbulence by applying them to a one-dimensional mixing model and a stable temperature profile. Our modelling experiment suggests a strong effect on sea surface temperature due to non-breaking wave-induced turbulent mixing.

1 Introduction

For many applications ocean waves can be adequately modelled using potential wave theory, meaning the water is presumed inviscid, incompressible and the motion adiabatic. As a result, the wave orbital motion, which in potential theory is irrotational, is often presumed incapable of producing turbulence, and therefore unable to contribute to ocean mixing. (A recent paper by Beyá et al., 2012 argued, based on experiments they carried out, that the deviations from irrotational motion are entirely negligible. We discuss this paper in Appendix B.) Yet real waves by definition do produce vorticity (for example Kinsman, 1965; Landau and Lifshitz, 1987), and moreover it may be shown that even potential waves can, by stressing vortex lines, amplify pre-existing vorticity (Phillips, 1961; Kinsman, 1965; Benilov et al., 1993). Nevertheless the effect was either ignored or thought too small to produce any significant effect, and it remains excluded from operational mixing models. However, recent work suggests that the effect is significant enough that its inclusion in geophysical models could greatly improve ocean as well as sediment suspension, weather and climate forecasts. Despite the encouraging results the existing approaches are in large part guesses, undeniably useful though they are, and a parameterisation based on the underlying physical principles is still needed.

Our aim is, ultimately, to produce a useful and efficient parameterisation for the wave-induced turbulence so that it can be applied to ocean mixing models and, by extension, to ocean circulation, weather and climate models. And because the generation of turbulence is a sink for wave energy, the same parameterisation should improve wave modelling itself by better accounting for swell dissipation. This has historically been poorly predicted (Rogers, 2002), although recent advances have improved the situation considerably (Ardhuin et al., 2010). Nevertheless, the physics is not yet settled, and while it seems likely that more than one process is responsible, the precise contributions are unknown. In this paper we assume that the bulk of the swell dissipation is due to turbulence production.

We briefly run through the theoretical and experimental justification for the mechanism in Sect. 2. In Sect. 3 we review the different approaches attempted and describe their strengths and weaknesses. We also suggest another way of introducing the effect; this is described in Sect. 4, along with its application to a one-dimensional mixing model and the results.

There are of course other mechanisms of turbulence generation involving waves, such as wave breaking (see for example Chalikov and Belevich, 1993) and Langmuir circulation (Sullivan and McWilliams, 2010). We do not discuss wave breaking in detail since it is a source independent of that produced by the wave orbital motion which concerns us here. Langmuir circulation, on the other hand, is connected, as it is driven by the Stokes drift, a property of the wave orbital motion – but it is not a complete description since the Stokes drift does not encompass the entire orbital motion. We briefly discuss this in Sect. 3, though we do not treat Langmuir circulation directly.

2 Wave-induced turbulence

As early as the first quarter of the 20th century, Jeffreys (1920) sought to explain observed eddy viscosity in the upper ocean by the wave orbital motion, which he noticed had a similar rate of depth attenuation. Later, Dobroklonskiy (1947) and Bowden (1950) produced the first useful parameterisations for eddy viscosity and turbulence production. Further theoretical work examining the order of the effect of the orbital motion on turbulence was published in the 1960s (Phillips, 1961; Kinsman, 1965). Meanwhile, observations from the 1970s (Yefimov and Khristoforov, 1971a, b) and possibly as early as the 1950s (Francis et al., 1953 observed higher than expected stresses near the surface in windless conditions and speculated that they may be due to a "hitherto unexpected complication ... caused by the free surface") seemed to confirm the existence of the mechanism, but the phenomenon had more recently been neglected by oceanographers, though not by experimentalists.

One reason for this neglect might be because as a mechanism for wave dissipation it is much smaller than that caused by wave breaking. Indeed, Phillips (1961) (see also Kinsman, 1965 for a more expansive treatment) tried to estimate wave dissipation caused by wave–turbulence interaction, and the conclusion that it was negligible could have had some influence among oceanographers. But the treatment assumed a Reynolds number defined using wavelength as the length scale rather than amplitude, which is the variable that controls the size of the water particle orbital motion; and wave dissipation due to turbulence was demonstrated all the same.

A further reason for the neglect could be that, especially from the 1960s, many important features of nonlinear wave dynamics were found to be well described by potential theory where viscosity and vorticity were neglected (see, for example, the discussion in Babanin and Chalikov, 2012). Because turbulence by definition requires vorticity, it might have seemed that since vorticity could be safely ignored in describing wave dynamics accurately enough that turbulence generated by them could be ignored too, but such an interpretation misses three vital points: firstly, that despite being small, the vorticity is not zero, and could still produce a measurable effect on the boundary layer if the waves were large enough; secondly, that irrotational waves can interact with and amplify pre-existing turbulence; and thirdly, that while the energy involved in turbulence production might be small relative to the wave energy, or even wave breaking, it does not necessarily follow that it is small with relation to ocean mixing.

To elaborate on that second point, Benilov et al. (1993) (this argument was revisited by Benilov, 2012) demonstrated by a direct instability analysis that vorticity grew with the passage of a linear potential plane wave, and that the only component that was stable was the one directed along the wave crest. Any treatment restricted to two-dimensional motion would miss this important aspect of the flow. Furthermore, Anis and Moum (1995), by considering the Reynolds-averaged equations of motion partitioned three ways into mean, orbital and turbulent components, differentiated between two mechanisms for turbulence production by the wave orbital motion – one that irrotational motion could account for and another that required rotational motion. They attempted to apply both to observed scalings of turbulent dissipation which the conventional mechanisms (such as wave breaking) could not adequately explain, but they did not determine the precise contribution of each term, and both scalings, though based on different mechanisms, could be used to describe the observed profiles.

Experiments such as those of Ölmez and Milgram (1992) and Milgram (1998) sought to investigate the passage of waves on a turbulent body of water, and demonstrated that the existence of turbulence, generated independently of the wave motion, led to a measurable attenuation of the surface waves. Taken together with the theoretical arguments that the waves should interact with the turbulence, and especially that they should transfer their energy to it, this finding highlights that turbulence production and swell attenuation in the absence of breaking are two sides of the same coin, a notion exploited by Bowden (1950), Babanin (2011) and by us too, in Sect. 4.1.

3 Existing parameterisations

Several approaches to parameterising non-breaking wave-induced mixing in numerical models have already been tried, and they have all been successfully used to improve model results. They fall into three broad categories: those that attempt to parameterise Langmuir circulation; those that are scaled with wind stress; and those that, like the present one, are based on the notion that a fundamental property of the wave orbital motion is capable of generating turbulence which can lead to mixing. Those based on wind stress include the parameterisations described by Jacobs (1978) and

Huang and Qiao (2010) (also Huang et al., 2011), and we discount these immediately since we are interested in capturing the effect of turbulence generated by swell in the absence of wind stress. Parameterising Langmuir circulation is usually done by scaling it with the Stokes drift. For example, Kantha and Clayson (2004) modify the Mellor–Yamada turbulence model by augmenting the turbulence production by adding the Stokes shear vectorially to the current shear in the turbulence production terms. In the kinetic energy equation this is done directly, however in the length-scale determining equation they rescale the Stokes-driven production in order to reflect the large size of the Langmuir cells in the turbulent macro length scale. Their model parameters were chosen in order to match the LES results of McWilliams et al. (1997). While studies such as these suggest that Langmuir circulation is an important mechanism in ocean mixing, we argue that it represents only a part of the dynamics. The Stokes drift is sometimes, as by Kantha and Clayson (2004) and Ardhuin and Jenkins (2006), treated as a vertical velocity shear, though it is really only one part of a motion which results in a mean transport. The rest of the motion is important: Benilov (2012) showed that vorticity was unstable for even linear waves, which generate no drift, and Anis and Moum (1995) described how irrotational wave motion could transport turbulent diffusion downward; this mechanism was also suggested by Ölmez and Milgram (1992) in order to explain their experiments. Even the analysis by Teixeira and Belcher (2002), who favoured Stokes drift as the primary mechanism for distortion of vorticity, nevertheless demonstrated that without it the waves affected the turbulence.

It is worth noting here that our proposed alternative also involves adding wave-induced production to the turbulent kinetic energy production in a two-equation model; however, we do not attempt to extract a value for the transport such as Stokes drift or mean shear in order to augment the shear velocity. Instead the production and its vertical scaling are independently determined. Of course in practice the difference in approach is notional rather than real, since any term can always be scaled by a parameter or by altering the empirical closure coefficients, but we argue that the method applied here involves quantities more directly related to the measurements and thus, since we make relatively few assumptions about the underlying physics, is likely to provide realistic scalings.

The proportion of mixing due to Langmuir circulation alone (that is, exclusive of other effects of non-breaking waves) is unknown, and while we do not know how prevalent it is in the ocean we do know that Langmuir cells are not always present. Since an efficient mixing due to wave motion as well as an efficient mechanism for wave attenuation have been observed in laboratory experiments with no reported Langmuir circulation (Dai et al., 2010; Babanin and Haus, 2009; Ölmez and Milgram, 1992), and this mechanism should in principle always exist, we make the assumption that Langmuir circulation is a minor contributor to the mixing. Thus we follow several other authors in applying a vertical distribution of turbulent kinetic energy that does not explicitly take into account Langmuir circulation, and it is to these approaches that we compare ours. On the other hand, our source of energy for mixing is derived from swell dissipation measurements, which means that even if Langmuir circulation is more prevalent than assumed here the total energy going into mixing will not be affected; in this case, our parameterisation can at least be useful as a first-order approximation. In the end, of course, neither a purely Langmuir nor a purely non-Langmuir formulation will perfectly capture the physics. Ideally, a Langmuir-based vertical scaling might be used in conjunction with the one proposed here, with the energy for turbulence production partitioned appropriately between the two mechanisms.

Previous approaches to non-breaking wave-induced turbulence have been applied to improve modelling of sea surface temperatures, sediment concentrations and climatic events. We have discovered some areas where the parameterisations could be improved upon, which we hope will lead to still more successful implementations. Below we describe two successful approaches and explain why we felt the need for devising yet another.

3.1 The B_v parameterisation of Qiao et al. (2004)

Qiao et al. (2004) recognised that ocean mixing is often underestimated in summer and postulated an additional source of mixing due to surface waves. Most prior attempts at implementing a wave-induced effect relied on wave breaking (see for example Chalikov and Belevich, 1993), but this has not been sufficient to explain the observed mixed-layer depths and surface temperatures. Qiao et al. (2004) sought to account for the disparity between models and observations by relating the new source of mixing to the wave spectrum. The new term was then directly added to the turbulent viscosity in the mixing scheme's turbulence model. Unlike turbulence generated by wave breaking, which is injected at the scale of the wave height, the non-breaking wave-induced turbulence is vertically distributed at the scale of the wavelength, comparable to the depth of the mixed layer.

Following Prandtl's mixing length hypothesis, they defined the wave-induced component of the turbulent viscosity ν_T as

$$B_\text{v} = \overline{l_z u'_z}. \tag{1}$$

This was developed by analogy with Prandtl's later definition of the kinematic turbulent viscosity,

$$\nu_\text{T} = c\sqrt{k}\,l \tag{2}$$

where k is the turbulent kinetic energy, l is a length scale (usually interpreted as the characteristic size of the large, energy-containing eddies) and c is a proportionality coefficient, usually related to the stability of the flow though sometimes set constant (see Wilcox, 1998).

In the equation for B_V the length l_z is related to the size of the water particle orbits due to the wave motion, and u'_z is the increment in vertical wave velocity over the distance l_z. Combining these definitions results in the following form for B_V (a full exposition is provided by Qiao et al., 2010):

$$B_V = \alpha \iint_{\kappa} E(\kappa) e^{2\kappa z} \mathrm{d}\kappa \frac{\partial}{\partial z} \left[\iint_{\kappa} \sigma^2 E(\kappa) e^{2\kappa z} \mathrm{d}\kappa \right]^{\frac{1}{2}} \quad (3)$$

where E is the wavenumber spectrum, $\boldsymbol{\kappa}$ is the wavenumber vector, σ is the angular frequency of the waves, z is depth and α is a tuning parameter, assumed to be of order one and set according to the observations. This formula embodies the assumption that the turbulence will be scaled by the wave motion and attenuate according to an exponential law. They applied it to different ocean mixing schemes as well as to model a more controlled laboratory experiment, reporting improved model fidelity in all cases.

The first published application was to the Princeton Ocean Model (Blumberg and Mellor, 1987), which uses the Mellor–Yamada turbulence model to drive mixing (Qiao et al., 2004, and also Qiao et al., 2010). This is a two-equation model, which means that it describes the evolution of the turbulence by way of two coupled transport equations: one for the turbulent kinetic energy and another for a quantity that may be related somehow to the length scale (Mellor and Yamada, 1982). Crucially, the turbulent viscosity in this kind of model is a function of both of these quantities, and in turn the transport equations both depend on the turbulent viscosity, as well as the turbulent kinetic energy and the length scale. Thus introducing the wave effect by adding directly to the turbulent viscosity adds a source of viscosity without corresponding quantities relating to the creation and destruction of turbulence energy, and this effectively amounts to modifying the turbulence scheme at a fundamental level. All turbulence models are calibrated empirically and therefore are open to modification in the light of new experimental data, but no recalibration seems to have been attempted. That notwithstanding, the addition of B_V did lead to a more accurate prediction of ocean mixing in the presence of waves. A more detailed description of the issue of modifying a two-equation model in this way is illustrated in Appendix A.

In a later application B_V was added to the turbulent viscosity in the KPP (K Profile Parameterisation) model (Wang et al., 2010), a popular mixing scheme in operational oceanography. It does not feature the kind of interdependence of parameters as two-equation turbulence models such as Mellor–Yamada do. The model produces a vertical profile for the turbulent viscosity, and this is used to determine the turbulent flux of a quantity. Similar to Qiao et al. (2004), the wave-induced turbulent viscosity was added to the KPP turbulent viscosity profile. Because it was unclear whether a wave effect had been implicitly included in the KPP model, they tuned B_V via the constant α, choosing a value which gave the most satisfactory match between modelled and observed data. Again, this pragmatic approach led to real improvements, significantly reducing errors in ocean simulations.

The "K profile" (K is the turbulent viscosity, just ν_T in our notation) is, for the purposes of the KPP model, divided into three separate regimes: two outer layers – a surface and bottom layer – and the ocean interior. The viscosity profile is matched at the boundaries between these layers, which are themselves defined based on a bulk Richardson number. Forgetting for the moment the bottom layer (in deep water this will not greatly concern us as the wave influence would be expected to vanish), the surface layer and the interior are formulated very differently. The former consists of the product of a cubic shape function and a velocity scale, which is a function of the Monin–Obukhov stability function, and the latter of simplified depth-dependent turbulent viscosities for heat, salinity and momentum. While merely adding B_V works, a preferable approach would be to reformulate KPP at a more fundamental level, reconciling the assumptions underpinning its development with the added influence of nonbreaking waves. This implies nothing less than replacing the Monin–Obukhov scaling with something more suitable to the ocean. Additionally, boundary conditions for the boundary layer, taking into account the wave-induced turbulence, would need to be adjusted, as would the Richardson number, which determines the boundary layer depth (although this last one would be done automatically if the Monin–Obukhov stability function were modified).

Since there does not appear to be a ready alternative to the Monin–Obukhov theory, for now we have to conclude that KPP is, strictly speaking, incompatible with the desire to model wave-induced turbulence. Indeed, this is not a new observation: D'Alessio (2002) considered Monin–Obukhov similarity theory and wave breaking, finding that when waves are breaking the Monin–Obukhov functions no longer describe the ocean boundary layer accurately. Yet he also remarked that despite this discrepancy, KPP was nevertheless successful as a mixing model. In his study, though, the deep mixing characterised by turbulence generated by the wave orbital motion was absent and, as he suggested, since wave breaking is confined to the near-surface layer, the difference could be insignificant.

A third application of the B_V parameterisation was to a coupled ocean–atmosphere climate model in simulations of the South Asian Summer Monsoon (Song et al., 2012). Here the mixing scheme used was that of Pacanowski and Philander (1981), which is much simpler than either a two-equation model or KPP. It consists of separate turbulent viscosities for momentum and scalars, which are empirically determined and dependent on a constant background value, in addition to a term based around the Richardson number. B_V was added to these viscosities, and once again the model results were much closer to the observations. Being a far less sophisticated mixing model, adding the wave-induced viscosity is unlikely to be as problematic from a theoretical standpoint, although we note that it was calibrated without waves and it is possible that including them might lead to different values of the constants, or a different form for the equations. On the other hand, given the controlled nature of the original calibrations, this may not make a large difference (Pacanowski and Philander, 1981).

Despite the issues we have illustrated, in all three studies big improvements were made in model performance. We note also that the B_v model was applied with considerable success in modelling a controlled laboratory experiment for temperature diffusion in a wave tank (Dai et al., 2010).

3.2 Pleskachevsky et al.'s (2011) wave shear frequency approach

Pleskachevsky et al. (2011) derived their model quite differently. Their approach dealt simultaneously with both a modification to the turbulent viscosity of the flow and an augmentation to the turbulence production by wave-induced shear. Both effects were incorporated in their mixing model, using a two-equation turbulence model. For geophysical models where horizontal gradients vary slowly and vertical velocities are small, the production term for free-shear flows may be parameterised as (see Eq. A1 for the full equation)

$$P = \tau_{ij} \frac{\partial U_i}{\partial x_j} \approx \nu_T M^2, \tag{4}$$

where M is the shear frequency. P thus represents the rate of production of turbulent kinetic energy by the mean current shear. They augmented the production by considering the addition to the shear from the waves, so P becomes $P + P_{\text{wave}}$. P_{wave} is the wave-induced shear production, parameterised as

$$P_{\text{wave}} = \nu_T E_{\text{AM}}^2 \overline{M}_{\text{wave}}^2 \tag{5}$$

and added to P. E_{AM} is the ratio of wave energy lost to dissipation (or, equivalently, the production of turbulence) and the total wave energy. The wave mean shear frequency, $\overline{M}_{\text{wave}}$, is defined a bit differently to the usual shear frequency M, involving not just horizontal shear but vertical shear as well, in order to take into account the orbital motion of the water under the influence of the waves. It is also averaged over a wavelength in order to match the scale of the mean currents (the wave motion is assumed periodic in a timescale greater than the turbulence, but less than the mean currents). The wave shear frequency is expressed as

$$M_{\text{wave}}^2 = \left(\frac{\partial U_{\text{wave}}}{\partial z}\right)^2 + \left(\frac{\partial W_{\text{wave}}}{\partial x}\right)^2. \tag{6}$$

The quantities U_{wave} and W_{wave} are component mean velocities of the wave motion, that is they represent the orbital motion with the turbulent fluctuations averaged out. The mean wave shear frequency is then the mean of M_{wave} over a wavelength.

To augment the turbulent viscosity in these expressions, they defined a wave-induced turbulent viscosity which they added to the turbulent viscosity ν_T; it is proportional to the wave mean shear frequency and the square of a length scale,

$$\nu_{\text{wave}} = l_{\text{wave}}^2 \overline{M}_{\text{wave}}. \tag{7}$$

In a similar vein to the B_v model, they assume that the length scale l_{wave} is equal to the amplitude of the water particle orbits, a function of depth. However, in contrast to the B_v model, Pleskachevsky et al. (2011) do not treat the full spectrum, and instead they work with bulk properties such as significant wave height, so they only deal with one value for l_{wave}.

One of the problems with this approach is the same with that of the first B_v implementation, which is that the wave-induced turbulent viscosity is added directly to the turbulent viscosity term in a two-equation model. However, unlike the B_v model, this wave shear model also includes a turbulence production term, which both creates an extra problem and an opportunity. On the one hand, now that we are not only directly augmenting the viscosity but producing turbulence, the viscosity is effectively being counted twice. On the other hand, with an explicit production term we should not need to invoke any explicit addition to the turbulent viscosity, so we ought to be able to rectify the problem by simply removing the viscosity term. Note that this in itself is not to say that the formula proposed for turbulent viscosity is incorrect, but instead that it is not appropriate in the context of a two-equation turbulence model, other than perhaps as a diagnostic tool to check that the production is indeed leading to the right value for the turbulent viscosity. In other kinds of models, where the turbulent viscosity is not explicitly dependent on the turbulent kinetic energy, it could still be a useful formulation.

We applied this model for the wave-induced turbulence into the GOTM one-dimensional mixing model in, so far as we know, the same manner as that described by Pleskachevsky et al. (2011). This meant using the parameters pre-selected in the "wave breaking" test case supplied by the model developers, and using the k–ϵ model after Rodi. Using their formulas, however, it became clear on running the model that the turbulence production due to the waves, Eq. (5), was entirely negligible, meaning that the mixing was completely due to the direct addition of the wave-induced turbulent viscosity – the very term we had intended to remove on theoretical grounds. It would appear that the ratio E_{AM}, which Pleskachevsky et al. (2011) set to 1.5×10^{-4}, is too small. They indicated, however, that the variable could take a range of values, anywhere from 1.0×10^{-3} to 1.0×10^{-7}, depending on wave steepness. As a test we tried the higher value, and indeed the wave-induced mixing, now without any direct augmentation to the turbulent viscosity, became very strong. This suggests that a recalibration might fix the model, though we elected to defer this task.

What we are left with, then, is a direct parameterisation for the turbulent viscosity. It is easily shown that this is in fact identical to the monochromatic form for B_v described by Qiao et al. (2010), provided that $\alpha = 1$. In deep water, this is

$$\nu_{wave} = B_v = A^3 \kappa \omega e^{3\kappa z}, \tag{8}$$

where A is the amplitude of the waves, and κ the wave number. This parameterisation was applied to a sediment suspension model and resulted in vastly improved modelling (Pleskachevsky et al., 2011); it has also been used to explain the under-prediction of mixed layer depth when waves are omitted from a mixing model (Toffoli et al., 2012).

3.3 Further comments

Despite some shortcomings we have discussed, the parameterisations explored in the previous section have led to improved model predictions, and it seems clear from previous studies that including some kind of extra turbulence due to the waves is preferable to leaving it out. Since the Pleskachevsky model as it stands is identical to B_v for monochromatic waves, the work described above effectively leaves us with only one model. This describes the turbulent viscosity rather than the production of turbulence, so it is not really suited for use in a two-equation model. We therefore feel that there is room for another method, one that calculates turbulence production and which will work naturally with these models, and many other Reynolds-averaged turbulence models. Since the generation of wave-induced turbulent kinetic energy must by definition involve a transfer of energy from the waves, the same parameterisation should predict a component of the wave dissipation. Thus in Sect. 4 we test an alternative model, developed in terms of wave dissipation through turbulence generation.

4 One-dimensional modelling of the stratified ocean

4.1 An amplitude-scaled turbulence production parameterisation

Babanin and Haus (2009) performed laboratory experiments with monochromatic waves and observed that the velocity spectra contained regions closely approximating the Kolmogorov interval. By measuring the spectrum at a depth just below that of the wave troughs, they determined a form for the dependence on amplitude of volumetric turbulent kinetic energy dissipation near the surface,

$$\epsilon = b a_0^p, \tag{9}$$

where a_0 is the surface wave amplitude, and $p = 3.0 \pm 1.0$ within the 95 % confidence limit, and $b = 300$. An earlier hypothesis (Babanin, 2006) on dimensional grounds supports the scaling with the cube of the amplitude, so p is set equal to 3. (This follows from a definition of a wave Reynolds number where the inertial forces scale with the square of amplitude,

$$Re = \frac{a V}{\nu} = \frac{a^2 \omega}{\nu}, \tag{10}$$

where V is the orbital velocity of the fluid particles, a is the amplitude of the motion (which scales exponentially with depth) and ν is the molecular viscosity of water.)

Since the value of $p = 3$ is assumed to be constant, the quantities in expression Eq. (9) are dimensional. The dimensionless form was described by Babanin (2011), and in it b becomes

$$b = b_1 \kappa \omega^3, \tag{11}$$

where b_1 is a dimensionless parameter initially assumed to be constant. Turbulence was observed not more than once out of every ten wave periods for the steepest waves in the experiment, which were just below the breaking threshold. The value of $b = 300$ was determined for these isolated patches of turbulence, so over time the average value would be less than a tenth of this. (This value and the intermittent nature of the turbulence were found to be consistent with numerical experiments by Babanin and Chalikov, 2012.) Since the waves in the experiment were steep it was further suggested that for ocean swell the number could be reduced further.

Ardhuin et al. (2009) had measured swell dissipation rates in the ocean and attributed them to a negative wind stress. By assuming instead that turbulence production in the ocean is the dominant sink for wave energy in swell, Babanin (2011) compared these swell dissipation rates with a depth-integrated form for the turbulence dissipation. The fit was good, except that it tended to over-estimate the swell dissipation, so a revised estimate of b was that it should be perhaps one twentieth of its laboratory-measured instantaneous value. Thus reckoned, b_1 was chosen to be 0.002. (We note that more recent measurements by Young et al., 2013 suggest a slightly smaller value of between 0.0014 and 0.0017, not too dissimilar from the results of Babanin, 2011, especially given that these values assume a constant value for b_1.) Assuming a steady state where the turbulence dissipation and production are equal, the final form for the turbulence production due to the wave orbital motion is

$$P_w = b_1 \kappa \omega_p^3 \frac{H^3}{8} e^{3\kappa z}, \tag{12}$$

where H is the wave height, and z is measured upwards from the mean surface. On closer inspection the data appears to exhibit a quadratic dependence on wave steepness, thus b_1 becomes

$$b_1 = 5 \left(\kappa \frac{H}{2}\right)^2. \tag{13}$$

This form turns out not to match the lab experiments – it is far too high, in fact. There are several possible reasons for this: firstly, in the ocean there is always turbulence, whereas in the lab the water was left to rest. Since the rate of production of turbulence depends on the turbulent viscosity we would expect that the rate of production of oceanic turbulence would be much higher than the lab experiment for otherwise equivalent waves. Secondly, it is possible that the wave orbital motion is not the only, nor even the primary, source for swell dissipation. Thirdly, we have assumed that the ocean waves are monochromatic. The tank experiments were for monochromatic waves, but it is not entirely clear how a realistic ocean spectrum would behave. A fourth reason is, perhaps obviously, that the quadratic dependence is only really valid within the range of the data used to find it. All that really remains to be said is that the dimensional coefficients to b_1 were deduced independently of the experiments and are consistent with them, so we do not anticipate the problem to lie there.

The formula was adapted from experimental data for monochromatic waves, and in using it we assume that the wave spectrum is narrow enough such that we can use characteristic wave parameters such as significant wave height and peak period instead of specifying the production based on a complete wave spectrum. In general, particularly for the case of swell with little wind stress, this is a good approximation; however we note that, due to the cubic amplitude dependence, in a random wave field it may be that the largest waves, and not the significant waves, dominate the mixing.

The formulation of b_1 could well be enhanced with more swell data, but in its favour this model is based solely on the properties of the waves (there is no wind dependence, for example), it is scaled with the complete orbital motion (rather than the Stokes drift or some other property) and it models the turbulence production explicitly – and exclusively – rather than the turbulent viscosity. We note that the model, with a different (and constant) b_1 based on much earlier estimations of swell dissipation, had been derived using different assumptions by Bowden (1950). We believe that, so far, this is the most likely formulation to succeed in a general ocean mixing model, especially, since it models turbulent kinetic energy production, when multiple sources of turbulence are introduced. There is a noteworthy deficiency, however, which is that the present model does not depend on turbulent viscosity. Indeed, in order to do so we would need either a more intimate knowledge of the physics which is thus far lacking, or a set of careful measurements showing how the rate of turbulence production changes with turbulent viscosity, also unavailable. One possible approach is to assume the validity of the wave mean shear approach (Eq. 6) and attempt to find an empirical closure coefficient. On the other hand, the observational data from Young et al. (2013) suggest that the parameterisation does, on the mean, a good job of modelling the rate of swell dissipation in the ocean, so for ocean modelling applications it should be adequate. As the model time step is decreased, however, the model would probably benefit from this kind of refinement.

4.2 The mixing model and numerical experiments

To simulate the ocean mixing we use the one-dimensional mixing model GOTM (Umlauf and Burchard, 2005; Umlauf et al., 2011). This model is easy to set up, modify and use with a variety of different turbulence closure schemes and inputs, and can be coupled to other models, such as circulation models. It is not, however, a general turbulence model because the momentum transport equations are simplified so that they can efficiently model the oceanic boundary layer (Burchard, 2002) (if this were not the case the model would also have to be three-dimensional). This places some limits on the kind of approach needed to introduce a wave-induced turbulence production term into the mixing scheme, in particular with the method of Pleskachevsky et al. (2011) which uses the wave mean shear frequency, $\overline{M}_{\text{wave}}$, and therefore requires a horizontal shear as well as a vertical one. This means that we must add turbulence production due to $\overline{M}_{\text{wave}}$ at a level higher, where it is completely parameterised based on mean wave properties rather than directly calculated from the flow.

While we do not anticipate that simply adding another source to the production term will be a problem, it is worth at least noting that this approach implicitly assumes that wave-induced turbulence is homogeneous on the scale of the waves. In their large eddy simulations Babanin and Chalikov (2012) showed that the turbulence is highly dependent on the gradients of the orbital velocities. Thus turbulence production occurs mostly in a narrow column immediately below the wave peaks. The turbulence may also be intermittent, as reported by Babanin and Haus (2009). This means that in order to include it in the GOTM model it must be presented in a space and time-averaged form, where the space and timescales are greater than the wavelength and period. In our application of the parameterisations this has been implicitly assumed, and is no different in that way to parameterising, say, turbulence from intermittent wave breaking.

The model setup was, just as with Pleskachevsky et al. (2011), based on the *wave breaking* scenario downloaded from the GOTM website. We removed the surface momentum flux which is meant to represent the effect of breaking waves and, in addition, changed the surface boundary condition from *tke injection* (intended for breaking waves) to *logarithmic law of the wall*. We tested two parameterisations: B_v (the only one remaining after we discarded the others considered in Sect. 3.3) and the new one described in Sect. 4.1.

In the context of a turbulence model, we assume that the wave orbital shear behaves in a similar way to the mean shear, so for the present parameterisation the wave-induced turbulence production Eq. (12) was added directly to the shear production term, which is calculated in the GOTM `production` subroutine. Because we are using a two-equation turbulence model this is the only modification that should be made, apart from adjusting closure coefficients, especially the one relating to production of dissipation. To test how sensitive the model would be to this we varied this coefficient as far as the numerics would allow without generating errors and the differences, while measurable, were not so great as to overturn the main conclusions of this study; hence, without any field or laboratory measurements to guide us, we kept the default value. The B_V model parameterises the turbulent viscosity directly, so is added to the viscosities calculated in the `kolpran` routine; we add the same value for B_V, with $\alpha = 1$, to the turbulent viscosities for momentum, heat and salinity. By eliminating all other sources of turbulence production this renders the turbulence transport equation irrelevant in the mixing model. More care would have to be taken if B_V were used and another source of turbulence production were present, since the turbulent viscosities would not simply be additive. As we have noted above, our parameterisation lacks dependence on turbulent viscosity, but we do not anticipate that this will be of great importance in ocean applications except at the very highest model resolutions.

The two-equation turbulence scheme used is the Rodi k–ϵ formulation. Tests with other two-equation models yielded results which differed a little, but compared with the difference between incorporating the wave-induced effect or omitting it entirely these differences were negligible. Apart from the waves there was no other forcing – no input or output of heat, no internal waves, surface wave breaking, biology, wind, currents or external sources of pressure.

The initial conditions consisted of prescribed salinity and temperature profiles which we sourced from the Global Temperature and Salinity Profile Program (GTSPP, 2006). We wanted to examine mixing due to non-breaking wave generated turbulence in isolation from other mixing mechanisms, so we sought a stable, stratified profile at mid-to-high latitudes in summer, so as to avoid the issues with convective mixing common in wintertime. It was also desirable to have something in deep water, well away from continents, as some of the wavelengths we tested were over 600 m. The profile we finally chose came from station 8541272, located some distance west of Chile at 115.9200° W, 45.8840° S, and measured at 03:24 UTC on 3 January 2010. The depth of the measured profile was 1813 m, and the model was set up to run with that depth of water (though for clarity we only show down to 250 m in the plots).

The results of the mixing on the temperature profile are shown in Figs. 1–4 (a similar effect may be seen on the salinity profiles, but we omit the plots). In each the dashed line is the original temperature profile, and the solid lines are, from darkest to lightest, in two-hour intervals, the temperature profiles as the water column is mixed by turbulence induced by the waves. We ran the experiments for several different wave heights and wave numbers. In Fig. 1 the simulations were performed using the parameterisation Eq. (12) for a set of wave numbers and wave heights which sit within the range of data used to develop the parameterisation. The wave heights are 3, 4 and 5 m, and the wave numbers 0.014 and 0.02, corresponding to wave lengths of 449 and 314 m

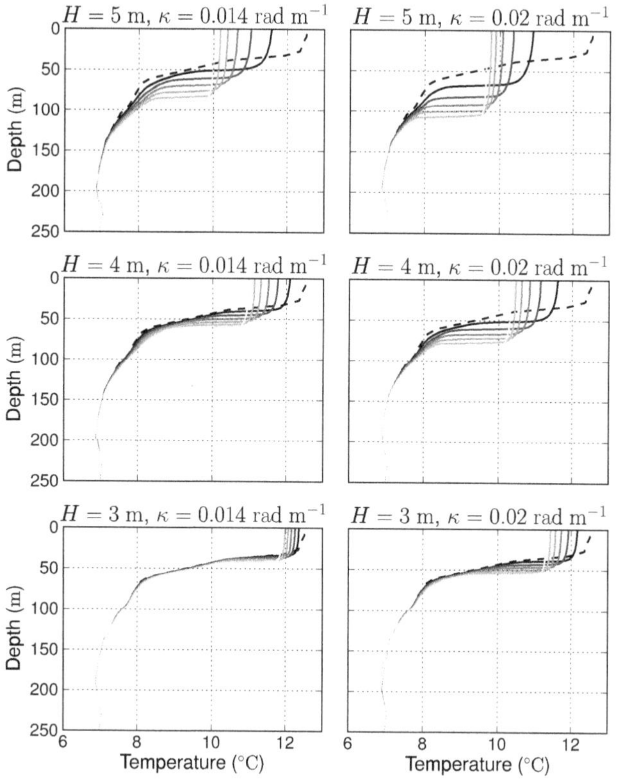

Fig. 1. Mixing in GOTM due to non-breaking wave-induced turbulence using the parameterisation for wave dissipation from Babanin (2011). The wave heights and wave numbers correspond to typical values from the data of Ardhuin et al. (2009). The dashed line is the initial temperature profile, and the solid lines for the profile at two-hour intervals, with lighter lines indicating more time having elapsed.

respectively. In Fig. 2 a wider range is used, where we have assumed that the parameterisation still holds. For this set of experiments the three wave heights are 2, 5 and 10 m and the two wave numbers 0.1 and 0.01 rad m^{-1}, corresponding respectively to wavelengths of 63 and 628 m. Note that ten-metre waves with a wave number of 0.1 are just above the breaking threshold, meaning that in any real situation there would be near-surface turbulence generated by wave breaking as well. Since we are investigating a source of turbulence in isolation (no currents, wind stress or wave breaking) these tests can be thought of as instructive, rather than a complete description of a column of ocean.

The results suggest that even for small H the wave motion may contribute substantially to the mixing. For wave heights of 5 m and a wave number of 0.02 it deepens the mixed layer by as much as a factor of four, and reduces the surface temperature by about three degrees after twelve hours. The largest part of the effect occurs in the first few hours, and significant cooling occurs after only minutes (not shown). Even the smaller, shorter waves have an impact on the mixed layer. Assuming an appropriately shallow mixed layer, an effect of

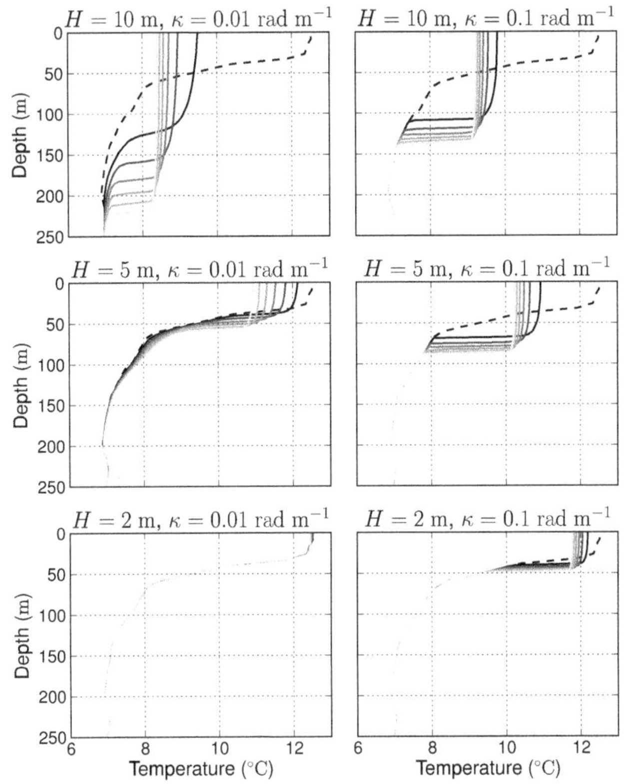

Fig. 2. As for Fig. 1 but for a wider range of wave heights and wave numbers.

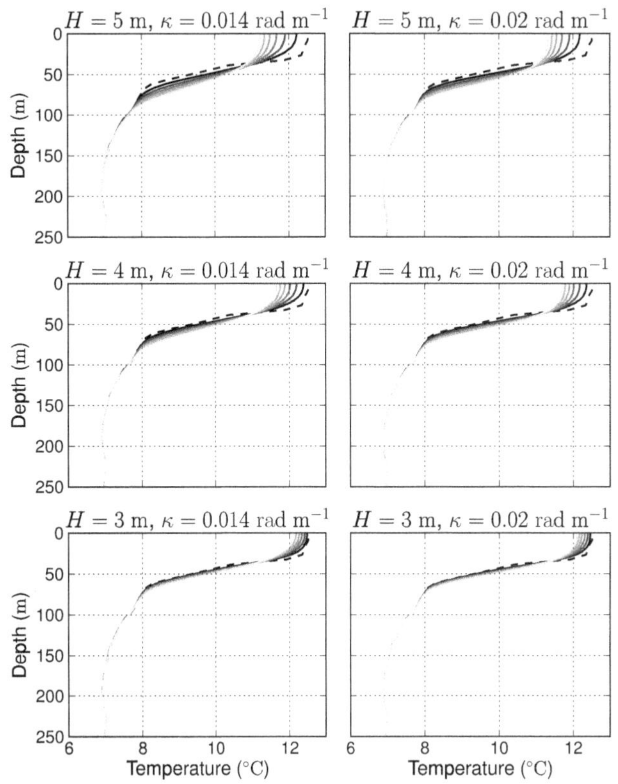

Fig. 3. Mixing in GOTM due to non-breaking wave-induced turbulence using the B_V parameterisation for monochromatic waves with $\alpha = 1$ from Qiao et al. (2010). The range of wave heights and wave numbers is as for Fig. 1.

this magnitude could strongly influence the intensity of tropical cyclones, especially where the sea surface temperature is close to the threshold for tropical cyclone generation. It may also be seen that the mixing extends much deeper than we would expect from wave breaking, which generates turbulence only to depths of the order of the wave height.

For the B_V turbulent viscosity parameterisation Eq. (8), the results corresponding to the wave numbers and heights in Fig. 1 are shown in Fig. 3, with results for the wider range of wave parameters plotted in Fig. 4. The mixing is weaker, especially for the steeper waves, and qualitatively different: instead of a clear thermocline the temperature gradient is reduced. (This degradation of the thermocline was observed and discussed in more detail by Martin et al., 2011.) Even so, the surface temperature decrease is of the same magnitude as the new parameterisation with smaller waves, which are not far outside the range of the data of Ardhuin et al. (2009), and especially for shorter times; this is encouraging given that the two parameterisations tested here are for different quantities (one is the rate of production of turbulence, the other of turbulent viscosity), and were developed along different theoretical lines and validated with different kinds of observational data. For much larger waves, though, the results diverge greatly, and for the short waves of $\kappa = 0.1$ the B_V turbulent viscosity is insufficient to affect the mixed layer depth, even for large wave heights – this is in no small part due to the initial depth of the mixed layer, and we would presumably have seen an effect if it were even shallower than it is.

The qualitative differences can be explored even further when we look at the plots for the turbulent viscosity in Fig. 5. For the B_V parameterisation (the dotted curve in the plot) this is calculated directly from the wave parameters, which here are constant, and therefore do not change with time, while for the parameterisation of turbulence production the turbulent viscosity is calculated from the transport equations for turbulent kinetic energy and dissipation (see Eq. A4 in Appendix A). Its value depends not just on the wave-induced turbulence production but also on the stratification of the water column (hence the small bulge between 50 and 100 m depth in the plots, which corresponds to the end of the thermocline where the lower stability leads to increased mixing).

The viscosities generated by the two-equation model are a very different shape to those generated by the B_V equation. While the B_V model scales exponentially from a maximum at the surface, the two-equation model calculates the turbulent viscosity such that it is zero at the surface and there is a maximum at some depth. This is what we would expect since the size of the eddies near the surface should be constrained by the surface, leading to a reduced ability to mix.

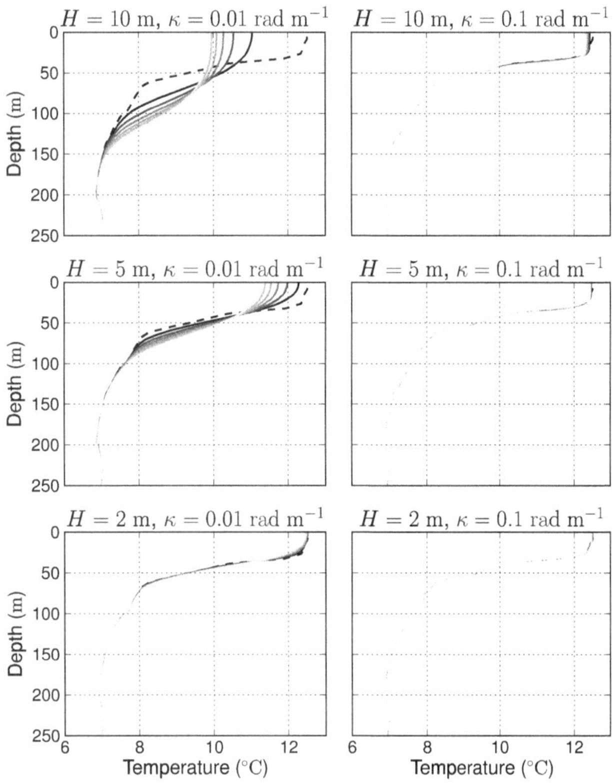

Fig. 4. As for Fig. 3 but for a wider range of wave heights and wave numbers.

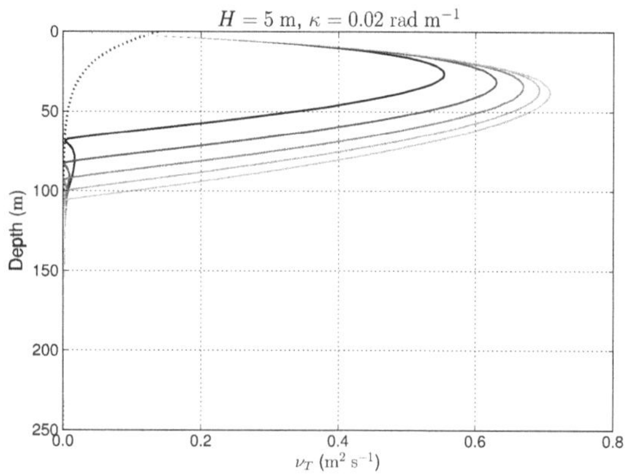

Fig. 5. Turbulent viscosity due to non-breaking wave-induced turbulence for a wave height of 5 m and a wave number of 0.02 rad s^{-1}. The dotted line is the constant-in-time B_V parameterisation, while the solid lines are calculated from the parameterisation based on wave dissipation (Babanin, 2011) using the k–ϵ model of Rodi (1987). The darkest line is that calculated two hours after the simulation started, with subsequent two-hour intervals indicated by progressively lighter lines.

It should be noted that this characteristic is a result of using a logarithmic surface boundary condition in the mixing model, which was primarily intended for vertical shear generated turbulence, but the result is in any case more realistic than a viscosity that increases continuously to the surface. Perhaps when more is known about the physics of this source of turbulence a more appropriate boundary condition may be formulated for inclusion in the mixing model.

Extrapolating from these academic tests to more realistic ocean modelling applications is not necessarily straightforward, and we offer the following observations to explain our results more clearly in that context, particularly with regard to those situations where the two parameterisations produce disparate results. In the present idealised tests the parameterisations are used alone with no other sources of turbulence (except for a very small background turbulence required by the k–ϵ model). In a full ocean simulation they will be used in conjunction with other sources of turbulence; for the B_V parameterisation this will mean that the value of B_V will be added to whatever turbulent viscosity already exists, whereas our new parameterisation will add to the rate of turbulence production. The net result will be quite different for two reasons: firstly, adding turbulence sources does not result in an analogous addition of turbulent viscosities – put simply, doubling production does not double viscosity (an inspection of the turbulence transport equations in Appendix A will make this clear). In more realistic applications, then, B_V may be more effective than it seems based on some of the results shown in the figures. The second reason is that because the B_V profile increases near the surface rather than decreasing to zero as is normally expected, it can help mix the very top layer of ocean into the deeper mixed layer even if mixing near the surface by currents alone is relatively inefficient.

5 Conclusions

While there are several different approaches available to modellers wishing to incorporate the effect of non-breaking wave-induced turbulence in their forecasting models, on closer inspection several of these cannot deal with situations of zero wind stress and here, in the context of wave-only mixing, had to be discarded from consideration. Leaving aside models invoking Langmuir circulation, which seems to represent only a part of the process, we identified two models which have been used successfully in modelling studies. One of these, the model proposed by Pleskachevsky et al. (2011), requires modification and at present represents a special case of the B_V model. This latter model, first proposed by Qiao et al. (2004), models not the turbulence production but the turbulent viscosity. As an alternative to these we have proposed using the turbulence dissipation measurements of Babanin and Haus (2009) and the swell dissipation measurements of Ardhuin et al. (2009) as a source of turbulence production. Our tests suggest that, in an environment without any other sources of turbulence, the two models give broadly similar results in some situations, though with some important qualitative differences, especially with regard to the position and slope of the thermocline. However in other, more extreme situations their behaviour diverges considerably. What is needed now is more laboratory data in order to refine the theory, which is still not completely understood, and a good data set of simultaneous field measurements against which model results can be compared.

For the broader question of modelling the ocean and the atmosphere, our rather academic simulation was designed to test the efficacy of this source of turbulence in deepening a shallow mixed layer, and we have deliberately left out other sources of turbulence to get a measure of its effect in isolation. Of course in the ocean several turbulence sources coexist, including tangential wind stress, internal wave breaking, currents and the other surface wave-related sources: wave breaking and Langmuir circulation (though this latter source likely has been implicitly included in our model, at least in terms of the total energy), and a physical description of the ocean–atmosphere system is impossible without accounting for large-scale three-dimensional effects. Be that as it may, our simulation results, in the absence of surface fluxes and other sources of turbulence, suggest that the turbulence generated by the wave orbital motion, as reckoned from assuming that it is the primary mechanism for swell decay, can deepen an initially shallow mixed layer and reduce sea surface temperatures by, potentially, several degrees. Even a much smaller change would be significant in many modelling and forecasting applications, which indicates that it ought to be featured in mixing models – especially when modelling scenarios where there exists persistent or large-amplitude swell. This also implies that a wave model should be a component in any coupled model system where ocean–atmosphere feedbacks are present. This finding aligns well with modelling studies by Jacobs (1978), Qiao et al. (2004, 2010), Wang et al. (2010), Huang and Qiao (2010), Huang et al. (2011) and Song et al. (2012). All showed improved correspondence between observations and models with the inclusion of turbulence attributed to the wave orbital motion, regardless of the details of the method chosen to parameterise it. We would anticipate that the model we have proposed, because of its more general approach, should aid future modelling studies.

As a final note, we would like to point out that from a modelling perspective the differences between the present model and other models such as B_V, and even models based on Langmuir circulation such as that by Kantha and Clayson (2004), are more theoretical than practical at this stage. They all invoke waves as a source of turbulent kinetic energy and a way of distributing it through the water column, so in a sense they are all equivalent even if they are motivated by different physical assumptions. While the scaling in the present model is derived from experiments and theoretical assumptions which may not be always be applicable in the open ocean, the source of energy is, uniquely, explicitly quantified through measured swell dissipation, providing at least a cap on the available energy for mixing. Our 1-D tests when restricted to the regime for the available swell dissipation data (Figs. 1 and 3) indicate that the resultant mixing from the present model is comparable, though greater, than that produced by the B_v model, which has been successfully applied in several different scenarios. Further research is needed to determine both the dominant physics of the mechanism and the partition between this and other possible sources of swell dissipation, such as that invoked by Ardhuin et al. (2009).

Appendix A

Two-equation models and artificially augmenting the turbulent viscosity

While there are many kinds of turbulence models, the so-called two-equation models occupy a special position in the modelling community. They are far more computationally efficient than large eddy simulations (which are too costly for many geophysical applications), and unlike algebraic and one-equation models they are "complete", which means that no prior knowledge of the flow need be had. They are the simplest such complete models, and their performance is not much hindered by their simplicity (Wilcox, 1998). There are a good number of formulations, but they all consist of two interdependent transport equations, one of which describes the evolution of the turbulent kinetic energy and the other that of a length-scale determining variable. The turbulent kinetic energy equation for an incompressible fluid takes the following form:

$$\frac{\partial k}{\partial t} + U_j \frac{\partial k}{\partial x_j} = 2\nu_\mathrm{T} S_{ij} \frac{\partial U_i}{\partial x_j} + \frac{g}{\rho} \frac{\nu_\mathrm{T}}{\sigma_\mathrm{T}} \frac{\partial \rho}{\partial x_j} + \frac{\partial}{\partial x_j}\left(\frac{\nu_\mathrm{T}}{\sigma_k}\frac{\partial k}{\partial x_j}\right) - \epsilon, \quad \text{(A1)}$$

where the rate-of-strain tensor is defined as

$$S_{ij} = \frac{1}{2}\left(\frac{\partial U_i}{\partial x_j} + \frac{\partial U_j}{\partial x_i}\right). \quad \text{(A2)}$$

Although some terms are more or less important depending on the application, this is the equation generally used (in simplified form) in geophysical turbulence modelling, and its essential features can be derived from the Reynolds stress equation. On the left-hand side we have the material derivative of the turbulent kinetic energy, k. On the right-hand side the first term is the production of turbulent kinetic energy due to shear stress in the fluid; the second term is the production due to buoyancy, where ρ is the potential density; the third term is a gradient diffusion term; and ϵ is the dissipation. ν_T is the turbulent viscosity for momentum, and for scalars determining the buoyancy (salinity and temperature) the diffusivity is proportional to ν_T, hence the scaling parameter σ_T. The closure coefficient σ_k is set empirically.

Dimensional arguments may be used to determine that

$$\epsilon \sim \frac{k^{\frac{3}{2}}}{l} \quad \text{(A3)}$$

where l is a length scale, usually interpreted to be the characteristic size of the largest eddies. Further dimensional analysis provides an expression for the turbulent viscosity:

$$\nu_{\mathrm{T}} = c_{\mu} \frac{k^2}{\epsilon}, \tag{A4}$$

where c_{μ} is another empirical parameter. From substituting Eq. (A3) into Eq. (A4), it becomes clear that l and ϵ are interchangeable quantities, since ν_{T} depends on k and only one of l or ϵ. It is also understood that they vary with the flow. The motivation behind a two-equation model is to construct a second transport equation in order to determine the length scale l, which in turn enables us to fully determine the turbulent viscosity and the turbulent kinetic energy transport. The primary difference between the various two-equation models boils down to the choice of the variable used for the second transport equation, be it, for example, the dissipation rate as in the k–ϵ models, the specific dissipation rate, ω, as in k–ω models, or the product kl, as in the Mellor–Yamada model found in the Princeton Ocean Model used by Qiao et al. (2004). Any variable can in principle be used as long as it can unambiguously determine the length scale, although all of these models for the second equation are empirical, so they can differ substantially in both form and behaviour.

The dissipation transport equation in the Rodi k–ϵ model used in the numerical experiments in Sect. 4.2 is (Rodi, 1987):

$$\frac{\partial \epsilon}{\partial t} + U_j \frac{\partial \epsilon}{\partial x_j} = \frac{\partial}{\partial x_j}\left(\frac{\nu_{\mathrm{T}}}{\sigma_\epsilon}\frac{\partial \epsilon}{\partial x_j}\right) + \frac{\epsilon}{k} \left[2 C_{1\epsilon} S_{ij} \frac{\partial U_i}{\partial x_j} + C_{3\epsilon} \frac{g}{\rho} \frac{\nu_{\mathrm{T}}}{\sigma_{\mathrm{T}}} \frac{\partial \rho}{\partial x_j} - C_{2\epsilon}\, \epsilon\right] \tag{A5}$$

where $C_{1\epsilon}$, $C_{2\epsilon}$, $C_{3\epsilon}$, σ_ϵ are empirical constants. Note the existence of production terms here too; the right-hand side is analagous to that of Eq. (A1). Naturally any additional turbulence production due to waves must be present here just as for the equation for k.

Now the main point we wish to make relates to Eq. (A4), which expresses the turbulent viscosity in terms of the turbulent kinetic energy k and the dissipation ϵ. Both of these quantities vary according to the rather complicated transport Eqs. (A1) and (A5), and both of the transport equations are functions of these quantities both explicitly and through the turbulent viscosity, so to simply add wave-induced turbulent viscosity to that obtained from other aspects of the flow (such as the current shear) not only amounts to effectively reformulating the equations, but it results in a kind of double counting of turbulent viscosity. Adding an extra source of turbulence production should be sufficient. This production term will appear in both the transport equations for turbulent kinetic energy and the dissipation (or another length-scale determining variable). Following Kantha and Clayson (2004), the only other change that one should wish to make would be to adjust the empirical coefficients. It is possible that this sort of approach might help to further improve the model presented in this paper, but it should be noted that in their parameterisation the augmentation to the length scale was necessary, since Stokes drift on its own adds relatively little to the current shear, and the primary production mechanism in their model, Langmuir circulation, involves length scales rather larger than the wave orbital motion. Hence we do not anticipate that in our model the adjustment will be as critical.

Appendix B

The experimental paper by Beyá et al. (2012)

When Beyá et al. (2012) attempted to replicate the experiment of Babanin and Haus (2009) they did not find any evidence of wave-induced turbulence. Based on their observations and other arguments they concluded that no such effect exists. Since that paper has been gaining traction we felt that it was appropriate to explain why it does not invalidate the hypothesis of non-breaking wave-induced turbulence.

Beyá et al. (2012) appear to have confused the papers by Babanin and Haus (2009) and Babanin (2006). They claim that Babanin and Haus (2009) established the Reynolds number threshold of 3000 for transition to fully developed turbulence, whereas it was actually done by Babanin (2006). Babanin and Haus (2009) investigated intermittent turbulence and could not confirm the threshold value either. Also, it was Babanin (2006) who conducted the laboratory experiments to illustrate the effect of fully developed turbulence on the dissolution of dye, but the experiments by Beyá et al. (2012) replicated the experimental conditions of Babanin and Haus (2009) where only intermittent turbulence was observed. Thus they did not and could not have seen the rapid dissolution of dye observed by Babanin (2006).

Secondly, Beyá et al. (2012) argue that the fact that Stokes waves are observed to high precision precludes the existence of vorticity, and hence turbulence. This is an apparent misreading of the basic theory of viscous wave motion where the viscosity, which leads to vorticity, reveals itself in the imaginary part of the solutions. This part leads to a slow decay of Stokes waves, but does not affect in any way the shape of such waves (see for example Kinsman, 1965).

Thirdly, the experiment conducted by Beyá et al. (2012) itself has serious shortcomings. In their experiment they observed dye in the presence of waves over 5 wave periods, for wave trains of 3 different steepnesses. The dye will disperse due to molecular diffusion even in the absence of turbulence, thus in order to determine the existence of turbulence a comparison between the rates of dissolution of dye with and without the presence of waves would be required. They made no such comparison, and indeed no measurement of molecular diffusion was attempted. Instead, the existence or non-existence of turbulence was decided by a more subjective criterion, being whether or not the experimenters had judged the dye to have sufficiently dispersed. The weak, intermittent turbulence observed by Babanin and Haus would only be expected to create a visual difference after several hundred periods (see also Dai et al., 2010). Beyá et al. (2012) only observed their dye over 5–10 wave periods, so they could not have seen any significant effect visually. On the other hand, they did observe a reduction of thickness of their dye line, but their explanation contradicts their own data. In their Fig. 7, vertical profiles of dye line thickness are shown at zero time and after 5 wave periods, for three wave steepnesses: 0.17, 0.21 and 0.24, from the surface down to 7 cm below the surface. The dye line became twice as thin at the 6–7 cm level over the 5 periods, but notably not at the surface and not for the 0.21 steepness case.

Beyá et al. (2012) explain the observed reduction of thickness by the effect of Stokes drift stretching the dye line. This is not supported by any estimates of expected thinning by this mechanism, but should it be the case then the thinning cannot be increasing away from the surface where the Stokes drift is decreasing exponentially. Furthermore, the Stokes drift could not have influenced the dye thickness at steepnesses 0.17 and 0.24 without affecting the dye at steepness 0.21. On the contrary, Fig. 7 of Beyá et al. (2012) much more likely reflects the impact of an intermittent turbulence in their experiment, and not the Stokes drift. Unless this figure is upside down, however, this turbulence could not have come from the waves and must have come from the bottom.

Finally, Beyá et al. (2012) comment that their observations of velocity time series do not "show any evidence of turbulence fluctuations", but they do not substantiate the claim. The original thesis (Beyá, 2011), however, provides more details on their observations of kinematics by acoustic doppler velocimeters (ADVs). The sampling frequency of the ADV was 50 Hz, which means that the Nyquist frequency of 25 Hz was below most of the frequency range of the Kolmogorov interval observed by Babanin and Haus (2009). More importantly, the Kolmogorov interval as described by Babanin and Haus only appeared over a fraction of the wave period and cannot be seen if the Fourier Transform is applied to the continuous time series. Beyá et al. (2012) did not attempt to locate segments of the intermittent turbulence, and therefore they could not have seen the Kolmogorov spectra and the turbulent oscillations even if their sampling frequency and resolution had been high enough.

Acknowledgements. The authors would like to thank Andrey Ganopolski and Dmitry Chalikov for useful discussions. We are also grateful to Alexander Benilov and an anonymous reviewer whose comments helped to improve the paper. This research was supported by the Australian Research Council (ARC) (grant nos. LP0883888 and DP1093517).

Edited by: Y. Troitskaya

References

Anis, A. and Moum, J. N.: Surface wave-turbulence interactions: scaling $\epsilon(z)$ near the sea surface, J. Phys. Oceanogr., 25, 2025–2045, 1995.

Ardhuin, F. and Jenkins, A. D.: On the Interaction of Surface Waves and Upper Ocean Turbulence, J. Phys. Oceanogr., 36, 551–557, 2006.

Ardhuin, F., Chapron, B., and Collard, F.: Observation of swell dissipation across oceans, Geophys. Res. Lett., 36, L06607, doi:10.1029/2008GL037030, 2009.

Ardhuin, F., Rogers, E., Babanin, A. V., Filipot, J.-F., Magne, R., Roland, A., van der Westhuysen, A., Queffeulou, P., Lefevre, J.-M., Aouf, L., and Collard, F.: Semiempirical Dissipation Source Functions for Ocean Waves, Part I: Definition, Calibration, and Validation, J. Phys. Oceanogr., 40, 1917–1941, 2010.

Babanin, A.: Breaking and Dissipation of Ocean Surface Waves, Cambridge University Press, 2011.

Babanin, A. V.: On a wave-induced turbulence and a wave-mixed upper ocean layer, Geophys. Res. Lett., 33, L20605, doi:10.1029/2006GL027308, 2006.

Babanin, A. V. and Chalikov, D.: Numerical investigation of turbulence generation in non-breaking potential waves, J. Geophys. Res., 117, C00J17, doi:10.1029/2012JC007929, 2012.

Babanin, A. V. and Haus, B. K.: On the existence of water turbulence induced by nonbreaking surface waves, J. Phys. Oceanogr., 39, 2675–2679, 2009.

Benilov, A. Y.: On the turbulence generated by the potential surface waves, J. Geophys. Res., 117, C00J30, doi::0.1029/2012JC007948, 2012.

Benilov, A. Y., McKee, T. G., and Safray, A. S.: On the vortex instability of linear surface wave, Numerical Methods in Laminar & Turbulent Flow VIII, Pineridge Press, UK, 1323–1334, 1993.

Beyá, J. F.: On the interaction of ocean waves and turbulence, Master's thesis, The University of New South Wales, 2011.

Beyá, J. F., Peirson, W. L., and Banner, M. L.: Turbulence beneath finite amplitude water waves, Exp. Fluids, 52, 1319–1330, 2012.

Blumberg, A. F. and Mellor, G. L.: A Description of a Three-Dimensional Coastal Ocean Circulation Model, in: Three Dimensional Coastal Ocean Models, edited by: Heaps, N. S., American Geophysical Union, 1–16, 1987.

Bowden, K. F.: The effect of eddy viscosity on ocean waves, Philosoph. Mag., 41, 907–917, 1950.

Burchard, H.: Applied Turbulence Modelling in Marine Waters, Springer Verlag, Berlin, Heidelberg, 2002.

Chalikov, D. V. and Belevich, M. Y.: One-dimensional theory of the wave boundary layer, Bound.-Lay. Meteorol., 63, 65–96, doi:10.1007/BF00705377, 1993.

Dai, D., Qiao, F., Sulisz, W., Han, L., and Babanin, A.: An Experiment on the Nonbreaking Surface-Wave-Induced Vertical Mixing, J. Phys. Oceanogr., 40, 2180–2188, 2010.

D'Alessio, J. D.: Concerning the influence of wave breaking on Monin–Obukhov similarity theory, Acta Mechanica, 157, 1–14, 2002.

Dobroklonskiy, S. V.: Eddy viscosity in the surface layer of the ocean and waves, Doklady Academii Nauk SSSR, 58, 1345–1348, 1947.

Francis, J. R. D., Stommel, H., Farmer, H. G., and Parson Jr., D.: Observations of Turbulent Mixing Processes in a Tidal Estuary, Tech. rep., Woods Hole Oceanographic Institution, Woods Hole, Massachusetts, submitted to the Office of Naval Research, 1953.

GTSPP: US National Oceanographic Data Center: Global Temperature–Salinity Profile Programme, US Department of Commerce, National Oceanic and Atmospheric Administration, National Oceanographic Data Center, Silver Spring, Maryland, http://www.nodc.noaa.gov/GTSPP/ (last access: 7 June 2011), 2006.

Huang, C. J. and Qiao, F.: Wave–turbulence interaction and its induced mixing in the upper ocean, J. Geophys. Res., 115, C04026, doi:10.1029/2009JC005853, 2010.

Huang, C. J., Qiao, F., Song, Z., and Ezer, T.: Improving simulations of the upper ocean by inclusion of surface waves in the Mellor–Yamada turbulence scheme, J. Geophys. Res, 116, C01007, doi:10.1029/2010JC006320, 2011.

Jacobs, C. A.: Numerical simulations of the natural variability in water temperature during BOMEX using alternative forms of the vertical eddy exchange coefficients, J. Phys. Oceanogr., 8, 119–141, 1978.

Jeffreys, H.: On turbulence in the ocean, Lond. Edinburgh Dublin Philos. Mag. J. Sci., 39, 578–586, 1920.

Kantha, L. H. and Clayson, C. A.: On the effect of surface gravity waves on mixing in the oceanic mixed layer, Ocean Model., 6, 101–124, 2004.

Kinsman, B.: Wind waves, Prentice-Hall, Englewood Cliffs, N.J., 1965.

Landau, L. D. and Lifshitz, E. M.: Fluid Mechanics, 2nd Edn., Butterworth-Heinemann, Oxford, 1987.

Martin, P. J., Rogers, E., Allard, R. A., Hogan, P. J., and Richman, J. G.: Results from tests of direct wave mixing in the ocean's surface mixed layer, Tech. Rep. NRL/FR/7322-03-10,045, Naval Research Laboratory, Washington, D.C., 2011.

McWilliams, J. C., Sullivan, P. P., and Moeng, C.-H.: Langmuir turbulence in the ocean, J. Fluid Mech., 334, 1–30, 1997.

Mellor, G. L. and Yamada, T.: Development of a Turbulence Closure Model for Geophysical Fluid Problems, Rev. Geophys. Space Phys., 20, 851–875, 1982.

Milgram, J. H.: Short wave damping in the simultaneous presence of a surface film and turbulence, J. Geophys. Res., 103, 15717–15727, 1998.

Ölmez, H. S. and Milgram, J. H.: An experimental study of attenuation of short water waves by turbulence, J. Fluid Mech., 239, 133–156, 1992.

Pacanowski, R. C. and Philander, S. G. H.: Parameterization of Vertical Mixing in Numerical Models of Tropical Oceans, J. Phys. Oceanogr., 11, 1443–1451, 1981.

Phillips, O. M.: A note on the turbulence generated by gravity waves, J. Geophys. Res., 66, 2889–2893, 1961.

Pleskachevsky, A., Dobrynin, M., Babanin, A. V., Günther, H., and Stanev, E.: Turbulent mixing due to surface waves indicated by remote sensing of suspended particulate matter and its implementation into coupled modeling of waves, turbulence and circulation, J. Phys. Oceanogr., 41, 708–724, 2011.

Qiao, F., Yuan, Y., Yang, Y., Zheng, Q., Xia, C., and Ma, J.: Wave-induced mixing in the upper ocean: Distribution and application to a global ocean circulation model, Geophys. Res. Lett., 31, L11303, doi:10.1029/2004GL019824, 2004.

Qiao, F., Yuan, Y., Ezer, T., Xia, C., Yang, Y., Lü, X., and Song, Z.: A three-dimensional surface wave–ocean circulation coupled model and its initial testing, Ocean Dynam., 60, 1339–1355, 2010.

Rodi, W.: Examples of calculation methods for flow and mixing in stratified fluids, J. Geophys. Res., 92, 5305–5328, 1987.

Rogers, W. E.: The US Navy's Global Wind-Wave Models: An Investigation into Sources of Errors in Low-Frequency Energy Predictions, Tech. Rep. 7320–02-10,035, Oeanography division, Naval Research Laboratory, Stennis Space Center, 2002.

Song, Y., Qiao, F., and Song, Z.: Improved Simulation of the South Asian Summer Monsoon in a Coupled GCM with a More Realistic Ocean Mixed Layer, J. Atmos. Sci., 69, 1681–1690, doi:10.1175/JAS-D-11-0235.1, 2012.

Sullivan, P. P. and McWilliams, J. C.: Dynamics of Winds and Currents Coupled to Surface Waves, Annu. Rev. Fluid Mech., 42, 19–42, doi:10.1146/annurev-fluid-121108-145541, 2010.

Teixeira, M. A. C. and Belcher, S. E.: On the distortion of turbulence by a progressive surface wave, J. Fluid Mech., 458, 229–267, 2002.

Toffoli, A., McConochie, J., Ghantous, M., Loffredo, L., and Babanin, A. V.: The effect of wave-induced turbulence on the ocean mixed layer during tropical cylcones: Field observations on the Australian North-West Shelf, J. Geophys. Res., 117, C00J24, doi:10.1029/2011JC007780, 2012.

Umlauf, L. and Burchard, H.: Second-order turbulence closure models for geophysical boundary layers, A review of recent work, Cont. Shelf Res., 25, 795–827, 2005.

Umlauf, L., Bolding, K., and Burchard, H.: GOTM Scientific Documentation. Version 4.2, available at: http://www.gotm.net/pages/documentation/manual/pdf/a4.pdf, last access: 23 March 2011.

Wang, Y., Qiao, F., Fang, G., and Wei, Z.: Application of wave-induced vertical mixing to the K profile parameterization scheme, J. Geophys. Res., 115, C09014, doi:10.1029/2009JC005856, 2010.

Wilcox, D. C.: Turbulence Modeling for CFD, 2nd Edn., DCW Industries, La Cañada, CA, 1998.

Yefimov, V. V. and Khristoforov, G. N.: Spectra and statistical relations between the velocity fluctuations in the upper layer of the sea and surface waves, Izvestiya, Atmos. Oceanic Phys., 7, 1290–1310, 1971a.

Yefimov, V. V. and Khristoforov, G. N.: Wave-related and turbulent components of velocity spectrum in the top sea layer, Izvestiya, Atmos. Oceanic Phys., 7, 200–211, 1971b.

Young, I. R., Babanin, A. V., and Zieger, S.: The Decay Rate of Ocean Swell Observed by Altimeter, J. Phys. Oceanogr., 43, 2322–2333, 2013.

Diagnostics on the cost-function in variational assimilations for meteorological models

Y. Michel

Météo-France and CNRS, CNRM-GAME, UMR3589, Toulouse, France

Correspondence to: Y. Michel (yann.michel@meteo.fr)

Abstract. Several consistency diagnostics have been proposed to evaluate variational assimilation schemes. The "Bennett-Talagrand" criterion in particular shows that the cost-function at the minimum should be close to half the number of assimilated observations when statistics are correctly specified. It has been further shown that sub-parts of the cost function also had statistical expectations that could be expressed as traces of large matrices, and that this could be exploited for variance tuning and hypothesis testing.

The aim of this work is to extend those results using standard theory of quadratic forms in random variables. The first step is to express the sub-parts of the cost function as quadratic forms in the innovation vector. Then, it is possible to derive expressions for the statistical expectations, variances and cross-covariances (whether the statistics are correctly specified or not). As a consequence it is proven in particular that, in a perfect system, the values of the background and observation parts of the cost function at the minimum are positively correlated. These results are illustrated in a simplified variational scheme in a one-dimensional context.

These expressions involve the computation of the trace of large matrices that are generally unavailable in variational formulations of the assimilation problem. It is shown that the randomization algorithm proposed in the literature can be extended to cover these computations, yet at the price of additional minimizations. This is shown to provide estimations of background and observation errors that improve forecasts of the operational ARPEGE model.

1 Introduction

Most operational data assimilation schemes in meteorology are loosely based on statistical linear estimation (e.g. Talagrand, 2010. The variational formulation (Le Dimet and Talagrand, 1986; Courtier et al., 1998; Rabier et al., 2000) has proven very effective in assimilating directly and massively satellite observations, which has been a source of huge progress in numerical weather prediction (Simmons and Hollingsworth, 2002; Rabier, 2005). The cost function to be minimized measures the distances of the atmospheric state to the observations and to the background, weighted by the inverse of their error covariances. These error statistics are unfortunately not very well known. Common methods to estimate observation errors include the innovation approach (Hollingsworth and Lonnberg, 1986; Daley, 1993), where observation error statistics are taken to be spatially uncorrelated and are deduced from the innovation covariance. Background error statistics are estimated within an existing data assimilation scheme, for instance using forecasts at different ranges valid at the same time (Parrish and Derber, 1992; Derber and Bouttier, 1999) or, more recently and inspired by the Ensemble Kalman Filter (e.g. Evensen, 2003), ensembles of perturbed variational assimilations (Fisher, 2003; Kucukkaraca and Fisher, 2006; Belo-Pereira and Berre, 2006).

Despite this recent progress, it is still necessary and useful to check for the consistency between the prescribed statistics and the a posteriori error statistics. Any inconsistency could help to point out imperfections in the specification of the statistics. Objective assessment of the quality of assimilation schemes can only be performed against independent observations that are not used in the estimation process (Talagrand, 1999). Yet internal consistency diagnostics can also prove useful. Two applications of these kinds of diagnostics have been proposed: hypothesis testing (Muccino et al., 2004) and variance tuning (Desroziers and Ivanov, 2001; Chapnik et al., 2006).

Hypothesis testing formally checks whether the obtained solution is consistent with the hypotheses made concerning the errors in the innovation vector. In the perfect case, the minimum value of the cost function has a χ^2_p distribution where p is the number of scalar observations assimilated in the system (Bennett et al., 1993; Talagrand, 1999). However, the values of the cost function were found to differ significantly from their expectation in a series of data assimilation experiments with an ocean model (Bennett et al., 2000). Muccino et al. (2004) also studied how various errors in the specified error covariance matrices were affecting the distribution of the value of the cost function at the minimum. A Kolmogorov–Smirnov test was used to identify departures from the χ^2 distribution, with significant skill in some cases.

When certain sets of observations are taken to have uncorrelated errors, it is possible to split the observation term further as the sum of sub-functions. The expectation of these sub-functions can still be expressed (Talagrand, 1999) as the trace of large matrices. Desroziers and Ivanov (2001) (hereafter DI01) proposed to estimate these traces using a randomization algorithm. They also proposed an iterative approach for tuning the variances. It enforces consistency between the observed value of the sub-parts of the cost functions and their theoretical expectations as computed by the randomized trace estimation. This method was later shown (Chapnik et al., 2004) to be similar to a maximum-likelihood approach. Operational implementation of the scheme in the global model ARPEGE[1] from Météo-France was achieved by Chapnik et al. (2006), and the tuning of observation error variances yielded a positive impact on both analyses and forecasts. The implementation in a limited area model was discussed by Sadiki and Fischer (2005). They proposed in particular the use of time or space averages to improve sampling. Finally, the ensemble of variational assimilations with perturbed observations as implemented at Météo-France and ECMWF (Kucukkaraca and Fisher, 2006) typically uses assimilations with perturbed backgrounds and observations. Desroziers et al. (2009) then showed that the previous statistical expressions were a direct by-product of such an ensemble.

More precisely, Dee (1995) and Dee and da Silva (1999) have introduced the maximum-likelihood approach for the estimation of covariance parameters in meteorology. The likelihood function to be optimized with respect to these covariance parameters is the sum of a quadratic form in the innovation vector and the log-determinant of a precision (inverse covariance) matrix. Because of the very large size of the innovation vector (currently of the order of 10^6 elements in the ARPEGE 4D-Var), a direct computation (using for instance Choleski decomposition) is not feasible. The approach that was favored in meteorology is to evaluate the derivatives (rather than the absolute values) of the likelihood function and then using stochastic trace estimation techniques, as demonstrated by Purser and Parrish (2003) in an idealized context. This stochastic approximation is nearly optimal in a well-define sense as proven by Stein et al. (2013). It is also possible that direct evaluation of the log-likelihood may become feasible even in high dimensions as shown by Aune et al. (2012). This required in particular (i) more efficient evaluation of the trace with probing vectors (Bekas et al., 2007) and (ii) Krylov subspace methods for matrix inversions (Saad, 2003). These approaches look very promising, but as recognized by Purser and Parrish (2003), it is still necessary to experiment with various aspects in meteorology including non-linear observation operators and introduction of a prior term (to stabilize the solution when the number of estimated parameters become large). This study rather builds upon the simpler but convenient approach taken in DI01 for variance tuning. However, it should be stressed that this approach is unlikely to be easily extended to the estimation of other paramaters such as localisation lengthscales (Houtekamer and Mitchell, 2001), or to the heteromorphic case where there is a need to compare log-likelihoods associated with covariances that use completely different sets of parameters (Purser and Parrish, 2003).

The aim of this paper is to present a different way of deriving the Bennett-Talagrand criterion. The sub-parts of the cost function at the minimum are expressed as quadratic forms in the innovation vector. This allows the application of the theory of quadratic forms in random variables as summarized in Mathai and Provost (1992) (hereafter M92). The statistical expectation, variances and cross-covariances between sub-parts of the cost-function at the minimum are shown to be related to the trace of matrices in the assimilation. This relies on the assumption that the innovation vector is normally distributed, but some extensions to the Non-Gaussian case have also been derived for some specific distributions. The derivation of these expressions is outlined in Sect. 2. Practical estimation of these terms based on the randomized trace algorithm is demonstrated in Sect. 3. The use of these formulas for variance tuning in the operational ARPEGE model is illustrated in Sect. 4.

2 Statistical distribution of the values of the terms in the cost-function at the minimum

2.1 Variational framework

The best estimate of the atmospheric state, $\boldsymbol{x}_\mathrm{a}$, is obtained as the solution of the minimization of the cost function:

$$
\begin{aligned}
J(\boldsymbol{x}) &= J_\mathrm{b}(\boldsymbol{x}) + J_\mathrm{o}(\boldsymbol{x}) \\
&= \frac{1}{2}\{(\boldsymbol{x}-\boldsymbol{x}_b)^\mathrm{T}\mathbf{B}^{-1}(\boldsymbol{x}-\boldsymbol{x}_b) \\
&+ \left[\boldsymbol{y}_\mathrm{o} - \mathcal{H}(\boldsymbol{x})\right]^\mathrm{T}\mathbf{R}^{-1}\left[\boldsymbol{y}_\mathrm{o} - \mathcal{H}(\boldsymbol{x})\right]\} \qquad (1)
\end{aligned}
$$

where any time index has been dropped for convenience. $J_\mathrm{b}(\boldsymbol{x})$ and $J_\mathrm{o}(\boldsymbol{x})$ stand for the background and observation

[1]"Action de Recherche Petite Echelle Grande Echelle", see Pailleux et al., 2000 for a historical description

terms of the cost-function respectively. **B** and **R** are the background and observation error covariance matrices, and $\mathcal{H}$ is the non-linear observation operator including model time integration in the 4D-Var framework. This cost-function is usually solved by a sequence of quadratic minimizations using linearized operators, an algorithm known in meteorology as incremental 4D-Var (Courtier et al., 1994). Here we are only interested in the final value of the cost function (at the minimum), but of course preconditioning in minimizing Eq. (1) is a crucial problem in practice (e.g. Lorenc, 1997; Gauthier et al., 1999). A Bayesian derivation of Eq. (1) is provided by Lorenc (1986).

Introducing the innovation vector

$$\boldsymbol{d} = \boldsymbol{y}_{\mathrm{o}} - \mathcal{H}(\boldsymbol{x}_{\mathrm{b}}) \tag{2}$$

the linearized cost-function (still written J) is expressed as:

$$\begin{aligned} \boldsymbol{\delta}_x &= \boldsymbol{x} - \boldsymbol{x}_b \\ J(\boldsymbol{\delta}_x) &= J_{\mathrm{b}}(\boldsymbol{\delta}_x) + J_{\mathrm{o}}(\boldsymbol{\delta}_x) \\ &= \frac{1}{2}\left\{\boldsymbol{\delta}_x^{\mathrm{T}}\mathbf{B}^{-1}\boldsymbol{\delta}_x + [\boldsymbol{d} - \mathbf{H}\boldsymbol{\delta}_x]^{\mathrm{T}}\mathbf{R}^{-1}[\boldsymbol{d} - \mathbf{H}\boldsymbol{\delta}_x]\right\} \end{aligned} \tag{3}$$

where **H** is the linearized observation operator and where every term is now quadratic.

A common derivation of the Bennett-Talagrand criterion casts the problem (3) into a single term using a generalized observation operator (Talagrand, 1999; Desroziers and Ivanov, 2001). This is not necessary for deriving the statistical moments of $\widetilde{J} = J(\boldsymbol{x}_{\mathrm{a}})$. An alternative way is to write down the solution of Eq. (3) in the standard form involving the data assimilation operator (or gain matrix) **K**:

$$\boldsymbol{x}_{\mathrm{a}} - \boldsymbol{x}_{\mathrm{b}} = \mathbf{K}\boldsymbol{d} \tag{4}$$

K is a representation of performing variational assimilation for the analysis increments. The background term at the minimum $\widetilde{J}_{\mathrm{b}}$ is therefore:

$$\widetilde{J}_{\mathrm{b}} = \frac{1}{2}\boldsymbol{d}^{\mathrm{T}}\mathbf{K}^{\mathrm{T}}\mathbf{B}^{-1}\mathbf{K}\boldsymbol{d} \tag{5}$$

and similarly, the observation term at the minimum $\widetilde{J}_{\mathrm{o}}$ is:

$$\widetilde{J}_{\mathrm{o}} = \frac{1}{2}\boldsymbol{d}^{\mathrm{T}}(\mathbf{I} - \mathbf{HK})^{\mathrm{T}}\mathbf{R}^{-1}(\mathbf{I} - \mathbf{HK})\boldsymbol{d} \tag{6}$$

From Eq. (3), it is also possible to express the observation term before the minimization $\widehat{J}_{\mathrm{o}}$ as

$$\widehat{J}_{\mathrm{o}} = \frac{1}{2}\boldsymbol{d}^{\mathrm{T}}\mathbf{R}^{-1}\boldsymbol{d} \tag{7}$$

and $\widehat{J}_{\mathrm{b}} = 0$. This explicitly shows that these terms are (halves of) quadratic forms in the innovation vector $\boldsymbol{d}$.

In general, observation errors are taken to be diagonal or block-diagonal, with each block corresponding to a given instrument, and observation errors are taken to be uncorrelated between instruments. Following Chapnik et al. (2004), this is written as a projection operator Π_i^{o} operating on the observation vector $\boldsymbol{y}$ and giving a subset $\boldsymbol{y}_i = \Pi_i^{\mathrm{o}}\boldsymbol{y}$ of the observations, with $i = 1, \ldots, P$ and where P is the number of such subsets. Up to a reordering, the matrix **R** can be written in the block-diagonal form:

$$\mathbf{R} = \begin{pmatrix} \mathbf{R}_1 & \mathbf{0} & \cdots & \mathbf{0} \\ \mathbf{0} & \mathbf{R}_2 & \cdots & \mathbf{0} \\ \vdots & \vdots & \ddots & \vdots \\ \mathbf{0} & \mathbf{0} & \cdots & \mathbf{R}_P \end{pmatrix}$$

where $\mathbf{R}_i = \Pi_i^{\mathrm{o}}\mathbf{R}\Pi_i^{\mathrm{o}\,\mathrm{T}}$ are observation error covariance matrices for the ith-subset of observations. The inverse covariance matrix $\mathbf{R}^{-1}$ is also block-diagonal, with elements $\mathbf{R}_i^{-1}$, and the observation cost-function can be split into P contributions J_{o}^i (either at the minimum or before the minimization):

$$\widetilde{J}_{\mathrm{o}}^i = \frac{1}{2}\boldsymbol{d}^{\mathrm{T}}(\mathbf{I} - \mathbf{HK})^{\mathrm{T}}\Pi_i^{\mathrm{o}\,\mathrm{T}}\mathbf{R}_{\mathbf{i}}^{-1}\Pi_i^{\mathrm{o}}(\mathbf{I} - \mathbf{HK})\boldsymbol{d} \tag{8}$$

$$\widehat{J}_{\mathrm{o}}^i = \frac{1}{2}\boldsymbol{d}^{\mathrm{T}}\Pi_i^{\mathrm{o}\,\mathrm{T}}\mathbf{R}_i^{-1}\Pi_i^{\mathrm{o}}\boldsymbol{d} \tag{9}$$

which are quadratic forms in the innovation vector.

In general the background error covariance matrix is not block-diagonal as the different variables are correlated. This was neglected in DI01, but it can be taken into account by introducing the parameter transform $\mathbf{K}_{\mathrm{p}}$ that changes variables into uncorrelated ones (Derber and Bouttier, 1999; Bannister, 2008). Different spatial autocovariance blocks $\mathbf{B}_i$ are specified for each unbalanced variable. This can be written as:

$$\begin{aligned} \mathbf{B} &= \mathbf{K}_{\mathrm{p}} \begin{pmatrix} \mathbf{B}_1 & \mathbf{0} & \cdots & \mathbf{0} \\ \mathbf{0} & \mathbf{B}_2 & \cdots & \mathbf{0} \\ \vdots & \vdots & \ddots & \vdots \\ \mathbf{0} & \mathbf{0} & \cdots & \mathbf{B}_S \end{pmatrix} \mathbf{K}_{\mathrm{p}}^{\mathrm{T}} \\ &= \mathbf{K}_{\mathrm{p}} \sum_{i=1}^{i=S} \Pi_i^{\mathrm{b}\,\mathrm{T}}\mathbf{B}_i\Pi_i^{\mathrm{b}}\mathbf{K}_{\mathrm{p}}^{\mathrm{T}} \end{aligned}$$

where usually there are $S = 3$–5 control variables (usually vorticity or streamfunction, unbalanced divergence or velocity potential, and some form of unbalanced temperature and humidity, see Bannister (2008) for a review and comparison between various implementations). $\mathbf{K}_{\mathrm{p}}$ is generally formulated as a lower-triangular non-singular matrix with left-inverse $\mathbf{K}_{\mathrm{p}}^{-1}$. The projection matrix Π_i^{b} is used to extract the ith-component of the background vector in unbalanced space and can be written explicitely with blocks being either the zero matrix **0** or the identity matrix **I** in ith position:

$$\Pi_i^{\mathrm{b}} = (\mathbf{0} \cdots \mathbf{0}\,\mathbf{I}\,\mathbf{0} \cdots \mathbf{0}).$$

Using this formulation allows us to split the background term into S contributions J_b^i that are, again, quadratic forms in $\boldsymbol{d}$:

$$\widetilde{J}_\mathrm{b} = \sum_{i=1}^{S} \widetilde{J}_\mathrm{b}^i$$
$$\widetilde{J}_\mathrm{b}^i = \frac{1}{2}\boldsymbol{d}^\mathrm{T}\mathbf{K}^\mathrm{T}\mathbf{K}_\mathrm{p}^{-\mathrm{T}}\Pi_i^{\mathrm{b}^\mathrm{T}}\mathbf{B}_i{}^{-1}\Pi_i^\mathrm{b}\mathbf{K}_\mathrm{p}^{-1}\mathbf{K}\boldsymbol{d} \tag{10}$$

There may be additional penalty terms in the cost function, related for instance to the coupling information in a limited area model (Guidard and Fischer, 2008) or to the weak constraint digital filter (Gauthier and Thépaut, 2001). These terms will not be considered explicitly here, but they can be put into similar quadratic forms. The digital filter expresses a constraint on the increments which are directly related to the innovation vector. The coupling information requires an extended definition of the information vector, and this issue may be addressed in future work.

2.2 Moments

The innovation vector is a random variable, so are the background and observation terms at the minimum, and it is possible to explicitly derive their expectation, but also their variances and covariances in the Gaussian case. Note that this Gaussian assumption can be relaxed for some classes of distributions; see e.g. Mathai et al. (1995) for the case of elliptically contoured distributions and Genton et al. (2001) for skew-normal ones.

2.2.1 Expectations

Without any loss of generality in this section (other than existence of second order moments), we will denote the first two moments of the innovation vector by:

$$\boldsymbol{\mu} = E(\boldsymbol{d})$$
$$\Sigma = \mathrm{Cov}(\boldsymbol{d})$$

where $\mathrm{E}(\cdot)$ is the expectation and $\mathrm{Cov}(\cdot)$ the covariance. Here, $\boldsymbol{\mu}$ and Σ are the actual (i.e. measured) mean and covariance of the innovation vector that may differ from their assumed values given in Eqs. (14) and (15). Then, it can be seen by applying the trace operator (M92, Corollary 3.2b.1) that the expectation of any quadratic form in $\boldsymbol{d}$ is:

$$E\left(\boldsymbol{d}^\mathrm{T}\mathbf{A}\boldsymbol{d}\right) = \mathrm{Tr}(\mathbf{A}\Sigma) + \boldsymbol{\mu}^\mathrm{T}\mathbf{A}\boldsymbol{\mu} \tag{11}$$

where $\mathrm{Tr}(\cdot)$ is the trace (which is a linear operator). This result is robust as it holds for any distribution of $\boldsymbol{d}$ with finite first-two moments (Ménard et al., 2000). Equation (11) can be readily applied to the previously derived quadratic forms. In particular the expectation of background and observation terms follows:

$$E\left(\widetilde{J}_\mathrm{b}\right) = \frac{1}{2}\mathrm{Tr}\left(\mathbf{K}^\mathrm{T}\mathbf{B}^{-1}\mathbf{K}\Sigma\right) + \frac{1}{2}\boldsymbol{\mu}^\mathrm{T}\mathbf{K}^\mathrm{T}\mathbf{B}^{-1}\mathbf{K}\boldsymbol{\mu} \tag{12}$$

$$E\left(\widetilde{J}_\mathrm{o}\right) = \frac{1}{2}\mathrm{Tr}\left[(\mathbf{I}-\mathbf{HK})^\mathrm{T}\mathbf{R}^{-1}(\mathbf{I}-\mathbf{HK})\,\Sigma\right] + \frac{1}{2}\boldsymbol{\mu}^\mathrm{T}(\mathbf{I}-\mathbf{HK})^\mathrm{T}\mathbf{R}^{-1}(\mathbf{I}-\mathbf{HK})\,\boldsymbol{\mu} \tag{13}$$

No hypothesis has been made on the distribution of the innovation yet. These equations are valid for any data assimilation scheme with possibly incorrect error covariance matrices (and hence $\mathbf{K}$).

Now, in a well specified system, we would have unbiased innovation and well-prescribed statistics:

$$\boldsymbol{\mu} = \mathbf{0} \tag{14}$$
$$\Sigma = \mathbf{D} \equiv \mathbf{HBH}^\mathrm{T} + \mathbf{R} \tag{15}$$

The gain matrix can also be written as:

$$\mathbf{K} = \mathbf{BH}^\mathrm{T}\left(\mathbf{HBH}^\mathrm{T}+\mathbf{R}\right)^{-1} = \mathbf{BH}^\mathrm{T}\mathbf{D}^{-1}$$

which allows us to write the following expressions:

$$\mathbf{B}^{-1}\mathbf{KD} = \mathbf{H}^\mathrm{T}$$
$$\mathbf{R}^{-1}(\mathbf{I}-\mathbf{HK})\mathbf{D} = \mathbf{I}$$

Thus, when $\Sigma = \mathbf{D}$, the expectation of background and observation terms take the form

$$E\left(\widetilde{J}_\mathrm{b}\right) = \tfrac{1}{2}\mathrm{Tr}\left[\mathbf{K}^\mathrm{T}\mathbf{H}^\mathrm{T}\right] = \tfrac{1}{2}\mathrm{Tr}(\mathbf{HK}) \tag{16}$$
$$\mathrm{E}\left(\widetilde{J}_\mathrm{o}\right) = \tfrac{1}{2}\mathrm{Tr}\left[(\mathbf{I}-\mathbf{HK})^\mathrm{T}\right] = \tfrac{1}{2}\mathrm{Tr}(\mathbf{I}-\mathbf{HK}) \tag{17}$$

as previously derived by Bennett et al. (1993), Talagrand (1999), and Desroziers and Ivanov (2001).

New results are obtained by applying a similar calculation with sub-terms of the background cost function at the minimum:

$$\begin{aligned} E\left(\widetilde{J}_\mathrm{b}^i\right) &= \frac{1}{2}\mathrm{Tr}\left[\mathbf{K}^\mathrm{T}\mathbf{K}_\mathrm{p}^{-\mathrm{T}}\Pi_i^{\mathrm{b}^\mathrm{T}}\mathbf{B}_i^{-1}\Pi_i^\mathrm{b}\mathbf{K}_\mathrm{p}^{-1}\mathbf{BH}^\mathrm{T}\right] \\ &= \frac{1}{2}\mathrm{Tr}\left[\mathbf{H}^\mathrm{T}\mathbf{K}^\mathrm{T}\mathbf{K}_\mathrm{p}^{-\mathrm{T}}\Pi_i^{\mathrm{b}^\mathrm{T}}\mathbf{B}_i^{-1}\Pi_i^\mathrm{b}\sum_{j=1}^{i=S}\Pi_j^{\mathrm{b}^\mathrm{T}}\mathbf{B}_j\Pi_j^\mathrm{b}\mathbf{K}_\mathrm{p}^\mathrm{T}\right] \\ &= \frac{1}{2}\mathrm{Tr}\left[\mathbf{H}^\mathrm{T}\mathbf{K}^\mathrm{T}\mathbf{K}_\mathrm{p}^{-\mathrm{T}}\Pi_i^{\mathrm{b}^\mathrm{T}}\Pi_i^\mathrm{b}\mathbf{K}_\mathrm{p}^\mathrm{T}\right] \\ &= \frac{1}{2}\mathrm{Tr}\left[\mathbf{K}_\mathrm{p}\Pi_i^{\mathrm{b}^\mathrm{T}}\Pi_i^\mathrm{b}\mathbf{K}_\mathrm{p}^{-1}\mathbf{KH}\right]. \end{aligned} \tag{18}$$

The expectation of any sub-term of the observation cost function at the minimum is derived in a similar way:

$$E\left(\widetilde{J}_\mathrm{o}^i\right) = \frac{1}{2}\mathrm{Tr}\left[\Pi_\mathrm{o}^{i\,\mathrm{T}}\Pi_\mathrm{o}^i(\mathbf{I}-\mathbf{HK})\right]. \tag{19}$$

Analogous expressions can be obtained for the sub-term of the observation cost function before the minimization (Eqs. 7 and 9).

2.2.2 Variances

In general, the variance of the quadratic forms will depend on the fourth-order moment of $\boldsymbol{d}$. If the innovation vector has a Gaussian distribution, then its fourth-order moment can be expressed as a function of the first two ones, and the variance of the quadratic forms can be expressed following M92 (theorem 3.2b.2):

$$\mathrm{Var}\left(\boldsymbol{d}^{\mathrm{T}}\mathbf{A}\boldsymbol{d}\right) = 2\mathrm{Tr}\left[(\mathbf{A}\Sigma)^2\right] + 4\boldsymbol{\mu}^{\mathrm{T}}\mathbf{A}\Sigma\mathbf{A}\boldsymbol{\mu}. \tag{20}$$

This formula can be readily applied to the previous quadratic forms (without assuming perfect statistics). Now, in the particular case of a system with well-specified statistics verifying Eqs. (14)–(15) we have:

$$\mathrm{Var}\left(\widetilde{J}_{\mathrm{b}}\right) = \frac{1}{2}\mathrm{Tr}\left[(\mathbf{HK})^2\right] \tag{21}$$

$$\mathrm{Var}\left(\widetilde{J}_{\mathrm{o}}\right) = \frac{1}{2}\mathrm{Tr}\left[(\mathbf{I}-\mathbf{HK})^2\right]. \tag{22}$$

The results do not seem to appear in the standard literature on data assimilation for the atmosphere, yet analogous formulas appear in Bennett et al. (2000) within the formalism of the representer method. They can be extended to sub-parts of the cost function (not shown).

Before deriving cross-covariances in the next section, one can point out that $\widetilde{J}_{\mathrm{b}}$ and $\widetilde{J}_{\mathrm{o}}$ should be positively correlated. Indeed, using the fact that $2\widetilde{J}$ has a χ_p^2 distribution, we can deduce that:

$$\begin{aligned}\mathrm{Var}\left(\widetilde{J}\right) &= p/2\\ &= \mathrm{Var}\widetilde{J}_{\mathrm{b}} + \mathrm{Var}\widetilde{J}_{\mathrm{o}} + 2\mathrm{Cov}\left(\widetilde{J}_{\mathrm{b}}, \widetilde{J}_{\mathrm{o}}\right).\end{aligned}$$

The matrices $\mathbf{HK}$ and $\mathbf{I}-\mathbf{HK}$ have eigenvalues between 0 and 1 (**?**), therefore:

$$\begin{aligned}&\mathrm{Tr}\left[(\mathbf{HK})^2\right] + \mathrm{Tr}\left[(\mathbf{I}-\mathbf{HK})^2\right]\\ &\qquad \leq \mathrm{Tr}(\mathbf{HK}) + \mathrm{Tr}(\mathbf{I}-\mathbf{HK}) = \mathrm{Tr}(\mathbf{I}) = p.\end{aligned}$$

Thus it is necessary that

$$\mathrm{Cov}\left(\widetilde{J}_{\mathrm{b}}, \widetilde{J}_{\mathrm{o}}\right) \geq 0. \tag{23}$$

In other words, in a system with well-specified statistics, $\widetilde{J}_{\mathrm{b}}$ and $\widetilde{J}_{\mathrm{o}}$ are positively correlated. This may sound like a counter-intuitive result, as one could expect the random fluctuations of J to be "split" (with anticorrelations) between $\widetilde{J}_{\mathrm{b}}$ and $\widetilde{J}_{\mathrm{o}}$, but this is not the case. Rather, the random fluctuations go in the same direction.

2.2.3 Covariances

The covariance between (sub)terms of the cost function can be calculated as follows (M92, theorem 3.2d.4):

$$\begin{aligned}\mathrm{Cov}\left(\boldsymbol{d}^{\mathrm{T}}\mathbf{A}_1\boldsymbol{d}, \boldsymbol{d}^{\mathrm{T}}\mathbf{A}_2\boldsymbol{d}\right) &= 2\mathrm{Tr}\left(\mathbf{A}_1\Sigma\mathbf{A}_2\Sigma\right)\\ &\quad + 4\boldsymbol{\mu}^{\mathrm{T}}\mathbf{A}_1\Sigma\mathbf{A}_2\boldsymbol{\mu}\end{aligned} \tag{24}$$

In particular, in a well-specified system:

$$\mathrm{Cov}\left(\widetilde{J}_{\mathrm{b}}, \widetilde{J}_{\mathrm{o}}\right) = \frac{1}{2}\mathrm{Tr}\left[\mathbf{HK}\left(\mathbf{I}-\mathbf{HK}\right)\right] \tag{25}$$

The magnitude of the correlation depends on how Tr($\mathbf{HK}$) and Tr[$(\mathbf{HK})^2$] differ. This highlights that it could be useful to estimate $\mathrm{Tr}\left[(\mathbf{HK})^2\right]$ in operational schemes. Also, further expressions can be derived for the cross-covariances between sub-parts of the cost function (not shown).

2.3 Distribution

Beyond the statistical moments, it may be of some interest to get information on the distribution of the sub-parts of the cost function at the minimum. The random variable $2\widetilde{J}$ has a variance that is double of its expectation, as expected for a χ_p^2-distributed variable. In view of Eqs. (16)–(21) and (17)–(22), this is generally not the case for $2\widetilde{J}_{\mathrm{o}}, 2\widetilde{J}_{\mathrm{b}}$ (nor their sub-parts). Thus they do not follow a χ^2 distribution – see also the necessary and sufficient conditions in the chapter 5 of M92. However, they are linear combinations of χ^2 distributions in the Gaussian case (M92).

Albeit the previous expressions may be thought to be restricted to variational assimilation, they would also apply to an ensemble variant of the Kalman filter provided that the associated cost function can be computed (Ménard and Chang, 2000; Talagrand, 2010).

3 Practical estimation and potential use

This section introduces the randomized trace algorithm to compute the previous statistical expressions. They could be used for hypothesis testing. Rather, we focus on their use when tuning global scaling factors for matrices $\mathbf{B}$ and $\mathbf{R}$.

3.1 Application of the randomized trace estimator

The previous expressions involve the computation of the trace of large matrices that are unavailable in large-size data assimilation schemes used in operations in meteorology. Following DI01, it is however possible to estimate them with a randomized trace estimator:

$$\mathrm{Tr}(\mathbf{A}) \simeq \frac{1}{M}\sum_{i=1}^{M}\eta_i^{\mathrm{T}}\mathbf{A}\eta_i \tag{26}$$

where η_i are independent random vectors whose components follow a standard $\mathcal{N}(0,1)$ normal distribution (Girard, 1991) or Rademacher (± 1 with probability 1/2) distribution (Hutchinson, 1989). The Hutchinson estimator has lower variance than Girard's but requires more samples to achieve a given relative precision (Avron and Toledo, 2011). In addition, using the Gaussian distribution is more natural in data assimilation and allows Tr($\mathbf{HK}$) to be computed as a by-product of an ensemble of variational assimilations

(Desroziers et al., 2009). It is shown next that randomized estimation of traces in new formulas (21,22, and 25) is possible through perturbed minimizations (as in DI01), yet at the cost of an additional minimization. Indeed, **K** (the assimilation algorithm) has to be applied three times.

Instead of the direct application of Eq. (26), DI01 take the approach of further introducing the square root forms of the error covariance matrices $\mathbf{B} = \mathbf{B}^{1/2}\mathbf{B}^{T/2}$ and $\mathbf{R} = \mathbf{R}^{1/2}\mathbf{R}^{T/2}$. Then, using the fact that the trace is invariant under cyclic permutations (not general ones), the randomized estimate (26) required in Eqs. (21), (22), and (25) follows:

$$\begin{aligned}
\mathrm{Tr}\left[(\mathbf{HK})^2\right] &= \mathrm{Tr}\left(\mathbf{HKHKR}^{1/2}\mathbf{R}^{T/2}\mathbf{R}^{-1}\right) \\
&= \mathrm{Tr}\left(\mathbf{R}^{T/2}\mathbf{R}^{-1}\mathbf{HKHKR}^{1/2}\right) \\
&\simeq \frac{1}{M}\sum_{i=1}^{M}\left(\mathbf{R}^{1/2}\boldsymbol{\eta}_i\right)^{T}\mathbf{R}^{-1}\mathbf{HKHK}\left(\mathbf{R}^{1/2}\boldsymbol{\eta}_i\right) \\
&= \frac{1}{M}\sum_{i=1}^{M}\delta\boldsymbol{y}_i^{\mathrm{o}\,T}\mathbf{R}^{-1}\mathbf{HKHK}\delta\boldsymbol{y}_i^{\mathrm{o}} \\
&= \frac{1}{M}\sum_{i=1}^{M}\delta\boldsymbol{y}_i^{\mathrm{o}\,T}\mathbf{R}^{-1}\mathbf{HKH}\left(\delta\boldsymbol{x}_{\mathrm{a}}^{*} - \delta\boldsymbol{x}_{\mathrm{a}|_{y\mathrm{o}}}\right) \\
&= \frac{1}{M}\sum_{i=1}^{M}\delta\boldsymbol{y}_i^{\mathrm{o}\,T}\mathbf{R}^{-1}\mathbf{H}\left(\delta\boldsymbol{x}_{\mathrm{a}}^{**} - \delta\boldsymbol{x}_{\mathrm{a}|_{y\mathrm{o}}}\right)
\end{aligned}$$

where

$$\begin{aligned}
\delta\boldsymbol{x}_{\mathrm{a}}^{*} &= \delta\boldsymbol{x}_{\mathrm{a}|_{y\mathrm{o}+\delta y_i^{\mathrm{o}}}} \\
\delta\boldsymbol{x}_{\mathrm{a}}^{**} &= \delta\boldsymbol{x}_{\mathrm{a}|_{y\mathrm{o}+\mathbf{H}(\delta x_{\mathrm{a}}^{*}-\delta x_{\mathrm{a}|y\mathrm{o}})}}
\end{aligned} \tag{27}$$

and $\delta\boldsymbol{y}_i^{\mathrm{o}} = \mathbf{R}^{1/2}\boldsymbol{\eta}_i$ are sets of perturbations of observations (accordingly to the error covariance matrix **R**). Thus, at least three $(2M+1)$ minimizations are necessary for the computation of this trace: first with original observations to give the unperturbed analysis increment $\delta\boldsymbol{x}_{\mathrm{a}|_{y\mathrm{o}}}$, then with perturbed observations to give the perturbed analysis increment $\delta\boldsymbol{x}_{\mathrm{a}}^{*}$, and finally the difference between those analysis increments is brought into observation space and used as a perturbation of observation in a third analysis giving the analysis increment $\delta\boldsymbol{x}_{\mathrm{a}}^{**}$. The trace computation then follows, provided that $\mathbf{R}^{-1}$ is available as an operator, which is easy if observations errors are taken uncorrelated as it is usually the case.

DI01 have provided an independent computation of $\mathrm{Tr}(\mathbf{KH}) = \mathrm{Tr}(\mathbf{HK})$. It is also possible to derive a similar computation of $\mathrm{Tr}\left[(\mathbf{HK})^2\right] = \mathrm{Tr}\left[(\mathbf{KH})^2\right]$, as:

$$\begin{aligned}
\mathrm{Tr}\left[(\mathbf{KH})^2\right] &= \mathrm{Tr}\left(\mathbf{KHKHB}^{1/2}\mathbf{B}^{T/2}\mathbf{B}^{-1}\right) \\
&= \mathrm{Tr}\left(\mathbf{B}^{T/2}\mathbf{B}^{-1}\mathbf{KHKHB}^{1/2}\right) \\
&\simeq \frac{1}{M}\sum_{i=1}^{M}\left(\mathbf{B}^{1/2}\boldsymbol{\epsilon}_i\right)^{T}\mathbf{B}^{-1}\mathbf{KHKH}\left(\mathbf{B}^{1/2}\boldsymbol{\epsilon}_i\right) \\
&= \frac{1}{M}\sum_{i=1}^{M}\delta\boldsymbol{x}_i^{\mathrm{b}\,T}\mathbf{B}^{-1}\mathbf{KHKH}\delta\boldsymbol{x}_i^{\mathrm{b}} \\
&= -\frac{1}{M}\sum_{i=1}^{M}\delta\boldsymbol{x}_i^{\mathrm{b}\,T}\mathbf{B}^{-1}\mathbf{KH}\left(\delta\boldsymbol{x}_{\mathrm{a}}^{\circ} - \delta\boldsymbol{x}_{\mathrm{a}|_{x\mathrm{b}}}\right) \\
&= -\frac{1}{M}\sum_{i=1}^{M}\delta\boldsymbol{x}_i^{\mathrm{b}\,T}\mathbf{B}^{-1}\left(\delta\boldsymbol{x}_{\mathrm{a}}^{\circ\circ} - \delta\boldsymbol{x}_{\mathrm{a}|_{x\mathrm{b}}}\right)
\end{aligned}$$

where

$$\begin{aligned}
\delta\boldsymbol{x}_{\mathrm{a}}^{\circ} &= \delta\boldsymbol{x}_{\mathrm{a}|_{x\mathrm{b}+\delta x_i^{\mathrm{b}}}} \\
\delta\boldsymbol{x}_{\mathrm{a}}^{\circ\circ} &= \delta\boldsymbol{x}_{\mathrm{a}|_{x\mathrm{b}+\delta x_{\mathrm{a}}^{\circ}-\delta x_{\mathrm{a}|_{x\mathrm{b}}}}}
\end{aligned} \tag{28}$$

and $\delta\boldsymbol{x}_i^{\mathrm{b}} = \mathbf{B}^{1/2}\boldsymbol{\epsilon}_i$ are sets of background perturbations. Thus, at least three minimizations are necessary for the computation of this trace: first with original background; second with a perturbed background, third with a background perturbed by differences of analysis increments. This last formula may be simplified when working with $\mathbf{B}^{1/2}$-preconditioning (e.g. Bannister, 2008 as it can be expressed in terms of the control variable. Also, those computations only involve analysis increments where observations or background have been perturbed, and they can still be computed when using a weakly non-linear algorithm.

Finally, another way to compute directly the expectations and covariances of the sub-parts of the cost function is the "simulated optimal innovation technique" introduced by Chapnik et al. (2006). In this approach, pseudo-observations and pseudo-backgrounds are generated with errors that are exactly consistent with the prescribed error statistics. Then, assimilation of these synthetic observations will result into sub-parts of the cost function that exactly follow the previous expressions. This is a Monte-Carlo approach to compute the expectations and covariances in the specific case where assimilation statistics are correctly specified.

3.2 A simplified framework

If one wants to compare the theoretical statistics to the observed ones, the obvious problem lies in the fact that there is only one realization of the observations (and thus of the innovation vector) in the real atmosphere. The expectation of the cost-function can be estimated from a single realization (Chapnik et al., 2004), yet for some observation types the use of space or time averages to build samples large enough is required (Sadiki and Fischer, 2005). Obviously, computing the variance of the cost-function requires many samples, thus the use of space or time averages. This kind of ergodicity assumption raises questions in itself, such that we will start investigating the use of these formulas first in a simplified framework where independent Monte-Carlo realizations of the innovation vector can be easily generated. Such an approach was also used in Muccino et al. (2004) with the same purpose.

For that purpose, a simplified one-dimensional variational assimilation framework has been set up in a similar way as in DI01. The goal is to analyze a single variable (say temperature) on a circular domain (say on an earth meridian), minimizing Eq. (3). A conjugate gradient algorithm with $\mathbf{B}^{1/2}$-preconditioning is used for the minimization. It involves a maximum of 100 iterations, which is sufficient for convergence unless correlated observation errors are specified (this is verified a posteriori). Background errors have homogeneous standard deviation $\sigma_\mathrm{b} = 1\,\mathrm{K}$. The correlation function c is chosen to belong to the Matérn class (Rasmussen and Williams, 2006) with "smoothness" parameter $\nu = 1$, corresponding to:

$$c(x) = \left(1 + \frac{\sqrt{3}x}{\ell}\right) \exp\left(-\frac{\sqrt{3}x}{\ell}\right)$$

where the length-scale is chosen to be $\ell = 250\,\mathrm{km}$; see Michel (2013) for details. The observation operator is a periodic linear interpolation. Observations are taken to be equally spaced, which allows easy inclusion of spatial correlations if needed, using the same model as $\mathbf{B}$ but in observation space. In order to roughly simulate the characteristics of operational data assimilation, the number of variables is taken as $n = 1000$ and there are $p = 500$ observations with precision $\sigma_\mathrm{o} = 1\,\mathrm{K}$. The fields are illustrated in Fig. 1. In this particular case, the value of the cost-function at the minimum is $\widetilde{J} = 245.8 \approx p/2$ as the specified error statistics are fully consistent with the ones of the innovation vector.

3.3 Validation of the randomized trace estimation

The distribution of the observation and background terms at the minimum have been obtained from running 10^4 independent realizations of the assimilation scheme. The distribution of the obtained values is shown in Fig. 2. The total cost function has a mean of $\simeq 250.11$ which is consistent with the theoretical expectation $p/2 = 250$ (within the bounds of a standard confidence interval with 10^4 realizations). The estimated variances of the observation and background terms are lower than their mean, which is consistent with the discussion in Sect. 2.

The trace estimation algorithm relies on randomization, thus its results depend on the sample size. The algorithm has been applied to compute $\mathrm{Tr}(\mathbf{HK})$, $\mathrm{Tr}\left[(\mathbf{HK})^2\right]$, and also independently $\mathrm{Tr}(\mathbf{KH})$ and $\mathrm{Tr}\left[(\mathbf{KH})^2\right]$ for various ensemble (sampling) sizes M between 1 and 100. The procedure has been applied 200 times in order to evaluate the statistical properties of the randomized trace estimator. Results are presented in Fig. 3. Almost identical results are obtained among ensemble sizes. Averaging over the ensemble sizes and the 200 realizations gives the estimations $\mathrm{Tr}(\mathbf{HK}) \simeq 80.15$, $\mathrm{Tr}(\mathbf{KH}) \simeq 80.19$, $\mathrm{Tr}\left[(\mathbf{HK})^2\right] \simeq 56.7$, $\mathrm{Tr}\left[(\mathbf{KH})^2\right] \simeq 56.86$ which are obviously consistent whenever Eq. (26) or Eq. (27) is used. The comparison with the independent realizations of the assimilation scheme (the simulated optimal innovation method, Fig. 2) is shown in Table 1.

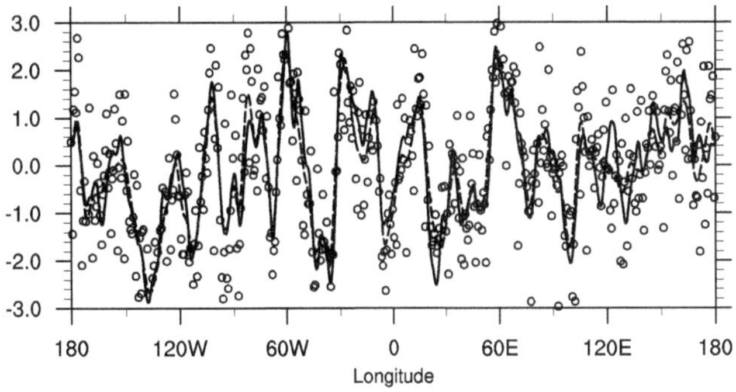

Fig. 1. True signal $\boldsymbol{x}_\mathrm{t}$ (solid black line), observations $\boldsymbol{y}_\mathrm{o}$ (circles) and the analysis $\boldsymbol{x}_\mathrm{a}$ (dashed black line) simulated with the prescribed statistics for background and observation errors (see text). The background $\boldsymbol{x}_\mathrm{b}$ is zero.

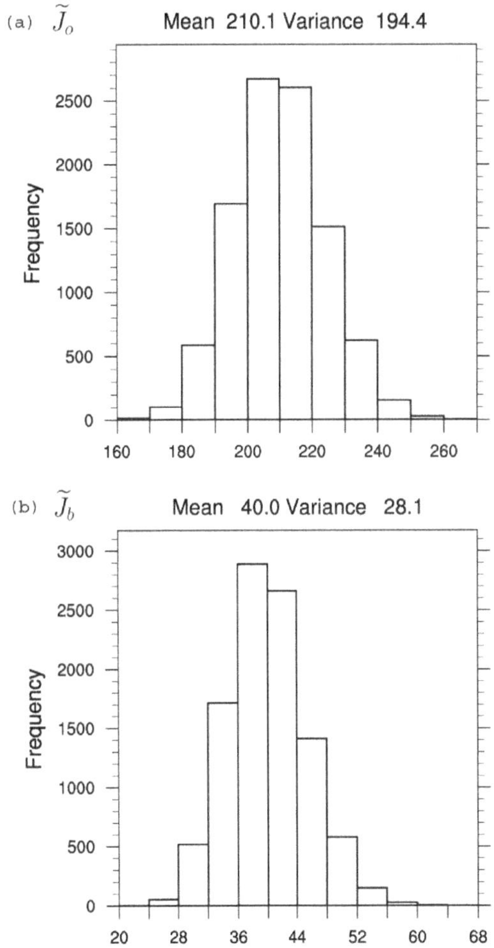

Fig. 2. Histograms obtained with 10^4 realizations of the assimilation scheme shown in Fig. 1 for the values of **(a)** the background cost function and **(b)** the observation cost function at the minimum.

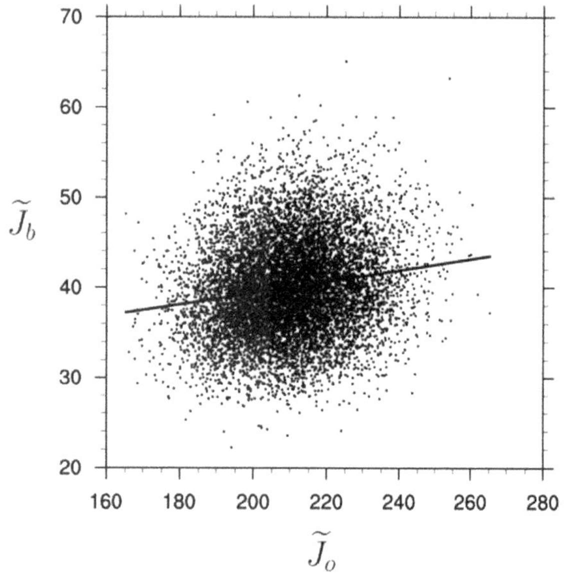

Fig. 3. Scatterplot of the values of the background cost function and the observation cost function at the minimum.

Table 1. Comparison of the simulated optimal innovation method with the randomized trace estimation, following Eqs. (16), (17), (21), and (22).

$E(\widetilde{J}_b)$	40.01	$\frac{1}{2}\mathrm{Tr}(\mathbf{HK})$	40.09
$\mathrm{Var}(\widetilde{J}_b)$	28.14	$\frac{1}{2}\mathrm{Tr}[(\mathbf{HK})^2]$	28.41
$E(\widetilde{J}_o)$	210.11	$\frac{1}{2}\mathrm{Tr}(\mathbf{I}-\mathbf{HK})$	209.91
$\mathrm{Var}(\widetilde{J}_o)$	194.38	$\frac{1}{2}\mathrm{Tr}[(\mathbf{I}-\mathbf{HK})^2]$	198.24

Differences between theoretical and estimated values arise from sample size. Moving to cross-correlation between the values of $\widetilde{J}_o$ and $\widetilde{J}_b$, the simulated optimal innovation method gives an estimate

$$\mathrm{Corr}\left(\widetilde{J}_o, \widetilde{J}_b\right) \simeq 0.164 \tag{29}$$

which is also in rather good agreement with the estimation from the trace algorithm:

$$\frac{\mathrm{Tr}[\mathbf{HK}(\mathbf{I}-\mathbf{HK})]}{\sqrt{\mathrm{Tr}\left[(\mathbf{HK})^2\right]\mathrm{Tr}\left[(\mathbf{I}-\mathbf{HK})^2\right]}} \simeq 0.156 \tag{30}$$

as also shown in Fig. 3; $\widetilde{J}_o$ and $\widetilde{J}_b$ are positively correlated.

Figure 4 shows how the trace estimator becomes more and more precise as the ensemble size goes larger. More precisely, the variance of the estimator (26) is (Chapnik et al., 2006):

$$\mathrm{Var}\left(\frac{1}{M}\sum_{i=1}^{M}\eta_i^{\mathrm{T}}\mathbf{A}\eta_i\right) = \frac{1}{M}\mathrm{Tr}\left[\left(\frac{\mathbf{A}+\mathbf{A}^{\mathrm{T}}}{2}\right)^2\right].$$

Thus, it is possible to estimate the precision of the randomized trace computation as a function of ensemble size

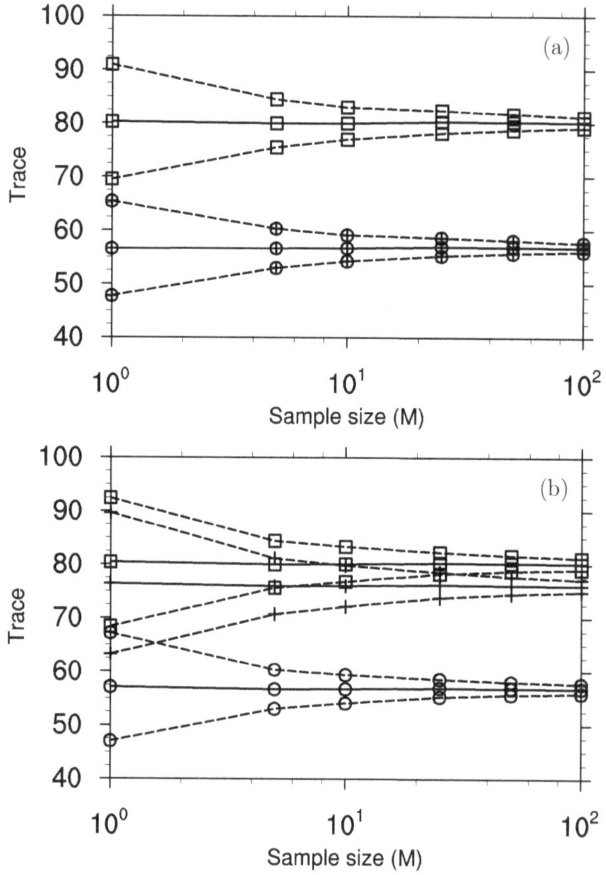

Fig. 4. Randomized trace estimation of $\mathrm{Tr}(\mathbf{HK})$ (**a**, squares), $\mathrm{Tr}[(\mathbf{HK})^2]$ (**a**, circles), $\mathrm{Tr}[(\mathbf{HK})^{\mathrm{T}}\mathbf{HK}]$ (**a**, plus signs) and $\mathrm{Tr}(\mathbf{KH})$ (**b**, squares), $\mathrm{Tr}[(\mathbf{KH})^2]$ (**b**, circles), $\mathrm{Tr}[(\mathbf{KH})^{\mathrm{T}}\mathbf{KH}]$ (**b**, plus signs), as a function of the ensemble size used for the randomization. Solid line: expectation; dashed lines: expectation plus or minus one standard deviation, as determined from a Monte-Carlo approach with 200 independent realizations.

M and as a function of traces $\mathrm{Tr}\left[(\mathbf{HK})^2\right]$, $\mathrm{Tr}\left[(\mathbf{HK})^{\mathrm{T}}\mathbf{HK}\right]$, $\mathrm{Tr}\left[(\mathbf{KH})^2\right]$ and $\mathrm{Tr}\left[(\mathbf{KH})^{\mathrm{T}}\mathbf{KH}\right]$ which themselves can be computed with the randomized approach proposed in Sect. 3.1. This also implies that while Eqs. (25) and (26) can be applied to estimate $\mathrm{Tr}(\mathbf{HK}) = \mathrm{Tr}(\mathbf{KH})$, both estimators have different precisions, as apparent in Fig. 4.

3.4 Variance tuning

DI01 proposed to adjust multiplicative scaling factors s^b and s^o for the covariance matrices $\mathbf{B}$ and $\mathbf{R}$, assuming that at least approximately

$$\mathbf{B}_t = s^b\mathbf{B} \tag{31}$$

$$\mathbf{R}_t = s^o\mathbf{R} \tag{32}$$

where $\mathbf{B}_t$ and $\mathbf{R}_t$ are the true covariance matrices. Then DI01 proposed to estimate the scaling factors as the ratio of the realization of the cost function at the minimum to their expectations:

$$s^{\mathrm{b}} = \frac{2\widetilde{J}_{\mathrm{b}}}{\mathrm{Tr}\,(\mathbf{HK})} \tag{33}$$

$$s^{\mathrm{o}} = \frac{2\widetilde{J}_{\mathrm{o}}}{\mathrm{Tr}\,(\mathbf{I}-\mathbf{HK})} \tag{34}$$

These equations can be seen as a fixed point relation that is solved iteratively.

Alternatively, if the conditions (31)–(32) are satisfied, then it is also possible to derive an alternative formulation for those coefficients. Indeed, the innovation vector $\boldsymbol{d}$ has covariance

$$\begin{aligned}\Sigma = \mathbf{HB}_{\mathrm{t}}\mathbf{H}^{\mathrm{T}} + \mathbf{R}_{\mathrm{t}} &= s^{\mathrm{b}}\mathbf{HBH}^{\mathrm{T}} + s^{\mathrm{o}}\mathbf{R} \\ &= \left(s^{\mathrm{b}} - s^{\mathrm{o}}\right)\mathbf{HBH}^{\mathrm{T}} + s^{\mathrm{o}}\mathbf{D} \qquad (35)\\ &= s^{\mathrm{b}}\mathbf{D} + \left(s^{\mathrm{o}} - s^{\mathrm{b}}\right)\mathbf{R}\,. \qquad (36)\end{aligned}$$

Using Eq. (35) in Eq. (12) with $\boldsymbol{\mu} = 0$ gives:

$$\begin{aligned}2\mathrm{E}\left(\widetilde{J}_{\mathrm{b}}\right) &= \mathrm{Tr}\left(\mathbf{K}^{\mathrm{T}}\mathbf{B}^{-1}\mathbf{K}\Sigma\right) \\ &= s^{\mathrm{b}}\mathrm{Tr}\left(\mathbf{K}^{\mathrm{T}}\mathbf{B}^{-1}\mathbf{KD}\right) + \left(s^{\mathrm{o}} - s^{\mathrm{b}}\right)\mathrm{Tr}\left[\mathbf{K}^{\mathrm{T}}\mathbf{H}^{\mathrm{T}}\mathbf{D}^{-1}\mathbf{R}\right] \\ &= s^{\mathrm{b}}\mathrm{Tr}\,(\mathbf{HK}) + \left(s^{\mathrm{o}} - s^{\mathrm{b}}\right)\mathrm{Tr}\,[\mathbf{HK}(\mathbf{I}-\mathbf{HK})] \qquad (37)\end{aligned}$$

Using Eq. (36) in Eq. (13) with $\boldsymbol{\mu} = 0$ gives

$$2\mathrm{E}\left(\widetilde{J}_{\mathrm{o}}\right) = s^{\mathrm{o}}\mathrm{Tr}\,(\mathbf{I}-\mathbf{HK}) + \left(s^{\mathrm{b}} - s^{\mathrm{o}}\right)\mathrm{Tr}\,[\mathbf{HK}(\mathbf{I}-\mathbf{HK})] \tag{38}$$

which was already been obtained by Chapnik et al. (2006) when discussing the first iteration of the algorithm designed by DI01. In view of Eqs. (21)–(22), the random fluctuations of $\widetilde{J}_{\mathrm{b}}$ and $\widetilde{J}_{\mathrm{o}}$ are small and the expectation can usually be replaced by a single realization. This turns Eqs. (37)–(38) into a *linear system* of two unknowns s^o, s^b which can be written as:

$$\begin{aligned}\alpha s^{\mathrm{b}} + (\alpha-\beta)\left(s^{\mathrm{o}} - s^{\mathrm{b}}\right) &= 2\widetilde{J}_{\mathrm{b}} \\ (p-\alpha)\,s^{\mathrm{o}} + (\alpha-\beta)\left(s^{\mathrm{b}} - s^{\mathrm{o}}\right) &= 2\widetilde{J}_{\mathrm{o}}\end{aligned}$$

and where the notations $\alpha = \mathrm{Tr}\,(\mathbf{HK})$, $\beta = \mathrm{Tr}\left[(\mathbf{HK})^2\right]$ have been introduced for simplification. If the system is not singular, then the unique solution is:

$$s^{\mathrm{b}} = 2\frac{(\beta-\alpha)\,\widetilde{J}_{\mathrm{o}} + (p-2\alpha+\beta)\,\widetilde{J}_{\mathrm{b}}}{\beta\,(p-2\alpha+\beta) - (\beta-\alpha)^2} \tag{39}$$

$$s^{\mathrm{o}} = 2\frac{\beta\widetilde{J}_{\mathrm{o}} + (\beta-\alpha)\,\widetilde{J}_{\mathrm{b}}}{\beta\,(p-2\alpha+\beta) - (\beta-\alpha)^2} \tag{40}$$

At the price of computing $\mathrm{Tr}\left[(\mathbf{HK})^2\right]$, the final solution of DI01's algorithm can be directly obtained, without any further iteration in the case where specified and true error covariances differ by a scaling factor as indicated by Eqs. (31)–(32). This may however not be the case when observation operators are non-linear or when this condition is not met.

In our setting, $\mathrm{Tr}\left[(\mathbf{HK})^2\right]$ is lower but comparable to Tr($\mathbf{HK}$) in magnitude (see Sect. 3.3 for details). Both traces are smaller by about one order of magnitude with respect to the number of observations, which is consistent with recent estimates in ARPEGE (Desroziers et al., 2009, Fig. 1). DI01's formulation neglects the term $(s^{\mathrm{o}} - s^{\mathrm{b}})\mathrm{Tr}\,[\mathbf{HK}(\mathbf{I}-\mathbf{HK})]$ (but the correct estimate is recovered through the iterations). This term is indeed likely to be negligible in Eq. (38) but not so in Eq. (37). Thus, more iterations of the scheme may be needed to recover the appropriate value of s^{b}, rather than s^{o}, as already noticed by DI01 (Tables 3 and 4). In contrast, the correct estimate can be obtained by solving the linear system with the direct solution given by Eqs. (39)–(40). The iterative tuning of the coefficients following DI01's approach is shown in Fig. 5 for overestimated or underestimated σ_{o} and σ_{b}. They confirm that generally, a single iteration is necessary for tuning σ_{o} (panels a, c, e, and g). In contrast, around four iterations can be necessary for recovering a useful value of σ_{b} especially when $s^{\mathrm{o}} - s^{\mathrm{b}}$ is large (panels b and f), as the term $(s^{\mathrm{o}} - s^{\mathrm{b}})\mathrm{Tr}\,[\mathbf{HK}(\mathbf{I}-\mathbf{HK})]$ cannot be neglected. The formulation (39)–(40) recovers the correct factor within one iteration (shown only in panels a and b).

This suggests that it may be useful to use the formulation (39)–(40) when trying to diagnose a multiplicative factor for the background term. This happens for instance for the inflation of variances in an ensemble of data assimilations (Raynaud et al., 2012). Extension of this formulation to the simultaneous tuning of several parameters (corresponding to subparts of the background and the observation cost-functions) is possible but more complicated, as a linear system of order $S+P$ (the number of sub-terms) has to be solved. This will be the topic of further studies.

4 Application in the ARPEGE 4D-Var

This section describes the application of the variance tuning factors to the operational model of Météo-France. The global model ARPEGE is a spectral model with currently 70 vertical levels and features a stretched geometry that allows higher resolution over Europe (about 10 km in physical space). The assimilation scheme is based on the multi-incremental formulation (Courtier et al., 1994) with two outer-loops.

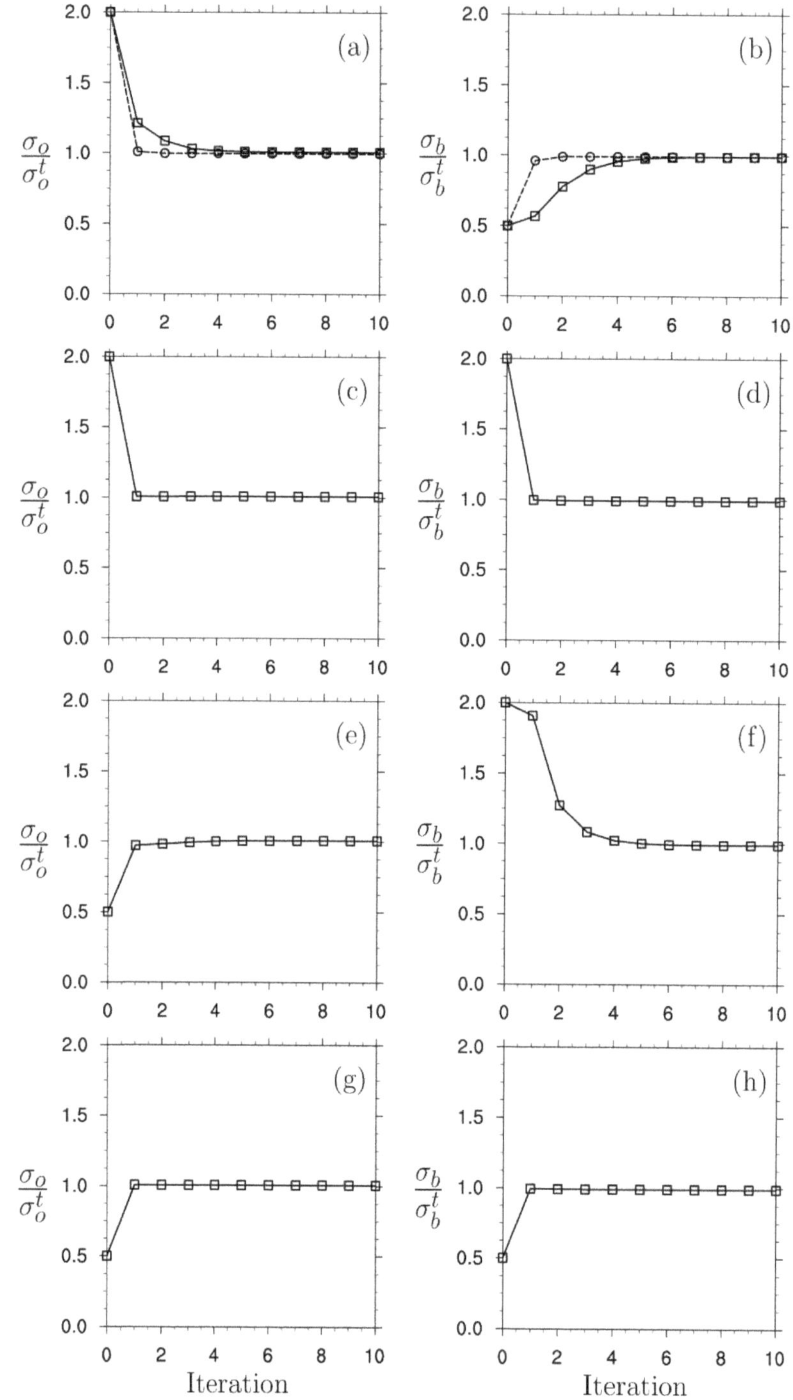

Fig. 5. Iterative tuning of multiplicative coefficients for the error covariance matrices. σ_o (left) and σ_b (right) are tuned simultaneously with **(a, b)** overestimated σ_o, underestimated σ_b; **(c, d)** overestimated σ_o and σ_b; **(e, f)** underestimated σ_o and overestimated σ_b; **(g, h)** underestimated σ_o and σ_b =. Solid lines and square marks: iterative tuning following DI01's approach; dashed lines and circle marks: new approach (shown only in **a, b**). Traces are estimated with the randomization method using 100 independent realizations.

4.1 Estimation of tuning factors

Randomized estimates of the traces are performed in the same way as in the idealized 1-D model using the assimilation with perturbed observations. For the background term, we estimate a single scaling factor with Eq. (39). Following the result of the previous section, it is clear that there is no need for such a refinement in the observation terms (e.g. inclusion of $(s^b - s^o)$ in Eq. 38). Observation scaling

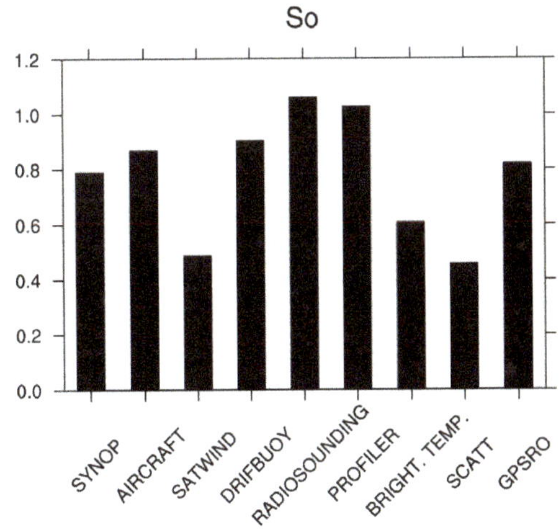

Fig. 6. Diagnosed σ_o's-tuning factors per observation type (September 2013).

factors are therefore estimated in the very same way as in Desroziers et al. (2009). Randomized estimates of the traces have been averaged over a one week period and use a sample size $M = 6$, giving the following values:

$$\mathrm{Tr}\,(\mathbf{HK}) \simeq 70\,000,$$

$$\mathrm{Tr}\left[(\mathbf{HK})^2\right] \simeq 57\,000.$$

An application of the preceding expressions shows that background error standard deviations are overestimated by a factor 0.85. Thus, the specified $\mathbf{B}$ matrix in ARPEGE is closer to its diagnosed value than in previous studies (Desroziers et al., 2009), but still $\mathbf{B} < \mathbf{B}_t$. The estimated values for the observation error standard deviations are shown in Fig. 6 for different observation types. Albeit previous studies have highlighted that observation errors are generally overestimated in data assimilation schemes (possibly because of approximate bias correction or neglect of spatial correlations in the errors), this was not the case anymore for some conventional observations (columns DRIFTBUOYS and RADIOSOUNDING have $S^o = \sqrt{s^o} > 1$). As explained by Desroziers et al. (2009), the obtained values for observation types AIRCRAFT, BRIGHT. TEMP. (satellite measurements) and GPSRO are suspiciously low and this is probably due to the fact that these observations indeed have correlated errors (thus, additional parameters describing these correlations should be included in the maximum-likelihood approach). In our study the corresponding coefficients will just be discarded for the tuning as in Desroziers et al. (2009).

4.2 Impact in terms of forecast performance

We have investigated the impact of tuning the background and observation variances in the global 4D-Var scheme. Figure 7 shows the differences between root-mean-square errors for the geopotential between the reference and the experiment (with tuned values of variances). The error is computed with respect to independent ECMWF analyses. The verification period is one month, which is generally sufficient to obtain reliable results. Bootstrap tests (not shown) indicate that the results are statistically significant during the first 36 h, with positive impact over the whole atmospheric column except for the surface. Other scores however point out to a degradation of the wind analyses in the Tropics with respect to the observations.

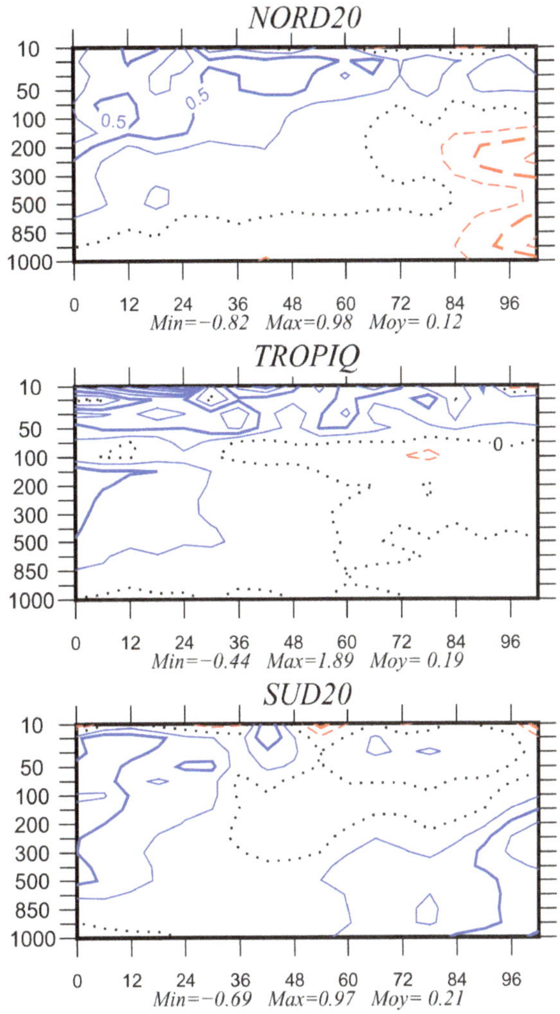

Fig. 7. Impact of tuning on forecast scores: differences between the root mean squared errors of the reference and of the experiment compared to independent ECMWF analyses (geopotential error in m), as a function of forecast range (in h). NORD20 is the domain on the globe that is north of 20° N, TROPIQ is the domain between 20° N and 20° S, and SUD20 is south of 20° S. Blue (red) indicates better (worse) performance. The verification period is one month (September 2013).

5 Conclusions

A posteriori diagnostics can help to detect misspecifications in the statistics of a data assimilation scheme. However, it is generally not possible to tell where the misspecification lies unless some further hypotheses are made (Talagrand, 2010).

Among those diagnostics, one is particularly simple. The value of the cost-function at the minimum should be close to its expected value $p/2$ where p is the number of observations (Bennett et al., 1993). It is possible to refine this diagnostic to sub-parts of the cost function (Talagrand, 1999). Despite care taken in the estimation of error covariance matrices, implementation of this diagnostic in early operational data assimilation schemes showed that the amplitude of the innovation vector was a priori overestimated by a factor 2–3 as reported by Talagrand (1999), possibly because of the misspecification of error correlations for satellite observations.

This paper introduces some additional results using the fact that the sub-parts of the cost function at the minimum can be expressed as quadratic forms of the innovation vector. For some specific distributions of the innovation vector, including the multivariate normal, expressions for the expectation, variances and cross-covariances can be given. In particular, the observation and background terms at the minimum are positively (generally weakly) correlated (when the innovation vector follows the expected statistics).

The expressions involve the trace of larges matrices such as $\mathrm{Tr}[(\mathbf{HK})^2]$. It is shown that these matrices can be estimated from a randomized method (when $\mathbf{K}$ is not explicitly available) by applying the assimilation scheme twice. The randomized method is shown to yield results that are consistent with the "simulated optimal innovation" method (Chapnik et al., 2006). Finally, it is advocated that the computation of $\mathrm{Tr}[(\mathbf{HK})^2]$ could be useful for variance tuning, in particular for the inflation factor of the background error covariance matrix. Further work will attempt to estimate additional covariance parameters following Purser and Parrish (2003) and also make use of recent advances in randomized linear algebra (Aune et al., 2012; Stein et al., 2013)

Acknowledgements. Part of this work was stimulated by a presentation given by M. Jardak at the International Conference on Ensemble Methods in Geophysical Sciences, Toulouse, November 2012. I am grateful to F. Rabier and to G. Desroziers for providing helpful comments on a previous version of this manuscript. This study benefited from the support of the RTRA STAE foundation within the framework of the FILAOS project.

Edited by: T. Gneiting

References

Aune, E., Simpson, D., and Eidsvik, J.: Parameter estimation in high dimensional Gaussian distributions, Stat. Comput., online first, 1–17, doi:10.1007/s11222-012-9368-y, 2012.

Avron, H. and Toledo, S.: Randomized algorithms for estimating the trace of an implicit symmetric positive semi-definite matrix, Journal of the ACM, 58, 8:1–8:34, 2011.

Bannister, R. N.: A review of forecast error covariance statistics in atmospheric variational data assimilation. II: Modelling the forecast error covariance statistics, Q. J. Roy. Meteor. Soc., 134, 1971–1996, 2008.

Bekas, C., Kokiopoulou, E., and Saad, Y.: An estimator for the diagonal of a matrix, Appl. Numer. Math., 57, 1214–1229, 2007.

Belo-Pereira, M. and Berre, L.: The Use of an Ensemble Approach to Study the Background Error Covariances in a Global NWP Model, Mon. Weather Rev., 134, 2466–2489, 2006.

Bennett, A., Leslie, L., Hagelberg, C., and Powers, P.: Tropical Cyclone Prediction Using a Barotropic Model Initialized by a Generalized Inverse Method, Mon. Weather Rev., 121, 1714–1729, 1993.

Bennett, A. F., Chua, B. S., Harrison, D. E., and McPhaden, M. J.: Generalized Inversion of Tropical Atmosphere-Ocean (TAO) Data and a Coupled Model of the Tropical Pacific. Part II: The 1995 La Nina and 1997 El Nino, J. Climate, 13, 2770–2785, 2000.

Chapnik, B., Desroziers, G., Rabier, F., and Talagrand, O.: Properties and first application of an error-statistics tuning method in variational assimilation, Q. J. Roy. Meteor. Soc., 130, 2253–2275, 2004.

Chapnik, B., Desroziers, G., Rabier, F., and Talagrand, O.: Diagnosis and tuning of observational error in a quasi-operational data assimilation setting, Q. J. Roy. Meteor. Soc., 132, 543–565, 2006.

Courtier, P., Thépaut, J.-N., and Hollingsworth, A.: A strategy for operational implementation of 4D-Var, using an incremental approach, Q. J. Roy. Meteor. Soc., 120, 1367–1387, 1994.

Courtier, P., Andersson, E., Heckley, W., Vasiljevic, D., Hamrud, M., Hollingsworth, A., Rabier, F., Fisher, M., and Pailleux, J.: The ECMWF implementation of three-dimensional variational assimilation (3D-Var). I: Formulation, Q. J. Roy. Meteor. Soc., 124, 1783–1807, 1998.

Daley, R.: Estimating observation error statistics for atmospheric data assimilation, Annales Geophysicae, 11, 634–647, 1993.

Dee, D. and da Silva, A.: Maximum-Likelihood Estimation of Forecast and Observation Error Covariance Parameters. Part I: Methodology, Mon. Weather Rev., 127, 1822–1834, 1999.

Dee, D. P.: On-line Estimation of Error Covariance Parameters for Atmospheric Data Assimilation, Mon. Weather Rev., 123, 1128–1145, 1995.

Derber, J. and Bouttier, F.: A reformulation of the background error covariance in the ECMWF global data assimilation system, Tellus A, 51, 195–221, 1999.

Desroziers, G. and Ivanov, S.: Diagnosis and adaptive tuning of observation-error parameters in a variational assimilation, Q. J. Roy. Meteor. Soc., 127, 1433–1452, 2001.

Desroziers, G., Berre, L., Chabot, V., and Chapnik, B.: A posteriori diagnostics in an ensemble of perturbed analyses, Mon. Weather Rev., 137, 3420–3436, 2009.

Evensen, G.: The Ensemble Kalman Filter: theoretical formulation and practical implementation, Ocean Dynam., 53, 343–367, 2003.

Fisher, M.: Background error covariance modelling, in: ECMWF Seminar on recent developments in data assimilation for atmosphere and ocean, ECMWF, 45–63, 2003.

Gauthier, P. and Thépaut, J.-N.: Impact of the digital filter as a weak constraint in the preoperational 4Dvar assimilation system of Météo-France, Mon. Weather Rev., 129, 2089–2102, 2001.

Gauthier, P., Charette, C., Fillion, L., Koclas, P., and Laroche, S.: Implementation of a 3D variational data assimilation system at the Canadian Meteorological Centre. Part I: The global analysis, Atmos. Ocean, 37, 103–156, 1999.

Genton, M. G., He, L., and Liu, X.: Moments of skew-normal random vectors and their quadratic forms, Stat. Probabil. Lett., 51, 319–325, 2001.

Girard, D.: Asymptotic optimality of the fast randomized versions of GCV and CL in ridge regression and regularization, Ann. Stat., 19, 1950–1963, 1991.

Guidard, V. and Fischer, C.: Introducing the coupling information in a limited-area variational assimilation, Q. J. Roy. Meteor. Soc., 134, 723–735, 2008.

Hollingsworth, A. and Lonnberg, P.: The statistical structure of short-range forecast errors as determined from radiosonde data. Part I: The wind field, Tellus A, 38, 111–136, 1986.

Houtekamer, P. and Mitchell, H.: A Sequential Ensemble Kalman Filter for Atmospheric Data Assimilation, Mon. Weather Rev., 129, 123–137, 2001.

Hutchinson, M. F.: A stochastic estimator of the trace of the influence matrix for Laplacian smoothing splines, Commun. Stat.-Simul. C., 18, 1059–1076, 1989.

Kucukkaraca, E. and Fisher, M.: Use of analysis ensembles in estimating flow-dependent background error variances, Tech. Rep. 492, ECMWF Technical Memorandum, 2006.

Le Dimet, F. X. and Talagrand, O.: Variational algorithms for analysis and assimilation of meteorological observations: theoretical aspects, Tellus A, 38, 97–110, 1986.

Lorenc, A. C.: Analysis methods for numerical weather prediction, Q. J. Roy. Meteor. Soc., 112, 1177–1194, 1986.

Lorenc, A. C.: Development of an operational variational assimilation scheme, J. Meteorol. Soc. Jpn, 75, 339–346, 1997.

Mathai, A. and Provost, S. B.: Quadratic forms in random variables: theory and applications, Marcel Dekker, Inc. New York, statistics, a series of textbooks and monographs, 1992.

Mathai, A., Provost, S., and Hayakawa, T.: Bilinear Forms and Zonal Polynomials, Springer-Verlag, New York, Lect. Notes Stat., 102, doi:10.1007/978-1-4612-4242-0, 1995.

Ménard, R. and Chang, L.-P.: Assimilation of Stratospheric Chemical Tracer Observations Using a Kalman Filter. Part II: χ^2-Validated Results and Analysis of Variance and Correlation Dynamics, Mon. Weather Rev., 128, 2672–2686, 2000.

Ménard, R., Cohn, S. E., Chang, L.-P., and Lyster, P. M.: Assimilation of Stratospheric Chemical Tracer Observations Using a Kalman Filter. Part I: Formulation, Mon. Weather Rev., 128, 2654–2671, 2000.

Michel, Y.: Estimating deformations of random processes for correlation modelling: methodology and the one-dimensional case, Q. J. Roy. Meteor. Soc., 139, 771–783, 2013.

Muccino, J. C., Hubele, N. F., and Bennett, A. F.: Significance testing for variational assimilation, Q. J. Roy. Meteor. Soc., 130, 1815–1838, 2004.

Pailleux, J., Geleyn, J.-F., and Legrand, E.: La prévision numérique du temps avec les modèles Arpege et Aladin: Bilan et perspectives = Numerical weather prediction with the models Arpege and Aladin; assessment and prospects, La Météorologie, 32–60, 2000 in French.

Parrish, D. and Derber, J.: The National Meteorological Center's Spectral Statistical-Interpolation Analysis System, Mon. Weather Rev., 120, 1747–1763, 1992.

Purser, R. J. and Parrish, D. F.: A Bayesian technique for estimating continuously varying statistical parameters of a variational assimilation, Meteorol. Atmos. Phys., 82, 209–226, 2003.

Rabier, F.: Overview of global data assimilation developments in numerical weather-prediction centres, Q. J. Roy. Meteor. Soc., 131, 3215–3233, 2005.

Rabier, F., Jarvinen, H., Klinker, E., Mahfouf, J.-F., and Simmons, A.: The ECMWF operational implementation of four-dimensional variational assimilation. I: Experimental results with simplified physics, Q. J. Roy. Meteor. Soc., 126, 1143–1170, 2000.

Rasmussen, C. E. and Williams, C.: Gaussian Processes for Machine Learning, the MIT Press, 2006.

Raynaud, L., Berre, L., and Desroziers, G.: Accounting for model error in the Météo-France ensemble data assimilation system, Q. J. Roy. Meteor. Soc., 138, 249–262, 2012.

Saad, Y.: Iterative Methods for Sparse Linear Systems, 2nd Edn., Society for Industrial and Applied Mathematics, Philadelphia, PA, USA, 2003.

Sadiki, W. and Fischer, C.: A posteriori validation applied to the 3D-VAR Arpege and Aladin data assimilation systems, Tellus A, 57, 21–34, 2005.

Simmons, A. and Hollingsworth, A.: Some aspects of the improvement in skill of Numerical Weather Prediction, Q. J. Roy. Meteor. Soc., 128, 647–677, 2002.

Stein, M. L., Chen, J., and Anitescu, M.: Stochastic Approximation of Score Functions for Gaussian Processes, Ann. Appl. Stat., 7, 1162–1191, 2013.

Talagrand, O.: A posteriori verification of analysis and assimilation algorithms, in: Workshop on diagnosis of data assimilation systems, 2–4 November 1998, ECMWF, Reading, UK, 1999.

Talagrand, O.: Data Assimilation: Making Sense of Observations, in: Evaluation of Assimilation Algorithms, Springer-Verlag Berlin Heidelberg, 217–240, 2010.

Permissions

All chapters in this book were first published in NPG, by Copernicus Publications; hereby published with permission under the Creative Commons Attribution License or equivalent. Every chapter published in this book has been scrutinized by our experts. Their significance has been extensively debated. The topics covered herein carry significant findings which will fuel the growth of the discipline. They may even be implemented as practical applications or may be referred to as a beginning point for another development.

The contributors of this book come from diverse backgrounds, making this book a truly international effort. This book will bring forth new frontiers with its revolutionizing research information and detailed analysis of the nascent developments around the world.

We would like to thank all the contributing authors for lending their expertise to make the book truly unique. They have played a crucial role in the development of this book. Without their invaluable contributions this book wouldn't have been possible. They have made vital efforts to compile up to date information on the varied aspects of this subject to make this book a valuable addition to the collection of many professionals and students.

This book was conceptualized with the vision of imparting up-to-date information and advanced data in this field. To ensure the same, a matchless editorial board was set up. Every individual on the board went through rigorous rounds of assessment to prove their worth. After which they invested a large part of their time researching and compiling the most relevant data for our readers.

The editorial board has been involved in producing this book since its inception. They have spent rigorous hours researching and exploring the diverse topics which have resulted in the successful publishing of this book. They have passed on their knowledge of decades through this book. To expedite this challenging task, the publisher supported the team at every step. A small team of assistant editors was also appointed to further simplify the editing procedure and attain best results for the readers.

Apart from the editorial board, the designing team has also invested a significant amount of their time in understanding the subject and creating the most relevant covers. They scrutinized every image to scout for the most suitable representation of the subject and create an appropriate cover for the book.

The publishing team has been an ardent support to the editorial, designing and production team. Their endless efforts to recruit the best for this project, has resulted in the accomplishment of this book. They are a veteran in the field of academics and their pool of knowledge is as vast as their experience in printing. Their expertise and guidance has proved useful at every step. Their uncompromising quality standards have made this book an exceptional effort. Their encouragement from time to time has been an inspiration for everyone.

The publisher and the editorial board hope that this book will prove to be a valuable piece of knowledge for researchers, students, practitioners and scholars across the globe.

List of Contributors

L. H. Baker
Department of Meteorology, University of Reading, Earley Gate, P.O. Box 243, Reading, RG6 6BB, UK

A. C. Rudd
Department of Meteorology, University of Reading, Earley Gate, P.O. Box 243, Reading, RG6 6BB, UK

S. Migliorini
Department of Meteorology, University of Reading, Earley Gate, P.O. Box 243, Reading, RG6 6BB, UK

R. N. Bannister
Department of Meteorology, University of Reading, Earley Gate, P.O. Box 243, Reading, RG6 6BB, UK

L. Siqueira
Rosenstiel School of Marine and Atmospheric Science, University of Miami, Miami, USA

B. Kirtman
Rosenstiel School of Marine and Atmospheric Science, University of Miami, Miami, USA

A. Turiel
Institut de Ciencies del Mar, CSIC, Barcelona, Spain

J. Isern-Fontanet
Institut Catala de Ciencies del Clima (IC3), Barcelona, Spain

M. Umbert
Institut de Ciencies del Mar, CSIC, Barcelona, Spain

E. Støylen
Department of Geosciences, University of Oslo, Oslo, Norway

I. Fer
Geophysical Institute, University of Bergen, Bergen, Norway now at: Norwegian Meteorological Institute, Oslo, Norway

J. I. Deza
Departament de Física i Enginyeria Nuclear, Universitat Politècnica de Catalunya, Colom 11, 08222, Terrassa, Barcelona, Spain

C. Masoller
Departament de Física i Enginyeria Nuclear, Universitat Politècnica de Catalunya, Colom 11, 08222, Terrassa, Barcelona, Spain

M. Barreiro
Instituto de Física, Facultad de Ciencias, Universidad de la República, Iguà 4225, Montevideo, Uruguay

A. Deluca
Centre de Recerca Matemàtica, Edifici Cc, Campus Bellaterra, 08193 Bellaterra, Barcelona, Spain
Departament de Matemàtiques, Universitat Autònoma de Barcelona, 08193 Bellaterra, Barcelona, Spain

Á. Corral
Centre de Recerca Matemàtica, Edifici Cc, Campus Bellaterra, 08193 Bellaterra, Barcelona, Spain
Departament de Matemàtiques, Universitat Autònoma de Barcelona, 08193 Bellaterra, Barcelona, Spain

G. A. Gottwald
School of Mathematics and Statistics, University of Sydney, Sydney, NSW 2006, Australia

R. Calif
EA 4539, LARGE laboratoire en Géosciences et Énergies, Université des Antilles et de la Guyane, 97170 P-á-P, France

F. G. Schmitt
CNRS, UMR 8187 LOG Laboratoire d'Océanologie et de Géosciences, Université de Lille 1, 28 avenue Foch, 62930 Wimereux, France

L. Tupikina
PIK Potsdam Institute of Climate Impact Research, P.O. Box 601203, 14412 Potsdam, Germany
Department of Physics, Humboldt-Universität zu Berlin, Newtonstr. 15, 12489 Berlin, Germany

K. Rehfeld
PIK Potsdam Institute of Climate Impact Research, P.O. Box 601203, 14412 Potsdam, Germany
Alfred-Wegener-Institut, Helmholtz-Zentrum für Polar- und Meeresforschung, Telegrafenberg A43, 14473 Potsdam, Germany

N. Molkenthin
PIK Potsdam Institute of Climate Impact Research, P.O. Box 601203, 14412 Potsdam, Germany
Department of Physics, Humboldt-Universität zu Berlin, Newtonstr. 15, 12489 Berlin, Germany

V. Stolbova
PIK Potsdam Institute of Climate Impact Research, P.O. Box 601203, 14412 Potsdam, Germany
Department of Physics, Humboldt-Universität zu Berlin, Newtonstr. 15, 12489 Berlin, Germany

N. Marwan
PIK Potsdam Institute of Climate Impact Research, P.O. Box 601203, 14412 Potsdam, Germany

J. Kurths
PIK Potsdam Institute of Climate Impact Research, P.O. Box 601203, 14412 Potsdam, Germany
Department of Physics, Humboldt-Universität zu Berlin, Newtonstr. 15, 12489 Berlin, Germany

M. Gorji Sefidmazgi
North Carolina A&T State University, Dept. of Electrical Engineering, Greensboro, USA

M. Sayemuzzaman
North Carolina A&T State University, Dept. of Energy and Environmental Systems, Greensboro, USA

A. Homaifar
North Carolina A&T State University, Dept. of Electrical Engineering, Greensboro, USA

M. K. Jha
North Carolina A&T State University, Dept. of Civil, Architectural and Environmental Engineering, Greensboro, USA

S. Liess
University of Minnesota, Department of Soil, Water and Climate, St. Paul, USA

S. V. Prants
Laboratory of Nonlinear Dynamical Systems, Pacific Oceanological Institute of the Russian Academy of Sciences, 43 Baltiyskaya st., 690041 Vladivostok, Russia

M. V. Budyansky
Laboratory of Nonlinear Dynamical Systems, Pacific Oceanological Institute of the Russian Academy of Sciences, 43 Baltiyskaya st., 690041 Vladivostok, Russia

M. Yu. Uleysky
Laboratory of Nonlinear Dynamical Systems, Pacific Oceanological Institute of the Russian Academy of Sciences, 43 Baltiyskaya st., 690041 Vladivostok, Russia

N. Gustafsson
Swedish Meteorological and Hydrological Institute, 60176 Norrköping, Sweden

J. Bojarova
Norwegian Meteorological Institute, P.O. Box 43 Blindern, 0313 Oslo, Norway

C. Mendoza
ETSI Navales, Universidad Politécnica de Madrid, Av. Arco de la Victoria 4, 28040 Madrid, Spain

A. M. Mancho
Instituto de Ciencias Matemáticas, CSIC-UAM-UC3M-UCM, C/ Nicolás Cabrera 15, Campus Cantoblanco UAM, 28049 Madrid, Spain

S. Wiggins
School of Mathematics, University of Bristol, Bristol BS8 1TW, UK

J. Marmain
Université de Toulon, UMR7294, CNRS/INSU, IRD - Mediterranean Institute of Oceanography (MIO), UM 110, 83957 La Garde, France
Aix Marseille Université, CNRS/INSU, IRD, Mediterranean Institute of Oceanography (MIO), UM 110, 13288 Marseille, France

A. Molcard
Université de Toulon, UMR7294, CNRS/INSU, IRD - Mediterranean Institute of Oceanography (MIO), UM 110, 83957 La Garde, France
Aix Marseille Université, CNRS/INSU, IRD, Mediterranean Institute of Oceanography (MIO), UM 110, 13288 Marseille, France

P. Forget
Université de Toulon, UMR7294, CNRS/INSU, IRD - Mediterranean Institute of Oceanography (MIO), UM 110, 83957 La Garde, France
Aix Marseille Université, CNRS/INSU, IRD, Mediterranean Institute of Oceanography (MIO), UM 110, 13288 Marseille, France

A. Barth
GeoHydrodynamics and Environment Research (GHER), AGO/MARE, Univeristy of Liège, Liége, Belgium

Y. Ourmières
Université de Toulon, UMR7294, CNRS/INSU, IRD - Mediterranean Institute of Oceanography (MIO), UM 110, 83957 La Garde, France
Aix Marseille Université, CNRS/INSU, IRD, Mediterranean Institute of Oceanography (MIO), UM 110, 13288 Marseille, France

K. Rehfeld
Potsdam Institute for Climate Impact Research, P.O. Box 601203, 14412 Potsdam, Germany
Department of Physics, Humboldt-Universität zu Berlin, Newtonstr. 15, 12489 Berlin, Germany
Alfred-Wegener-Institut Helmholtz-Zentrum für Polar- und Meeresforschung, Telegrafenberg A43, 14473 Potsdam, Germany

N. Molkenthin
Potsdam Institute for Climate Impact Research, P.O. Box 601203, 14412 Potsdam, Germany
Department of Physics, Humboldt-Universität zu Berlin, Newtonstr. 15, 12489 Berlin, Germany

J. Kurths
Potsdam Institute for Climate Impact Research, P.O. Box 601203, 14412 Potsdam, Germany
Department of Physics, Humboldt-Universität zu Berlin, Newtonstr. 15, 12489 Berlin, Germany

A. M. Balk
Department of Mathematics, University of Utah, Salt Lake City, UT 84112, USA

N. Molkenthin
PIK Potsdam Institute of Climate Impact Research, P.O. Box 601203, 14412 Potsdam, Germany
Department of Physics, Humboldt-Universität zu Berlin, Newtonstr. 15, 12489 Berlin, Germany

K. Rehfeld
PIK Potsdam Institute of Climate Impact Research, P.O. Box 601203, 14412 Potsdam, Germany
Alfred-Wegner Institute for Polar and Marine Research, Telegrafenberg A43, 14473 Potsdam, Germany

V. Stolbova
PIK Potsdam Institute of Climate Impact Research, P.O. Box 601203, 14412 Potsdam, Germany
Department of Physics, Humboldt-Universität zu Berlin, Newtonstr. 15, 12489 Berlin, Germany

L. Tupikina
PIK Potsdam Institute of Climate Impact Research, P.O. Box 601203, 14412 Potsdam, Germany
Department of Physics, Humboldt-Universität zu Berlin, Newtonstr. 15, 12489 Berlin, Germany

J. Kurths
PIK Potsdam Institute of Climate Impact Research, P.O. Box 601203, 14412 Potsdam, Germany
Department of Physics, Humboldt-Universität zu Berlin, Newtonstr. 15, 12489 Berlin, Germany

A. Carrassi
Institut Català de Ciències del Clima (IC3), Barcelona, Spain

R. J. T. Weber
Institut Català de Ciències del Clima (IC3), Barcelona, Spain
Department of Physics, University of Heidelberg, Heidelberg, Germany

V. Guemas
Institut Català de Ciències del Clima (IC3), Barcelona, Spain
Centre National de Recherches Météorologiques/Groupe d'Etude de l'Atmosphère Météorologique, UMR3589, Toulouse, France

F. J. Doblas-Reyes
Institut Català de Ciències del Clima (IC3), Barcelona, Spain
Institució Catalana de Recerca i Estudis Avançats (ICREA), Barcelona, Spain

M. Asif
Institut Català de Ciències del Clima (IC3), Barcelona, Spain

D. Volpi
Institut Català de Ciències del Clima (IC3), Barcelona, Spain
Department of Mathematics, University of Reading, Reading, UK

M. Ghantous
Swinburne University of Technology, Hawthorn, Victoria, Australia

A. V. Babanin
Swinburne University of Technology, Hawthorn, Victoria, Australia

Y. Michel
Météo-France and CNRS, CNRM-GAME, UMR3589, Toulouse, France